The New Era in American Mathematics, 1920–1950

The New Era in
American Mathematics,
1920–1950

KAREN HUNGER PARSHALL

PRINCETON UNIVERSITY PRESS

PRINCETON & OXFORD

Requests for permission to reproduce material from this work should be sent to permissions@press.princeton.edu

Published by Princeton University Press
41 William Street, Princeton, New Jersey 08540
6 Oxford Street, Woodstock, Oxfordshire OX20 1TR

press.princeton.edu

All Rights Reserved

ISBN 978-0-691-19755-5
ISBN (pbk.) 978-0-691-23524-0
ISBN (e-book) 978-0-691-23381-9

British Library Cataloging-in-Publication Data is available

Editorial: Susannah Shoemaker, Diana Gillooly, and Kristen Hop
Production Editorial: Nathan Carr and Michelle Scott
Cover Design: Layla Mac Rory and Lauren Smith
Production: Danielle Amatucci
Publicity: Alyssa Sanford and Charlotte Coyne
Cover image: G. H. Hardy's "calculus" and "score card" for ranking mathematicians in the United States and England. Veblen Papers, Library of Congress

This book has been composed in Arno and Helvetica Neue LT Std

10 9 8 7 6 5 4 3 2 1

For Brian and Mom

CONTENTS

"We are evidently on the verge of important steps forward," declared Roland Richardson in 1926.[1] The Canadian-born, Yale-trained analyst, and Brown University professor was forty-seven years old and just beginning his fifth year as Secretary of the American Mathematical Society (AMS). It was the AMS, comprised of those who valued and were committed to the production of original mathematical research, that constituted the inclusive "we" in his assessment.

In four short years, Richardson had been instrumental in helping AMS Presidents—analyst Gilbert Bliss of the University of Chicago, Princeton geometer-turned-algebraic-topologist-turned-differential-geometer Oswald Veblen, and Harvard analyst George Birkhoff—to effect, oversee, and document a number of key changes. Together, they had worked toward the legal incorporation of the AMS; they had effected a massive membership drive that had brought the Society's numbers from 770 when Richardson began as Secretary in 1921 to 1,692 in 1926 for an almost 220% increase; they had organized a capital campaign that had set the AMS's publication enterprise on a sound financial footing; and they had successfully made the case for the inclusion of mathematicians in the Rockefeller Foundation–funded postdoctoral fellowship program of the National Academy of Science's National Research Council.[2] It had been, in Bliss's view, a "critical period in the development of our American mathematical school," while Birkhoff felt that he and Veblen, in particular, had "worked together to 'boost' mathematics into the position

1. Roland Richardson to Gilbert Bliss, Luther Eisenhart, Edward Huntington, Dunham Jackson, and Robert Moore, undated but early 1926, Box 4RM75, Folder: Richardson, Roland George Dwight (1922–1947), Moore Papers.

2. For the membership numbers, see Roland Richardson, "The Twenty-seventh Annual Meeting of the American Mathematical Society," *BAMS* 27 (1921), 245–265 on p. 246 and "The Thirty-third Annual Meeting of the Society," *BAMS* 33 (1927), 129–152 on p. 131.

it deserves to have."[3] Richardson was thus not alone among politically savvy 1920s mathematical leaders in seeing these as auspicious changes that clearly delineated that decade in the evolution of research-level mathematics in the United States.

Research, however, was but one aspect of professional mathematical life. Teaching was another. Since its founding in 1888 as the New York Mathematical Society, the AMS had been primarily devoted to, as its founders expressed it, "preserving, supplementing, and utilizing the results of their mathematical studies," that is, to research.[4] Yet, it had always been the case that its members, as they pursued their research and increasingly valued it as a means of determining professional stature and aiding professional advancement, spent much of their time in classrooms, both undergraduate and graduate. When push came to shove, though, the AMS, while appreciative both "of the importance of" work geared toward the teaching of undergraduate mathematics and of "its value to mathematical science," had no real interest in fostering such work in and of itself and welcomed the creation of a new organization to that end.[5]

The Mathematical Association of America (MAA) came into existence at the end of December 1915 thanks largely to the efforts of Herbert Slaught, professor of mathematics at the University of Chicago, overseer of all things pedagogical in its Department of Mathematics, and the then editor of the more undergraduate teaching–oriented *American Mathematical Monthly*. Indeed, the *Monthly* became the official journal of the new MAA. Like Slaught and many other mathematicians, the MAA's first President, the Göttingen-trained, University of Missouri analyst Earle Hedrick, was a member of both the AMS and the MAA in recognition of his dual professional mission. Over the course of its first ten years, the MAA firmly established itself as the AMS's faithful partner in encouraging the mathematical endeavor in the United States, even though their respective areas of concern remained largely separate.[6]

3. Gilbert Bliss, "A Letter from the President," *BAMS* 28 (1922), 16 and Birkhoff to Veblen, 17 February, 1927, Box 2, Folder: Birkhoff, George D., Veblen Papers, respectively.

4. Thomas Fiske, Edward Stabler, and Harold Jacoby in a circular dated November 1888, as quoted in Raymond Archibald, *A Semicentennial History of the American Mathematical Society, 1888–1938* (New York: American Mathematical Society, 1938; reprint ed., New York: Arno Press, 1980), p. 4.

5. Frank Cole, "The April Meeting of the Society in New York," *BAMS* 21 (1915), 481–493 on p. 482.

6. I treat the founding of the MAA and its first decade in "The Stratification of the American Mathematical Community: The Mathematical Association of America and the

If the Roaring Twenties started off strong for the American mathematical research community, they certainly did not end that way. When the U.S. stock market crashed in October 1929, a worldwide Depression ensued. University budgets were slashed. Trained mathematicians, like so many others, had their salaries cut or, worse, found themselves out of work, while mathematicians-in-training faced a grim job market.

The long climb out of the financial hole throughout the 1930s coincided with unsettling political developments in Europe. Adolph Hitler's rise to power was followed by the Nazis' assault on the Jews of Germany and prompted a Jewish exodus. In particular, beginning in the spring of 1933, the American mathematical community was confronted with a first wave of foreigners seeking positions in the United States. There seemed no way for all of them to be absorbed, but Veblen, by then at the Institute for Advanced Study, Richardson, and others, nevertheless worked hard in an effort to accommodate as many of them as possible.

Had it been the 1890s or the pre–World War I decades of the new century when so many aspiring American mathematicians like Hedrick had gone abroad, and especially to Germany, for their advanced mathematical training, these émigrés might well have been welcomed with unequivocally open arms. After all, that would have represented a reversal. The European mathematical mountain would have been coming to the American Muhammad rather than vice versa. Still, some like analyst Griffith Evans, hired by the University of California in 1934 to revivify its program in mathematics, saw things differently in the Depression's depths. "A generation ago we were in need of direct stimulation and there was plenty of room," he allowed. "[N]ow," however, "we could well interchange."[7] In other words, from the perspective he shared with others in the community, American mathematicians were no longer students of the Europeans. They had been producing fundamental mathematical research that was filling the pages of their journals at home and, to a lesser extent, abroad. In so doing, they had, in their view at least, become mathematical peers, even if it seemed that the Europeans did not yet see it that way. It was another auspicious change from the situation at the century's turn.

American Mathematical Society, 1915–1925," in *A Century of Advancing Mathematics*, ed. Stephen Kennedy et al. (Washington, D.C.: Mathematical Association of America, 2015), pp. 159–175.

7. Griffith Evans to Oswald Veblen, 16 January, 1934, Box 4, Folder: Courant, Richard 1923–38, Veblen Papers.

By the end of the 1930s, the American mathematical community was readying itself to highlight its talents as host, in 1940, of its first International Congress of Mathematicians (ICM). As its planning drew to a close and as it anticipated the event itself—one its leaders felt sure would establish once and for all its mathematical equality with Europe—Hitler's ongoing aggression resulted in September 1939 in the outbreak of World War II. An American ICM would have to be postponed indefinitely. Not only that, a second, even larger, wave of European mathematicians came crashing westward toward a community that thought it had already reached its saturation point. This renewed emigration, together with initiatives, among them, the creation of the *Mathematical Reviews* on American shores, left Richardson with the strong sense that "the center of gravity of mathematics has moved more definitely toward America."[8]

The manifestation of that shift would only become more widely evident after the war's end. The four years from 1941 to 1945 found the American mathematical community focused on its standing not internationally but at home. It faced challenges on at least two fronts: how to mobilize itself effectively for the war effort and how to engage with a nascent but then soon fast-growing scientific bureaucracy that threatened to ignore the contributions it could make and, in so doing, fail to support it.

Relative to the first, it recognized that, while its technical expertise, the purview of the AMS, could contribute in key ways to the nation's military success, perhaps its chief contribution would come through teaching, the purview of the MAA. After all, basic mathematics was essential to the effectiveness of gunners in the air and soldiers on the ground. Young men from all walks of life, they would need instruction. It would not be glamorous work, but it would be critical, and the country's mathematicians already knew how to do it well.

Meeting the second challenge, however, would require mathematical leaders to hone a new set of skills. Richardson, Bliss, Veblen, Birkhoff, and others in the 1920s and 1930s had ultimately been successful in persuading both individual donors and private foundations that research-level mathematics merited their support. Richardson's successor, in 1941, as AMS Secretary, the University of Pennsylvania topologist John Kline and the AMS Presidents with whom he served in the 1940s—analyst Marston Morse of the Institute

---

8. Roland Richardson to "Dear Colleagues," 25 April, 1939, Box 11, Folder: Richardson, R.G.D. 1939, Veblen Papers.

for Advanced Study and Harvard's algebraically oriented analyst Marshall Stone, among others—found themselves thrust into Washington politics as well as into the geopolitics of the postwar era. Their success or failure thus depended on their ability to master the political game.

When the National Defense Research Committee (f. 1940) and then the Office of Scientific Research and Development (f. 1941) were created in Washington to coordinate the anticipated mobilization of science in the nation's defense, American mathematicians found themselves overlooked by the physicists, chemists, and engineers at their helms. It became imperative to get the message across in Washington that mathematics was just as fundamental to a war effort as the other sciences, especially since large sums of money soon began to flow from Federal coffers in support of war-related research. Moreover, some in Congress had begun advocating for the Federal support of science even in peacetime. Mathematicians needed to be part of that conversation as well. Kline, Morse, and Stone, among others, worked capably to these ends.

The American mathematical research community emerged from the war with a new sense of self. "The United States," Kline stated categorically in November 1945, "has assumed world leadership in mathematics."[9] If he was right, that status brought with it new responsibilities. Research would have to remain strong, but, as in the case of the Germany to which Americans had flocked prior to the First World War, research agendas at home would have to spark interest abroad as well as domestically. The community might also be expected to take on, or might actually be called upon to tackle, broader initiatives.

In the closing five years of the 1940s, American mathematicians not only poured themselves into their research, as evidenced by the inundation of their journals, but also renewed their plans for hosting an ICM, this time in 1950. A decade earlier, they had aimed to show that they were Europe's mathematical peer; in 1950, they were the field's presumptive world leader. Unfortunately, their planning took place just as the Cold War's chill began to set in. It would be incumbent upon the world's leader actually to bring the world's mathematicians together, despite extra-mathematical geopolitics, yet it was precisely those politics that risked preventing the attendance of colleagues both from formerly enemy nations and of Communist or socialist political persuasions. Such geopolitics also affected American initiatives, and

9. John Kline, "Rehabilitation of Graduate Work," *AMM* 53 (1946), 121–131 on p. 131.

especially those of Marshall Stone, to reestablish the International Mathematical Union (IMU), a body founded in 1920 that had ultimately succumbed in the 1930s to post–World War I animosity. Politics are almost always messy, and those in which American mathematicians engaged in the late 1940s were no exception. Still, in 1950, the American mathematical community was not only host to the largest and most geographically diverse ICM since the inception of ICMs in Zürich in 1897, but the groundwork had also been laid for the IMU's re-formation in 1951.

Told in this way, the story of the American mathematical research community over the three decades from 1920 to 1950 is a rousing success. Indeed, as the quotations from their correspondence and speeches attest, that is how the community's leaders saw it. It could, however, have been otherwise. At many points, outcomes could have been different. Successes could have been failures. In the 1920s, private foundations like the Rockefeller Foundation might not have been persuaded by the mathematicians' arguments that their field merited external support just as much as, say, medicine did. Some universities, like the University of California in Berkeley, might not have been convinced in the depths of the Depression in the 1930s that the time was actually ripe to build up their departments of mathematics. American mathematicians might not have exhibited the agility over the course of all three decades to take their research in new directions, to embrace and create new approaches, and to attract others internationally to their way of thinking.

This book follows members of the American mathematical community over the course of the thirty-year period from 1920 to 1950 as they developed strategies to deal with the challenges they faced. At the same time, it finds them at work: producing original research, training and nurturing future researchers, engaging in collegiate teaching, building professional infrastructure, participating in the international life of their field, serving their country in wartime, and positioning themselves in the postwar world. Between 1920 and 1950, they witnessed the dramatic growth of their community from 770 to some 4,411 strong for a more than 570% increase.[10] Given a group this large, it will be impossible to mention all of the men and women of various ethnicities who comprised it or to highlight every program that contributed to it. The book thus necessarily aims to be representative not comprehensive.

10. William Whyburn, "The Annual Meeting of the Society," *BAMS* 57 (1951), 109–152 on p. 113.

Part one treats the 1920s, a decade, as noted, when Richardson viewed his community as "on the verge of important steps forward." One of those steps involved, as detailed in chapter one, the production of original research. Americans worked in many areas of contemporaneous mathematics—algebraic and point-set topology, algebraic and differential geometry, algebra, and analysis—but some areas—among them, number theory, mathematical logic, and applied mathematics—engaged them less. This chapter focuses on who was doing what sort of mathematics where in an effort to give not only a flavor of American mathematical research in the 1920s but also a sense of the research community's geographical distribution. Chapters six and ten in part two on the 1930s and part three on the 1940s, respectively, pick up and weave these technical threads further, tracking mathematical developments, shifts in research foci, changing personnel, and the geographical dispersion of programs supportive of mathematics at the research level.

Part one's second chapter focuses on the building and strengthening of mathematical infrastructure in the 1920s. In addition to advances on philanthropic fronts that provided support for mathematical publications, postdoctoral fellowships were created that gave newly minted Ph.D.s the opportunity to establish individual research agendas—and thereby significantly to increase the United States' mathematical output—before being thrust into teaching-intensive faculty positions. Community leaders also recognized the importance of championing the mathematical research achievements that their infrastructure aimed to foster. To be competitive with the other sciences, like physics and chemistry with their Nobel Prizes, those leaders recognized that prizes needed to be created, honors bestowed, and other forms of recognition devised for mathematics as well.

Yet, academic competition does not just take place at home. American mathematicians of the 1920s, many of whom had received all or part of their advanced training in Europe in the decades before World War I, had a strong sense of the wider mathematical world and of their evolving place in it. But how could they change European perceptions of them as students? One way would be for key Europeans to come to the United States in order to experience their mathematical community firsthand. Was it time for American mathematicians actually to host an ICM? Another way would be for Americans to go abroad as mathematical peers, giving lectures on their latest work, attending meetings, interacting one-on-one. Two-way international travel was a much more viable option by the 1920s thanks to the inclusion of mathematics in the postdoctoral travel grants programs of the International

Education Board and the National Research Council, both underwritten by the Rockefeller Foundation. Chapter three develops these and other, more outward-looking, international themes.

The "Threadbare Thirties" and the lead-up to the United States' entry into World War II followed the Roaring Twenties and serve to frame the book's second part.[11] A decade characterized in the minds of some like Evans as one in which the American mathematical community had finally pulled even with Europe, it was complicated by two major extra-mathematical events: the Depression and Hitler's Nazi regime. Chapter four looks at the American mathematical research community in the aftermath of the stock market crash and analyzes the extent to which it was able to sustain the momentum it had built up over the course of the twenties. Publication was affected as the AMS and the MAA both had to cut costs, but the AMS, in particular, launched another fund drive, this time targeting academic institutions for support for its journal enterprise. If universities in particular were going to place increasing emphasis on the publication of original research in matters of academic advancement, then, the AMS's ultimately successful argument went, they should help offset the cost of that publication. That such an argument actually worked is almost astonishing, given its timing during the Depression. Also counterintuitive perhaps, not only did a number of programs in the Northeast, the Midwest, the West, and the South maintain or even strengthen their programs in mathematics at this time of financial hardship, but new journals were founded and sustained and a new type of institution, the Institute for Advanced Study in Princeton, was created *ex nihilo* from philanthropic dollars.

This historical juncture also presented the American mathematical research community with a very different challenge: how to cope with the influx of would-be mathematical émigrés who came in two distinct waves over the course of the decade. Chapters five and seven examine both the close relationship that key American mathematicians developed with the various societies created to aid European refugees and the strategies they employed in accepting and placing them. The two chapters also consider the reception those émigrés received in the communities into which they found themselves transplanted. Chapter seven, in particular, looks at the placement process in

11. Ivan Niven, "The Threadbare Thirties," in *A Century of Mathematics in America*, ed. Peter L. Duren et al., 3 vols. (Providence: American Mathematical Society and London: London Mathematical Society, 1988–1989), 1: 209–229.

the broader context of an American mathematical research community positioning itself to assume a leadership role internationally. That leadership manifested itself both in the organization of an ultimately aborted International Congress of Mathematicians and in creating a home base for a new international reviewing journal, the *Mathematical Reviews*, under the editorship of émigré mathematician and historian of mathematics, Otto Neugebauer.

The book's third and final part provides evidence for Richardson's perception that the "center of gravity of mathematics … moved more definitely toward America" as the 1940s opened as well as for Kline's view that "[t]he United States ha[d] assumed world leadership in mathematics" by the middle of the decade. Chapter eight follows American mathematicians as they first mobilized for and then actively engaged in the war effort. At Brown, Richardson, on stepping down after almost two decades as AMS Secretary, immediately poured himself into the creation of a new program of advanced instruction and research in mechanics that aimed, once and for all, to begin to fill the United States' institutional lacuna in applied mathematics. Mathematicians at Brown and, indeed, at institutions large and small from coast to coast, also engaged in teaching relatively elementary mathematics to the country's Armed Forces. Others contributed to the solution of specific problems posed by the military in contexts such as the Ballistics Research Laboratory at the Aberdeen Proving Ground in Maryland; the Applied Mathematics Panel of the National Defense Research Committee with its pockets of expertise at Columbia, Princeton, Brown, Berkeley, and elsewhere; in England as part of the Eighth Army's Operations Research Service; and at Los Alamos in New Mexico as part of the Manhattan Project. As they engaged in this war work, they also managed to sustain their research momentum and other professional activities.

After the war, as they reestablished their prewar rhythms in research and publication as well as undergraduate and graduate training—matters introduced in chapter nine and detailed in chapter ten—they also began actively to engage in what became the politics of postwar American science. The problem was that it was unclear initially just what such politics would look like and how they would be played. Chapter nine opens by examining the failed efforts in 1945 and 1946 to establish a Research Board for National Security within the Federal government and proceeds to consider the successful creation of the Office of Naval Research (ONR). As Congress continued to debate the possibility of a Federally funded National Science Foundation, the realization of which would only come in 1950 as discussed in the book's coda, the

ONR set an example for the Federal funding of mathematical research under the sage guidance of Chicago-trained algebraist, Mina Rees. These political debates were set against backdrops of postwar program building—at, for example, the University of Chicago and, in the South, at Tulane and Louisiana State University—and of what developed into a postwar publication crisis as the nation's journals were flooded with new mathematical results. The book closes with a coda that sketches the initial contours of a postwar era in which the American mathematical community, through its engagement in the politics both of finally hosting the first postwar ICM and of setting up a new International Mathematical Union, began playing an active leadership role in mathematics internationally.

———

When David Rowe and I published our book, *The Emergence of the American Mathematical Research Community, 1876–1900,* in 1994, we saw the evolution of research-level mathematics in terms of a periodization in which the general structure-building of American science as a whole, of which mathematics was a part, preceded the period from 1876 to 1900 in which an actual *mathematical research* community emerged.[12] Although our book and our research focused on that emergent period, we posited two more periods: one, from 1900 to 1933 in which American mathematics at the research level consolidated and grew and another from 1933 to roughly 1960 defined by the influx of European mathematical émigrés, the development of various areas of applied mathematics, and the institutionalization of Federal funding for science in general and mathematics in particular. Here—and building on research done by many in the intervening twenty-six years but by, in particular, Reinhard Siegmund-Schultze on the impact on American mathematics of both the Rockefeller Foundation and the European émigrés—I argue for a refinement of our earlier periodization.[13]

12. Karen Hunger Parshall and David Rowe, *The Emergence of the American Mathematical Research Community, 1876–1900: J. J. Sylvester, Felix Klein, and E. H. Moore,* HMATH, vol. 8 (Providence: American Mathematical Society and London: London Mathematical Society, 1994), pp. 427–428.

13. Reinhard Siegmund-Schultze, *Rockefeller and the Internationalization of Mathematics between the Two World Wars: Documents and Studies for the Social History of Mathematics in the 20th Century* (Basel: Birkhäuser Verlag, 2001) and *Mathematicians Fleeing from Nazi Germany: Individual Fates and Global Impact* (Princeton: Princeton University Press, 2009).

An American mathematical research community *had* emerged from the American scientific community as a whole by 1900, as, arguably, had an American research community in physics and other areas. Thereafter, it *did* consolidate and grow, even if this did not initially detract from the continuing allure of mathematical Europe. Programs, especially at Chicago and Harvard, consolidated by making strong hires, thereby continuing to attract American mathematical aspirants. Others, at, for example, Princeton and several of the land grant universities, grew. These developments were reflected in the more than doubling of the AMS's membership between 1900 and 1920.[14]

The processes that resulted in the research community's emergence around 1900, however, were largely the same as those that shaped the consolidation and growth of the first two decades of the twentieth century: research, publication, and the training of future researchers; institution-building within the context of the American university; and self-governance within the context of the AMS.[15] While these processes still characterized the decades from 1920 to 1950, those thirty years witnessed fundamental changes, among them, the actual capitalization of American mathematics by sources—private donors, private foundations, and the Federal government—outside the college and university context, the accommodation and assimilation of European mathematical émigrés into what was already a fully formed and highly functioning research community, and the conviction that the American mathematical research endeavor equaled that of Europe. To the extent that periodizations are helpful, the years from 1900 to 1920 might, indeed, be termed a period of consolidation and growth, but, as I argue in the pages that follow, those from 1920 to 1950 were distinct from it in representing a "new era" in which the American mathematical research community emerged as a major player on the international mathematical stage.

14. The membership numbered 357 in 1900. See "Members of the Society," *BAMS* 7 (1901), 9–33 on p. 33.

15. In *American Mathematics 1890–1913: Catching Up to Europe* (Washington, D.C.: MAA Press, 2017), Steve Batterson contends that a major change took place in 1913, namely, that by that date the American mathematical community was on "the verge of parity" with mathematical Europe (p. 197). The argument I present here suggests that that date is premature by roughly a quarter-century.

ACKNOWLEDGMENTS

The idea for this book began to take shape in 2006 as I was casting about for a new project following the publication of my biography of James Joseph Sylvester. That June, the now-lamented Peter Neumann (1940–2020) invited me to speak in the Oxford History of Mathematics Forum that he hosted in his rooms in Queen's College. The topic was "work in progress," so I took the opportunity to try out some early thoughts on what a book on the American mathematical research community in the first half of the twentieth century might look like. The discussion that ensued convinced me that while the idea was sound, the project would likely be much bigger and more complex than I had initially thought.

In 1994, David Rowe and I had published our book, *The Emergence of the American Mathematical Research Community, 1876–1900: James Joseph Sylvester, Felix Klein, and E. H. Moore.* As our subtitle indicated, that process depended on a small number of key figures. The new project, were it to materialize, would be different. Many more people would be involved; the mathematics produced would be more diverse; the forces shaping the community would be more complex. How to rein it in?

As I continued to mull things over and to pursue various avenues of research, life intervened. Victor Katz and I conceived of, wrote, and then published (in 2014) our book, *Taming the Unknown: A History of Algebra from Antiquity to the Early Twentieth Century*, and I served a three-year stint as Associate Dean of the Social Sciences in the College of Arts and Sciences. On returning to my departments after a year's research leave, my then History Department chair, Paul Halliday, invited me to give our annual Robert Cross Memorial Lecture, an event that uniquely brings the whole department together to focus on and discuss the current work of one of its own. I have never worked harder on a talk! I chose to speak on the American mathematics project. It was time to move it off the back burner. Putting together that talk effectively allowed me to outline the argument of the present book, an argument honed thanks to the perspicuous critiques—over multiple drafts—of my History Department colleague, Joe Kett.

One of the university's Sesquicentennial Fellowships in the spring of 2016 allowed me actually to begin writing the book. I spent an incredible month of that semester as a professeur invité at the Université Pierre et Marie Curie in Paris as the guest of my friend and colleague, Catherine Goldstein. From my commodious apartment within the university's walls and inspired by my view of the Institut du monde arabe and the Seine, I completed a first chapter, the book's second, as I participated in the vibrant history of mathematics community that Catherine animates at Paris VI. The project was "officially" under way.

That was a good thing, since I began a three-year tour of duty as chair of the History Department in the fall of 2016. Progress during those three years was slow but steady. The light at the end of that particular tunnel was a year's research leave, one semester the result of the chaired professorship with which the university had honored me in 2016 and the other for "service rendered" as department chair. At the end of my leave, the book was complete, owing in no small part to Virginia's appreciation and support of its faculty's research.

In addition to Peter, Paul, Joe, and Catherine, many people encouraged and helped me bring this project to a close. First, I owe a huge debt of gratitude to all of the archivists and their staffs who worked with me to extract materials—photographic and otherwise—from their holdings, but, in particular, to Carol Mead at the University of Texas at Austin's Archives of American Mathematics and to Heather Riser in Special Collections at the University of Virginia. Carol warmly welcomed me on my numerous visits to Austin to plumb the depths of the expanding collection she oversaw, and Heather allowed me to serve as "honorary archivist," bringing order to the previously uncatalogued but, for my project, much-needed Gordon T. Whyburn Papers. Special thanks also go to Michele Melancon, Assistant University Archivist at Louisiana State University. Although we have never met, Michele answered my (out of the blue) e-mail query about the mathematics faculty at LSU in the 1940s and early 1950s with a stunning and painstakingly extracted spreadsheet of names, dates, and titles that ultimately allowed me to compare and contrast two nascent research-level mathematics programs in the South (see chapter nine).

Second, friends and colleagues read my manuscript in whole or in part and offered constructive criticism and other food for thought. My husband and mathematics colleague, Brian Parshall, and my former mathematics colleague, Jim Rovnyak, scrutinized especially the mathematical exposition, while my friends, historians of science Joe Dauben and Albert Lewis, poured over the manuscript as a whole as did my mother, "Mike" Hunger. If Brian and Jim helped to root out mathematical obscurities, and Joe and Albert pushed me to address questions I had not thought to ask, my mother had her eagle eye out for what she euphemistically calls stylistic "snags." Albert also "held my hand" and served as my technical expert and advisor in the final stages of readying everything for the press as I tried, during a pandemic and often in extreme frustration, to secure images and permissions that met the press's specifications. Judy Green, fellow historian of mathematics formerly of Marymount University, and Elizabeth Meyer, my History Department friend and colleague, actually shared with me photographs from their own private collections thus sparing me the at times overwhelming task of tracking down copyright holders or otherwise determining "fair use." Other friends, particularly my colleague in history, Tico Braun, and our fellow members of the Wednesday Seminar, helped me keep things in perspective with their lively conversations about things both academic and otherwise at our weekly get-togethers.

Last, but certainly not least, I thank everyone with Princeton University Press who helped me see this project into hard covers.

Karen Hunger Parshall
Charlottesville, VA
28 January, 2021

ABBREVIATIONS USED IN THE FOOTNOTES AND TABLES FOR AMERICAN MATHEMATICAL PUBLICATIONS

AJM	American Journal of Mathematics
AM	Annals of Mathematics
AMM	American Mathematical Monthly
AMS	Annals of Mathematical Statistics
BAMS	Bulletin of the American Mathematical Society
BSL	Bulletin of Symbolic Logic
DMJ	Duke Mathematical Journal
JASA	Journal of the American Statistical Association
JMP	Journal of Mathematics and Physics
JSL	Journal of Symbolic Logic
MI	The Mathematical Intelligencer
MR	Mathematical Reviews
MTOAC	Mathematical Tables and Other Aids to Computation
NAMS	Notices of the American Mathematical Society
NMM	National Mathematics Magazine
PAAAS	Proceedings of the American Academy of Arts and Sciences
PNAS	Proceedings of the National Academy of Sciences of the United States of America
QAM	Quarterly of Applied Mathematics
TAMS	Transactions of the American Mathematical Society

PART I

1920–1929: "We are evidently on the verge of important steps forward."

—Roland Richardson, undated but early 1926

1

Surveying the 1920s Research Landscape

In December 1918, Edward Van Vleck was "crazy to get back into real scientific work."[1] The University of Wisconsin mathematician had turned fifty-four just months after the United States had entered World War I in 1917 and had engaged in the war effort as an instructor for the Student Army Training Corps (SATC) on his home campus in Madison. With his usual nine hours of teaching a week augmented by two additional four-hour classes of freshman algebra targeted at SATC students, his "war work," not surprisingly, had "absorbed all of [his] spare time and energy." He had been completely diverted from the research in analysis that he had been faithfully pursuing since his days in Göttingen as a doctoral student of Felix Klein.[2]

Van Vleck was, in some sense, a member of the "first generation" of research mathematicians in the United States.[3] Although he had done graduate work at the Johns Hopkins University before earning his Göttingen degree, he, like many other American mathematical aspirants born in the 1860s, had recognized that the kind of training he sought was largely unavailable in the United

1. Edward Van Vleck to George Birkhoff, 9 December, 1918, HUG 4213.2, Box 4, Folder: Correspondence, 1918–1919, S–Z, Birkhoff Papers. The quotation that follows is from Van Vleck to Birkhoff, 4 May, 1918, op. cit.

2. Thomas Archibald, Della Dumbaugh, and Deborah Kent treat the involvement of American mathematicians in the war effort in "A Mobilized Community: Mathematicians in the United States during World War I," in *The War of Guns and Mathematics: Mathematical Practices and Communities in France and Its Western Allies around World War I*, ed. David Aubin and Catherine Goldstein, HMATH, vol. 42 (Providence: American Mathematical Society, 2014), pp. 229–271.

3. David Rowe and I consider that generation in general and Van Vleck in particular in *Emergence*.

States in the early 1890s. He thus went abroad and returned with a personal mathematical research agenda as well as a dual sense of his academic mission. He was a *teacher* of undergraduate as well as graduate students, but he was also an active *researcher*. After 1904 and thanks to its then president, the geologist Charles Van Hise, the University of Wisconsin to which Van Vleck had moved in 1906 was also coming to share this ethos. It was one of the state universities that had begun to respond to changes in American higher education under way at least since 1876 with the founding of Hopkins in Baltimore. In fits and starts, other institutions followed suit into the opening decades of the twentieth century.

In many ways, World War I had served as a wake-up call to those in academe but, perhaps more importantly, to others in newly created philanthropies as well as to some within the Federal government. They had begun to recognize the value of original research for the welfare of the nation; they increasingly saw the need to support research financially. Savvy university administrators witnessed and steadily responded to this trend over the course of the 1920s and 1930s. They followed the money. Maybe the philanthropies were on to something. Maybe research *should* be more vigorously encouraged within the universities. Maybe faculties *should* be formed and sustained on the basis of research productivity and graduate training, first, and undergraduate teaching, second.

The war had also served as a break in business as usual. In its aftermath, there was a sense within the scientific community more broadly, but within the mathematical community, in particular, of entering into "a new era in the development of our science."[4] "Every nerve should be strained to get our research back on its feet," in Roland Richardson's view.[5] He was apparently not alone in this conviction. He and other American mathematicians poured themselves into their work in the 1920s, but what did that mean? What were their main research interests? Where were those interests fostered? What, in short, was the lay of the American mathematical research landscape in the 1920s?

4. Roland Richardson to Oswald Veblen, 19 December, 1923, Box 10, Folder: Richardson, R.G.D. 1923, Veblen Papers. Daniel Kevles analyzes this attitude among American physicists in *The Physicists: The History of a Scientific Community in Modern America* (Cambridge: Harvard University Press, 1987), especially pp. 75–138.

5. Richardson to George Birkhoff, 31 December, 1918, HUG 4213.2, Box 4, Folder: Correspondence, 1918–1919, M–R, Birkhoff Papers.

FIGURE 1.1. Oswald Veblen (1880–1960) (ca. 1915). (Photo from Wikimedia Commons.)

Mathematicians in Colleges and Universities

"Mathematical research is done almost entirely by university and college teachers," Princeton's Oswald Veblen patiently explained in 1924 to Vernon Kellogg, an entomologist and the permanent secretary of the National Research Council (NRC).[6] Yet, he continued, "[a] mathematics department in an American university has to deal with an enormous mass of freshmen, a very large number of sophomores, and with extremely small numbers of juniors, seniors and graduate students." Veblen was certainly in a position to know.

His father had been a professor of mathematics and physics at the University of Iowa, where the young Veblen had pursued his undergraduate studies. After a year at Harvard to earn a second B.A.—and presumably to supplement

6. Veblen to Kellogg, 11 February, 1924, Box 7, Folder: Kellogg, Vernon 1924–28, Veblen Papers. The next quotation is also from this letter. On the National Research Council and its role in mathematics, see the next chapter.

the more limited offerings that had been available to him in Iowa City—he proceeded to the University of Chicago in 1900, where his uncle, the iconoclastic economist and sociologist, Thorstein Veblen, happened then to be on the faculty.[7] As a graduate student, Veblen imbued an ethos of research, research, research under his doctoral advisor E. H. Moore. His 1903 Ph.D. was followed by two years at Chicago as an associate in mathematics and, in 1905, by a preceptorship at Princeton.[8] All the while, he churned out new results in what was then his main field, geometry. Veblen had thus experienced firsthand American higher mathematics education at levels from the so-so to the very best and had fully embodied the teacher-researcher mindset.

Moreover, from his highly privileged position as President of the American Mathematical Society from January 1923 through December 1924, he had become "rather acutely conscious of the fact that the needs of mathematical research have not yet been brought to the attention of those," like Kellogg, "whose position enables them to have a view of the strategy of Science."[9] But if Veblen laid blame for this state of affairs, it was at the feet of the mathematicians themselves, for they "have too easily assumed that an outside world which cannot understand the details of their work is not interested in its success." In 1924, having embraced the role of mathematical leader in the research as well as in the political sense, Veblen had many reasons to reject that

7. Thorstein Veblen was best-known for the 1899 book, *The Theory of the Leisure Class*, in which he coined the phrase "conspicuous consumption." In 1918, however, he published *The Higher Learning in America: A Memorandum on the Conduct of Universities by Business Men* (New York: B. W. Huebsch, 1918; reprint ed., New York: The Viking Press, 1935), where he argued that World War I would leave "American men of learning in a strategic position . . . in that . . . they command those material resources without which the quest for knowledge can hope to achieve little along the modern lines of inquiry" (p. 52). As the first part of the present book will document, Veblen's nephew, Oswald, and other mathematicians sought to capitalize on what they viewed as the "strategic position" that the American mathematical community had gained in the 1920s.

8. Conceived by Princeton University President (from 1902–1910) Woodrow Wilson, the preceptorial system served as a means of reorienting the Princeton faculty from teaching to teaching and research through the appointment of talented young scholars to serve as intellectual guides for undergraduate students. The basics were learned in lecture courses, while preceptors and their charges met in small groups of from two to six to discuss common readings of a more advanced nature. As Wilson described them, preceptors were "men who are older and more mature and whose studies have touched them with an enthusiasm for the subjects they are teaching." See Woodrow Wilson, "The Preceptorial System at Princeton," *Educational Review* 39 (1910), 385–390 on p. 389.

9. Veblen to Kellogg, 11 February, 1924, Box 7, Folder: Kellogg, Vernon 1924–28, Veblen Papers. The quotation that follows is also from this letter.

assumption (see the next chapter), but he also appreciated the need clearly to articulate how mathematicians, as distinct from other types of scientists, fit into the modern college and university.

Since the beginnings of higher education in the United States, mathematics had been a key, required component of the undergraduate, liberal arts curriculum.[10] By the 1920s, however, America's universities—as opposed to its four-year colleges—had produced a cadre of college and university professors who were trained to do original research but who were hired largely to teach undergraduates. They populated a wide array of institutions.

The colonial colleges—Harvard, Yale, Princeton, Columbia, Pennsylvania, Brown, and others—had, over the course of the final quarter of the nineteenth century and into the opening decades of the twentieth, begun to reorient themselves toward undergraduate *and* graduate instruction. Owing to their relatively long histories and to their traditionally collegiate focus, some of these schools experienced more difficulty than others in redefining themselves as actual universities in which faculties were expected actively to engage in research and publication. The same was true of some of the state-supported schools—like the Universities of Michigan, Iowa, Wisconsin, Kansas, Texas, and California at Los Angeles. After the 1862 Morrill Act provided funding for them, moreover, the Federal land-grant universities—such as the University of California in Berkeley, the University of Illinois, the Massachusetts Institute of Technology (MIT), and the Ohio State University—realized their more practical orientation at both the undergraduate and graduate levels. These types of schools were supplemented, in the so-called Gilded Age that followed the U.S. Civil War, by privately endowed women's colleges—especially Pennsylvania's Bryn Mawr—and other institutions—such as Hopkins, Clark University, and the University of Chicago—that set new standards particularly for graduate education and the production of original research.[11] Faculty members at both colleges

10. Ubiratan D'Ambrosio, Joseph Dauben, and I consider the place of mathematics in the seventeenth- and eighteenth-century North American curriculum in our chapter "Mathematics Education in America in the Premodern Period," in *Handbook on the History of Mathematics Education*, ed. Alexander Karp and Gert Schubring (New York: Springer Verlag, 2014), pp. 175–199.

11. Cornell University represents an interesting hybrid. Founded in 1865 thanks to the private benefaction of telegraph tycoon, Ezra Cornell, it was also named New York State's land-grant college. For a nice overview of the emergence of the research university per se in the United States, see Roger Geiger, *To Advance Knowledge: The Growth of American Research Universities, 1900–1940* (New York: Oxford University Press, 1986).

and universities were coming to define themselves in terms of teaching *and* research.

For American mathematicians, this dual personality was both like and unlike that of their European counterparts. American and European mathematicians strove to do research and to publish the fruits of their labors, but in Europe—and especially in Germany and France where a system of *Gymnasien* and *lycées*, respectively, provided instruction at the freshman and sophomore levels—mathematicians were not involved in more introductory teaching.[12] Yet, in the United States, as Veblen explained to Kellogg, "[a] man with good mathematical gifts and normal personal qualities has little trouble in obtaining as good a position as is available under our system," "[b]ut when he obtains it he has a teaching schedule of from nine to fifteen hours a week as compared with three hours a week for his colleague in the Collège de France."[13] "Moreover," Veblen went on, "he becomes tremendously interested in this teaching; he sees the manifold ways in which it could be improved, and he plays his part in the committees and other administrative devices which are trying to do the obvious tasks of the university in a better way." The American mathematician was thus able to spend only a relatively "small fraction" of time on research, given that a certain "sense of responsibility" dictated that he respond "in a normal way to his environment."

A contradictory state of affairs had thus resulted in mathematics, although, at least as Veblen saw it, not at all in astronomy and much less so in the laboratory sciences. In mathematics, he explained, "we recognize ability in scientific research as *a basis for* university appointments but not as a *primary occupation for* the appointees." Astronomers, however, were often associated with observatories where observation and research defined their primary occupations, and although some physicists taught, they were also often responsible for maintaining research laboratories, whether in an academic or in an industrial setting.[14] Veblen and many of his contemporaries believed that the time had

12. I give a comparative look at the situations for mathematicians in "Training Research Mathematicians circa 1900: The Cases of the United States, Germany, France, and Great Britain," in *A Global History of Research Education: Disciplines, Institutions, and Nations*, ed. Kuming "Kevin" Chang and Alan Rocke, vol. 34(1) (Oxford University Press, 2021), pp. 65–83.

13. Veblen to Kellogg, 11 February, 1924. The quotations that follow in this and the next two paragraphs are also from this letter (with my emphases).

14. John Lankford treats American astronomy in *American Astronomy: Community, Careers, and Power, 1859–1940* (Chicago: University of Chicago Press, 1997), especially pp. 125–181; on physics, see Kevles, *The Physicists*, pp. 60–90.

come for colleges, but especially universities, to reverse the order of their priorities for mathematics, making research paramount and teaching secondary although still important.

They envisioned a system—with an implied hierarchy—in which those "who have shown in their own environments that their impulse to research is a vital one" would be "freed from all other obligations and thenceforth paid for devoting their energies to research." Those whose "impulse to research" was less "vital" would focus on teaching. Indeed, this tension was already reflected in the existence of two mathematical societies: the AMS, founded in 1888, served the needs of the researchers, while the Mathematical Association of America, created in 1915, aimed at those engaged in undergraduate teaching.[15] These two sets of mathematicians were by no means disjoint, but Veblen's was an idealistic vision of the future of research-level mathematics that collided with the reality of college and university life at many, if not most, institutions in the 1920s.

Consider, for example, John Kline's experiences at Yale following his 1916 Pennsylvania Ph.D. under University of Chicago–trained Robert L. Moore. At Penn, Kline had internalized the research mantra thanks to Moore—a mentee of Veblen and fellow student with him of E. H. Moore—and had taken it with him to Yale as an instructor during the 1918–1919 academic year. He was shocked by the attitudes he encountered there.

At a faculty meeting early in the second semester, the department chair, mathematical astronomer Ernest Brown, announced that there would be no more than one new entering graduate student and that not even that candidate was certain. When elder statesman and European-trained James Pierpont noted that the department used to graduate several first-rate Ph.D.s a year but that "lately we have had only a few men and they mostly a poor lot," Brown replied that, in his view, that "was due to the fact . . . that the money Yale had to put out in fellowships and scholarships was very small as compared with Chicago, Harvard, and Princeton."[16] When Pierpont pressed the issue, agreeing that that was likely part of the problem but questioning whether it was the whole of it, Brown constituted a committee of the "younger men" to study the situation and to make recommendations.

15. AMS-MAA relations at this time are treated in the next chapter as well as in my "The Stratification of the American Mathematical Community."

16. John Kline to R. L. Moore, 9 February, 1919, Box 4RM74, Folder: Kline, John Robert (1918–1921), Moore Papers. The quotations that follow in this and the next two paragraphs are also from this letter (with my emphases).

As one of those "younger men," Kline got right to work canvassing his colleagues, but his findings dismayed him. William Longley, an assistant professor who had earned his Ph.D. at the University of Chicago in 1906 likely under the mathematical astronomer Forest Moulton, initially "seemed interested in doing something for [the] encouraging of research here" but then suggested that "the decline in graduate students was because pure mathematics was a drudge on the market, [that] the pure mathematician had nothing that anyone else wanted and that perhaps we had been following false gods in patterning [ourselves] after the Germans in our highly specialized mathematics." Longley also offered the opinion "that most men . . . are enthusiastic research men when in graduate school but when they got out into teaching and got away from this influence, they gradually returned to their *normal* selves and a *correct* balance of things." Egbert Miles, another assistant professor and another Chicago Ph.D. but one who had earned his degree under Oskar Bolza in 1910, made Kline "still sorer." Miles "felt that pure mathematics was a subject which had no place in our university life at present, that we were at present engaged in building up a great industrial nation and that it was the business of the mathematician not to delve into pure science but to do effective teaching and apply mathematics to industrial problems."

Kline next moved on to the members of what he pejoratively termed "the teaching gang." One of that number held "that it is our business to look after the interests of the men who are going to be primarily interested in teaching, that there has been a false evaluation and that heads of departments have been unjust in making promotion depend only on research." In sum, James Whittenmore, like Kline an instructor but unlike him a European-trained mathematician who had nevertheless not taken a Ph.D., thought that the two of them "were the only ones of the younger men who had any interest in doing research." In Kline's view, "if that was the attitude of the rest of the bunch, I should not be surprised if harm had already been done along the research lines."

Clearly, not all members of the younger generation were of a mind relative to the desirability and value of doing original research. Yale's Department of Mathematics, unlike those at Chicago, Harvard, and Princeton, was thus not in a position as the 1920s opened to make a strong push into research, even though Pierpont, for one, hoped to convince the Yale administration "to strengthen the Department of Mathematics in the sphere of Research in Pure Mathematics" by making sufficient funds available to lure George

Birkhoff from Harvard.[17] That initiative failed. Birkhoff, then regarded as one of America's best mathematicians, spent his career at Harvard. For Yale, as for numerous other schools, a strong research reorientation had evolved only by the 1940s.[18]

Kline left Yale after one year for an instructorship at the University of Illinois. There, he found a department much different from the one he had left on the East Coast. There, the geometers Edgar Townsend and Arthur Coble and the algebraists James Shaw and George Miller, among others, had been fostering what Kline deemed "a good research atmosphere."[19] Although Coble had just narrowly edged out Kline's advisor, R. L. Moore, for an Illinois professorship, Kline had been "asked for suggestions of good men" and had been actively campaigning to get Moore's name back in the running should a new senior position open up. Kline felt, moreover, that the primacy of research was fully appreciated at Illinois, whereas it had not been at Yale. As he put it to Moore, "[c]ouldn't we make this a centre if you came here"? The University of Illinois, one of the newer land-grant institutions, had already embraced, at least in mathematics, a more modern research ethos by 1920.

At another land-grant, the Ohio State University, that transition was proving a bit more difficult. Kline's academic brother, Raymond Wilder, had finished his Ph.D. under Moore at the University of Texas in Austin in 1923 and had accepted an associate professorship at Ohio State a year later. After settling into the routine there, he wrote to Moore to convey his impressions of the place. He was candid. "[A]s you no doubt would guess, the dept. needs new life," he told Moore. "Outside of Kuhn, Bohannan & Weaver—*dead wood*. . . . Of course, I am speaking of the dept. as it stands without MacDuffee. The latter is a good one—seems to have good ideas, and we've already formed

---

17. E. H. Moore to James Pierpont, 6 April, 1923, Box 2: Correspondence, 1921–1925 'J-Z,' Folder: Pierpont, James, Richardson Papers.

18. Harold Dorwart, who was a graduate student at Yale in the 1920s, reminisced in rosy terms about his student days there. Still, even his account of the 1920s pointed to the period from the mid-1930s to the mid-1940s as the epoch when department chair, Oystein Ore, actually "recruited many fine mathematicians to the department" with support, that had been lacking earlier, of the higher administration. See Harold L. Dorwart, "Mathematics at Yale in the Nineteen Twenties," in *A Century of Mathematics in America*, ed. Peter Duren et al., 2: 87–97 on p. 94.

19. Kline to R. L. Moore, undated but likely the fall of 1919, Box 4RM75, Folder: Kline, John Robert, Letters by or to Kline, undated, in Moore Papers. The quotations that follow are also from this letter.

a 'dynamite squad' or 'flying wedge' consisting of our two selves. It's a case of stand together or drop into oblivion."[20] Harry Kuhn had earned his doctorate at Cornell in finite group theory under the direction of George Miller in 1901; Rosser Bohannan, chair of the department, had taken degrees in engineering from the University of Virginia in 1876 before proceeding for post-graduate studies abroad at Cambridge and Göttingen in the 1880s; James Weaver was a 1916 Ph.D. in geometry under Maurice Babb at Penn; and Cyrus MacDuffee had earned a doctorate in 1921 under Leonard Dickson at Chicago. It had been under Bohannan that the Ohio State department had begun hiring Ph.D.s and had started to offer more advanced courses, among them some graduate-level seminars.[21]

Despite the "dead wood," Wilder thought that the department at Ohio State did have "some good points, chief of these being *freedom.*" As a case in point, he was teaching both a freshman and an advanced course that he could "run as [he] please[d]." Moreover, he hoped to teach his special field of topology in the second quarter and had "two graduate students—likely looking boys, one an M.A. already—intending to take it."

All in all, though, the department needed improvement. "MacDuffee expressed it very well," Wilder told Moore, " 'I don't want to say anything about any members of the dept., *but,* there aren't enough vertebrae in Kuhn, Weaver, & Rasor put together to make one spinal column.' "[22] He and MacDuffee therefore had "to reform not only the character of the work in the dept., but the attitude of the adminstration toward" the group as a whole.

That, in fact, was the self-appointed task of many in the 1920s—like Veblen, Kline, Wilder, and others—in departments of mathematics in all manner of colleges and universities around the United States. These

20. Wilder to Moore, 15 October, 1924, Box 86-36/8, Folder 6: General Correspondence R. L. Moore, Wilder Papers (his emphasis). The quotations in the next paragraph are also from this letter with his emphasis.

21. For an overview of the history of the Ohio State Mathematics Department, see https://math.osu.edu/about-us/history.

22. Wilder to Moore, 22 December, 1924, Box 86-36/8, Folder 6: General Correspondence R. L. Moore, Wilder Papers (his emphasis). The next quotation is also from this letter. A differential geometer interested in the calculus of variations, Samuel Rasor had earned his M.S. at Ohio State in 1902, and although he had done additional coursework at the University of Chicago in 1906 and at Berlin during the 1910–1911 academic year, he never took a doctorate. He had nevertheless moved up the ranks at Ohio State, becoming a full professor in 1913 and serving in that post until his retirement in 1943.

mathematicians sought to convince their administrators to allow the pendulum to swing from teaching to research relative to professional advancement. Although in the 1920s it was not yet clear whether that swing would occur, a not insignificant number of America's mathematicians endeavored to pursue their research and graduate instruction as they dutifully taught their undergraduate classes and served their institutions. In so doing, they contributed to a number of areas that filled the pages of journals at home and appeared side by side with European research in journals abroad. Veblen captured at work the "most active and successful investigators" among them in a 1928 snapshot that well reflected where American mathematicians were deemed, by at least some of their contemporaries, to be making the most important advances (see fig. 1.2).[23]

A Recognized American Specialty: Analysis Situs

Analysis situs, or what would today be called topology, was considered in the 1920s perhaps the most distinctive of the American mathematical research specialities. In fact, as Göttingen's Richard Courant saw it in 1927, it was "[t]he one mathematical field in which America has had perhaps the greatest success."[24] It came, however, in two flavors. Combinatorial, that is, algebraic, topology treated space as comprised, in some sense, of "visible" building blocks that were stuck together in particular ways. It asked just how those building blocks were "combined," or, in other words, what were their "combinatorial" properties? This type of topology—acknowledged by Courant— was fostered primarily at Princeton initially under Veblen's leadership. The other kind—point-set topology and ignored by Courant—considered space microscopically as a collection of "invisible" points. It focused largely on continuity considerations from an axiomatic point of view and was developed as an American speciality thanks to the efforts principally of R. L. Moore, at Penn until his move in 1920 to the University of Texas in Austin. Each of these types of topology sought to isolate those properties of spaces that

23. Oswald Veblen, "Report for Mathematics to the Trustees of the National Research Fund," 17–18 June, 1928, part C, Box 26, Folder: NAS National Research Fund (1928), Veblen Papers. Fig. 1.2 is a retyped version of the original that preserves as much as possible its layout, spacing, etc.

24. Courant to Augustus Trowbridge, 27 April, 1927, in Siegmund-Schultze, *Rockefeller and the Internationalization of Mathematics between the Two World Wars*, pp. 272–274 on p. 272 (Siegmund-Schultze's translation).

Table I.

Algebra and Theory of Numbers.

L. E. Dickson (Chicago)
J. H. M. Wedderburn (Princeton)
E. T. Bell (C.I.T.)
H. S. Vandiver (Texas)
H. H. Mitchell (Penna.)
H. F. Blichfeldt (Stanford)
W. A. Manning (Stanford)
G. A. Miller (Illinois)
A. J. Kempner (Colorado)
O. Ore (Yale)

Function Theory.

W. F. Osgood (Harvard)
G. D. Birkhoff (Harvard)
T. H. Gronwall (Columbia)
Norbert Wiener (M.I.T.)
J. Tamarkin (Brown)
J. F. Ritt (Columbia)
Einar Hille (Princeton)
O. D. Kellogg (Harvard)
Dunham Jackson (Minnesota)
J. A. Shohat (Michigan)
C. N. Moore (Univ. of Cincinnati)
W. A. Hurwitz (Cornell)
A. Pell Wheeler (Bryn Mawr)
R. D. Carmichael (Illinois)
J. L. Walsh (Harvard)
M. H. Stone

Calculus of Variations

G. A. Bliss (Chicago)
M. Morse (Harvard)

General Analysis & Theory of Functionals

E. H. Moore (Chicago)
G. C. Evans (Rice Inst.)
T. H. Hildebrandt (Michigan)
L. M. Graves (Chicago)

Analysis Situs

J. W. Alexander (Princeton)
S. Lefschetz (Princeton)
O. Veblen (Princeton)
G. D. Birkhoff (Harvard)
M. Morse (Harvard)
R. L. Moore (Texas)
E. W. Chittenden (Iowa)
J. R. Kline (Penna.)

Algebraic Geometry.

S. Lefschetz (Princeton)
A. B. Coble (Illinois)
V. Snyder (Cornell)
F. R. Sharpe (Cornell)
J. L. Coolidge (Harvard)

Differential Geometry

L. P. Eisenhart (Princeton)
O. Veblen (Princeton)
W. C. Graustein (Harvard)
E. Kasner (Columbia)
J. D. Struik (M.I.T.)
T. Y. Thomas (Princeton)
E. P. Lane (Chicago)
E. B. Stauffer (Kansas)

Dynamics and Relativity

G. D. Birkhoff (Harvard)
L. P. Eisenhart (Princeton)
G. Y. Rainich (Michigan)
E. W. Brown (Yale)
H. Bateman (C.I.T.)
Paul Epstein (C.I.T.)
F. D. Murnaghan (Johns Hopkins)

FIGURE 1.2. Veblen's list of America's "Research Mathematicians" (1928). (Typed Facsimile of the document in Veblen Papers, Library of Congress.)

are preserved under homeomorphism, that is, under the action of a continuous, one-to-one and onto map with continuous inverse. Each thus also dealt with the properties of geometrical figures that remain invariant under such a map.[25] As students in the classes of E. H. Moore at Chicago in the first decade of the twentieth century, Veblen and Moore had both been influenced by the foundational, postulate-theoretic agenda that the elder Moore had then embraced.[26] Interestingly, each initially attacked his own brand of topology from, so to speak, the ground up.

Veblen had come to the field from work, in the opening decade of the twentieth century, on the foundations, first, of geometry in general and, then, of projective geometry in particular. By 1912, his focus had shifted to an exploration of ideas that Henri Poincaré had only incompletely developed in a series of papers published between 1895 and 1904 on the concept of the connectivity of a space and on what Poincaré termed "analysis situs." With Princeton student James Alexander, for example, Veblen co-authored a paper on "Manifolds of N Dimensions" in 1913 that explicitly aimed "to establish some of the fundamental definitions and theorems as rigorously as possible, so as to furnish an introduction to the memoirs of Poincaré."[27] This paper

25. For these general characterizations of algebraic and point-set topology, see Solomon Lefschetz, Review of Oswald Veblen's *Analysis Situs*, *Bulletin des sciences mathématiques*, ser. 2, 46 (1922), 421–424 on pp. 421–422; Oswald Veblen, *The Cambridge Colloquium: Part II: Analysis Situs* (New York: American Mathematical Society, 1922; 2d. ed., 1931), p. 5; and James Alexander, "Some Problems in Topology," in *Verhandlungen des Internationalen Mathematiker-Kongresses Zürich 1932*, ed. Walter Saxer, 2 vols. (Zürich and Leipzig: Orell Füssli, 1932), 1: 249–257 on p. 249.

Ioan James provided an idiosyncratic account of the American topological scene in his essay, "Combinatorial Topology Versus Point-set Topology," in *Handbook of the History of General Topology*, ed. Charles Aull and Robert Lowen, 3 vols. (Dordrecht: Kluwer Academic Publishers, 1997–2001), 3: 809–834. There, he asserted that "point-set topology seems to have become separated from the rest of topology around the middle of the twentieth century" (p. 809). Here, I show that, at least in the United States, the two types of topology were fairly separate from the start.

26. See the section on algebraic research below. On Moore's role in American mathematics in the decades around 1900, see my "E. H. Moore and the Founding of a Mathematical Community in America: 1892–1902," *Annals of Science* 41 (1984), 313–333, reprinted in *A Century of Mathematics in America*, ed. Peter Duren et al., 2: 155–175 and Parshall and Rowe, *Emergence*, chapters 6, 9, and 10. On the doctoral research of Veblen and R. L. Moore, see Parshall and Rowe, *Emergence*, pp. 383–387.

27. Oswald Veblen and James Alexander, "Manifolds of N Dimensions," *AM* 14 (1912–1913), 163–178 on p. 164. The quotation that follows is also on this page. For the set-up in the next paragraph, see pp. 164–165. The quotations in the next paragraph are on p. 164.

marked Alexander's publication debut as a topologist and set him on the research path he would continue to pursue throughout his career.

To fix the ideas and establish some terminology, consider Euclidean n-space and take $n + 1$ points not all in the same $(n - 1)$-space as well as the 1-, 2-, . . . , $(n - 1)$-dimensional simplexes of which they are the vertices. These constitute a finite region in n-space called an n-dimensional simplex, that is, "that one among the regions into which n-space is subdivided by $n + 1$ linearly independent $(n - 1)$-spaces which does not contain a point at infinity." For example, "the interior of a triangle in a plane is a two-dimensional simplex, and the linear segment joining two points is a one-dimensional simplex." The $n + 1$ points are called the vertices, and the points on the boundary are not part of the simplex.

Now, consider a set of objects in one-to-one correspondence with the points in an n-dimensional simplex together with its boundary. The objects corresponding to the points of the simplex constitute an n-cell and the objects corresponding to the boundary of the simplex form the n-cell's boundary. Finally, consider the set C_n of cells consisting of α_i i-cells for $0 \leq i \leq n$. C_n is called a complex if every i-cell, for $i > 0$, is made up entirely of cells of dimensions less than i and if every i-cell, for $i < n$, is on the boundary of some $(i + 1)$-cell. The ordered set of points in the various cells of a complex C_n is a manifold M_n provided: 1) every point is interior to some n-cell, 2) if two n-cells have a point in common, there is an n-cell contained within each of them, and 3) for any two points p and q in C_n, there is always a chain of overlapping n-cells that connects an n-cell about p to an n-cell about q.

As Veblen and Alexander noted, Poincaré had shown that it was possible to characterize any oriented, n-dimensional manifold M_n in terms of certain matrices from which are derivable a set of $n - 1$ positive integers P_i, which he called the Betti numbers and which are invariants of M_n.[28] The P_i satisfy both the duality relation (now named after Poincaré)

$$P_i = P_{n-i}$$

and the so-called generalized Euler theorem

$$\sum_0^n (-1)^i \alpha_i = 1 + (-1)^n + \sum_1^{n-1} (-1)^i (P_i - 1),$$

---

28. Veblen and Alexander, pp. 163–164. Jean Dieudonné gives a modern technical discussion both of Poincaré's work and of Veblen and Alexander's 1913 paper in *A History of Algebraic and Differential Topology, 1900–1960* (Boston: Birkhäuser Boston, 1989), pp. 15–35 and pp. 41–42, respectively.

where α_i is the number of i-cells into which M_n may be dissected. If, however, M_n is non-oriented, then the numbers P_i do *not* satisfy the duality relation but do satisfy

$$\sum_0^n (-1)^i \alpha_i = 1 + \sum_1^{n-1} (-1)^i (P_i - 1).$$

In their paper, Veblen and Alexander showed how to simplify things so that "certain systems of linear equations reduced modulo 2" led to matrices in just zeros and ones. From those matrices, they derived $n - 1$ constants R_i which satisfied *both* Poincaré duality *and* the generalized Euler theorem, regardless of whether or not the manifold was oriented. Although more general than Poincaré's set-up, theirs, as they realized, unfortunately did not yield any invariants of the manifold different from those already determined by Poincaré's methods.

Three years after the publication of this joint work and just before he joined the American war effort, Veblen gave the fifth AMS Colloquium Lectures on his evolving thoughts on analysis situs.[29] In particular, he aimed to present merely "an introduction" for his American audience "to the problem of discovering the n-dimensional manifolds and characterizing them by means of invariants."[30] He had an even higher aspiration in the published version of the lectures, which appeared only in 1922 due to his wartime involvement.[31] Ever intent on the clarity and precision he had been honing since his student days at Chicago, he took on the challenge of providing a "more formal," "systematic treatise on the elements of [this type of] Analysis Situs." In writing it, Veblen introduced the ideas by treating the cases of $n = 1$ and $n = 2$ before tackling the general case.

Veblen's work initiated a research focus on algebraic topology at Princeton that flourished beginning in the 1920s.[32] Although his own interests shifted into the not-unrelated area of differential geometry over the course of that decade (see the next section), Veblen's student and, beginning in 1916, his colleague, Alexander, as well as their colleague, after his 1925 move from the University of Kansas, Solomon Lefschetz, continued to churn out new results

29. On the establishment of the AMS Colloquium Lectures, see the next chapter.

30. Veblen, *Analysis Situs*, p. vi. The next quotation is also from this page.

31. For more on the latter, see David Alan Grier, "Dr. Veblen Takes a Uniform: Mathematics in the First World War," *AMM* 108 (2001), 922–931.

32. Saunders Mac Lane briefly characterizes this group in "Topology and Logic at Princeton," in *A Century of Mathematics in America*, ed. Peter Duren et al., 2: 217–221.

in the rapidly evolving field. In 1926, for example, Alexander significantly generalized the results that he and Veblen had obtained (mod 2) in their 1913 paper to results (mod n), while Lefschetz proved his famous fixed point theorem, a result that provided an actual formula for counting the number of fixed points of a continuous transformation of manifolds.[33] They, but especially Veblen and Lefschetz, also trained members of a next generation of algebraic topologists that included, in the decade of the 1920s, Veblen's student, Philip Franklin at MIT, and Lefschetz's student, Paul Smith of Columbia's Barnard College.[34] These young mathematicians were complemented by others like the University of Iowa's Edward Chittenden, who had earned his Ph.D. in 1912 at Chicago under E. H. Moore for a thesis on Moore's brand of general analysis.[35]

By 1930, then, the time was already ripe for the new overview of results that Lefschetz provided on the occasion of his AMS Colloquium Lectures at Brown University. As he explained, while Poincaré had left "the foundations" of combinatorial analysis situs "in a rather unstable equilibrium," "[i]t is largely to Veblen and Alexander that we owe the remedy for this state of affairs, and the present improved situation." In fact, as Lefschetz saw it, "[a] date marks the transition: 1922, when there appeared Veblen's excellent *Cambridge Colloquium Lectures: Analysis Situs*, which has deservedly become the standard work on the subject."[36]

In his own Colloquium volume, entitled simply *Topology*, Lefschetz pushed beyond Veblen's work to deal with what he termed the "new phases

---

33. James Alexander, "Combinatorial Analysis Situs," *TAMS* 28 (1926), 301–329 and Solomon Lefschetz, "Intersections and Transformations of Complexes and Manifolds," *TAMS* 28 (1926), 1–49, respectively. The AMS awarded its third Bôcher Prize to Alexander in 1928 for this paper; Lefschetz had won the second prize in 1924 for an earlier paper in an algebraic-geometric vein. This prize per se is treated in the next chapter. Dieudonné treats Alexander's work in a technical context in *A History of Algebraic and Differential Topology*, pp. 36–59, while Robert Brown discusses Lefschetz's development of the theory of fixed points in "Fixed Point Theory," in *History of Topology*, ed. Ioan James (Amsterdam; Elsevier Science B.V., 1999), pp. 271–299 on pp. 275–280.

34. The Englishman, Henry Whitehead, also studied at Princeton beginning in 1929 and earned his Ph.D. there in 1932. See chapter three.

35. Reinhard Siegmund-Schultze provides a historical contextualization of this work in "Eliakim Hastings Moore's 'General Analysis,'" *Archive for History of Exact Sciences* 52 (1998), 51–89.

36. Solomon Lefschetz, *Topology*, AMS Colloquium Publications, vol. 12 (New York: American Mathematical Society, 1930), p. iii. The next quotation is also from this page.

of the subject." Among those were the fixed point theory that he had been developing as well as new results on duality. Relative to the latter, there were then two types of duality relations: "those discovered by Poincaré which exist between the various connectivity indices of a manifold, and those due to Alexander in which the invariants of a surrounding residual space also enter."[37] In reviewing his former advisor's book, Paul Smith highlighted the fact that Lefschetz's "discovery that these two types of relations are special cases of a third more general type, is revealed in a set of formulas of striking symmetry and generality." Smith closed on a boosteristic note that reflected Courant's view of the strength of American algebraic topology. "Analysis situs," he acknowledged, "is a comparatively young science," but "[i]t is pleasant to reflect that much of what has been accomplished has been the work of American mathematicians, and to that work the present volume is a distinguished contribution."

The other branch of topology—point-set analysis situs—grew out of the set-theoretic work on which Georg Cantor had embarked beginning in the 1870s. Cantor concerned himself with deep, fundamental questions about the real line that involved concepts like limits, convergence, and continuity.[38] He tackled them through a whole new theory of sets that rested on formalized notions such as open and closed sets and the set of limit points of a set. At the hands of Maurice Fréchet, Frigyes Riesz, and Felix Hausdorff into the 1910s, these ideas were extended and developed into a general theory of topological spaces independent of any particular metric. It was this kind of analysis situs that ultimately attracted Veblen's slightly younger colleague, R. L. Moore.

Moore's promising start at Chicago was followed by a decade during which the newly minted Ph.D. cast about for both a productive line of research and

37. Paul Smith, "Lefschetz on Topology," *BAMS* 37 (1931), 645–648. The quotations that follow are on pp. 646 and 648, respectively. Alexander presented what is now called "Alexander duality" in "A Proof and Extension of the Jordan-Brouwer Theorem," *TAMS* 23 (1922), 333–349. In addition to this important work, Alexander discovered the Alexander horned sphere, a particular embedding of a sphere in Euclidean three-space that cuts it into two regions, one of which is not simply connected. He also applied topological methods to the theory of knots. See James Alexander, "An Example of a Simply Connected Surface Bounding a Region Which is Not Simply Connected," *PNAS* 10 (1924), 8–10 and "Topological Invariants of Knots and Links," *TAMS* 30 (1928), 275–306, respectively.

38. Joseph Dauben treats Cantor and his work in detail in *Georg Cantor: His Mathematics and Philosophy of the Infinite* (Cambridge: Harvard University Press, 1979).

a suitable position.[39] After short stints on the faculties at the University of Tennessee and at Northwestern as well as at Princeton with Veblen, Moore accepted a call in 1911 to the University of Pennsylvania, where he began to find his academic and professional footing. By 1916, he had published what proved to be a seminal paper "On the Foundations of Plane Analysis Situs,"[40] and he had directed the thesis research of his first student, John Kline. In a sense, these two events defined the subsequent course of Moore's career as a researcher and teacher.[41]

His 1916 paper harkened back philosophically and mathematically both to David Hilbert's 1899 *Grundlagen der Geometrie* and to Cantor's work in the 1870s and 1880s on the theory of point sets.[42] In it, Moore took the two notions of "point" and "region" as primitive: if S is a class of elements called "points," then a "region" is a class of subclasses of points, that is, a region is a class of what Moore termed "point-sets." He then stated a number of axioms in terms of these primitives and, from them, developed the topology of the Euclidean plane, giving topological characterizations of such notions as the simple arc and the simple closed curve. At the same time, he was careful to provide examples that demonstrated the independence of the axioms one from the others.[43] This postulate-theoretic mode of reasoning—learned during his student days at Chicago—characterized much of Moore's subsequent research as well as the eponymous style of teaching that he developed in which students independently test conjectures and derive mathematical theorems from a set of axioms.[44]

39. Raymond Wilder, "The Mathematical Work of R. L. Moore," *Archive for History of Exact Sciences* 26 (1982) 73–97 on p. 77; reprinted in *A Century of Mathematics in America*, ed. Peter Duren et al., 3: 265–291 on pp. 269–270. There is one full-length biography of Moore: John Parker, *R. L. Moore: Mathematician and Teacher* (n.p.: Mathematical Association of America, 2005).

40. R. L. Moore, "On the Foundations of Plane Analysis Situs," *TAMS* 17 (1916), 131–164.

41. Compare Albert Lewis, "The Beginnings of the R. L. Moore School of Topology," *Historia Mathematica* 31 (2004), 279–295 and Wilder, "The Mathematical Work of R. L. Moore," pp. 79–80 (or pp. 272–273 in the reprinted edition).

42. See, for example, Jerome Manheim, *The Genesis of Point Set Topology* (London: The Macmillan Company, 1964).

43. Wilder, "The Mathematical Work of R. L. Moore," pp. 79–82 (or pp. 272–275 in the reprinted edition.)

44. David Zitarelli discusses this technique and its reception in "The Origin and Early Impact of the Moore Method," *AMM* 111 (2004), 465–486. For more on the role of postulate theory in the United States in the opening decade of the twentieth century, see the section on algebra below.

Like Veblen before him, Moore codified his ideas in the context of AMS Colloquium Lectures. Speaking in Boulder, Colorado, in 1929, he laid out his point-set brand of analysis situs. When the printed volume appeared in 1932—two years after Lefschetz's account of algebraic topology—Moore had greatly extended the results derivable from the axiomatic set-up for point-set topology that he had presented in his 1916 paper, among them the Moore-Kline theorem that gives necessary and sufficient conditions under which a closed set is a subset of an arc.[45] As Harry Gehman of the State University of New York at Buffalo saw it, "this book will undoubtedly be an excellent text from which to obtain an insight into the nature of the problems considered by the school of mathematicians headed by Professor Moore."[46] In addition to Moore, that school ultimately consisted in the 1920s of Moore's students—such as Kline at Penn, Raymond Wilder first at Ohio State but then at Michigan, and Gordon Whyburn ultimately at the University of Virginia—and students of these students—like Kline's students, Gehman, Leo Zippin first at Penn State but later at Queens College in New York, and William Ayres at Michigan.[47]

By the 1920s, then, the United States sustained two largely disjoint schools of topology. One was associated with Princeton and was actively spearheaded by Lefschetz after 1925. The other was linked with R. L. Moore especially after he settled in Austin in 1920. Led by strong-minded advocates, these two topological camps fairly quickly found themselves in competition. Kline, as a professor in Philadelphia and a topologist of the point-set variety, felt this

45. R. L. Moore, *Foundations of Point Set Theory*, AMS Colloquium Publications, vol. 13 (New York: American Mathematical Society, 1932), pp. 317–322. (Moore and Kline had proven their theorem in "On the Most General Closed Point-set Through Which It Is Possible To Pass a Simple Continuous Arc," *AM* 20 (1919), 218–223.) A main line of research that emerged from Moore's approach in this book was "the search for necessary and sufficient conditions for the metrizability of topological spaces" (Lynn Steen, "Conjectures and Counterexamples in Metrization Theory," *AMM* 79 (1972), 113–132 on p. 113).

46. Harry Gehman, "Moore on Point Sets," *BAMS* 39 (1933), 479–483. Gehman did, however, point out a number of "minor inaccuracies" (p. 481) at the same time that he leveled a number of criticisms at the book.

47. The Moore school has received much historical consideration, for example, Lewis, pp. 285–288, as well as Ben Fitzpatrick, "Some Aspects of the Work and Influence of R. L. Moore" and F. Burton Jones, "The Beginning of Topology in the United States and the Moore School," the latter both in *Handbook of the History of General Topology*, ed. Charles Aull and Robert Lowen, 1: 41–61 and 97–103, respectively. Fitzpatrick and Jones counted themselves among the members of the "Moore school."

rivalry particularly keenly, given that his closest mathematical colleagues were the combinatorial topologists in Princeton.

In March 1925, for example, Lefschetz ran into Kline and Gehman at the AMS meeting in New York City and learned that Gehman was applying for an NRC fellowship to study with his mathematical "grandfather," R. L. Moore in Austin.[48] As Moore related in a letter to Kline, Lefschetz promptly wrote to tell him that he had "strongly urged" Gehman to go to Princeton first in order to "get all he could from the local analysis situs gang before going to you."[49] As Lefschetz saw it, that "would be a very excellent thing for both gangs, the local and yours," since then Gehman could "go down to Texas and thus establish the bridge, etc." Lefschetz's query—"What do you think of it?"—drew a sharp rebuke from Moore. "As to what I think of it," Moore sniped to Kline, "it doesn't sound *sincere* to me. If Gehman wants to go into the Princeton line of analysis situs—let him *go*, with his eyes *open*. But don't let him go with his eyes half-*shut*, led by some pretense that in that way he will be better prepared to come down here. He has started on a definite line of work. If he wants to continue that line let him do it. If he *doesn't*, let him do *that*."

Kline could not have agreed more. "This whole matter makes me mad," he told Moore.[50] Lefschetz and others had recognized a strong student in Gehman and were trying to win him over to their point of view. As Kline put it to Moore, the Princetonians hoped to convince Gehman "that our line was . . . highly specialized," while theirs was less so. Moreover, they announced that they "did not appreciate our type of Analysis Situs," so "it was for our good to have the two bridged etc. etc."[51] The rivalry inherent in this

48. These fellowships and the role that they played in mathematics are considered in the next chapter.

49. Moore to Kline, 7 March, 1925, Box 4RM74, Folder: Kline, John Robert (1925–1928), Moore Papers. The quotations that follow are also from this letter with Moore's emphases.

50. Kline to Moore, 10 March, 1925, Box 4RM74, Folder: Kline, John Robert (1925–1928), Moore Papers. The quotations that follow are also from this letter.

51. Indeed, this perception of the point-set approach was not limited to members of the combinatorial camp. Albert Bennett, for example, was a Princeton-trained algebraist who took a position at Brown in 1927 after stints at the University of Texas and Lehigh University. When John Kline's student, William Ayres, was on the job market following an International Education Board fellowship year in Vienna, Bennett mentioned him to Roland Richardson as a possible hire, but with a caveat: "as with a number of people working on R. L. Moore's form of analysis situs, his interests are probably rather narrow." See Bennett to Richardson, 21 March, 1929, Box 3: Correspondence, 1926–1930 'A-B,' Folder: Bennett, Albert Arnold, Richardson Papers.

exchange continued unabated into the 1930s and served to spur the further development of both camps in the interwar period (see chapter six).

Geometries, Differential and Algebraic

Americans also pursued two types of geometry in the 1920s—differential and algebraic—but, in this instance, there was no rivalry and little overlap between the respective practitioners. By and large, the differential geometers were motivated by the still-recent discovery and ongoing development of Einstein's general theory of relativity, while the algebraic geometers drew primarily from a nineteenth-century tradition imported to American shores by students of Felix Klein as well as by James Joseph Sylvester and his friend and mathematical confidant, Arthur Cayley.[52] As characterized by Harvard's self-described "modern" geometer, Julian Coolidge, the differential geometers studied the properties of figures as revealed by the differential calculus and worked more abstractly in terms of groups of one-to-one, analytic transformations. The algebraic geometers, on the other hand, were concerned with uncovering the properties of figures in terms of algebraic relations that linked their coordinates or their equations and worked with birational groups of one-to-one, algebraic transformations.[53]

Although differential geometry had adherents at Chicago in Ernest Lane, at Columbia in Edward Kasner, and at Harvard in William Graustein, the American center in the field in the 1920s was Princeton where Veblen and Luther Eisenhart attracted both graduate students and postdoctoral fellows to their vibrant intellectual environment. That Veblen was also a leader in algebraic topology attests to how closely related that flavor of topology is

52. Samson Duran treats American geometrical research in the period from 1888 to 1920, that is, in the period immediately preceding the one considered here, in "Des géométries états-suniennes à partir de l'étude de l'*American Mathematical Society*: 1888–1920" (unpublished doctoral dissertation, Université Paris-Sud (Orsay), 2019).

53. Julian Coolidge, *A History of Geometrical Methods* (Oxford: University Press, 1940; reprint ed., New York: Dover Publications, Inc., 1963), p. viii. Coolidge also isolated two other types of geometry in his book: synthetic geometry, as exemplified by Euclid's *Elements*, and topology. Coolidge's characterization of himself as a "modern" geometer reflected his broad interests: in non-Euclidean geometry, which he categorized as synthetic geometry, as well as in algebraic geometry. See Jaques Cattell, ed., *American Men of Science: A Biographical Dictionary*, 7th ed. (Lancaster: The Science Press, 1944), p. 357 and Dirk Struik, "Julian Lowell Coolidge," *AMM* 62, (1955), 669–682, especially pp. 674–682.

to the differential flavor of geometry. As for algebraic geometry, it was fostered somewhat more diffusely in the 1920s by, among others, Klein's students Virgil Snyder at Cornell and Henry White at Vassar,[54] Cambridge-trained Frank Morley at Hopkins and Charlotte Angas Scott at Bryn Mawr, Morley's student, Arthur Coble, at both Hopkins and the University of Illinois, and Julian Coolidge at Harvard. Of these, only White at Vassar, an undergraduate women's college, trained no doctoral students in the field, although that made him no less of a research participant. While still in Kansas, moreover, Lefschetz also engaged in geometric research of an algebraic bent prior to turning his attentions more exclusively to topology.

Of these two geometric research streams, differential geometry was unquestionably the more exciting and the more avant-garde in the 1920s, as leading mathematicians in Europe and the United States tried effectively to mathematize the general theory of relativity.[55] Yet, it also came in different flavors. The more classical versions, which extended and developed late-nineteenth-century work, also had their representatives in the United States.

For example, in 1906, Ernest Wilczynski published the first American text on projective differential geometry, a subarea that had grown out of work particularly by Gaston Darboux and Georges Halphen in France.[56] After he assumed the professorship at the University of Chicago in 1910 that he would hold until ill health forced him from the classroom in 1923, Wilczynski not only actively pursued his own research in this area but also produced almost two dozen Ph.D.s in it. Among them, Ernest Lane took over for his advisor

54. Parshall and Rowe discuss Snyder's and White's training under Klein in *Emergence*, pp. 202–229.

55. Tracy Thomas gives an overview of work in this field in the 1920s to 1938 in "Recent Trends in Geometry," in *Semicentennial Addresses of the American Mathematical Society*, ed. The Committee on Publications (New York: American Mathematical Society, 1938), pp. 98–135. Other efforts to mathematize the general theory of relativity as well as quantum mechanics had their roots in mathematical physics. Among the American contributors to these research strains were two Russian-born Americans, Paul Epstein at the California Institute of Technology and George Rainich at the University of Michigan. See, for example, Paul Epstein, "On the Evaluation of Certain Integrals in the Theory of Quanta," *PNAS* 12 (1926), 629–633 and George Rainich, "Electrodynamics in General Relativity," *TAMS* 17 (1925), 106–136.

56. Ernest Wilczynski, *Projective Differential Geometry of Curves and Ruled Surfaces* (Leipzig: B. G. Teubner Verlag, 1906). Duran treats Wilczynski's work in some detail in chapter six of his dissertation.

in 1923 and continued to churn out graduate students.[57] As Saunders Mac Lane pejoratively characterized it, "Chicago had become in part a Ph.D. mill in mathematics" in the 1920s.[58] Moreover, what Mac Lane termed its "inheritance principle" in hiring—that is, the replacement of faculty members by their former students without particular regard for the evolution of newer and more exciting mathematical ideas—resulted in a certain stagnation there in geometry as well as in the calculus of variations (see below) despite a prodigious output of new Ph.D.s in these fields. Those doctorate holders nevertheless left Chicago to populate American colleges and universities desirous of ostensibly better credentialed, more research-oriented faculties.

At Columbia, Edward Kasner also worked in differential geometry along more classical lines, exploring the purely geometric properties of the trajectories—defined in terms of certain differential equations—of particles moving in general positional fields of force. When 1919 brought the confirmation of Einstein's prediction that light rays bend when passing close to a large gravitational mass like the Sun, Kasner redirected his techniques to the problem of teasing out the more "purely mathematical aspects of . . . relativity theory, based as it is, on regarding the space-time continuum as a four-dimensional Riemannian manifold."[59] In particular, in a flurry of work presented to the AMS in 1921 (some of which was not actually published until 1925), Kasner studied the mathematical ramifications of Einstein's cosmological equations, finding, for example, that an Einstein space that was not itself Euclidean could not be embedded in a five-dimensional Euclidean space.[60]

57. At the University of Kansas, Ellis Stouffer had also been a student of Wilczynski and pursued his advisor's brand of differential geometry there. See, for example, Ellis Stouffer, "Singular Ruled Surfaces in Space of Five Dimensions," *TAMS* 29 (1927), 80–95. Stouffer, however, had few students compared to his academic "brother," Lane.

58. Saunders Mac Lane, "Mathematics at the University of Chicago: A Brief History," in *A Century of Mathematics in America*, ed. Peter Duren et al., 2: 127–154 on p. 138. The quotation that follows is on p. 141. Mac Lane was a graduate student at Chicago for one year, 1930–1931, but left to earn his doctorate at Göttingen. He returned to Chicago as an instructor for the 1937–1938 academic year, moved to Harvard, and then returned to the Chicago faculty in 1947 thanks to the efforts of Marshall Stone (see chapter nine).

59. Jesse Douglas, "Edward Kasner, 1878–1955," *Biographical Memoirs*, vol. 31 (Washington, D.C.: National Academy of Sciences, 1958), pp. 179–209 on p. 195. Douglas gives a nice technical overview of Kasner's work in this tribute.

60. Edward Kasner, "The Impossibility of Einstein Fields Immersed in Flat Space of Five Dimensions," *AJM* 43 (1921), 126–129.

William Graustein at Harvard also pursued differential geometry from a classical point of view and, in the 1920s, drew on that background to treat questions arising from the mathematization of Einstein's theory. Graustein was intrigued by the implications of the application to differential geometry of the tensor calculus that Italians Gregorio Ricci-Cubastro and Tullio Levi-Civita had developed in the opening years of the twentieth century and that the physicists were then employing.[61] As his colleague and biographer, Julian Coolidge, explained, while "[m]any geometers threw themselves entirely into the new work," "Graustein was more cautious." "[H]e recognized the advantages in the new notations, new points of view and new techniques, especially when more than three dimensions were involved. But what attracted him most was the invariant or covariant character of the new processes, and that led him to the idea of developing methods on the more classical lines."[62] Graustein's efforts resulted in the paper "Méthodes invariantes dans la géométrie infinitésimale," which, although published only in 1929, won the Royal Academy of Belgium's 1925 prize "for an important contribution to infinitesimal [that is, differential] geometry."[63] Graustein had succeeded in producing fruitful, new techniques for determining "what sort of things are invariant under the transformations of differential geometry," and he laid them out, this time in English by invitation of the AMS, at its meeting in April 1930.[64]

It was at Princeton, however, that Luther Eisenhart bridged the old and the new differential geometry and, with Veblen, inspired novel research

61. The same was true of Graustein's mathematical neighbor, Dirk Struik. Struik, later perhaps better known for his work as a historian of mathematics, had been a student of Jan Schouten at Delft and had done postdoctoral work with Levi-Civita in Rome before settling in the United States at MIT. There, he was a colleague and collaborator (in differential geometry) of Norbert Wiener in addition to pursuing his own differential geometric work. See, for example, Dirk Struik and Norbert Wiener, "A Relativistic Theory of Quanta," *JMP* 7 (1927–1928), 1–23 and Dirk Struik, "On Sets of Principle Directions in a Riemannian Manifold of Four Dimensions," op. cit., 193–197.

62. Julian Coolidge, "William Caspar Graustein–In Memoriam," *BAMS* 47 (1941), 343–349 on p. 345.

63. William Graustein, "Méthodes invariantes dans la géométrie infinitésimale," *Mémoires de l'Académie royale de Belgique (Classe des Sciences)* 11 (1929), 1–96. For the quotation, see "Notes," *BAMS* 32 (1926), 176–186 on p. 177.

64. William Graustein, "Invariant Methods in Classical Differential Geometry," *BAMS* 36 (1930), 489–521. The quotation appears in Coolidge, "William Caspar Graustein–In Memoriam," p. 345.

directions. Prior to Einstein's work, Eisenhart had continued to pursue the research line stemming from the doctoral work on the "Infinitesimal Deformation of Surfaces" that he had completed in 1900 at Hopkins under the guidance of Sylvester's student, Thomas Craig.[65] This had culminated, in some sense, in 1923 with the publication of *Transformations of Surfaces*, in which Eisenhart gave the first unified, book-length treatment of the research that had been done up to that time on the generalization of three-dimensional differential geometry to n dimensions.[66] Interestingly, this was one of the first mathematical monographs to be published through a subvention provided by the National Research Council (see the next chapter).

Three years earlier, Eisenhart, with his graduate-level background in mathematics, physics, and astronomy, had already begun embracing the new Einsteinian physics from a mathematical point of view. He was thus the obvious person to introduce American mathematicians to those ideas at a special, afternoon-long symposium held in conjunction with the April 1920 meeting of the AMS at Columbia. Eisenhart, who spoke on the "Geometric Aspects of the Einstein Theory," shared the stage with physicist Leigh Page of Yale, who discussed "The Physical and Philosophical Significance of the Principle of Relativity and Einstein's Theory of Gravitation." Some fifty mathematicians were present to hear their remarks.[67]

By October, Eisenhart had written to Einstein himself, inviting him to come to Princeton to lecture for a semester on his evolving ideas. Although that initial invitation was declined, Einstein did visit Princeton the following May to give the Stafford Little Lectures. Their published English version

65. When Eisenhart was a student at Hopkins, Craig was hard at work on a book on the theory of surfaces, and Eisenhart consistently took his courses. Eisenhart also took a number of physics and astronomy courses as a graduate student, although astronomer Simon Newcomb was not teaching at the time. Eisenhart's doctoral committee consisted, however, of Craig and Newcomb. It seems safe to say that Craig was Eisenhart's doctoral advisor, although, unfortunately, Craig died on 8 May, 1900, just a month before Eisenhart officially graduated. See *Johns Hopkins University Circulars* 17–19 (1897–1900), especially, "Degrees Conferred June 12, 1900," 19 (June 1900), 84–85 on p. 84. Eisenhart published his dissertation as "Infinitesimal Deformation of Surfaces," *AJM* 24 (1902), 173–204.

66. Luther Eisenhart, *Transformations of Surfaces* (Princeton: Princeton University Press, 1923) was favorably reviewed by William Graustein in "Eisenhart's Transformation of Surfaces," *BAMS* 30 (1924), 454–460.

67. Frank Cole, "The April Meeting of the American Mathematical Society in New York," *BAMS* 26 (1920), 433–444 on p. 435. Eisenhart published his remarks in "The Permanent Gravitational Field in the Einstein Theory," *AM* 22 (1920), 86–94.

"became the classic Einsteinian introduction to general relativity in the English-speaking world and served as an implicit declaration by Princeton University of its claim to be the center of relativity research in America."[68]

Eisenhart and Veblen began to set up that center as early as the 1921–1922 academic year when they offered their joint seminar on "The Theory of Relativity" and began to publish papers on their emergent ideas. In the first of those, joint work on "The Riemannian Geometry and Its Generalizations," they laid the groundwork for what they termed the geometry of paths. As they explained, "[o]ne of the simplest ways of generalizing Euclidean Geometry is to start by assuming (1) that the space to be considered is an n-dimensional manifold in the sense of Analysis Situs, and (2) that in this space there exists a system of curves called paths which, like the straight lines in a euclidean space, serve as a means of finding one's way about."[69] These paths, defined as the solutions of a particular system of differential equations, generated, in Eisenhart and Veblen's view, "a more natural" geometry in terms of which to mathematize space than that then-recently developed by Hermann Weyl and Arthur Eddington because, under certain conditions, it reduces to Riemannian geometry. One problem then became to determine "under what conditions the geometry of paths is Riemannian." The exploration of that and other questions launched Eisenhart and Veblen on a research agenda in the geometry of paths, in particular, and in differential geometry, more generally, that occupied not only them but also a string of students and postdoctoral fellows—Tracy Thomas, Harry Levy, Morris Knebelman, Joseph Thomas, Aristotle Michal, Jesse Douglas, and Henry Whitehead, among others—as well as new faculty members—Howard "Bob" Robertson and beginning in 1930 John von Neumann and Eugene Wigner—throughout the 1920s and into the 1930s.[70] In particular, Michal engendered a so-called "Pasadena school" of differential geometry applied to physics on the West Coast at the

68. Jim Ritter, "Geometry as Physics: Oswald Veblen and the Princeton School," in *Mathematics Meets Physics: A Contribution to Their Interaction in the 19th and the First Half of the 20th Century*, ed. Karl-Heinz Schlote and Martina Schneider (Frankfurt: Verlag Harri Deutsch, 2011), pp. 146–179 on p. 153.

69. Luther Eisenhart and Oswald Veblen, "The Riemannian Geometry and Its Generalizations," *PNAS* 8 (1922), 19–23 on p. 19. The next two quotations appear on pp. 20 and 20–21, respectively.

70. For more on the work particularly of Joseph Thomas and Jesse Douglas in the 1920s, see chapter three. Ritter gives the full story of the Princeton research center in the geometry of paths in the article cited above.

California Institute of Technology (Caltech) beginning in 1929 (see chapter six).[71]

To promote their agenda, Eisenhart published two more synthetic texts in the 1920s. His *Riemannian Geometry* of 1926 provided an advanced introduction to the subject that incorporated an exposition of recent results including some of his own, while *Non-Riemannian Geometry*, the topic of his 1925 AMS Colloquium Lectures, gave a systematic treatment of the new mathematics that was evolving, especially at Princeton, from the geometry of paths.[72] Veblen, too, contributed to the codification of this work in his 1927 treatment of *Invariants of Quadratic Differential Forms* as well as in the *Foundations of Differential Geometry* that he co-authored with his Ph.D. student, Henry Whitehead, in 1932.[73]

Veblen had conceived of creating within Princeton's Department of Mathematics a mathematical research group that, in pooling the individual strengths of its members and working collaboratively, would serve to focus international attention on mathematics in the United States and to "advance the position and role of American mathematics in the new post-war world."[74] Together with Eisenhart, he achieved that goal in the 1920s with the generation of a new brand of differential geometry that found itself in active competition with rival schools in the Netherlands under Jan Schouten and in France under Élie Cartan.

The 1920s also witnessed the continued development of algebraic geometry on American shores. As Veblen explained in a 1926 sketch of the contours of the American mathematical landscape, again for the NRC's Vernon Kellogg, "[t]he development of mathematics on [an] extensive scale in this country was brought about by a series of waves of interest in new subjects," and the first of those, thanks to the influence of Sylvester and Cayley, had been algebraic geometry.[75] By the 1920s, however, much of that work was

71. Tracy Thomas, "Recent Trends in Geometry," p. 120.

72. Luther Eisenhart, *Riemannian Geometry* (Princeton: Princeton University Press, 1926) and *Non-Riemannian Geometry*, AMS Colloquium Publications, vol. 8 (New York: American Mathematical Society, 1927).

73. Oswald Veblen, *Invariants of Quadratic Differential Forms*, Cambridge Tracts in Mathematics and Mathematical Physics, no. 24 (Cambridge: University Press, 1927) and (with Henry Whitehead) *The Foundations of Differential Geometry*, Cambridge Tracts in Mathematics and Mathematical Physics, no. 29 (Cambridge: University Press, 1932).

74. Ritter, p. 152.

75. Veblen to Kellogg, 7 April, 1926, Box 7, Folder: Kellogg, Vernon 1924–28, Veblen Papers.

beginning to look dated in comparison with what was coming out of Germany informed by the algebraic insights of Emmy Noether and others. Still, from their more shielded vantage point, America's algebraic geometers felt that the time was ripe to survey their field, and they did so in 1928 under the auspices of no less than the National Academy of Sciences. Their aim? To aid "investigators in this field" as well as to serve "a wider circle."[76] As the 1920s came to a close, they had no reason to doubt that their approach would have anything but a bright future.

The survey's authors—Virgil Snyder, Arthur Coble, Arnold Emch, Solomon Lefschetz, Francis Sharpe, and Charles Sisam—reflected the changing demographics of American algebraic geometry. Snyder had returned from Göttingen to take up, in 1895, the teaching position at Cornell that he would hold for his entire career. There, he taught many in his classrooms—among whom was his future Cornell colleague, Francis Sharpe—and trained in his style of geometric research almost forty graduate students, one of whom was Colorado College's Charles Sisam. Coble, who, as noted, had done his doctoral work under Morley at Hopkins in 1902, taught with his colleague Emch at the University of Illinois for all but one year of the 1920s. Together he and Emch, like Snyder, produced like-minded graduate students throughout their long careers. Finally, Lefschetz earned his Ph.D. at Clark University under Sylvester's student and successor at Hopkins, William Story, leaving Kansas for Princeton in 1925. Whereas in many regards his co-authors on the survey perpetuated algebraic geometry's past, he reflected its future with his dual interests in algebraic geometry and algebraic topology.

Readers of the collaborative survey that these men wrote found lengthy lists of results and extensive bibliographies of mostly nineteenth-century works, at least in the first fourteen of the volume's seventeen chapters. Those were the chapters written by Snyder, Coble, Emch, Sharpe, and Sisam. The largely nineteenth-century mathematicians who inspired them and their fellow American algebraic geometers into the 1920s were men like Julius Plücker, Felix Klein, Max Noether, Alexander von Brill, and Alfred Clebsch in Germany, Cayley, Sylvester, and George Salmon in the British Isles, Luigi Cremona, Guido Castelnuovo, Federigo Enriques, and Gino Fano in Italy, and Gaston Darboux and Georges Halphen in France. Theirs were the techniques

---

76. Virgil Snyder, Arthur Coble, Arnold Emch, Solomon Lefschetz, Francis Sharpe, and Charles Sisam, *Selected Topics in Algebraic Geometry*, Bulletin of the National Research Council, no. 63 (Washington, D.C.: National Research Council of the National Academy of Sciences, 1928), p. 3.

that the Americans continued to employ. Theirs was the approach that the Americans continued to play out.

For example, Charlotte Angas Scott, an 1885 D.Sc. from the University of London who actually did her doctoral work under Cayley at Cambridge,[77] moved to the United States to take a position on the first faculty at Bryn Mawr in 1885. Modeled on Hopkins, Bryn Mawr was the only women's college in the United States that offered graduate training, albeit in a limited number of subjects deemed key.[78] One of those, however, was mathematics, and Scott crafted and animated a program in the field until her retirement in 1924. At the same time, she continued to pursue algebraic geometric research that focused on such matters as the intersections and singularities of plane algebraic curves. As fellow Briton Francis Macaulay described her, Scott was "an enthusiastic searcher and propounder of new ideas" as well as a gifted "interpreter of the work of others, adding simplifications and extensions of her own."[79] She shared those insights in the course of training seven graduate students, two in the 1920s. One of the latter, Marguerite Lehr, ultimately succeeded Scott on the Bryn Mawr faculty.

Among Scott's "interpretations" was Max Noether's so-called Fundamental Theorem: "Given two algebraic curves in the same plane, $f = 0$, $\phi = 0$. Every curve which has at least the multiplicity $r_i + s_i - 1$ at every point, distinct or clustering, common to the two curves, where f has the multiplicity r_i and ϕ the multiplicity s_i, has an equation of the form $F \equiv \phi' f + f' \phi = 0$, where f' has the multiplicity $r_i - 1$ at least, and ϕ' the multiplicity $s_i - 1$ at least."[80] Although Noether had given a justification of this in 1873, it had

77. Scott had been an undergraduate at all-female Girton College, Cambridge and had even come in eighth on the infamous Mathematical Tripos after being given permission, as a woman, to take it. Since Cambridge did not officially grant degrees to women in the 1880s, she took both her 1882 B.Sc. and her 1885 D.Sc. at London (Patricia Kenschaft, "Charlotte Angas Scott (1858–1931)," in *Women in Mathematics: A Biobibliographic Sourcebook*, ed. Louise Grinstein and Paul Campbell (Westport: Greenwood Press, 1982), pp. 193–203).

78. For more on that program, see my "Training Women in Mathematical Research: The First Fifty Years of Bryn Mawr College (1885–1935)," *MI* 37 (2) (2015), 71–83 as well as Jemma Lorenat's "'Actual Accomplishments in This World': The Other Students of Charlotte Angas Scott," *MI* 42 (2020), 56–65.

79. Francis Macaulay, "Dr. Charlotte Angas Scott," *Journal of the London Mathematical Society* 7 (1932), 230–240 on p. 232.

80. Max Noether, "Über einen Satz aus der Theorie der algebraischen Funktionen," *Mathematische Annalen* 6 (1873), 351–359 on p. 351. Coolidge gives an English treatment of the theorem in *A History of Geometrical Methods*, p. 205.

not been deemed particularly satisfying. This motivated Scott to provide in a paper published in 1899 what was later termed the theorem's "best proof."[81]

Twenty-six years later, Vassar's Henry White was still working along these lines. In *Plane Curves of the Third Order*, a book like Eisenhart's *Transformations of Surfaces* published with an NRC subvention, he aimed to provide an introduction to what he viewed as the "rich and attractive field" of cubic curves and, in so doing, to provide "a stepping-stone to many extensive and beautiful treatises on special themes, and a stimulus to further exploration."[82] The book mainly treated the invariant theory of the cubic—in the style of Clebsch and Gordan that White had studied and reported on as a graduate student in Klein's seminar at the end of the 1880s[83]—-but it also explored the explicitly geometrical properties of cubic curves from an algebraic point of view. What are their inflection points? Describe and analyze their tangents. "Can a pentagon be inscribed in a cubic so that every point where a side meets the opposite diagonal shall be a point on the curve?" White dealt with these and other questions in what Charles Sisam appreciatively termed "the most natural and logical manner," that is, "by establishing and using Noether's fundamental theorem."[84]

This example—from Noether's 1873 result to Scott's turn-of-the-twentieth-century reproving of it to White's 1925 continued exploration of it in the particular context of cubic curves—illustrates well not only the nineteenth-century inspiration for much of American algebraic geometry in the 1920s but also the perpetuation of that classical style by an active community of practitioners. Harvard's Julian Coolidge also continued in this vein in his 1931 text, *A Treatise on Algebraic Plane Curves*, although Snyder criticized the work for its effort to treat "[a] great many, perhaps too many, points of view."[85] In exasperation, Snyder described "[t]he expansion of the field during the last

81. Charlotte Angas Scott, "A Proof of Noether's Fundamental Theorem," *Mathematische Annalen* 52 (1899), 593–597. Coolidge gives the characterization in *A History of Geometrical Methods*, p. 205 (note †).

82. Henry White, *Plane Curves of the Third Order* (Cambridge: Harvard University Press, 1925), pp. vi-vii. The quotation that follows is on p. 136.

83. Parshall and Rowe detail White's presentations in Klein's seminar in *Emergence*, pp. 223–229 and 255–257.

84. Charles Sisam, "White on Cubic Curves," *BAMS* 32 (1926), 555–556 on p. 555.

85. Julian Coolidge, *A Treatise on Algebraic Plane Curves* (Oxford: Clarendon Press, 1931) as reviewed by Virgil Snyder, "Coolidge on Algebraic Curves," *BAMS* 38 (1932), 163–165 on p. 163. The quotation that follows is also on this page.

half-century" as "simply appalling." He thus gave expression to an insider's view of an epoch in the history of algebraic geometry much later characterized by mathematician Jean Dieudonné as that of "development and chaos," namely, the period from the mid-nineteenth century to 1920.[86] That was precisely the era among the last representatives of which were Coolidge, Scott, White, Snyder and his survey co-authors, and others like Morley.

At the same time that it contributed to the cacophony characteristic of this pre-1920 period, Solomon Lefschetz's work also suggested some of the new research directions of what Dieudonné styled a next epoch of "new structures in algebraic geometry." In, for example, the influential 1924 monograph, *L'analysis situs et la géométrie algébrique*, that he wrote just before leaving Kansas, Lefschetz applied the evolving techniques of algebraic topology to classical algebraic geometry and thereby revealed the latter's "essentially topological nature."[87] In a 1926 letter to Hermann Weyl, he had confessed his hope of having at least begun the process of "bring[ing] the theory of Algebraic Surfaces under the fold of Analysis and An[alysis] Situs." As he saw it, "[t]here is a great need to unify mathematics and cast off to the wind all unnecessary parts leaving only a skeleton that an average mathematician may more or less absorb. Methods that are extremely special should be avoided."[88] Lefschetz thus foresaw a future for algebraic geometry in which new and very different techniques would supplant those of the past. His work, in fact, influenced one of that future's European shapers.

The Dutchman Bartel van der Waerden had studied algebra at the feet of Emmy Noether in Göttingen in the early 1920s. By the middle of the decade, he had begun a project of "algebraizing algebraic geometry *à la*" Noether that had ultimately and interestingly led him to Lefschetz's 1924 work.[89] Classical algebraic geometry had dealt with the analysis of equations with coefficients

86. Jean Dieudonné, *History of Algebraic Geometry* (Monterey: Wadsworth, Inc., 1985), pp. 27–58. Dieudonné deals with the next period in the subject's development on pp. 59–90.

87. Solomon Lefschetz, *L'analysis situs et la géométrie algébrique* (Paris: Gauthier-Villars et Cie, 1924) and compare Dieudonné, *History of Algebraic Geometry*, p. 70 for the quotation.

88. Lefschetz to Weyl, 30 November, 1926, Archiv der ETH Zürich, HS 91:659 as quoted in Norbert Schappacher, "A Historical Sketch of B. L. van der Waerden's Work in Algebraic Geometry: 1926–1946," in *Episodes in the History of Modern Algebra (1800–1950)*, ed. Jeremy Gray and Karen Hunger Parshall, HMATH, vol. 32 (Providence: American Mathematical Society and London: London Mathematical Society, 2007), pp. 245–283 on p. 262 (Schappacher's translation).

89. Schappacher, pp. 250–261 on p. 250.

in the fields of rational, real, or complex numbers. With the advent of modern algebra in the opening decades of the twentieth century—and its emphasis in the work of Noether and others on structures like groups, rings, and fields—it became natural to ask whether the results of the classical theory could be extended to equations with coefficients in an arbitrary field. As Dieudonné explained, "to be able to develop algebraic geometry over an arbitrary field in the same manner" that Lefschetz had developed the classical version, "it was necessary to invent purely algebraic tools" to replace those topological tools honed to treat such topological concepts as continuity and connectivity.[90] Van der Waerden did just that, especially in his famous series of papers entitled "Zur algebraischen Geometrie" that ran to some twenty installments over the almost four decades from 1933 to 1971.[91] His work suggested a new approach to algebraic geometry that drew on both algebraic *and* topological ideas and methods and that was developed in parallel by Oscar Zariski in the United States (see chapter six).

The American geometrical scene of the 1920s, like its topological counterpart, was thus both subdivided and lively. Yet, whereas the two topologies were, in some sense, young, the two geometries had much longer histories. Work from their nineteenth-century classical periods continued to attract the attention and to define the agendas of active twentieth-century researchers especially in the Northeast and Midwest. Yet, as Harvard's Birkhoff saw it in his assessment of "Fifty Years of American Mathematics" on the occasion of the AMS's semicentennial in 1938, the areas of algebraic and classical differential geometry actually "seemed most vital fifty years ago" and were more than somewhat spent by the 1920s.[92] Be that as it may, in differential as well as in algebraic geometry, American mathematicians like Eisenhart, Veblen, and Lefschetz were taking their fields in fresh, new directions and were being recognized for their efforts on the international stage.

Algebraic Research

If work in algebraic geometry had represented a first wave of serious mathematical research in the United States in the late nineteenth century, "finite group theory and its applications to algebraic equations," according to Veblen,

90. Dieudonné, *History of Algebraic Geometry*, p. 70.

91. Schappacher considers this work in detail on pp. 264–278.

92. George Birkhoff, "Fifty Years of American Mathematics," in *Semicentennial Addresses of the American Mathematical Society*, ed. The Committee on Publications, pp. 270–315 on p. 308.

had come in on a second, "even more intense wave" that had originated in Europe and had come ashore on the other side of the Atlantic beginning in the 1880s.[93] Frank Nelson Cole, a student in Klein's classes in Leipzig who returned to take his doctorate at Harvard in 1886, was initially inspired in his group-theoretic work by Klein's innovative approach to the icosahedron and fifth-degree polynomial equations.[94] He returned to the United States to pursue those interests from positions first at Michigan and then at Columbia from 1895 until his retirement in 1926. Klein also directed the doctoral work of two German students, Oskar Bolza and Heinrich Maschke, following his move to Göttingen in 1885. They both ultimately landed jobs in 1892 at Chicago, where Bolza reprised the course on the theory of substitution groups that he had earlier taught at Hopkins, and where Maschke continued his work on the theory of finite linear groups. Their example may well have spurred their colleague, E. H. Moore, actively to take up research in finite group theory in the 1890s. Moore promptly directed the dissertation research of his first Ph.D. student, Leonard Dickson, in that area.[95]

Also in Germany, but in Leipzig, Sophus Lie attracted the Danish student Hans Blichfeldt, as well as George Miller, who had already studied with Cole at Michigan. Both young men attended Lie's lectures, but Blichfeldt actually earned his doctoral degree under the Norwegian's supervision, while Miller continued his mathematical peregrinations in order to take advantage of Camille Jordan's presence in Paris. Miller followed his European sojourn with posts first at Cornell, then at Stanford, and finally at Illinois in 1906. For his part, Blichfeldt settled at Stanford and spent a long career there that ended only with his retirement in 1938.[96]

These and other Americans made significant contributions to finite group theory in the 1890s through the 1910s that culminated, in some sense, with

93. Veblen to Kellogg, 7 April, 1926.

94. Felix Klein, *Vorlesungen über das Ikosaeder und die Auflösung der Gleichungen vom fünften Grade* (Leipzig: B. G. Teubner Verlag, 1884) as well as Parshall and Rowe, *Emergence*, pp. 192–196, 203–204, and 349–350.

95. Parshall and Rowe treat this early group-theoretic work at Chicago in *Emergence*, pp. 374–382. See also Karen Hunger Parshall, "Defining a Mathematical Research School: The Case of Algebra at the University of Chicago, 1892–1945," *Historia Mathematica* 31 (2004), 263–278.

96. For more on the life and work of Blichfeldt and Miller, see Leonard Dickson, "Hans Frederik Blichfeldt, 1873–1945," *BAMS* 53 (1947), 882–883 and Henry Brahana, "George Abram Miller," *Biographical Memoirs*, vol. 30 (Washington, D.C.: National Academy of Sciences, 1957), pp. 257–312, respectively.

the book, *Theory and Applications of Finite Groups,* co-authored by Miller, Blichfeldt, and Dickson and published in 1916 just before the United States' entry into World War I.[97] By the 1920s, however, it was perceived that the field no longer "occup[ied] the whole horizon" of American mathematical research "as it once did."[98] Although by then it shared the stage with both topology and differential geometry, it continued to represent a well-defined sphere of American research thanks largely to Miller's efforts.[99]

Miller's approach to group theory was, despite his direct exposure to European ideas, most influenced by the lessons he had learned over the course of the two years he had spent as an instructor at Michigan under Cole's tutelage. In the fall of 1893, Miller had just arrived in Ann Arbor, and Cole had just returned from the Mathematical Congress held in conjunction with the World's Columbian Exposition in Chicago. There, he had essentially laid out the research program of determining and classifying all finite simple groups, that is, all nontrivial finite groups that contain as normal subgroups only the trivial group and the group itself. Cole had acknowledged that "in the absence of a general method, something may be accomplished by the tentative, step-by-step process, especially within moderate limits where the labor involved is not incommensurate with the value of the result."[100] "Step by step" characterized well the approach to finite groups of Miller, his students, and others into and through the 1920s.

For example, in 1900, thanks to the work of Otto Hölder in Germany, Cole in the United States, and William Burnside in England, all of the finite simple groups of order up to 1092 had been determined. In that year, however, Miller together with George Ling, then an instructor at Wesleyan, extended those results. By fully exploiting the numerology of the theorems that Ludwig Sylow

97. George Miller, Hans Blichfeldt, and Leonard Dickson, *Theory and Applications of Finite Groups* (New York: John Wiley & Sons, 1916).

98. Veblen to Kellogg, 7 April, 1926.

99. Miller's collected works run to almost 2,500 pages in five quarto volumes. Volume four covers the years from 1916 to 1929 and comprises some 450 pages. In all, Miller published over 130 papers during this fourteen-year period. See George Miller, *The Collected Works of George Abram Miller,* 5 vols. (Urbana: University of Illinois, 1935–1959).

100. Frank Cole, "On a Certain Simple Group," in *Mathematical Papers Read at the International Mathematical Congress Held in Conjunction with the World's Columbian Exposition, Chicago 1893,* ed. E. H. Moore et al. (New York: Macmillan & Co., 1896), pp. 40–43 on p. 40. Parshall and Rowe discuss this congress and the mathematics expounded there in *Emergence,* pp. 309–327 as well as in "Embedded in the Culture: Mathematics at the World's Columbian Exposition," *MI* 15 (2) (1993), 40–45.

had established in 1872, they demonstrated that there are no other simple groups of order less than or equal to 2000.[101]

By 1922, Miller was revisiting the question of low-order simple groups. It had long been known that the alternating group on 7 letters, that is, the group A_7 of even permutations of a set with seven elements, is a simple group of order 2520. Yet, in the search for *all* finite simple groups, it was natural to ask whether A_7 was the *only* finite simple group of that order. In a letter to Cole, an extract of which appeared in the AMS's *Bulletin*, Miller gave a proof by contradiction that, indeed, no other simple group of order 2520 besides A_7 could exist.[102]

Another natural, big-picture question was, what makes one group essentially different from another? Or, in other words, what types of elements or internal structures do individual groups have that fundamentally differentiate them? Step by step, Miller approached this question, too, in the 1920s. He considered such cases as "subgroups of index p^2 contained in a group of order p^m," "groups generated by two operators of order three whose product is of order three," "groups generated by two operators of order three whose product is of order six," etc., etc.[103] By 1929, he had also determined all the abstract groups of order 72.[104] Perhaps not surprisingly, these types of questions characterized the work of those who came under Miller's group-theoretic sway.

Among those was his colleague at Illinois, Henry Brahana. Although Brahana had earned his Ph.D. under Veblen at Princeton for a thesis in topology in 1920, his move to Urbana in that year prompted a shift in his research direction thanks to the presence there of both Coble, the algebraic geometer, and Miller, the group theorist. By the end of the decade, Brahana was writing papers like "Certain Perfect Groups Generated by Two Operators of Orders Two and Three" that clearly reflected Miller's influence.[105] So, too,

101. George Miller and George Ling, "Proof That There Is No Simple Group Whose Order Lies Between 1092 and 2001," *AJM* 22 (1900), 13–26.

102. George Miller, "The Simple Group of Order 2520," *BAMS* 28 (1922), 98–102.

103. George Miller, "Subgroups of Index p^2 Contained in a Group of Order p^m," *AJM* 48 (1926), 253–256; "Groups Generated by Two Operators of Order Three Whose Product Is of Order Three," *PNAS* 13 (1927), 24–26; and "Groups Generated by Two Operators of Order Three Whose Product Is of Order Six," *op. cit.* 13 (1927), 170–174, respectively.

104. George Miller, "Determination of All the Abstract Groups of Order 72," *AJM* 51 (1929), 491–494.

105. Henry Brahana, "Certain Perfect Groups Generated by Two Operators of Orders Two and Three," *AJM* 50 (1928), 345–356.

were Miller's Illinois graduate students. Harry Bender, for example, wrote a 1923 doctoral dissertation on the "Sylow Subgroups in the Group of Isomorphisms of Prime Power Abelian Groups" and continued to push these sorts of group-theoretic ideas at his alma mater under Miller's watchful eye for five more years, first as an instructor and then as an associate.[106]

And, Miller had trained a number of graduate students even before his move to Illinois. At Stanford, he supervised the 1904 doctoral work of William Manning, who, although a member of Stanford's Department of Applied Mathematics after earning his Ph.D., continued to maintain his group-theoretic interests. Manning focused on particular classes of primitive permutation groups, that is, groups G (initially identified by Évariste Galois) that act on a set X (where $|X| > 2$) such that G preserves no nontrivial partition of X.[107] He was still thinking about such groups more than twenty years later.[108] When Stanford's two mathematics departments merged under Blichfeldt as department chair in 1927, Manning began to train students in the theory of primitive groups, making Stanford an American group-theoretic focal point in the late 1920s and into the 1930s.[109] His first student, Marie Weiss, followed closely in her advisor's footsteps, working on "Primitive Groups Which Contain Substitutions of Prime Order p and of Degree $6p$ or $7p$" before winning NRC fellowships for the two academic years 1928–1930 and ultimately taking a position in 1935 on the faculty at H. Sophie Newcomb College, the women's branch of Tulane University.[110]

106. Harry Bender, "Sylow Subgroups in the Group of Isomorphisms of Prime Power Abelian Groups," *AJM* 45 (1923), 223–250. "Associate" was a then not uncommon category of temporary employment. Bender left Illinois for an actual assistant professorship at the University of Akron in 1928.

107. To get the flavor of Manning's work, see, for example, these two papers which comprised the results in his dissertation: "The Primitive Groups of Class $2p$ Which Contain a Substitution of Order p and Degree $2p$," *TAMS* 4 (1903), 351–357 and "On the Primitive Groups of Class $3p$," *TAMS* 6 (1905), 42–47.

108. William Manning, "The Primitive Groups of Class 14," *AJM* 51 (1929), 619–652 is just one example.

109. For more on the history of Stanford mathematics, see Halsey Royden, "A History of Mathematics at Stanford," in *A Century of Mathematics in America*, ed. Peter Duren et al., 2: 237–277.

110. Marie Weiss, "Primitive Groups Which Contain Substitutions of Prime Order p and of Degree $6p$ or $7p$," *TAMS* 30 (1928), 333–359. Judy Green and Jeanne LaDuke discuss her life and career in *Pioneering Women in American Mathematics: The Pre-1940 PhD's*, HMATH, vol. 34 (Providence: American Mathematical Society and London: London Mathematical Society, 2009), p. 310.

The work of Weiss, Bender, Manning, Brahana, and others reflected a group-theoretic program introduced to the United States by Frank Cole in the 1890s and perpetuated, particularly by Miller, after the First World War. As the description above might suggest and, in fact, as E. T. Bell characterized it with tongue in cheek in his retrospective on "Fifty Years of Algebra in America, 1888–1938," that program already seemed "to have been pushed to the limit of human endeavor and even slightly beyond" by the mid-1920s.[111] Americans nevertheless continued to pursue research and to train graduate students in this vein through the 1920s and into the 1930s.

Another American algebraic focal point in the 1920s—the theory of linear associative algebras—had been defined largely via the program at Chicago just after the turn of the twentieth century.[112] As an assistant professor back at his alma mater by 1900, and following a European mathematical tour as well as positions in Berkeley and Austin, Leonard Dickson had briefly embraced the postulate-theoretic agenda of his former advisor and then colleague, E. H. Moore. In 1902, Moore had discovered that the axioms for geometry that Hilbert had presented in his *Grundlagen der Geometrie* three years earlier were not actually independent, despite the German's claim to the contrary. This discovery briefly led Moore and Dickson as well as Moore's students, Veblen and R. L. Moore, and Moore's brother-in-law, the Cornell-trained John Wesley Young, down the postulate-theoretic path of determining systems of axioms for various mathematical constructs that were both mutually independent and consistent, that is, not mutually contradictory. Moore considered groups; Veblen, R. L. Moore, and Young reconsidered geometry; and Dickson thought about fields and linear associative algebras.[113]

In particular, in his 1903 paper on "Definitions of a Linear Associative Algebra by Independent Postulates," Dickson considered a set *A* of elements

---

111. E. T. Bell, "Fifty Years of Algebra in America, 1888–1938," in *Semicentennial Addresses of the American Mathematical Society*, ed. The Committee on Publications, pp. 1–34.

112. Strong hints of it had appeared earlier: in the 1870s at Harvard in the work of Benjamin Peirce and in the 1880s at Hopkins owing to James Joseph Sylvester and Charles Peirce. On this earlier history, see Karen Hunger Parshall, "Joseph H. M. Wedderburn and the Structure Theory of Algebras," *Archive for History of Exact Sciences* 32 (1985), 223–349, especially pp. 241–261.

113. Parshall and Rowe, *Emergence*, pp. 382–387. Christopher Hollings gives a technical analysis of (mostly) American postulate-theoretic work on groups per se in " 'Nobody Could Possibly Misunderstand What a Group Is': A Study in Early Twentieth-Century Group Axiomatics," *Archive for History of Exact Sciences* 71 (2017), 409–481.

consisting of linear combinations $a = \sum_{i=1}^{n} a_i e_i$ of linearly independent quantities e_i and scalars a_i in some field F, where the e_i's are assumed to satisfy the multiplication $e_i e_j = \sum_{k=1}^{n} \gamma_{ijk} e_k$, for $\gamma_{ijk} \in F$ and $1 \leq i, j \leq n$. Given this set-up, the sum and difference of two elements—a as above and $b = \sum_{i=1}^{n} b_i e_i$—are defined to be $a \pm b = \sum_{i=1}^{n} (a_i \pm b_i) e_i$, and their (associative) product is given by $ab = \sum_{i,j=1}^{n} a_i b_j e_i e_j = \sum_{i=k}^{n} u_k e_k$, where $u_k = \sum_{i,j=1}^{n} \gamma_{ijk} a_i b_j$. Such a set A is called a linear associative algebra (or a hypercomplex number system if the field of scalars is restricted to the real numbers $\mathbb{R}$ or to the complex numbers $\mathbb{C}$), and Dickson formulated a defining set of four independent axioms.[114]

Shortly after Dickson did this work, the young Scots mathematician Joseph Wedderburn brought a Carnegie fellowship to Chicago. There, he not only spurred Dickson to do additional work on linear associative algebras but also produced ground-breaking research on their structure theory. In particular, Wedderburn proved his so-called "principal theorem," namely, every linear associative algebra A (over a field F of characteristic zero like $\mathbb{R}$ and $\mathbb{C}$) can be expressed as the direct sum of a semi-simple subalgebra S and a maximal nilpotent invariant subalgebra N. He also showed that a semi-simple algebra is the direct sum of simple algebras and that a simple algebra can be realized as the tensor product of a full matrix algebra and a division algebra, that is, a linear associative algebra in which division by any nonzero element is possible.[115] Wedderburn thus demonstrated that the classification of linear associative algebras ultimately reduces to the classification of division algebras. Both he and Dickson were still at work developing this area of mutual interest in the 1920s.

After earning his doctorate at Edinburgh University in 1908, Wedderburn was lured back to the United States by a call to Princeton to serve—like Veblen whom he had met at Chicago and who became his lifelong friend and

114. Leonard Dickson, "Definitions of a Linear Associative Algebra by Independent Postulates," *TAMS* 4 (1903), 21–26.

115. Joseph Wedderburn, "On Hypercomplex Number Systems," *Proceedings of the London Mathematical Society*, 2d ser., 6 (November 1907), 77–118 and compare Parshall, "Wedderburn and the Structure Theory of Algebras" as well as "In Pursuit of the Finite Division Algebra Theorem and Beyond: Joseph H. M. Wedderburn, Leonard E. Dickson, and Oswald Veblen," *Archives internationales d'Histoires des Sciences* 35 (1983), 274–299. Wedderburn stated his results in general, that is, regardless of the underlying base field F. As would soon become clear thanks to the work of Ernst Steinitz, F, in fact, has to be at least perfect. Today, the hypothesis is that F be separable, but that notion was not at Wedderburn's disposal in 1907.

colleague—as a preceptor. Wedderburn's algebra-theoretic work was interrupted, however, by Great Britain's entry into World War I and his service from 1914 to 1919 as an officer in the British Army.[116] That he managed to pick up his research thread almost immediately on his return is evidenced by the paper "On Division Algebras" that he presented to the AMS in February of 1920.[117]

As early as 1905, Dickson had defined the notion of a cyclic algebra, that is, an algebra A "defined by the relations $xy = y\theta(x)$ and $y^n = g$, where $\theta(x)$ is a polynomial in x which is rational" in the base field F over which A is defined and where $g \in F$ is not the norm of any rational polynomial in x.[118] These algebras have dimension n^2 over F, and, for suitable choices of θ and g, Dickson had noted that they are division algebras. In 1914, Wedderburn established a sufficient condition for a cyclic algebra to be a division algebra, a result he had extended by 1920 to central division algebras, that is, division algebras with center equal to F. In particular, he showed that every central division algebra of dimension 9 over its base field F is cyclic and that Dickson's cyclic algebras are actually special cases of an even more general type of algebra, a so-called crossed product algebra.[119] By the mid-1920s, Wedderburn had also had some success in extending his structure theory to infinite-dimensional algebras, and he had begun work on what would ultimately be his 1934 AMS Colloquium volume, *Lectures on Matrices*.[120]

Dickson was even more prolific. After the completion of his book on the theory of groups with Miller and Blichfeldt, he poured himself into what

116. Karen Hunger Parshall, "New Light on the Life and Work of Joseph Henry Maclagan Wedderburn (1882–1948)," in *Amphora: Festschrift für Hans Wussing zu seinem 65. Geburtstag*, ed. Menso Folkerts et al. (Basel: Birkhäuser Verlag, 1992), pp. 523–537.

117. Joseph Wedderburn, "On Division Algebras," *TAMS* 22 (1921), 129–135.

118. Joseph Wedderburn, "A Type of Primitive Algebra," *TAMS* 15 (1914), 162–166 on p. 162. Dickson originally defined these algebras in the abstract of a talk that he gave at the summer meeting of the AMS in Williamstown, Massachusetts, in 1905 (Frank Cole, "The Twelfth Summer Meeting of the American Mathematical Society," *BAMS* 12 (1906), 53–63 on p. 61).

119. Wedderburn, "On Division Algebras," pp. 133–134. The definition of a crossed product algebra is somewhat involved; see, for example, Israel Herstein, *Noncommutative Rings*, Carus Mathematical Monographs, no. 15 (n.p.: Mathematical Association of America, 1968), pp. 107–108.

120. Joseph Wedderburn, "Algebras Which Do Not Possess a Finite Basis," *TAMS* 26 (1924), 395–426 as well as *Lectures on Matrices*, AMS Colloquium Publications, vol. 17 (New York: American Mathematical Society, 1934).

was ultimately his three-volume compilation of number-theoretic results, the *History of the Theory of Numbers*.[121] The insights he gained in doing this encyclopedic work, together with his algebra-theoretic research, inspired two book-length forays in the 1920s: *Algebras and Their Arithmetics* (1923) and the substantially extended *Algebren und ihre Zahlentheorie* (1927). In these studies, Dickson considered linear associative algebras *A* (with identity) over the field of rational numbers. He aimed to develop "for the first time a general theory of the arithmetics of algebras, which furnishes a direct generalization of the classic theory of algebraic numbers" of such nineteenth-century German greats as Peter Lejeune Dirichlet, Ernst Kummer, and Richard Dedekind.[122] To that end, he exploited Wedderburn's structure theory, which, as Dickson recognized, implied that to study the arithmetic of an algebra *A* is to study the arithmetic of its semi-simple part *S*. Dickson proceeded to show how to construct the integral elements and the units in *A* as well as how to determine the properties of unique factorization from the analogous elements in *S*.[123] Indicative of his sense of the importance of this work, he lectured on it in 1924 in a venue no less auspicious than the International Congress of Mathematicians in Toronto (see chapter three).

121. Leonard Dickson, *History of the Theory of Numbers*, 3 vols. (Washington, D.C.: Carnegie Institution of Washington, 1919, 1920, 1923; reprint ed., New York: Chelsea Publishing Company, 1992). See also Della Dumbaugh Fenster, "Leonard Dickson, History of the Theory of Numbers," in *Landmark Writings in Western Mathematics, 1640–1940*, ed. Ivor Grattan-Guinness (Amsterdam: Elsevier Press, 2005), pp. 833–843 and "Why Dickson Left Quadratic Reciprocity Out of His History of the Theory of Numbers," *AMM* 106 (1999), 618–627.

122. Leonard Dickson, *Algebras and Their Arithmetics* (Chicago: University of Chicago Press, 1923; reprint ed., New York: Dover Publications, Inc., 1960), p. vii. See also Leonard Dickson, *Algebren und ihre Zahlentheorie* (Zürich: Orell Füssli Verlag, 1927). For a concise account of the nineteenth-century number-theoretic analog, see Victor Katz and Karen Hunger Parshall, *Taming the Unknown: A History of Algebra from Antiquity to the Early Twentieth Century* (Princeton: Princeton University Press, 2014), pp. 388–399. Dickson's work on the arithmetics of algebras was honored by the AMS with the first Frank Nelson Cole Prize in Algebra in 1928. See the next chapter for more on this prize.

123. Della Dumbaugh Fenster contextualizes this research historically in "Leonard Eugene Dickson and His Work in the Arithmetics of Algebras," *Archive for History of Exact Sciences* 52 (1998), 119–159 and "American Initiatives toward Internationalization: The Case of Leonard Dickson," in *Mathematics Unbound: The Evolution of an International Mathematical Research Community, 1800–1945*, ed. Karen Hunger Parshall and Adrian Rice, HMATH, vol. 23 (Providence: American Mathematical Society and London: London Mathematical Society, 2002), pp. 311–333.

Dickson was also more prolific than Wedderburn in the production of new researchers who actively added to the store of knowledge about linear associative algebras in the 1920s. For example, Olive Hazlett, one of his numerous women students, earned her Chicago Ph.D. in 1915 for a classification of all (not necessarily associative) nilpotent algebras of dimension four or less over the field of complex numbers.[124] From positions at Mount Holyoke College beginning in 1918 and then at Illinois starting in 1925, she not only considered questions associated with division algebras but also followed her advisor into the theory of the arithmetics of algebras. Like him, she spoke on her ideas about the latter at the Toronto Congress. There, in a postulate-theoretic spirit, she offered a definition for the notion of an integral element in an algebra A different from the one Dickson had given in *Algebras and Their Arithmetics* and explored the ramifications of her new formulation.[125] Like her advisor, too, she had the distinction of being "starred" in *American Men of Science*, in her case in the fourth edition of 1927, the second female mathematician so recognized.[126]

Another of Dickson's students, the same Cyrus MacDuffee who had impressed Raymond Wilder at Ohio State, steadily pursued into the 1920s the panoply of ideas he had identified in his 1921 dissertation on the theory of algebras as well as in his related research on the theory of matrices. Like his

124. Olive Hazlett, "On the Classification and Invariantive Characterization of Nilpotent Algebras," *AJM* 38 (1916), 109–110. On Dickson as a mentor of doctoral students, see Della Dumbaugh Fenster, "Role Modeling in Mathematics: The Case of Leonard Eugene Dickson (1874–1954)," *Historia Mathematica* 24 (1997), 7–24. Of Dickson's sixty-seven doctoral students, eighteen or 27% were women.

125. Olive Hazlett, "On the Arithmetic of a General Associative Algebra," *Proceedings of the International Mathematical Congress Held in Toronto, August 11–16, 1924*, ed. John Fields, 2 vols. (Toronto: University of Toronto Press, 1928), 1: 185–191 and "The Arithmetic of a General Algebra," *AM* 28 (1926), 92–102.

126. Some 1,000 natural and exact scientists received this designation in recognition of the importance of their work, with the 1,000 stars being distributed proportionally among the different sciences based on the total number of scientists in each of the given fields. In the first edition of 1906, for example, of the 1,000 stars, 175 went to chemists, 150 each to physicists and zoologists, 100 each to botanists and geologists, 80 to mathematicians, etc. See James McKeen Cattell, ed., *American Men of Science*, 2d ed. (Lancaster: The Science Press, 1910), pp. vi-vii. See also Stephen Visher, *Scientists Starred, 1903–1943, in American Men of Science: A Study of Collegiate and Doctoral Training, Birthplace, Distribution, Background, and Developmental Influences* (Baltimore: Johns Hopkins University Press, 1947). The first female mathematician to be "starred" was analyst Anna Pell Wheeler of Bryn Mawr, in the 1921 third edition.

FIGURE 1.3. Olive Hazlett (1890–1974). (Photo courtesy of the private collection of Judy Green.)

academic "sister," Hazlett, he, too, found interesting his advisor's work on the arithmetics of algebras and sought to push it further. In "An Introduction to the Theory of Ideals in Linear Associative Algebras," for example, MacDuffee noted that "[w]ith the development of the number theory of linear algebras, it was natural that attempts should be made to extend to these domains of integrity the theory of ideal numbers."[127] Still, it was a hard problem.

The German mathematician, Adolph Hurwitz, had "investigated the number theory of quaternions by using right and left ideals, and ha[d] found that they are powerless to introduce unique factorization into this algebra." Similarly, his Swiss contemporary, Andreas Speiser, had explored "the properties of right, left and two-sided ideals in semi-simple algebras" but had ruefully remarked "that some of the most remarkable properties of ideals are 'but foreign adjuncts which are essentially restricted to algebraic number fields.' " As MacDuffee explained, "[a]lthough it is historically true that ideals were introduced into algebraic number theory to establish unique factorization," that was only their "secondary function." "Primarily they establish the property that every two numbers have a greatest common divisor expressible linearly in terms of the numbers. In algebraic fields this property implies unique factorization but in the general linear algebra it does not—hence the success of the ideal theory in algebraic fields and its partial failure in the more general domain." It was in the context of that "more general domain" that MacDuffee developed "a correspondence between ideals and matrices whose elements are rational integers" in order that the multiplication of ideals, "which causes so much difficulty in non-commutative domains," could be replaced by matrix multiplication.

Perhaps Dickson's strongest student, however, was Adrian Albert. In his 1928 doctoral dissertation, Albert followed directly in the footsteps of both Wedderburn and Dickson in considering the classification of division algebras. He first pushed Wedderburn's immediately postwar results to the next dimension, showing that every central division algebra of dimension 16 over its base field F, while not necessarily cyclic, is a crossed product algebra.[128] This was the first of a flurry of papers in 1929 and 1930 in which Albert

127. Cyrus MacDuffee, "An Introduction to the Theory of Ideals in Linear Associative Algebras," *TAMS* 31 (1929), 71–90 on p. 71. The quotations that follow in the next paragraph are also on this page.

128. Adrian Albert, "A Determination of All Normal Division Algebras in Sixteen Units," *TAMS* 31 (1929), 253–260.

considered successively higher square dimensions in his quest for the general result.

If group theory as practiced in the United States was more than somewhat old-fashioned in the 1920s, the theory of algebras was anything but. In fact, in Birkhoff's assessment, "there ha[d] been a great algebraic advance in the direction of a unified theory of linear associative algebra and their arithmetics" in the United States in the immediately postwar years.[129] That advance was due, in no small part, to the work of Wedderburn and Dickson as well as of Dickson's students, especially Albert.[130] These mathematicians actively engaged in research that also attracted the attention of some of the best algebraists on the other side of the Atlantic and that would, in the 1930s, bring American mathematicians even more fully into competition internationally (see chapter six).

Research in Analysis

Americans also pursued at least one other major area of mathematical research in the 1920s—analysis—and as with other areas, the principal loci of activity were widely recognized. At Harvard, William Osgood, Maxime Bôcher, "and their followers" like George Birkhoff had "created a function-theoretic current," in Veblen's view, that "is one of the most important elements in the mathematical stream," while at Chicago, work on the calculus of variations was "initiated by Bolza and continued by Bliss."[131] Osgood had received graduate

---

129. Birkhoff, "Fifty Years of American Mathematics," p. 292.

130. Prolific in students though Dickson undoubtedly was, not all of his contemporaries viewed him as an enlightened advisor. Derrick Norman Lehmer, a number theorist who, like Dickson had been a student of E. H. Moore at Chicago, took his 1900 Ph.D. to Berkeley where he was ultimately promoted to a full professorship in 1918. When his son, Derrick Henry Lehmer, the future number theorist and computing pioneer, was trying to decide on graduate schools, Lehmer *père* confided to Roland Richardson that "from what I hear of the methods of turning out doctors [at Chicago] . . . it is no place for a man with ideas of his own. Dickson does not want him to think his own thoughts apparently. He will be required to drop all the problems which have interested him and work out a special case of some of Dickson's researches. Thus [*sic*], I will confess, seems to me to be a very stupid attitude to take toward any show of originality." Lehmer to Richardson, 4 February, 1928, Box 5: Correspondence, 1926–1930 'H-M,' Folder: Lehmer, Derrick Norman, Richardson Papers.

131. Veblen to Kellogg, 7 April, 1926. Another of these "followers" was Charles Moore, who earned his Ph.D. under Bôcher in 1905 and who worked from his position at the University of Cincinnati on, among other things, the summability of series. See, for example, Charles Moore,

training in Germany under Klein and Max Noether, ultimately earning his Erlangen doctorate under the latter in 1890; Bôcher was also German-trained and took his degree under Klein in 1891; Bolza, as noted, a German student of Klein, ultimately found a position at Chicago in 1892. The tradition of analysis in the United States was thus directly imported from Germany in the early 1890s, and it was carried into the twentieth century by members— like Birkhoff and Gilbert Bliss—of America's second mathematical research generation.[132]

Birkhoff had had the best mathematical training that the United States could offer at the turn of the twentieth century. Born in Michigan into a doctor's family, he began his undergraduate training at Chicago, but moved to complete both his B.A. and M.A. degrees at Harvard before returning to Chicago to take his Ph.D. under E. H. Moore in 1907. Influenced especially by Bôcher and Moore, in many regards Birkhoff, who would not venture across the Atlantic until 1926 (on this first trip, see chapter three), was just as much a student of Poincaré, owing to his avid mathematical reading. Indeed, it was in 1912, the year that he left a preceptorship at Princeton for a beginning professorship at Harvard, that Birkhoff proved Poincaré's so-called "Last" or "Geometric Theorem": consider "a continuous one-to-one transformation T [that] takes the ring [that is, the annulus] R, formed by concentric circles C_a and C_b of radii a and b respectively $(a > b > 0)$, into itself in such a way as to advance the points of C_a in a positive sense, and the points of C_b in the negative sense, and at the same time to preserve areas. Then there are at least two invariant points."[133] His proof of this special case of the three-body problem in dynamical systems cemented his reputation as an American mathematical force to be reckoned with.

By the 1920s, Birkhoff was actively pursuing a wide range of analytic topics as well as training graduate students across the field. One of his interests stemmed from the doctoral dissertation he had written at Chicago, inspired

---

"On the Application of Borel's Method to the Summation of Fourier's Series," *PNAS* 11 (1925), 284–287. By 1938, Moore had codified his work in *Summable Series and Convergence Factors*, AMS Colloquium Publications, vol. 22 (New York: American Mathematical Society, 1938).

132. Parshall and Rowe examine the work of students in the so-called "*Wanderlust* generation" of the 1880s and 1890s in *Emergence*, pp. 189–259.

133. George Birkhoff, "Proof of Poincaré's Geometric Theorem," *TAMS* 14 (1913), 14–22 on p. 14. Birkhoff presented his result before the AMS on 26 October, 1912, his first fall on the Harvard faculty. Compare Marston Morse, "George David Birkhoff and His Mathematical Work," *BAMS* 52 (1946), 357–391.

by the work of his Harvard professor, Bôcher, on the asymptotic behavior of solutions of ordinary linear differential equations, boundary-value problems, and Sturm-Liouville theory.[134] He lectured on these ideas at Harvard in the fall of 1920 to an audience that included his 1922 Ph.D. student, Rudolph Langer.

Langer published two papers in 1923 that had constituted his dissertation and that drew on results that Birkhoff had presented in his own doctoral work. One considered a class of differential equations different from that initially studied by Birkhoff and explored the expansion problem associated with that class.[135] The other, "The Boundary Problems and Developments Associated with a System of Linear Differential Equations of the First Order," was joint with Birkhoff. As the co-authors explained, roughly three-quarters of the material in their paper stemmed directly from Birkhoff's 1920 course, although it had been reorganized to a large extent by Langer and aimed to lay a matrix-theoretic foundation for the theory as a whole. In the paper's closing quarter, Langer, working within that framework, considered a system composed of a homogeneous differential vector equation, together with appropriate boundary conditions, and demonstrated convergence under suitable constraints. In so doing, he provided an interesting generalization of the expansions that Birkhoff had given some fifteen years earlier.[136]

At essentially the same time that Birkhoff and Langer were producing these classical results, Birkhoff was also engaged in functional-analytic research of

134. George Birkhoff, "On the Asymptotic Character of the Solutions of Certain Linear Differential Equations Containing a Parameter," *TAMS* 9 (1908), 219–231 and "Boundary Value and Expansion Problems of Ordinary Linear Differential Equations," *TAMS* 9 (1908), 373–395. Parshall and Rowe give more details in *Emergence*, p. 392. Birkhoff followed this work in 1911 with a paper on the "General Theory of Linear Difference Equations," *TAMS* 12 (1911), 243–284, that influenced the 1911 doctoral research in analysis of his first Ph.D. student, Robert Carmichael. Carmichael continued his mathematical explorations from the position at the University of Illinois that he held from 1915 until his retirement in 1947. See, for example, Robert Carmichael, "On the Expansion of Certain Analytic Functions in Series," *AM* 22 (1920), 29–34. By the 1930s, however, Carmichael had become better known for his work in number theory and group theory than for his research in analysis.

135. Rudolph Langer, "Developments Associated with a Boundary Problem not Linear in the Parameter," *TAMS* 25 (1923), 155–172.

136. George Birkhoff and Rudolph Langer, "The Boundary Problems and Developments Associated with a System of Linear Differential Equations of the First Order," *PAAAS* 58 (1923), 49–128. Langer continued working in this and related areas from positions first at Dartmouth and Brown and then, from 1927 on, at the University of Wisconsin.

a more abstract nature with his relatively new Harvard colleague, Oliver Kellogg. Their paper, "Invariant Points in Function Space," was inspired both by the axiomatic approach of Kellogg's advisor, David Hilbert, and by the general analysis that Birkhoff's mentor, E. H. Moore, had begun to develop in 1906. As Moore had put it, "[t]he existence of analogies between central features of various theories implies the existence of a general theory which underlies the particular theories and unifies them with respect to those central features."[137] For their part, Birkhoff and Kellogg proposed a "general program of functional analysis concerning existence theorems" that seemed "more effective than the obvious treatment by direct abstraction" that Moore, for example, had advocated.[138] As Birkhoff later explained, "[t]he ordinary implicit equations of analysis can be written in the form of $f = T(f)$ where f is the 'point' in function space whose existence is to be established and $g = T(f)$ for any f is a transformed point in the same functional space. The desired existence theorem merely affirms that the transformation T of the space into a subspace admits of a fixed point."[139] Birkhoff and Kellogg established the existence both "of invariant points in a region of n-space which is convex toward an interior point, under a continuous, one-valued transformation which carries points of the region into points of the region" and "of the inverses of points on the hypersphere in n-space (n odd) with respect to a parametric transformation containing the identity."[140] This then allowed them to infer analogous theorems for function spaces, "first by a method of interpolation, and second, by a transition through a Hilbert space."

One Harvard student who drew inspiration from both the abstract and the more classically oriented strains of analysis was Birkhoff's 1926 Ph.D., Marshall Stone. Like Langer, Stone initially generalized some of Birkhoff's dissertation results[141] and continued to mine that vein from the instructorship at Columbia that he accepted in 1925. By 1929, he had not only returned to take up an instructorship at his alma mater, but his interests had also shifted

137. E. H. Moore, "Introduction to a Form of General Analysis," in *The New Haven Mathematical Colloquium* (New Haven: Yale University Press, 1910), pp. 1–150 on p. 1.

138. George Birkhoff and Oliver Kellogg, "Invariant Points in Function Space," *TAMS* 23 (1922), 96–115. For the quotation, see Birkhoff, "Fifty Years of American Mathematics," p. 297.

139. Birkhoff, "Fifty Years of American Mathematics," p. 297.

140. Roland Richardson, "The February Meeting of the American Mathematical Society," *BAMS* 28 (1922), 233–244 on p. 236. The quotation that follows is also on this page.

141. See, for instance, Marshall Stone, "A Comparison of the Series of Fourier and Birkhoff," *TAMS* 28 (1926), 695–761.

to more properly functional-analytic considerations motivated by then-recent mathematical discussions of the new quantum theory by Hermann Weyl, John von Neumann, and others. In a series of three short notes published between 1929 and 1930 in the *Proceedings of the National Academy of Sciences*, Stone developed the kernel of what would become his 600-page, 1932 AMS Colloquium volume on *Linear Transformations in Hilbert Space and Their Applications to Analysis*.[142] That book, described as "one of the great classics of twentieth-century mathematics,"[143] was initially inspired by Stone's exposure to some of von Neumann's "early and still incomplete work" on self-adjoint operators on Hilbert space, that is, linear operators T from a complex Hilbert space into itself such that T equals its adjoint T^*.[144] As Stone explained, he then developed "independently" and "without further knowledge of [von Neumann's] progress along the same or similar lines" the ideas he presented in his massive tome.[145] In particular, it extended from the context of bounded to unbounded operators Hilbert's spectral theorem, a result that, loosely speaking, provides conditions under which an operator or its associated matrix can be diagonalized, that is, represented as a diagonal matrix relative to some basis.

If Stone's functional-analytic work may be seen to have stemmed directly from Moore's general analysis, so, too, did the research of his older contemporary, Theophil Hildebrandt. A 1910 Chicago Ph.D. under Moore, Hildebrandt, like Birkhoff, took his advisor's work in general analysis as a starting point from which he explored both functional analysis and the theory of integration over the course of a long career at the University of Michigan. In 1923, for example, and drawing directly from contemporaneous work of Moore

142. Marshall Stone, "Linear Transformations in Hilbert Space. I. Geometrical Aspects," *PNAS* 15 (1929), 198–200; "Linear Transformations in Hilbert Space. II. Analytical Aspects," op. cit., 15 (1929), 423–425; "Linear Transformations in Hilbert Space. III. Operational Methods and Group Theory," op. cit., 16 (1930), 172–175, and *Linear Transformations in Hilbert Space and Their Applications to Analysis*, AMS Colloquium Publications, vol. 15 (New York: American Mathematical Society, 1932).

143. George Mackey, "Marshall Harvey Stone 1903–1989," *NAMS* 36 (3) (1989), 221–223 on p. 221.

144. See, specifically, John von Neumann, "Mathematische Begründung der Quantenmechanik," *Nachrichten von der Gesellschaft der Wissenschaften zu Göttingen, Mathematisch-Physikalische Klasse* (1927), 1–57 or John von Neumann, *Collected Works*, ed. Abraham Taub, 6 vols. (New York: Pergamon Press, 1961), 1: 151–207.

145. Stone, *Linear Transformations in Hilbert Space*, pp. iv–v. Jean Dieudonné discusses the work of von Neumann (and to a much lesser extent Stone) in historical context in *History of Functional Analysis* (Amsterdam: North-Holland Publishing Company, 1981), pp. 172–183.

and another of his students, Herman Smith, Hildebrandt gave the first general proof of the principle of uniform boundedness for what would come to be called Banach spaces, a special kind of vector space (with complete metric) named in honor of the Polish mathematician Stefan Banach.[146] By 1925, Hildebrandt's continuing exploration of general spaces had led him to a consideration and exposition of the (Heine-)Borel Theorem: given a subset S of Euclidean n-space, S is closed and bounded if and only if S is compact. As he saw it, "[t]he attempts to derive th[is] theorem in increasingly general situations has [sic] led to interesting new properties and characterizations of spaces."[147] Likely for that reason, he chose it as the topic of the invited address he gave before the joint meeting of the AMS and the American Association for the Advancement of Science in Kansas City in December 1925. In 1929, the resulting paper won the second Chauvenet Prize of the Mathematical Association of America, an award then given once every three years for the best expository article appearing in an American mathematical publication.[148]

Another strand of research in analysis that occupied American mathematicians in the 1920s was potential theory or, broadly speaking, the study of harmonic functions.[149] Also imported into the United States from Germany—largely by Bôcher and Kellogg—it was championed by Kellogg following his move from the University of Missouri to fill the potential-theoretic void created by Bôcher's death in 1918. In the 1920s, one of the questions that particularly intrigued Kellogg—as well as his contemporaries like MIT's Norbert Wiener and Bôcher's student, Griffith Evans, then at the Rice Institute

146. Theophil Hildebrandt, "On Uniform Limitedness of Sets of Functional Operations," *BAMS* 29 (1923), 309–315. The theorem was proven independently four years later by Banach and Hugo Steinhaus in "Sur le principe de la condensation de singularités," *Fundamenta mathematicae* 9 (1927), 50–61. Compare also E. H. Moore and Herman Smith, "A General Theory of Convergence," *AJM* 44 (1922), 102–121, where they defined the notions—that Hildebrandt used to great advantage in his proof—of what are now called Moore-Smith sequences and Moore-Smith convergence.

147. Theophil Hildebrandt, "The Borel Theorem and Its Generalizations," *BAMS* 32 (1926), 423–474 on p. 454.

148. For more on this prize in the context of infrastructure-building for the American mathematical endeavor in the 1920s, see the next chapter.

149. A function $u(x_1, x_2, \ldots, x_n)$ defined in some region R of Euclidean n-space is called harmonic in R if it is 1) continuous, 2) has, considered as a function of each variable x_i singly, continuous first derivatives $\partial u / \partial x_i$ and finite second derivatives $\partial^2 u / \partial x_i{}^2$, and 3) if the expression $\Delta u = \frac{\partial^2 u}{\partial x_1{}^2} + \cdots + \frac{\partial^2 u}{\partial x_n{}^2}$ vanishes identically on R. See Constantin Carathéodory, "On Dirichlet's Problem," *AJM* 59 (1937), 709–731 on p. 710.

in Houston—was the so-called Dirichlet problem, that is, given a particular partial differential equation, find a (harmonic) function that solves it in the interior of a given region while taking on prescribed values on that region's boundary. By 1926, there was such a strong sense that the subject was undergoing "a period of remarkable development" that the AMS invited Kellogg to speak on "Recent Progress on the Dirichlet Problem" at its January meeting.[150] There, he aimed both to contextualize the myriad contributions of American and European mathematicians to the problem's solution and to indicate open problems that might spur further American research. This was followed by the advanced overview of the entire area that he gave in his 1929 book, *Foundations of Potential Theory*.[151]

Although much more analytic work could be singled out for mention— the algebraic approach to function theory of Columbia's Joseph Ritt, work in approximation theory by Minnesota's Dunham Jackson and Harvard's Joseph Walsh, results of UCLA's Earle Hedrick on partial differential equations and on functions of complex variables, research on the theories of real and complex functions as well as on special functions by Columbia's Thomas Gronwall, and results on ordinary differential equations by Bryn Mawr's Anna Pell Wheeler, among others[152]—perhaps the final, major research focus of American analysts to consider in some detail is the calculus of variations, that

150. Oliver Kellogg, "Recent Progress on the Dirichlet Problem," *BAMS* 32 (1926), 601–625 on p. 624. See Birkhoff, "Fifty Years of American Mathematics," pp. 300–301 for a brief technical discussion of the potential-theoretic work of Kellogg, Wiener, and Evans and compare Marcel Brelot, "Norbert Wiener and Potential Theory," *BAMS* 72 (1966), 39–41 and Norbert Wiener, "The Dirichlet Problem," *JMP* 3 (1924), 127–146.

151. Oliver Kellogg, *Foundations of Potential Theory* (New York: Frederick Ungar Publishing Company, 1929). Kellogg died prematurely of a heart attack while climbing in 1932 thus cutting his mathematical career short. For a heartfelt look at the man and his work, see George Birkhoff, "The Mathematical Work of Oliver Dimon Kellogg," *BAMS* 39 (1933), 171–177.

152. See, for example, Joseph Ritt, "Elementary Functions and Their Inverses," *TAMS* 27 (1925), 68–90; Dunham Jackson, *The Theory of Approximation*, AMS Colloquium Publications, vol. 11 (New York: American Mathematical Society, 1930); Joseph Walsh, "The Approximation of Harmonic Functions by Harmonic Polynomials and by Harmonic Rational Functions," *BAMS* 35 (1929), 499–544 and "On Approximation by Rational Functions to an Arbitrary Function of a Complex Variable," *TAMS* 31 (1929), 477–502; Earle Hedrick, "On the Derivatives of Non-analytic Functions," *PNAS* 14 (1928), 649–654; Thomas Gronwall, "On the Zeros of the Function $\beta(z)$ Associated with the Gamma Function," *TAMS* 28 (1926), 391–399; and Anna Pell Wheeler, "Linear Ordinary Self-adjoint Differential Equations of the Second Order," *AJM* 49 (1927), 309–320. Alan Gluchoff highlights Gronwall's particularly interesting career in "Pure Mathematics Applied in Early Twentieth-Century America:

is, the study of the conditions under which a given integral takes on a maximum or a minimum. Indeed, if Harvard was one American center of analysis in the 1920s, another was the University of Chicago precisely in this subfield. Inaugurated there by Oskar Bolza, it was perpetuated after Bolza's departure for Germany in 1910 by Gilbert Bliss, Bolza's first and most prominent student and his successor—after stints at Minnesota, Missouri, Chicago, and Princeton—on the Chicago faculty. As he pursued the research agenda he had begun in his 1900 doctoral dissertation, Bliss also populated the American mathematical research community with over twenty new researchers in the calculus of variations in the years from 1920 through 1930, among whom were Lawrence Graves, ultimately Bliss's colleague and successor at Chicago, and Edward J. "Jimmy" McShane, who transplanted the field to the University of Virginia after his move there in 1935.

In the 1920s, one of Bliss's research foci concerned the second variation of an integral along a curve. The problem was to determine a curve C joining the points (x_1, y_1) and (x_2, y_2) defined by $y = y(x)$ for $x_1 \leq x \leq x_2$ in the xy-plane that minimizes an integral

$$J(C) = \int_{x_1}^{x_2} f(x, y(x), y'(x)) dx.$$

As Bliss explained, "[i]n order to obtain conditions which must be satisfied by a minimizing arc,"[153] it was necessary to "consider the values of the integral along the curves of a family of the form

$$\bar{y} = y(x) + \alpha \eta(x), \quad (x_1 \leq x \leq x_2),$$

where α is a constant to be varied at pleasure and $\eta(x)$ is a function which vanishes at x_1 and x_2." Since all of the curves of this family pass through the endpoints of C, it is clear that

$$J(\alpha) = \int_{x_1}^{x_2} f(x, y + \alpha \eta, y' + \alpha \eta') dx$$

The Case of T. H. Gronwall, Consulting Mathematician," *Historia Mathematica* 32 (2005), 312–357.

In his list (see fig. 1.2), Veblen also singled out the analytic work of Einar Hille at Yale and Jacob Tamarkin at Brown (principally on integral equations and Fourier series), Wallie Hurwitz at Cornell (on divergent series, among other topics), and James Shohat at Michigan (on the so-called moment problem).

153. Gilbert Bliss, "Some Recent Developments in the Calculus of Variations," *BAMS* 26 (1920), 343–361 on p. 345.

"must have a minimum for $\alpha = 0$, so that by the usual theory of maxima and minima the conditions

$$J'(0) = \int_{x_1}^{x_2} (f_y\eta + f_{y'}\eta')dx = 0, \tag{1.1}$$

$$J''(0) = \int_{x_1}^{x_2} (f_{yy}\eta^2 + 2f_{yy'}\eta\eta' + f_{y'y'}\eta'^2)dx = 0, \tag{1.2}$$

must be satisfied for every choice of the function $\eta(x)$ vanishing at x_1 and x_2." Equations (1.1) and (1.2) are the first and second variations, respectively, of the integral J along the curve C. Bliss argued that "the theory of the second variation in its entirety could be viewed with success from the standpoint of the minimum problem of the second variation, a minimum problem within a minimum problem," and he explored that claim especially in the lecture he gave in 1924 at the Toronto International Congress.[154]

Not all work on the calculus of variations in the 1920s had a Chicago connection, though. Roland Richardson, Secretary of the AMS from 1921 to 1940 and indefatigable proponent of mathematics in the United States, reprised in the 1920s questions in the area that had occupied him as early as 1910. Richardson had earned his Ph.D. under James Pierpont at Yale in 1906, had accepted a position at Brown the next year, and had spent the 1908–1909 academic year studying in Göttingen with Klein and especially Hilbert. On the latter's recommendation, he turned his attention to a conjecture that the German master had made involving the calculus of variations in the context of certain boundary-value problems with a finite number of isoperimetric conditions.[155] By 1928, he had extended this work (although somewhat imperfectly) to consider "properties enjoyed by the individual proper functions as extrema for variational problems involving an infinite number of isoperimetric conditions."[156] In the 1920s, through the 1930s, and into the 1940s, although he tried to keep his hand in mathematics per se,

154. For the quotation, see ibid. p. 358. For the lecture, see Gilbert Bliss, "The Transformation of Clebsch in the Calculus of Variations," *Proceedings of the International Mathematical Congress Held in Toronto, August 11–16, 1924*, ed. John Fields, 1: 589–603.

155. Raymond Archibald, "R.G.D. Richardson 1878–1949," *BAMS* 56 (1950), 256–265 on pp. 257–258.

156. Roland Richardson, "A Problem in the Calculus of Variations with an Infinite Number of Auxiliary Conditions," *TAMS* 30 (1928), 155–189. For the quotation, see Archibald, "R.G.D. Richardson," p. 259.

FIGURE 1.4. Roland Richardson (1878–1949) in 1940 on the occasion of his "retirement"
after nineteen years as AMS Secretary. He is holding his copy of the testimonial he
received from the AMS and standing beside the silver tea set given to him by his
grateful colleagues. (Photo courtesy of Brown University Digital Depository.)

Richardson was almost exclusively focused on mathematical institution-
building both within the AMS and at Brown (on the latter, in particular, see
chapter eight).

Also outside the immediate circle defined by Bliss and his colleagues at
Chicago, Birkhoff was active in the calculus of variations like he was in many
other areas of analysis in the 1920s. Unlike Bliss, Birkhoff had come to the

field from his study of dynamical systems, since, as he explained, "dynamical trajectories may be regarded in many cases as geodesics along which the arc length is an extremum."[157] His was a fundamentally topological approach.

Birkhoff had followed his stunning proof of Poincaré's Last Geometric Theorem with a massive—and ultimately award-winning (see the next chapter)—paper in the AMS's *Transactions* in 1917 on "Dynamical Systems with Two Degrees of Freedom." There, among other ideas, he introduced the notion of the minimax method for establishing the existence of periodic motions of dynamical systems.[158] In codifying his work in this area in the book, *Dynamical Systems*, published in 1927 and based on the AMS Colloquium Lectures he had given in 1920, Birkhoff "created a new branch of mathematics separate from its roots in celestial mechanics."[159]

One of Birkhoff's students at this juncture, Marston Morse, followed his advisor's calculus-of-variations lead, producing a dissertation on "Certain Types of Geodesic Motion of a Surface of Negative Curvature."[160] By 1925, Morse had extended this work to consider the "Relations Between the Critical Points of a Real Function of n Independent Variables" which drew fundamentally from Birkhoff's development of the minimax principle.[161] The ideas that Morse presented in this paper formed the basis of what is now called Morse theory, an area in differential topology that studies differentiable functions on a manifold in order to analyze its topology. By the end of the 1920s from the position he had taken at Harvard in 1926, Morse was hard at work developing what he called the calculus of variations in the large, an analog of Morse

157. Birkhoff, "Fifty Years of American Mathematics," p. 299.

158. George Birkhoff, "Dynamical Systems with Two Degrees of Freedom," *TAMS* 18 (1917), 199–300 on pp. 239–257.

159. George Birkhoff, *Dynamical Systems*, AMS Colloquium Publications, vol. 9 (New York: American Mathematical Society, 1927) and David Aubin, "George David Birkhoff, *Dynamical Systems* (1927)," in *Landmark Writings in Western Mathematics, 1640–1940*, ed. Ivor Grattan-Guinness, pp. 871–881. The quotation is on p. 871.

160. Marston Morse, "A One-to-one Representation of Geodesics on a Surface of Negative Curvature," *AJM* 43 (1921), 33–51. The publication of Morse's dissertation was delayed by his participation in World War I. Raoul Bott gives a nice discussion of Morse's main results in "Marston Morse and His Mathematical Works," *BAMS* 3 (1980), 907–950 on pp. 915–918. See also Everett Pitcher, "Marston Morse 1892–1977," *Biographical Memoirs*, vol. 65 (Washington, D.C.: National Academy of Sciences, 1994), pp. 222–240 for a focus more specifically on Morse theory and Morse's supporting work in the calculus of variations.

161. Marston Morse, "Relations Between the Critical Points of a Real Function of n Independent Variables," *TAMS* 27 (1925), 345–396.

theory for functionals as opposed to functions (see chapter five).[162] Morse's work, like that of his contemporary Jesse Douglas (see chapter three), testified to the fact that research in the calculus of variations was alive and well in the 1920s.

Areas of Lesser American Interest

Although Americans in the 1920s were clearly engaged in research that covered a broad swath of the contemporaneous mathematical landscape, there were a number of well-established areas in which, for one reason or another, they showed relatively less interest. For example, while Leonard Dickson produced his massive *History of the Theory of Numbers* and promptly moved into not unrelated research on the arithmetic of algebras, few of his contemporaries shared his number-theoretic interests. Three who did were Harry Vandiver, E. T. Bell, and Oystein Ore.[163] Vandiver, after 1924 at the University of Texas in Austin, did important, although ultimately unsuccessful, work toward a proof of Fermat's Last Theorem, while Caltech's Bell developed a theory of what he termed "arithmetical paraphrases," which served to unify and extend seemingly disparate number-theoretic results due particularly to Joseph Liouville.[164] Ore, who actually became better known later for his work in ring theory and especially on lattices (see chapter six) than for

162. Marston Morse, "The Foundations of a Theory of the Calculus of Variations in the Large," *TAMS* 30 (1928), 213–274; "The Foundations of the Calculus of Variations in the Large in *m*-Space (First Paper)," *TAMS* 31 (1929), 379–404; and "The Foundations of a Theory of the Calculus of Variations in the Large in *m*-Space (Second Paper)," *TAMS* 32 (1930), 599–631. It was for the second of these papers that Morse won the AMS's Bôcher Memorial Prize in 1933.

163. Another, the English-born, Göttingen-educated student of noted number-theorist Edmund Landau, Aubrey Kempner, settled at the University of Colorado and published on, among other topics, "Polynomials and Their Residue Systems," *TAMS* 22 (1921), 240–266 and 267–288.

164. See, for example, Harry Vandiver, "On Fermat's Last Theorem," *TAMS* 31 (1929), 613–642, a paper on the partial basis of which Vandiver was awarded the first Frank Nelson Cole Prize in Number Theory in 1931, and E. T. Bell, "Arithmetical Paraphrases," *TAMS* 22 (1921), 1–30 and 198–219, for which he was awarded the AMS's Bôcher Memorial Prize in 1924. (Like the Bôcher Prize, the Cole Prize is discussed in the next chapter.) Bell later extended his ideas in *Algebraic Arithmetic*, AMS Colloquium Publications, vol. 7 (New York: American Mathematical Society, 1927). See also Christopher Hollings, "A Tale of Mathematical Myth-Making: E. T. Bell and the 'Arithmetization of Algebra,'" *BSHM Bulletin: Journal of the British Society for the History of Mathematics* 31 (2016), 69–80.

his research in number theory, came to Yale in 1927 from the University of Oslo as a direct result of a faculty-finding mission to Europe by James Pierpont in 1926.[165] Pierpont, the same mathematician who had questioned why Yale's graduate enrollment had dwindled to one by 1919, sought specifically to import a European-trained, research-driven mathematician to Yale and, in so doing, to begin moving Yale into a research position more competitive with those of Chicago, Harvard, and Princeton.[166]

Dickson, Vandiver (when still an instructor at Cornell), Penn's Howard Mitchell, and Illinois's Gustav Wahlin actually felt that the time was right already in 1923 to survey the field of algebraic numbers for the NRC with Vandiver and Wahlin producing a follow-up report the year after Ore's arrival in 1928.[167] As they explained, they aimed not only "to bring up to date the extensive report on the theory of algebraic number fields" that Hilbert had published in the 1890s but also "to deal with the literature, not cited in Hilbert's report, on fields of functions and related topics, such as [Kurt] Hensel's p-adic numbers and modular systems."[168] Perhaps not surprisingly, they, and particularly Dickson and Vandiver, had done work in the latter areas and so sought to highlight and contextualize it as well as the results of other American or American-trained mathematicians within the broader sweep of largely German number-theoretic research.

Postulate-theoretic work—like that animated by E. H. Moore at the turn of the twentieth century and in which both Veblen and R. L. Moore had

165. For a sense of Ore's number-theoretic research at the time, see Oystein Ore, "Abriß einer arithmetischen Theorie der Galoisschen Körper," *Mathematische Annalen* 100 (1928), 650–673. Ore continued this line of research into the 1930s with, for example, his short book, *Les corps algébriques et la théorie des idéaux* (Paris: Gauthier-Villars, 1934). In Norwegian, Ore's first name is spelled Øystein. After moving to the United States, he used the simplified Oystein. I will follow that convention here.

166. By 1933, Yale had also lured Einar Hille, the American-born son of Swedish immigrants, from Princeton to New Haven. This effectively signaled the beginning of Yale's ascent into the competitive mathematical research ranks.

167. Leonard Dickson, Howard Mitchell, Harry Vandiver, and Gustav Wahlin, *Algebraic Numbers*, Bulletin of the National Research Council, no. 28 (Washington, D.C.: National Research Council of the National Academy of Sciences, 1923) and Harry Vandiver and Gustav Wahlin, *Algebraic Numbers II*, Bulletin of the National Research Council, no. 62 (Washington, D.C.: National Research Council of the National Academy of Sciences, 1928).

168. Dickson, Mitchell, Vandiver, and Wahlin, p. 3 and David Hilbert, "Die Theorie der algebraischen Zahlkörper," *Jahresbericht der Deutschen Mathematiker-Vereinigung* 4 (1894–95), 175–535.

engaged—defined yet another American research strain that persisted into the 1920s, although its American heyday had been in the century's opening two decades. It attracted, among others, Harvard's Edward Huntington, Berkeley's Benjamin Bernstein, and Princeton's Alonzo Church and gradually evolved into work on both Boolean algebra and symbolic logic.[169] In particular, the Lithuanian-born Bernstein had done his undergraduate work at Hopkins before moving to Berkeley to prepare the doctoral dissertation on "A Complete Set of Postulates for the Logic of Classes Expressed in Terms of the Operation 'Exception'" that he defended under Mellen Haskell's direction in 1913. Bernstein stayed on at Berkeley and, by 1929, had worked his way up the ranks to professor. In 1925, for example, he published a paper in the AMS's *Transactions* in which he drew not only from his own earlier work but also from that of Huntington and the English mathematician and philosopher Alfred North Whitehead.[170] There, as Bernstein explained, he "determine[d] all the operations with respect to which the elements of a boolean algebra form a group in general and an abelian group in particular." (Marshall Stone would also actively pursue the theory of Boolean algebras, but in the 1930s. See chapter six.)

Bernstein's younger contemporary, Church was similarly interested in axioms and their ramifications. In a paper as early as 1925, he considered the problem of redundancy in axiomatic systems, citing results in particular of E. H. Moore and James (Sturdevant) Taylor, another of Haskell's students who, like Bernstein, worked on Boolean algebras.[171] In the doctoral dissertation Church completed under Veblen in 1927, however, he explored "Alternatives to Zermelo's Assumption," that is, alternatives to what is now generally termed the axiom of choice.[172] Following a two-year NRC research

169. Michael Scanlon, "Who Were the American Postulate Theorists?" *JSL* 56 (1991), 981–1002; Hollings, "Nobody Could Possibly Misunderstand What a Group Is"; and Bell, "Fifty Years of Algebra in America," pp. 15–19. On the development of symbolic logic as a well-defined research field in the United States in the 1930s, see chapter four. Developments in Boolean algebra in 1930s America are treated in chapter six.

170. Benjamin Bernstein, "Operations with Respect to Which the Elements of a Boolean Algebra Form a Group," *TAMS* 26 (1924), 171–175 (see p. 171 for the quotation) as well as Benjamin Bernstein, "Complete Sets of Representations of Two-Element Algebras," *BAMS* 30 (1924), 24–30.

171. Alonzo Church, "On Irredundant Sets of Postulates," *TAMS* 27 (1925), 318–328.

172. Alonzo Church, "Alternatives to Zermelo's Assumption," *TAMS* 29 (1927), 178–208. Gregory Moore treats the history of this axiom in *Zermelo's Axiom of Choice: Its Origins, Development, and Influence* (New York: Springer-Verlag, 1982).

fellowship that took him to Hilbert in Göttingen and to L.E.J. Brouwer in Amsterdam, Church returned to an assistant professorship at his alma mater in 1929 and trained an impressive string of doctoral students in symbolic logic in the 1930s (see chapter four).

The area of lesser interest in the 1920s that nevertheless represented if not an elephant in the "room" then an elephant in the landscape of American mathematics was applied mathematics. In the closing quarter of the nineteenth century, mathematics of a more applied nature had actually been well represented in the work in celestial mechanics of, for example, Benjamin Peirce and George William Hill, as well as in the mathematical physics of Josiah Willard Gibbs.[173] Yet, as a result of what they had brought back from their trips abroad, and especially from Göttingen, American mathematicians of the 1890s and opening decades of the twentieth century established doctoral programs—at the University of Chicago, at Harvard, at Princeton, and elsewhere—focused on *pure* mathematics and, consequently, embarked on *purist* research programs.[174]

As early as 1894, Emory McClintock, an actuary at the Mutual Life Insurance Company of New York and President of what was then the New York Mathematical Society, had bemoaned, in his retiring presidential address, the fact that "our young mathematicians" have been "most thoroughly instructed" in "the pure science," and so it was that aspect of the mathematical endeavor— and not applied aspects—to which "he will ... most likely confine his efforts" in pursuing original mathematical research.[175] Robert Woodward, professor of mechanics and later of mechanics and mathematical physics at Columbia and the Society's fifth President, was even more pointed on this score when he stepped down from his AMS post at the close of 1899. He acknowledged and applauded the American community's support of pure mathematics, but, while certainly not "urging the cultivation of pure mathematics less," he argued that the AMS should most definitely support "the pursuit of applied mathematics more."[176] Similar concerns were raised again in 1916 by the English-born, Yale mathematical astronomer and thirteenth

173. Parshall and Rowe, *Emergence*, pp. 23–40.

174. Ibid., especially chapter 10.

175. Emory McClintock, "The Past and Future of the Society," *BAMS* 1 (1895), 85–94 on p. 93. I provide a fuller contextualization of this and other calls for nurturing applied mathematics within the American mathematical community in "Perspectives on American Mathematics," *BAMS* 37 (2000), 381–405.

176. Robert Woodward, "The Century's Progress in Applied Mathematics," *BAMS* 6 (1900), 133–163 on p. 163.

AMS President, Ernest Brown, as Americans grimly watched a Europe at war from the other side of the Atlantic.[177] Still, despite these and ongoing calls (see the next chapter) for a greater emphasis on applied topics, only a few American mathematicians—among them, Forest Moulton, at Chicago until 1926, Frances Murnaghan at Hopkins, Harry Bateman at Caltech, and Norbert Wiener at MIT—actively pursued such research in the 1920s, even though it was well cultivated especially in Great Britain and Germany. Wiener, in particular, worked with engineering colleagues at MIT such as Vannevar Bush to advance and develop various practical applications of his work in harmonic analysis and analytic number theory.[178] World War II and the war work that some mathematicians did before and during that conflict would serve to focus greater American attention on applied mathematics during and after the 1940s (see chapters eight and ten).

———

American mathematicians in the 1920s were clearly engaged in research in many of the areas of then contemporaneous mathematics. And, whereas at the close of the nineteenth century and up to World War I, they had largely received their post-baccalaureate training in Europe, by the 1920s, they were staying at home at least for their initial higher mathematical education (compare chapter three). They had recognized that the United States had come to support not just viable but strong graduate programs in mathematics from coast to coast.

This fact was clearly reflected in an informal survey conducted in 1924 by Raymond Hughes, President of Miami University of Ohio.[179] There,

177. Ernest Brown, "The Relation of Mathematics to the Natural Sciences," *BAMS* 23 (1917), 213–230.

178. See, for example, Forest Moulton, *Differential Equations* (New York: The Macmillan Company, 1930); Hugh Dryden, Frances Murnaghan, and Harry Bateman, *Hydrodynamics*, Bulletin of the National Research Council, no. 84 (Washington, D.C.: National Research Council of the National Academy of Sciences, 1932); and Norbert Wiener's appendix, "Fourier Analysis and Asymptotic Series," to Vannevar Bush, *Operational Circuit Analysis* (New York: Wiley, 1929), pp. 366–379. Wiener discusses his interests in the applications of mathematics in *I Am a Mathematician: An Autobiography* (Garden City: Doubleday & Co., 1956).

179. Hughes solicited experts in the various academic fields to rank-order programs in their disciplines. Some forty mathematicians, among them Birkhoff, received his questionnaire, and roughly half responded to it. This survey was also discussed in Karen Hunger Parshall, " 'A New Era in the Development of Our Science': The American Mathematical Research Community, 1920–1950," in *A Delicate Balance: Global Perspectives on Innovation and Tradition in the History*

Chicago—under E. H. Moore as it had been in the 1890s and into the first decade of the twentieth century—was unequivocally deemed to have the leading graduate program in mathematics in the United States. Harvard was a close—but distinct—second thanks to the presence on its faculty not only of Moore's student, Birkhoff, but also of the German-trained Bôcher and Osgood, while Princeton, especially owing to another Moore student, Veblen, but also to Hopkins-trained Eisenhart, was a more distant but still clearly delineated third. These top three were followed by the programs at Illinois, Columbia, Cornell, Yale, Wisconsin, Hopkins, Michigan, Berkeley, Penn, and Minnesota.[180] These were some—but by no means all—of the schools that fostered American mathematical research in the 1920s.

Interestingly, these rankings roughly mirrored data that Richardson later compiled on American Ph.D. production in mathematics. Chicago had produced by far the most mathematics Ph.D.s (177) by 1924, followed by Hopkins (77), Yale (63), Harvard (59), Cornell and Columbia (45 each), Penn (32), Princeton (26), Illinois (25), Berkeley (23), and Wisconsin (14).[181] These students had fanned out across the country. Some built or otherwise actively participated in new graduate programs. Many firmly staked their research claims. This was the new lay of the American mathematical research landscape.

---

of Mathematics: A Festschrift in Honor of Joseph W. Dauben, ed. David Rowe and Wann-Sheng Horng (Basel: Birkhäuser Verlag, 2015), pp. 275–308 on pp. 279–280.

180. Raymond Hughes to Birkhoff, 29 October, 1924, HUA 4213.2, Box 4, Folder: Correspondence 1924 K–M, Birkhoff Papers. The University of Chicago received seventeen first-place and two second-place ratings; Harvard got thirteen first-place and six second-place rankings; and Princeton garnered five first-place, seven second-place, five third-place, and two fourth- or fifth-place ratings. Hughes combined the fourth- and fifth-place rankings, hence it is not clear whether Princeton received two fourth-place rankings, two fifth-place rankings or one fourth-place and one fifth-place ranking.

181. Roland Richardson, "The Ph.D. Degree and Mathematical Research," *AMM* 43 (1936), 199–215; reprinted in *A Century of Mathematics in America*, ed. Peter Duren et al., 2: 361–378, Table 1 on p. 203 (or see p. 366 in the reprinted edition). In 1924, Michigan (12) and Minnesota (4) had been giving the Ph.D. in mathematics for less than a decade.

Perhaps not surprisingly, the 1920s was also a key decade in the growth of American physics. Spencer Weart gives a statistical overview that documents a similarly steady increase in Ph.D. output in that field in "The Physics Business in America, 1919–1940: A Statistical Reconnaissance," in *The Sciences in the American Context: New Perspectives*, ed. Nathan Reingold (Washington, D.C.: Smithsonian Institution, 1979), pp. 295–358, especially on p. 296.

2

Strengthening the Infrastructure of American Mathematics

The American mathematical community was hard at work in the 1920s not only on research and program-building but also on better establishing and aligning itself within what appeared to be an emergent postwar scientific establishment. As a result of the involvement of scientists of all stripes—but especially of chemists and physicists—during the United States' limited involvement in World War I, it was becoming clear that scientists, if suitably organized and supported, could pursue their research and, at the same time, play a more prominent role in national affairs. But just exactly what form should that organization and support take? How should scientists position themselves to bring their contributions—both realized and potential—into the wider public consciousness?

These were new sorts of questions, but they had already been addressed successfully in the 1910s—and particularly within the privileged context of the National Academy of Sciences (NAS)—by the chemists and physicists, men like chemist Arthur Noyes of MIT, physicist Robert Millikan of the University of Chicago, and astrophysicist George Ellery Hale of the Mt. Wilson Observatory in California.[1] In so doing, these men had revived and modernized a category—the politically savvy scientist—that their nineteenth-century predecessors like Joseph Henry and Louis Agassiz had tried to

---

1. Kevles, *The Physicists*, especially pp. 91–138. Compare, too, Parshall, " 'A New Era,' " for an overview of the comparative situation in mathematics. Hale was incorrectly identified in that article (pp. 276–277) as President of the NAS and on the faculty at Caltech. While he was a trustee of what became Caltech, his affiliation was at the nearby observatory funded by the Carnegie Institution of Washington, and while he was a member of the National Academy, he was never its President.

define.[2] They had also assumed new roles as research scientists, fund-raisers, and valued governmental advisers. When postwar *mathematical* leaders began to consider the same questions in the 1920s, they found themselves playing catch-up. How could mathematicians establish a competitive position in a quickly evolving American *scientific*—as opposed to the more narrowly *mathematical*—landscape? One thing seemed certain. The answer to this question would depend on a more complete professionalization through the establishment of mathematics as a field worthy of external financial support as well as on the improvement of mathematicians' ability to produce, publish, and publicize new research results.

"Corporatizing" Research-Level Mathematics

In 1921, Gilbert Bliss began a two-calendar-year term as the sixteenth President of the AMS. As evidenced by the fact that he had earned a star in the second edition of James McKeen Cattell's *American Men of Science* in 1910, Bliss had quickly succeeded in establishing his mathematical reputation. His election to the National Academy of Sciences in 1916 and his assumption of the AMS presidency five years later provided further testament to his standing in the American mathematical research community, while his term in office demonstrated his adaptability as a leader.

At its annual meeting in New York City in December 1919, the members of the AMS Council had adopted a series of resolutions aimed at "the more extensive development of pure *and applied* mathematics in America" in explicit recognition of "the importance of such development to the nation."[3] Indeed, calls for the support of applied mathematics had come to represent a recurring theme in the rhetoric of an American mathematical community that, for better or worse, had developed over the course of the first two decades of the new century along largely purist lines.

On the Council members' minds in 1919 was nothing less than the "future needs of the Society," and to that end, it appointed a seven-person committee to consider the problem. Without question, central to that future—both

---

2. On Henry and Agassiz and their roles in nineteenth-century American science, see, for example, Albert Moyer, *Joseph Henry: The Rise of an American Scientist* (Washington, D.C.: Smithsonian Institution Press, 1997) and Edward Lurie, *Louis Agassiz: A Life in Science* (Chicago: University of Chicago Press, 1960).

3. Frank Cole, "The Twenty-sixth Annual Meeting of the American Mathematical Society," *BAMS* 26 (1920), 241–259 on p. 242 (my emphasis).

pure and applied—was the success of the AMS's publication ventures.[4] The Council recognized that in order to spur research and to communicate new findings most broadly, America's mathematicians required, in addition to the space provided by their research journals, the means to publish their work in book form. Since even journal space was becoming scarce, money would clearly be required to effect the needed change in publication support. (See Fig. 4.3 for data on pages published in American mathematical journals from 1920 to 1939.) A critical and foundational step, at least as they saw it, should thus be the incorporation of the AMS in order to provide legal protection for its assets and thereby to establish a firm foundation on which it could seek charitable donations and bequests. When Hopkins professor and then AMS President Frank Morley failed to set the necessary machinery in motion, the Council explicitly charged a committee in February 1921 to incorporate the Society.[5] This mandate, as well as other financial concerns, consumed the two years of Bliss's presidency.

While the committee on incorporation, chaired by Harry Tyler of MIT and rounded out by William Osgood and Columbia's David Smith, set about its work, Bliss launched what amounted to the beginning of a capital campaign for the AMS. The Society's financial situation had steadily worsened since the end of the war and seemed to be coming to a head in 1921. Nearly all of its income went into the publication of its *Bulletin* and *Transactions*, but printing costs had tripled, and it had become unclear whether the *Bulletin* could be sustained at its volume of some 500 pages per year or whether publication of the *Transactions* could continue at all (see the next section).[6] By contrast, the MAA, which supported its only journal, the *American Mathematical Monthly*, completely through membership dues, was on a relatively sound financial footing through the 1920s.[7]

4. Archibald, *Semicentennial History*, p. 29.

5. Ibid., p. 13.

6. On the state of the Society, see Roland Richardson, (Draft of a) "Report of the Secretary for the Years 1921–25," undated, p. 2, HUG 4213.2, Box 5, Folder: R, Birkhoff Papers and compare the printed version, Roland Richardson, "Report of the Secretary to the Council for the Years 1921–1925," *BAMS* 32 (1926), 203–211 on pp. 203–204.

7. For example, at the close of 1924, the MAA had a balance of $2,500, a slight increase over the close of the previous year. William Cairns, Secretary-Treasurer of the MAA, Minutes of the Meeting of the Outgoing Board of Trustees, 31 December, 1924, Box 4RM114, 2004–170, Folder: Headquarters: Board of Governors Minutes 1920–1928, MAA Records. The "real wealth" of $2,500 in 1924 dollars is equivalent to $39,042 in 2021 dollars. See https://www.measuringworth.com (where "real wage or real wealth" is defined as

Although skeptical at first about fund-raising, Bliss was finally swayed by the arguments particularly of Earle Hedrick, German-trained analyst in transit between 1919 and 1920 from the University of Missouri to UCLA, and a fellow member with Bliss of the committee that had been appointed to advise on the Society's future needs.[8] In 1921, and now as AMS President, Bliss personally assumed the chair of a committee that aimed to raise revenue, first, by increasing membership, and so income through dues and, second, by attracting new subscribers to the Society's *Transactions*. The drive proved successful. Targeting especially college teachers of mathematics, the AMS grew by just over 400 new members to over 1,000 strong, and the *Transactions* gained over 100 new subscribers between October 1920 and the end of Bliss's presidency in December 1922.[9]

In order to keep up the momentum during what he styled "this critical period in the development of our American mathematical school,"[10] Bliss, in essentially his last act as AMS President, appointed new standing committees on endowment, chaired by Julian Coolidge, and on policy and budget, led by Veblen, the AMS President-Elect. He also named a new committee on membership under Hopkins's Abraham Cohen, with Wellesley's Clara Smith as its "permanent secretary."[11] In an age increasingly characterized

---

"measur[ing] the purchasing power of an income or wealth by its relative ability to buy a (fixed over time) bundle of goods and services such as food, shelter, clothing, etc." This will be used as the basis of wage and wealth conversions hereinafter).

8. Archibald, *Semicentennial History*, p. 29.

9. The significantly larger MAA had almost 1,500 members by the end of that same year. Ibid., pp. 30 and 44 provide the data on the AMS, and *Register of Officers and Members for the Academic Year 1922–1923* (Lancaster and Providence: Mathematical Association of America), p. 43 gives data on the MAA.

10. Bliss, "A Letter from the President," p. 16.

11. Roland Richardson, "The Twenty-ninth Annual Meeting of the Society," *BAMS* 29 (1923), 97–116 on p. 99 and Archibald, *Semicentennial History*, p. 30. After her death in May 1943, the AMS honored Smith for her "devotion to the Society," citing her tireless efforts on this committee over the course of "the dozen years beginning with 1922, when it was necessary greatly to increase the membership of the Society in order to find financial support for its publications." The fact that the membership was doubled, the AMS Council publicly stated, "was due to her laborious and painstaking assembling of data regarding teachers of college mathematics." See Temple Hollcroft, "The Summer Meeting in New Brunswick," *BAMS* 49 (1943), 823–834 on p. 828. Born in 1865, Smith had earned her Ph.D. at Yale for a thesis on Bessel functions under Pierpont's supervision in 1904 and, after short stays elsewhere, moved to Wellesley College in 1908. She remained on the faculty there until her retirement as Helen Day Gould

by philanthropy, the AMS was poised to become an active fund-raiser for mathematics.[12]

As would soon become apparent, however, whereas the AMS's sixteenth President, Bliss, had been a somewhat reluctant convert to such an idea, its seventeenth President, Veblen, represented the first of a new breed of mathematician. He was a *politically savvy* mathematical leader, that is, a first-rate researcher who also recognized the need and had the skills to adapt mathematics to America's changing philanthropic and institutional environments.[13] Veblen, like Bliss, was starred in the second edition of *American Men of Science*. Like Bliss, too, he had been elected (in his case in 1919) to the National Academy at a young age. The two men had proceeded apace in making their names in the American mathematical community.

As AMS President, Veblen hit the ground running, thanks in no small part to the path that Bliss had cleared for him. By May 1923 and after two years of work, Tyler's committee on incorporation had finally succeeded in its charge, and at the end of October, "the unincorporated body turned over the conduct of its affairs to the corporation known as the American Mathematical Society" in what AMS Secretary Richardson solemnly described as "one of the milestones in the progress of the Society."[14] An initial board of thirty-one trustees was named—among whom were two women, Clara Smith and Bryn Mawr's Anna Pell (later Wheeler)—with Veblen elected its

---

Professor in 1934. For more on her life and contributions, see Green and LaDuke, especially pp. 20–24, 26, 79–80, and 289.

12. Olivier Zunz treats this aspect of twentieth-century American history in *Philanthropy in America: A History* (Princeton: Princeton University Press, 2012).

13. I characterize the Veblen of the 1920s as a "mathematician as scientific statesman" in " 'A New Era,' " p. 277. Without the broader historical contextualization provided there, Steve Batterson termed Veblen a "statesman of mathematics" in "The Vision, Insight, and Influence of Oswald Veblen," *NAMS* 54 (2007), 606–618 on pp. 615–616 (the quotation that follows is on p. 616). He says, without providing references, that it was Veblen's humanitarian endeavors relative to the refugees fleeing Europe in the 1930s (see chapters five and seven) that earned him this "unusual appellation." Here, I will use less gendered terms such as "politically savvy" and "leader" in referring to this new breed of mathematician.

14. Roland Richardson, "Incorporation of the American Mathematical Society," *BAMS* 30 (1924), 1–3 on p. 1. The actual articles of incorporation were published in *American Mathematical Society: List of Officers and Members, 1923–1924* (New York: American Mathematical Society, 1924) (and *BAMS* 30 (1924)), p. 47 and in Archibald, *Semicentennial History*, pp. 13–14. For more details, see Roland Richardson, "The October Meeting of the Society," *BAMS* 30 (1924), 4–11 on pp. 4–5. The MAA had already incorporated in 1920.

chair and Richardson its secretary. The latter two men—together with Herbert Hawkes, Dean of Columbia College, Robert Henderson, actuary with the Equitable Life Assurance Society of the United States, and incorporation committee chair, Tyler—were then chosen to constitute an executive committee that would be responsible for conducting the board's business through May 1924. At that time, a new, elected board numbering five trustees would be seated, and, thereafter, a five-member board would be elected every two years. The AMS's initial "corporatization" was complete. Now, it had to begin literally to capitalize on its new legal status.

Coolidge's committee on endowment had already been hard at work over the course of the first ten months of 1923. By the end of the year, and thanks to its efforts, the AMS membership—despite the fact that "they are nearly all college and university men of very limited means"—had contributed $25,000 or one fourth of the $100,000 endowment that Veblen and his colleagues hoped to raise.[15] This represented donations from one-seventh of the Society's membership, a rate of return deemed "beyond the most sanguine expectations."[16] The rest, though, would have to come from elsewhere, and Veblen and Coolidge had a number of ideas on that score.

First, Coolidge wrote to some fifty well-placed "local agents," among them Veblen at Princeton, soliciting their personal appeal to colleagues who had not yet contributed, as well as to others of their acquaintance who might be enjoined to donate.[17] Second, Coolidge targeted "a small minority in the Society who are more fortunately placed than their fellows" and urged them to give at a "really substantial rate."[18] He had already received two four-figure contributions, and he envisioned a whole class of member-donors at that level.

And, there was at least one other possible constituency that could be approached, namely, the "teachers of mathematics" defined as those at the

15. The quotation is from a general appeal letter from Veblen dated 10 March, 1924, in Box 3, Folder: Coolidge, Julian L., 1923, Veblen Papers. This endowment campaign is discussed in both Archibald, *Semicentennial History*, pp. 30–31 and Loren Butler Feffer, "Oswald Veblen and the Capitalization of American Mathematics: Raising Money for Research 1923–1928," *Isis* 89 (1998), 474–497, especially pp. 480–482.

16. Oswald Veblen to Julian Coolidge, 1 December, 1923, Box 3, Folder: Coolidge, Julian L., 1923, Veblen Papers. The quotation is in Archibald, *Semicentennial History*, p. 31.

17. Coolidge to Veblen, 3 December, 1923, Box 3, Folder: Coolidge, Julian L., 1923, Veblen Papers.

18. The quotations are from the undated copy of this more targeted appeal that accompanied Coolidge to Veblen, 27 February, 1924, Box 3, Folder: Coolidge, Julian L., 1923, Veblen Papers.

some 1,200 institutions that offered courses in analytical geometry or above in both the United States and Canada.[19] As early as January 1924, the idea was floated that the AMS and MAA somehow join forces in such a membership drive. This had the unintended consequence of rekindling—at least between Veblen, Richardson, and the then MAA President Herbert Slaught—the discussion from 1915 that had ultimately resulted in the MAA's creation.[20] Should there be two professional societies, one focused on research and one on teaching? Or, were those not two sides of the same professional coin and best handled under the aegis of a single organization?

In 1915, when the MAA was created, the answer to the latter question was "no." Regardless of their individual and collective aspirations toward research, the fact of the matter was that American mathematicians were college and university *teachers*. While the AMS had fostered research since its founding in 1888, it had done precious little to promote the actual teaching mission of its members. That had been the explicit objective of the founders of the MAA.[21]

By the 1920s, however, it was becoming ever clearer that universities were beginning to stress research, both in hiring their faculties and in assessing individual merit. Had it been a mistake organizationally to separate the two aspects—teaching and research—of the professional persona of the American mathematician? In January 1924, Richardson, for one, was "not ready to either endorse or reject" the single society idea. It was simply unclear to him whether the AMS and the MAA could "be better run as one organization or as two cooperating."[22] A month later, the question had been decided. Slaught and Veblen had come to agree that "it would now be quite out of the question to attempt any amalgamation of the two mathematical organization[s]" and that "we shall accomplish much more as formally distinct

19. Roland Richardson to Oswald Veblen, George Birkhoff, William Cairns, Abraham Cohen, Arnold Dresden, [Harry ?] Everett, William Roever, Clara Smith, and Herbert Slaught, 16 December, 1924, Box 10, Folder: Richardson, R.G.D. 1924, Veblen Papers.

20. Veblen to Richardson, 25 January, 1925, and Richardson to Veblen, 28 January, 1924, both in Box 10, Folder: Richardson, R.G.D. 1924, Veblen Papers and Slaught to Veblen, 5 February, 1924, and Veblen to Slaught, 18 February, 1924, both in Box 12, Folder: Slaught, H. E. 1923–30, Veblen Papers.

21. For more on the history of the MAA, see David Zitarelli, "The Mathematical Association of America: Its First 100 Years," in *A Century of Advancing Mathematics*, ed. Stephen Kennedy, et al., pp. 135–157.

22. Richardson to Veblen, 28 January, 1924.

organizations which ... coöperate very closely."[23] That cooperation manifested itself immediately in 1924 and 1925 as the two societies undertook their proposed joint membership drive, but theirs would be an uphill battle.

As Coolidge and Veblen realized, the mathematical community, no matter how broadly construed, had largely been tapped out, so they would have to focus on other, richer sources in order to reach their endowment goal. Clearly, the philanthropies represented obvious targets, and Veblen suggested that they "regard the Rockefeller Foundation, the General Education Board, and the Laura Spellman Rockefeller Fund as a unit for the purposes of" the fund drive.[24] Coolidge agreed that it was "the height of unwisdom" to approach more than one of these related Rockefeller philanthropies because of their administrative interconnectedness. Since he had already heard through a series of personal connections that the Foundation had been advised that the AMS was "one of the societies most deserving of help," he argued that following up on that lead might be "the best approach to the Rockefellers."[25] It was all a tricky business, and the mathematicians were learning as they went along.

Industries represented another possible target of opportunity. There, Veblen and Coolidge naturally focused on those corporate concerns with arguably the closest ties to mathematics: electric companies, insurance companies, and publishing houses. Yet, successfully approaching these sources presented at least two problems. It was widely viewed as illegal for corporations to make gifts outright, and it was hard for companies to appreciate how supporting the AMS might actually benefit *them*. Relative to the latter problem, Coolidge had initially proposed the "very radical" idea of contracting, "for not too small a quid pro quo," with firms to direct them to sources capable of satisfying their mathematical needs.[26] The AMS Council approved not only this idea but also the ultimately much more successful one of creating "sustaining" memberships for companies at a cost of no less than $100 a year. For a contribution of at least $500, a company would earn the additional designation of "patron." Such memberships would, among other privileges, give

23. Veblen to Slaught, 18 February, 1924.

24. Veblen to Coolidge, 1 December, 1923.

25. Coolidge to Veblen, 3 December, 1923.

26. The quotations are from Coolidge to Veblen, 6 March, 1924, Box 3, Folder: Coolidge, Julian L. 1924, Veblen Papers. See also Roland Richardson to the Members of the Council of the American Mathematical Society, 29 March, 1924, Box 10, Folder: Richardson, R.G.D. 1924, Veblen Papers.

the companies the right to nominate a limited number of their employees for AMS membership free of dues.[27]

Veblen, Coolidge, and Richardson on the East Coast, together with Wisconsin's Arnold Dresden west of the Appalachians, beat the corporate bushes in earnest. Coolidge's letter to Samuel Insull of Commonwealth Edison in Chicago was typical of their written strategy. After calling Insull's attention to the fact that the AMS fostered "the interests of mathematics and mathematical physics" in the United States, he coyly added that "I need not point out to you the vital importance of these sciences to every feature of our civilization. I have often heard engineers say," he continued, "that they used little mathematics in their work; but it is abundantly evident that without the mathematics of the past there would be no engineering whatever today. May we not go a step further and say that the development of the engineering of the future will depend in no small measure upon the cultivation of the mathematics of the present?"[28] Assuming that he had captured Insull's attention with this opening salvo, Coolidge proceeded to explain that the AMS had embarked upon a capital campaign and that, so far, the membership—for the most part "teachers on very limited salaries"—had already contributed a quarter of the goal. "Have they not the right," he argued, "to turn now and ask help from the great engineering corporations, whose business would never have existed but for mathematical science? Am I wrong in including the Commonwealth Edison Company in this category?"

Apparently not. By the end of 1924, Coolidge and his colleagues had succeeded in securing fourteen sustaining members among which four were patrons. There was one individual, E. W. Rice, Jr., President until 1922 of General Electric; two institutions of higher education, Dartmouth and the University of Washington; two publishers, Allyn and Bacon and Ginn and Company, both of Boston; four insurance companies, John Hancock and New England Life Insurance, both of Boston, as well as Metropolitan Life of New York and Union Central Life of Cincinnati; and five electric companies or power concerns, Babcock and Wilcox of New York, Insull Interests of Chicago (among which was Commonwealth Edison), and General Electric, Western Electric, and Westinghouse Electric, the latter three in New York and the latter

27. During the decade of the 1920s, individual annual dues to the AMS were $6.00. Archibald, *Semicentennial History*, p. 29.

28. Coolidge to Samuel Insull, 19 March, 1924, Box 3, Folder: Coolidge, Julian L. 1924, Veblen Papers. The quotations that follow in this paragraph are also from this letter.

four patrons.[29] The concentration of corporate support from the Boston area testified to Coolidge's untiring efforts on behalf of the Society, that from in and around New York City reflected Veblen's dedication, and that in the Midwest owed largely to Dresden's persuasive powers.[30]

While Coolidge's written argument—together with his legwork—may have sufficed to win over these industrial concerns to the AMS's cause, the Society actually aimed for more. It saw its campaign in terms not just of immediate financial outcomes but of longer-range and broader educational goals. It aimed at nothing less than educating "the public concerning the basic character of mathematics *in our present civilization* and the importance of mathematical research in *advancing that civilization*."[31] Mathematics as the foundation of science and technology and hence as the foundation of modern society, this had been at the core of Coolidge's argument to Insull, but it was also at the heart of another initiative launched during Veblen's AMS presidency.

The Josiah Willard Gibbs Lectureship was named in honor of the prominent, late-nineteenth-century, Yale mathematician, chemist, and physicist. To be "of a popular nature on topics in mathematics or its applications," the Gibbs lectures, it was hoped, would serve a dual purpose. First, they would cross over into fields "adjacent" to mathematics like electrical engineering and actuarial science and, in so doing, attract the interest and sympathies of their practitioners. Second, they would expose the purist members of the AMS to applied mathematics as well as, perhaps, spark interest among them in such topics. If these educational processes were successful, it would be easier and more natural in the future to make the case for mathematics.[32]

The first two Gibbs lectures were given in 1924, one by Columbia physicist and AMS life member Michael Pupin on the aesthetics of then-modern

29. *American Mathematical Society: List of Officers and Members, 1923–1924*, p. 7.

30. Archibald, *Semicentennial History*, p. 32.

31. Ibid., p. 31 (my emphasis) and compare Parshall, "'A New Era,'" p. 281.

32. Roland Richardson, "The Josiah Willard Gibbs Lectureship," *BAMS* 29 (1923), 385. Feffer dicusses the role the Gibbs lectures played in Veblen's strategy in "Oswald Veblen and the Capitalization of American Mathematics," pp. 481–482. The participation of mathematicians in the American Association for the Advancement of Science (AAAS), a national organization founded in 1848, also had, theoretically, this same crossover effect, since its annual meetings brought scientists of all stripes together under the same roof. Physicists and other scientists could easily attend talks of mathematicians and vice versa, but, in practice, the division of the AAAS into specialized sections (mathematics defines Section A) perhaps tended more to foster compartmentalization and specialization.

physics and the other by actuary Robert Henderson on life insurance as both a social science and a mathematical problem.[33] Both were in the spirit of intellectual crossover and communication between mathematics and other fields, but it is ultimately unclear just how many physicists or insurance professionals might have been in the audiences to hear and take away these broader messages, or how many mathematicians the lectures might have swayed into pursuing applications to physics or the actuarial sciences. Likely very few.

The inaugural Gibbs lectures were delivered just before Veblen's term as AMS President came to an end in December 1924. His two years at the AMS's helm had been, in the eyes of his colleagues, nothing short of "momentous."[34] His friend and confidant, Harvard's George Birkhoff, succeeded him in the presidency and furthered the initiatives that he had so ably fostered.

Birkhoff's professional trajectory had been remarkably similar to Veblen's. Both men had come from middle-class families. Both had earned bachelor's degrees at Harvard before moving on to complete doctoral work under E. H. Moore at Chicago.[35] Both had served as Princeton preceptors and were starred in the second edition of *American Men of Science*. Both quickly secured professorships at their respective institutions. Both were elected to the National Academy, Birkhoff in 1918 one year ahead of Veblen. They were strong mathematicians—with Birkhoff perhaps the stronger of the two—who shared a vision for American mathematics and who worked—with Veblen definitely the more intense of the activists—in concert to help realize it.

As AMS President, Birkhoff picked up seamlessly where Veblen had left off. In particular, he convinced Coolidge, who assumed the presidency of the MAA in January 1925,[36] to stay on as chair of the AMS's committee on the endowment. Coolidge remained in that post until the end of 1925, when

---

33. Michael Pupin, "From Chaos to Cosmos," *Scribner's Magazine* 76 (1924), 3–10 and Robert Henderson, "Life Insurance as a Social Science and as a Mathematical Problem," *BAMS* 31 (1925), 227–252.

34. Archibald, *Semicentennial History*, p. 30.

35. Birkhoff, however, had apparently embraced the research ethos while in high school and had then had a double dose of it, from both his mentor at Harvard, Bôcher, and from Moore. See Garrett Birkhoff, "Mathematics at Harvard, 1836–1944," in *A Century of Mathematics in America*, ed. Peter Duren et al., 2: 3–58 on p. 25.

36. During the hectic AMS fund-raising year of 1924, Coolidge had also served as MAA Vice President. He is a prime example of both the cooperation between and the interconnectedness of the constituencies represented by the two organizations.

FIGURE 2.1. George D. Birkhoff (1884–1944). (Photo by David Bachrach, Jr.)

the capital campaign—begun under Bliss in 1921, spectacularly developed under Veblen, and carried to its conclusion under Birkhoff—was declared "a complete success."[37] The membership had contributed $25,000 to the cause; another $30,000 had been raised from "others interested in mathematics"; some thirty-seven sustaining and/or patron-level members had been secured. As Richardson declared in the report he prepared for the AMS Council, if the corporate sponsorship could "be kept near its present level, the Society is in a financial condition even more favorable than if the sum originally named [of $100,000] had been procured for the treasury." Indeed, a new committee was formed under the leadership of the 1912 Harvard Ph.D. and student of Maxime Bôcher, Tomlinson Fort (then of Hunter College in New York

37. Richardson, "Report of the Secretary to the Council for the Years 1921–1925," p. 205. The quotations that follow in this paragraph are also on this page.

City), to work to assure precisely that. In the first half of the 1920s, the AMS had been *legally* incorporated at the same time that fund-raising for research-level mathematics had been *fundamentally* incorporated into its professional mission.

Raising Money to Enable Research

Clearly, the AMS's main objectives were the encouragement and support of mathematical research, and it had largely realized them by providing venues for face-to-face but especially for printed communication. Indeed, the bulk of its financial resources had been directed to the latter. But what about support for the creative process itself, that is, the process that actually generates the new research to be published? The universities, and to a lesser extent the colleges, were placing a growing emphasis on research in the 1920s. If only philanthropic dollars could be pried loose in support of that noble endeavor as well. Veblen, a mathematician with political savvy, was also the leader in the quest for dollars explicitly to enhance mathematical research during his two years in the AMS presidency.

As the AMS's representative in the Division of Physical Sciences of the National Research Council starting in 1920, Veblen had almost immediately come to appreciate the impact that the NRC could have on the development of American science, in general, and American mathematics, in particular. Scarcely five years old, the NRC had been founded as an arm of the National Academy of Sciences in 1916 for the purpose of coordinating civilian and military scientific and technical resources, given what was then the likely entrance of the United States into World War I.[38] Two years later and after the war's close, President Woodrow Wilson charged the Academy by Executive Order to make the NRC permanent in recognition of the role it had managed to play even during the United States' limited wartime engagement.

A reimagined NRC emerged in 1919. Funded through an endowment to the NAS of $5,000,000 made possible through the beneficence of the Carnegie Corporation of New York, it aimed explicitly "to promote research in the mathematical, physical, and biological sciences, and in the application of these sciences to engineering, agriculture, medicine, and other useful arts, with the object of increasing knowledge, of strengthening the national

---

38. For the account that follows of the NRC and its engagement in mathematics, compare Parshall, " 'A New Era,' " pp. 275–279.

defense, and of contributing in other ways to the public welfare."[39] It partially realized this goal through its National Research Fund, which, in mathematics at least, made small grants to "the leading research mathematicians" in order to give them "more control over their own time and energy by freeing them from some of their obligations in routine teaching and administration."[40] For example, fifteen mathematicians garnered such support during the 1928–1929 academic year. Among them, Marston Morse at Harvard received $2,000 a year for three years in order to reduce his teaching commitment from nine hours to four-and-a-half hours a week, while E. T. Bell at Caltech was awarded $3,000 a year for two years to provide for two assistants.[41]

Another way that the NAS and its NRC fostered American scientific productivity was through an innovative, Rockefeller-funded, postdoctoral fellowship program by means of which "the more able young scientists" would "acquire 'momentum' in their research, before settling down to permanent positions."[42] These fellowships were thus intended to give postdoctoral students time to focus *solely* on their research—in conjunction with recognized experts in their particular field—for a year or two prior to taking permanent

39. "The National Research Council," *Science* 49 (1919), 458–462 on p. 458 and Albert Barrows, "General Organization and Activities" in *A History of the National Research Council, 1919–1933* (Washington, D.C.: National Research Council, 1933), pp. 7–11 on p. 7. See also "Research Fellowships in Mathematics," *AMM* 31 (1924), 168–169. The "real wealth" of $5M in 1919 dollars is equivalent to $77.2M in 2021 dollars (https://www.measuringworth.com). Of the $5M endowment, roughly a third funded the construction of a building to house the NAS and the NRC, while the rest went to the operating costs of the NRC (Kevles, *The Physicists*, p. 150).

40. "Report for Mathematics to the Trustees of the National Research Fund," undated but around 1928, Box 26, Folder: NAS National Research Fund 1928 (file 1), Veblen Papers. This fund did not survive the economic depression of the 1930s.

41. The other "leading mathematicians" so rewarded in that round were Dickson at Chicago, Lefschetz, Alexander, and Wedderburn at Princeton, Osgood and Walsh at Harvard, Bliss, E. H. Moore, and Lane at Chicago, Wiener at MIT, Kline at Penn, Kasner at Columbia, and Manning at Stanford. See Box 26, Folder: NAS National Research Fund 1928 (file 1), Veblen Papers.

42. Siegmund-Schultze discusses Rockefeller support for the NRC in *Rockefeller and the Internationalization of Mathematics*, pp. 29–30. For the quotation, see Floyd Richtmyer, "Division of Physical Sciences" in *A History of the National Research Council, 1919–1933* (Washington, D.C.: National Research Council, 1933), pp. 12–16 on p. 15. Alexi Assmus details the impact of postdoctoral fellowships in "The Creation of Postdoctoral Fellowships and the Siting of American Scientific Research," *Minerva: A Review of Science, Learning, and Policy* 31 (1993), 151–183.

positions that, at least in academe, would require them to divide their energies between research, teaching, and service.

Initially, however, that fellowship program supported only the fields of chemistry and physics. Men like Millikan, who had had the ear of the higher-ups in the Rockefeller philanthropies since before the war, had already successfully made the case for those two fields by linking them to the medical sciences that were so central to Rockefeller philanthropic concerns.[43] By thus assuring external support for their research, the physicists and chemists had firmly anchored their fields in a fast-evolving philanthropic environment. Veblen recognized that, by contrast, the mathematicians were adrift and needed to be brought swiftly to anchor as well. By 1923, the timing seemed right for them to make their move, since Veblen found himself well-positioned not only as a member of the NRC's postdoctoral fellowship board and the chair of its Division of Physical Sciences but also as the President of the AMS. A mathematical trifecta.

Veblen proceeded strategically. Simon Flexner, a pathologist by training, served as both fellowship board chair and Director of the Rockefeller Institute of Medical Research. A close adviser to the Rockefellers and a highly placed member of their philanthropic team, Flexner had been instrumental in sealing the deal for chemistry and physics and, hence, in successfully arguing for the expansion of their support for medicine into those two scientific fields via the NRC postdoctoral fellowship program. It was Flexner, then, who would have to be persuaded that mathematicians equally merited such funding.

Veblen laid out his argument—one in spirit not unlike that Coolidge had employed in his approach to Samuel Insull and effectively one by descent—in a letter to Flexner in October 1923. While medicine may depend critically upon chemistry and physics, Veblen argued, chemistry and physics depend just as critically upon mathematics.[44] Turned on its head, strength in mathematics is required for strength in chemistry and physics, and that, in turn, is necessary for strength in the medical sciences. If this argument prevailed, as Veblen hoped it would, mathematics would establish itself on the equal footing with chemistry and physics that he felt it deserved relative to philanthropic support.

43. Robert Kohler, *Partners in Science: Foundations and Natural Scientists 1900–1945* (Chicago: University of Chicago Press, 1991), p. 138.

44. Compare Feffer, pp. 477–479.

Veblen did not craft this argument purely for opportunistic reasons. He firmly believed that mathematics and the sciences, especially physics, had important interconnections. This was clearly reflected in his building, with his colleague Eisenhart, of the research group at Princeton in the 1920s that sought to establish the foundations of general relativistic dynamics and the new quantum theory on the mathematical areas of topology and the geometry of paths. Veblen also tried to convince the mathematical community as a whole of the importance of such applications on the occasion of his retiring address as AMS President in 1924.[45] There, he sounded yet again a call for the active encouragement and development of more applied mathematics within the largely purist American mathematical community. He stated the case baldly. "The foundations of geometry *must* be studied *both* as a branch of physics *and* as a branch of mathematics."[46] Like those who had made similar arguments before him, Veblen, too, largely failed to convince American mathematicians of the desirability of such a research turn.

If his argument proved unsuccessful relative to the American mathematical community as a whole, it scored a major coup for mathematics in the context of the Rockefeller Foundation. By December 1923, the Foundation had announced that it would extend its postgraduate fellowship funding from physics and chemistry to mathematics.[47] Jubilant, Veblen wrote to his right-hand man, Richardson, to share the good news. "You have done," Richardson wrote in reply, "a momentous service to the Society and to mathematics in general."[48]

The two men immediately set to work deciding how best to get not only the word out but also first-rate candidates to apply. The Foundation had specified that the new program in mathematics would run for an initial five-year period beginning only on 1 July, 1925, but Veblen told Richardson that he

45. Oswald Veblen, "Remarks on the Foundations of Geometry," *BAMS* 31 (1925), 121–141 and compare the discussions in Feffer, "Oswald Veblen and the Capitalization of American Mathematics," pp. 74–75 and Parshall, " 'A New Era,' " p. 278.

46. Veblen, "Remarks on the Foundations of Geometry," p. 121 (my emphases).

47. Interestingly, the biologists, who had been arguing, although not with one voice, for Rockefeller support for three years, were also added to the fellowship program in 1923 as were the astro- and geophysicists. Kohler, pp. 106–107.

48. Richardson to Veblen, 19 December, 1923, Box 10, Folder: Richardson, R.G.D. 1923, Veblen Papers. The Rockefeller philanthropies—through both the NRC and the International Education Board—as well as the John Simon Guggenheim Foundation provided postgraduate fellowships in mathematics specifically for study abroad. See the next chapter.

had "every reason to believe that mathematical fellowships will be granted this year if suitable candidates turn up."[49] Again, he was right.

Veblen and Richardson took full advantage of the December 1923 meeting of the AMS in New York to "give the thing as much publicity and emphasis as possible,"[50] and Veblen called on the AMS membership to help identify "talented and well-equipped young men and women" as well as to encourage them to present their applications for funding as soon as possible.[51] He also convinced Flexner that the fellowship selection committee should be immediately enlarged by the inclusion of another mathematician. To that end, he secured the services of his predecessor as AMS President, Bliss; his successor, Birkhoff, would soon fill out the number to three.[52] In all, four NRC fellows were selected in mathematics for the class of 1924–1925, with eight more being chosen by 1 October, 1925.[53]

It is perhaps not surprising that in this earliest round connections were important. Two of the four—Lawrence Graves and James (Henry) Taylor— had earned Ph.D.s under Bliss at Chicago in 1924; one, Harry Levy, had taken his in the same year under Eisenhart at Princeton; only the fourth, Joseph Thomas, a 1923 Ph.D. at Penn under Frederick Beal, had no direct institutional advocate on the mathematics selection committee.

As it matured, the program fortunately became less incestuous and more inclusive of the mathematical talent being produced throughout the United States.[54] By the end of its second year, Richardson could already state unequivocally to Birkhoff that he could "see very considerable growth during the past twelve months. The National Research Fellowships have stimulated

49. Veblen to Richardson, 15 December, 1923, Box 10, Folder: Richardson, R.G.D. 1923, Veblen Papers. Compare Parshall, " 'A New Era,' " pp. 278–279.

50. Veblen to Richardson, 15 December, 1923, Box 10, Folder: Richardson, R.G.D. 1923, Veblen Papers.

51. Roland Richardson, "The Thirtieth Annual Meeting of the Society," *BAMS* 30 (1924), 199–216 on p. 203.

52. *A History of the National Research Council*, p. 14 and *National Research Fellowships 1919–1938* (Washington, D.C.: National Research Council, 1938), pp. 5–7. Veblen served on the fellowship selection board from 1923 to 1937, Bliss from 1924 to 1937, and Birkhoff from 1925 to 1937.

53. Compare "Notes," *BAMS* 30 (1924), 376–378 and Wilbur Tisdale to Oliver Kellogg, 1 October, 1925, "Memorandum" on "Research Fellowships in Physics, Chemistry and Mathematics," Box 14, Folder: Tisdale, W. E. 1923–24, Veblen Papers.

54. From 1919 to 1938, 112 mathematicians held NRC fellowships. *National Research Fellowships 1919–1938*, pp. 22–28.

production."[55] Indeed, significant numbers of the mathematical generation that earned the Ph.D. in the late 1920s and into the 1940s—and who then went on to make fundamental contributions to the American mathematical research endeavor—got an initial "leg up" in research as a result of the fellowship opportunities afforded by philanthropies like the Rockefeller Foundation through its support of the NRC (see the next chapter).

The Foundation's promotion of research, especially in the 1920s and 1930s, paralleled changes taking place within higher education at roughly the same time. As American colleges, but especially universities, came more fully to embrace the research ethos, American mathematicians in particular worked to pursue their research agendas. By the end of the 1920s, tangible signs of the reorientation had begun to appear.

The year 1926 opened with good news out of both Harvard and Princeton.[56] In Massachusetts, Birkhoff had been named the Walter Channing Cabot Fellow, a position he would hold for four years and that would provide for a deputy specifically to relieve him "of part of his teaching so that he [might] devote more time to research." This fellowship had been created in 1906 by Cabot's widow and in his memory specifically "to encourage literary achievement and original research work."[57] In New Jersey, Veblen had been named the first incumbent of the Henry Burchard Fine Professorship of Mathematics, thanks to a gift of $200,000 from 1876 Princeton alumnus and Director of International Harvester, Thomas D. Jones, and his niece, Gwethalyn Jones, both of Chicago. The perks for Veblen were even better. He would be freed from all formal teaching while holding the chair, although he would be expected "to be accessible to graduate students" and "to give a number of lectures each year" on his current research.[58]

The import of such private benefaction was not lost on the American mathematical community. When he heard the news of Birkhoff's fellowship, Richardson described it as "a dream come true" and as "a great thing for American mathematics" that "comes not a bit too soon."[59] The grand old

55. Richardson to Birkhoff, 3 July, 1926, HUG 4231.2, Box 6, Folder: P–R, Birkhoff Papers.

56. "Notes," BAMS 32 (1926), p. 183. The quotation in the next sentence is from this announcement.

57. See "Cabot Fellowship to Prof. Royce," The Harvard Crimson (15 April, 1911).

58. Oswald Veblen to Earle Hedrick, 27 March, 1926, Box 6, Folder: Hedrick, E. R. 1924–32, Veblen Papers. The "real wealth" of $200,000 in 1926 dollars is equivalent to $3M in 2021 dollars (see https://measuringworth.com).

59. Richardson to Birkhoff, 14 January, 1926, HUG 4231.2, Box 6, Folder: P–R, Birkhoff Papers.

man of American mathematics and the dissertation adviser of both Veblen and Birkhoff, E. H. Moore, took an even wider view of the situation. "Altogether we seem," he wrote, "to be approaching, or rather in the beginning, of an interesting period of scientific development in this country having its focus in the idea of research and research professorships, in a sense deeper than ever heretofore."[60] Not to be outdone by Harvard and Princeton, both of which were nipping at Chicago's heels in mathematics, the E. H. Moore Distinguished Service professorship was established there in 1929 and awarded to the university's deepest and most productive mathematician of the second generation, Leonard Dickson, in order to afford him more time to pursue his algebraic and number-theoretic agendas. Such dedicated support—like the Carnegie-funded National Research Fund for the support of the nation's leading scientific researchers—represented explicit acknowledgment by the end of the 1920s that, at least in certain exemplary cases, the emphasis should be on *research* and teaching as opposed to *teaching* and research.[61]

Support for mathematics also came in the very concrete form of physical spaces in which research could be optimally pursued. In the mid-1920s, and so roughly contemporaneously with their support through the NRC of postdoctoral research, the Rockefeller philanthropies, this time via the International Education Board, provided money to construct two European mathematics institutes for the fostering of international contacts as well as of mathematics in relation to other sciences, especially physics. Thanks to infusions of Rockefeller funds, the Institut Henri Poincaré opened in Paris in 1928, and the Mathematical Institute in Göttingen followed one year later in 1929.[62]

Whether these developments abroad sparked similar initiatives at home or merely occurred with them in parallel is unclear. What *is* clear is that Chicago and Princeton both constructed new homes for their mathematicians at more or less the same time. Chicago's Eckhart Hall was completed in 1930, Fine Hall at Princeton a year later in 1931. Both of these new structures had been made possible by gifts that were in hand before the stock

60. Moore to Birkhoff, 15 April, 1926, HUG 4231.2, Box 6, Folder: K–O, Birkhoff Papers.

61. Yale's Sterling Professorships also date from the 1920s. Established thanks to the allocation of $5M from a $15M bequest from Yale alumnus and lawyer John Sterling, the first was awarded in 1920. A year later, the first mathematical scientist, Ernest Brown, a mathematical astronomer and the thirteenth President of the AMS, was named a Sterling Professor. The next, Oystein Ore, would only follow in 1931. The "real wealth" of $5M in 1920 dollars is equivalent to $66.6M in 2021 dollars (see https://measuringworth.com).

62. For more on these developments, see Siegmund-Schultze, *Rockefeller and the Internationalization of Mathematics*, pp. 143–156 on Göttingen and pp. 156–177 on Paris.

market crashed in October 1929. In Chicago's case, the money had come largely from Bernard A. Eckhart, owner-founder of a successful milling company as well as a one-time Illinois state politician and Chicago municipal activist. In that of Princeton, the same Thomas Jones, who had given the money for the Fine chair, underwrote the building project. Bliss was primarily responsible for the planning at Chicago and shared architectural and other plans with Veblen, who guided the process at Princeton. Both men sought to provide suitably commodious homes—right down to the furnishings and complete with well-equipped libraries and congenial common rooms for afternoon teas—wherein mathematical research would flourish. These were dedicated, state-of-the-art buildings for the pursuit of mathematical teaching and research.[63]

At least in the minds of some philanthropy and university administrators, then, mathematics needed to play a greater and ever more visible role within the academy both nationally and internationally. To do that, the ablest mathematicians needed relief from the heavy burdens of teaching and more dedicated time for research, and their mathematics faculties needed conducive physical spaces in which to conduct their work. At home, other university presidents took note of these developments and sought to follow the examples of Chicago, Harvard, and Princeton into the 1930s, even as the Great Depression severely limited funding (see chapter four). Higher education was changing, specifically around the role of research in establishing institutional pecking order. These changes would have significant repercussions not only in mathematics but also in other academic fields.

Sustaining Support for Publication

If the completion of the AMS's capital campaign, the inclusion of mathematics in the fellowship programs of the NRC, and manifestations of greater

---

63. Bliss to Veblen, 14 May, 1929, Box 3, Folder: G. A. Bliss 1923–40, Veblen Papers as well as Series II: Administration, Course Materials and Writing, 1892–1975, Box 15, Folder 7, Department of Mathematics Records, University of Chicago. For more on, especially Fine Hall, see "A Memorial to a Scholar-Teacher," *Princeton Alumni Weekly* 32 (30 October, 1931), pp. 111–113 (which includes the text of remarks given by Veblen on the occasion of the building's dedication); William Aspray, "The Emergence of Princeton as a World Center for Mathematical Research, 1896–1939," in *History and Philosophy of Modern Mathematics*, ed. William Aspray and Philip Kitcher (Minneapolis: University of Minnesota Press, 1988), pp. 346–366; and Feffer, "Oswald Veblen and the Capitalization of American Mathematics," pp. 492–493.

university support marked successes for the American mathematical endeavor in the 1920s, another was the development of strategies to secure the financial resources actually to sustain mathematical publications. As early as 1915, the MAA had taken over, as its official journal, the financially ailing *American Mathematical Monthly* and, in a shrewd business move, had included a subscription to the journal as a part of its membership dues thereby providing the financial support that the journal had previously lacked. The MAA had thus assured the existence of at least one publication venue aimed at advancing "the interests of mathematics in the collegiate and advanced secondary fields."[64]

A year later, an MAA committee—chaired by E. H. Moore and including Veblen as well as Raymond Archibald of Brown and Alexander Ziwet of the University of Michigan—devised a plan to collaborate with the *Annals of Mathematics*, then underwritten by Princeton University and on the editorial board of which Veblen served. At this moment, the *Annals*, founded by astronomer Ormond Stone at the University of Virginia in 1884, was one of four, home-grown, research-oriented journals open to American mathematicians.[65] Another, the *American Journal of Mathematics*, had been founded by Sylvester at Hopkins in 1878 and had been underwritten financially from the start by the university in explicit acknowledgment of its commitment to research and the training of future researchers.[66] The other two were

64. William Cairns, "The Mathematical Association of America," *AMM* 23 (1916), 1–6 on p. 2. I give a concise account of the first twenty-five years of the *American Mathematical Monthly* in "*The American Mathematical Monthly* (1894–1919): A New Journal in the Service of Mathematics and Its Educators," in *Research in History and Philosophy of Mathematics*, ed. Maria Zach and Elaine Landry (Basel: Birkhäuser Verlag, 2016), pp. 193–204.

65. David Eugene Smith and Jekuthiel Ginsburg detail the early history of the *Annals* in *A History of Mathematics in America before 1900*, Carus Mathematical Monographs, no. 5 (Chicago: The Open Court Publishing Company, 1934), pp. 116. Compare Parshall and Rowe, *Emergence*, pp. 411–412. There was a fifth journal, the *Journal of Mathematics and Physics* launched by the Massachusetts Institute of Technology in 1922. It, however, was "founded to give an outlet for papers in pure and applied mathematics by members of the Institute" and was thus not open to the broader mathematical community. See "Massachusetts Institute of Technology, Department of Mathematics, Report to the Visiting Committee," March 13, 1935, Box 12: Correspondence, 1935–1939, Folder: Massachusetts Institute of Technology, Richardson Papers.

66. On Sylvester and the *American Journal*, see, for example, Karen Hunger Parshall, "America's First School of Mathematical Research: James Joseph Sylvester at the Johns Hopkins University 1876–1883," *Archive for History of Exact Sciences* 38 (1988), 153–196 and Karen

supported by the AMS: the *Bulletin*, begun in 1891 for the publication of shorter articles and research announcements as well as book reviews and news of the profession, and the *Transactions*, first published in 1900 for the presentation of full-scale research results.[67] In 1916, and in a creative move, the MAA proposed providing the *Annals* with an annual subvention of $300 to allow it to increase its size from 200 to 300 pages annually, the additional 100 pages being "devoted to expository articles."[68]

As Veblen knew, the *Annals* already had a policy that did not preclude the publication of expository articles. The MAA's financial gesture could thus be seen as encouragement for qualified authors to undertake such writing and thereby to reach—and, in so doing, nurture—a broader mathematical reading public than would be possible with a narrowly focused piece of specialized research. That broader reading public might then be inspired to provide even more generous financial support for the activities of the American mathematical community. Moreover, the arrangement had the potential further to define the *Monthly*'s publication sphere, for if the agreement were not renewed at the end of the initial three-year period, the Board of Editors of the *Annals* was directed thereafter to "conduct the *Annals* as a journal devoted primarily to research, yielding the field of historical and expository articles (not necessarily absolutely but principally) to the publications of the Association."[69] The MAA was being true to its spirit of cooperation with the AMS, to its broader definition of research as the product of "a great deal of labor of a purely investigational sort," and to its inherent sense of more effective community outreach.[70]

Those commitments were further reinforced both through the MAA's renewal of the subvention to the Princeton journal and through the creation of a new expository publication venture—the Carus Mathematical Monograph Series—underwritten by Mary Hegeler Carus at the level of $1,500 a year for

---

Hunger Parshall, *James Joseph Sylvester: Jewish Mathematician in a Victorian World* (Baltimore: Johns Hopkins University Press, 2006), pp. 239–248.

67. Archibald provides historical accounts up to 1938 of the *Bulletin* and the *Transactions* in *Semicentennial History*, pp. 48–55 and 56–65, respectively.

68. Earle Hedrick and William Cairns, "First Summer Meeting of the Association," *AMM* 23 (1916), 273–288 on p. 288.

69. Ibid., p. 288.

70. Earle Hedrick gives this more expansive sense of research in "A Tentative Platform for the Association," *AMM* 23 (1916), 31–33 on p. 33. Compare my "The Stratification of the American Mathematical Community," pp. 164–166.

five years beginning in January 1922.[71] Hegeler Carus, the daughter of German immigrants and later the wife of Paul Carus, had been the first woman to earn a bachelor of science degree (in 1882) from the University of Michigan. Her father, who had made his fortune in zinc smelting, also founded the Open Court Publishing Company in 1887, and hired another German immigrant, Paul Carus, to serve initially as a tutor to his children but then as the press's first managing editor. Hegeler Carus's donation in her husband's memory acknowledged his—and her—interest in mathematics through the publication of book-length manuscripts that aimed "to popularize mathematics by making accessible at nominal cost the best thought and keenest researches in this field set forth in expository form comprehensible to teachers and students of mathematics and to other readers of mathematical intelligence." By 1925, the first in the series, Bliss's *The Calculus of Variations*, had been published for the MAA in Chicago by the Open Court Publishing Company. By September, almost 900 copies had been sold to MAA members with 400 more purchased thanks to the press's promotional strategies.[72] Other volumes, *Analytic Functions of a Complex Variable* by Northwestern's David Curtiss and *Mathematical Statistics* by Iowa's Henry Rietz, soon followed in 1926 and 1927, respectively.[73] It seemed clear that the Carus Mathematical Monograph series was filling a need of mathematically inclined America.

71. "Notes," *BAMS* 28 (1922), 72–82 on p. 74 (the quotation given below may also be found here); Herbert Slaught, "The Carus Mathematical Monographs," *AMM* 30 (1923), 151–155; and David Eugene Smith, "Mary Hegeler Carus," *AMM* 44 (1937), 280–283. The "real wealth" of $1,500 in 1922 dollars is $22,500 in 2018 dollars (https://www.measuringworth.com). William Cairns, MAA Secretary-Treasurer, Minutes of the Meeting of the Incoming Board of Trustees, 1 January, 1925, Box 4RM114, 2004–170, Folder: Headquarters: Board of Governors Minutes, 1920–1928, MAA Records elaborates the eight-point publication agreement between Carus and the MAA. It included the stipulation that the MAA would "receive the returns from sales to its members and keep all such receipts," together with the annual donation from Carus, in a separate account from which would be paid "all clerical and other expense appropriated to authors as assistance in preparing manuscripts" as well as "whatever honoraria may be agreed upon to be paid to authors in lieu of all royalties."

72. William Cairns, MAA Secretary-Treasurer, Minutes of the Meeting of the Board of Trustees, 9–10 September, 1925, Box 4RM114, 2004–170, Folder: Headquarters: Board of Governors Minutes, 1920–1928, MAA Records.

73. See the bibliography for the full references. In 1927, Hegeler Carus increased the amount of her gift from $1,500 to $2,500. Carl Boyer, "The First Twenty-five Years," in *The Mathematical Association of America: Its First Fifty Years*, ed. Kenneth May (Washington, D.C.: The Mathematical Association of America, 1972), pp. 24–54 on p. 28. This series continues to appear, although now under the AMS/MAA Press imprint.

In addition to the appearance of the first Carus monograph, the year 1925 also witnessed the establishment, thanks to the generosity of then MAA President Julian Coolidge of the MAA's Chauvenet Prize for mathematical exposition. Named in honor of William Chauvenet, a nineteenth-century mathematical worthy recognized by his peers as one of the original fifty members of the National Academy of Sciences, the $100 Chauvenet Prize represented a further testament to the commitment that the Association had made through its financial support of the *Annals* and was first awarded to Bliss for his *Annals* article on "Algebraic Functions and Their Divisors."[74] Subsequent awards were made roughly every three years into the 1950s. All of these various projects and initiatives attest to the fact that the MAA was actively and successfully engaged in efforts both to provide financial support for mathematics at the collegiate level and to foster a broader appreciation of the mathematical endeavor.

At the research level, the AMS was also at work. An increase in 1920 in both AMS dues—from $5.00 to $6.00—and in the subscription rate to the *Bulletin*—from $5.00 to $7.00—had forestalled, at least for a year, the publication cost woes of that journal.[75] The subscription-raising initiative pursued by Bliss in 1921 and 1922 had served as a stop-gap in stabilizing, again at least temporarily, the finances of the Society's other journal, the *Transactions*. The AMS's capital campaign had also effectively solved the short-term problems caused by rising printing costs in the mid-1920s. Sustaining the publication of research journals at a level commensurate with the growing output of the American mathematical community would remain a challenge throughout the 1920s and into the 1930s (see chapter four, particularly fig. 4.3).

In the five years from 1921 through 1925 alone, both the *Bulletin* and the *Transactions* had been enlarged by 20% in an ultimately inadequate effort to accommodate the demand for printed pages.[76] Since the Ph.D. was conferred

74. Gilbert Bliss, "Algebraic Functions and Their Divisors," *AM* 26 (1924), 95–124. Cairns recounts the founding and defines the prize in Minutes of the Meeting of the Board of Trustees, 9–10 September, 1925. The role of prizes in the American mathematical society is discussed further in the section "Championing American Achievements" later in this chapter.

75. Archibald, *Semicentennial History*, pp. 28–29.

76. Richardson, "Report of the Secretary to the Council for the Years 1921–1925," p. 206. The size of the *Bulletin* was increased yet again—to 750 pages per volume—in 1926. "Notes," *BAMS* 32 (1926), 565–570 on p. 565. Archibald documents the dramatic rise in the number of pages per volume for the *Bulletin* and the *Transactions* in *Semicentennial History*, pp. 55 and 64, respectively. The *Transactions*, in particular, grew from 462 pages in 1920 to 950 pages in 1930.

almost solely on the basis of the production of a publishable piece of research, and since the number of Ph.D.s produced steadily rose through the 1920s and into the 1930s, journal capacity represented a key component of the community's growth and development.[77] Solutions such as dues and subscription rate adjustments as well as member donations to the cause were all dependent on the relatively static personal financial resources of the membership and, so, were simply untenable in the long run. As with the capital campaign, it appeared that the ultimate solution would hinge on external sources.

Indeed, the AMS had already had some success in that regard as early as 1921, when the *Transactions* had found itself in a compromised position. It had an embarrassment of riches in the sense of accepted papers for its twenty-second volume but a dearth of riches in the sense of actual money to pay for their publication. At that critical juncture, the AMS leadership—in addition to launching Bliss's subscription drive—approached the NRC for assistance, and it stepped in with an emergency $600 contribution that allowed the Society to bring volume twenty-two out on time, even if there was still a surfeit of accepted papers. Another donation—this time from an anonymous source with Harvard mathematician Edward Huntington as the intermediary—allowed for the publication of an extra volume, numbered twenty-three, as well as a volume twenty-four, both in 1922. The intercalated volume twenty-three was then distributed gratis to all those who already had subscriptions for what had become volume twenty-four, and the problem of the large backlog was solved, at least temporarily.[78]

The success of this approach to the philanthropically funded NRC was perhaps not unexpected. In 1920, the NRC's Division of Physical Sciences had appointed a committee to investigate the feasibility of establishing a Revolving Book Fund for the publication of technical scientific books in the full range of the sciences.[79] By providing subventions to help offset the high cost

77. Of course, potential Ph.D.s also took courses and seminars, but then, unlike now, additional hurdles such as qualifying examinations were not in place. As noted in the previous chapter, Roland Richardson gives data on the number of mathematics Ph.D.s produced in the United States in "The Ph.D. Degree and Mathematical Research," Table 1 on p. 203 (or see p. 366 in the reprinted edition).

78. Archibald, *Semicentennial History*, p. 61.

79. Prior to the NRC's creation, the Carnegie Institution of Washington, founded in 1902, had already recognized this need and had underwritten the publication of books in a variety of scientific fields. Della Fenster details its involvement in mathematical publication per se in "Funds for Mathematics: Carnegie Institution of Washington Support for Mathematics from 1902–1921," *Historia Mathematica* 30 (2003), 195–216.

of printing, the Fund helped assure that important scientific work, even if commercially unviable, appeared in print. Although a journal is not a book, a single issue of a journal could be construed as one. The NRC apparently agreed.

The NRC, in fact, had earmarked $1,500 (coincidentally the same amount that Mary Hegeler Carus had given in that same year for the MAA's monograph series) explicitly for mathematics with Veblen as chair of that particular Revolving Fund committee from 1921 to 1922.[80] Not only well aware of the inner workings of the NRC but also in a position actively to advocate there for mathematics, Veblen was the right person at the right time. In 1922, for example, he was involved in obtaining subventions for monographs in geometry by Eisenhart and Henry White (recall the previous chapter), both of which carried the acknowledgment "published with the cooperation of the National Research Council."[81] A year later, in 1923, as chair of the Division of Physical Sciences and AMS President, Veblen made it a top priority "to try to help find funds for the Revolving Fund for the publication of Mathematical Books."[82] The American mathematical community had entered, in the 1920s, into a highly productive period. It was critical to get its discoveries out in either journal or book form.

These successes with the NRC may have prompted Veblen, in the fall of 1924 as his term as AMS President was drawing to a close, to open discussions about possible support with Wickliffe Rose, head of the Rockefeller-funded General Education Board (GEB). What he learned both heartened him and spurred him to action. Rose allowed that Veblen, on behalf of America's mathematicians, was not the first to approach the GEB regarding funding for technical publications. The biologists and others had already made similar requests, which the GEB was then considering. The question was, however, if the GEB were to get into the business of supporting scientific publication, how best should that support be administered? Directly from the GEB to each of the different scientific societies? Or in some other, more indirect way?

80. Birkhoff took over as chair of the Revolving Fund committee in 1922, serving in that role until 1936. Archibald, *Semicentennial History*, p. 15.

81. The books in question are Eisenhart's *Transformations of Surfaces* and White's *Plane Curves of the Third Order*.

82. Veblen to Birkhoff, 26 September, 1923, Box 29, Folder: N.R.C. Miscellaneous Correspondence, etc., 1921–24, Veblen Papers.

By October 1924, the GEB seemed close to making a decision. It was considering the allocation of a lump sum of $10,000 to the National Academy of Sciences to be used specifically for scientific publication. In that way, it could support the sciences without, in the first instance, being directly involved in what could quickly have become the administrative nightmare of dealing individually with the full panoply of scientific societies and without, in the second instance, appearing overtly to meddle in decision-making best left to the experts themselves. Veblen, the savvy politician, promptly met with the relevant Academy committee to argue the case for the inclusion of mathematics in such a publication fund, should it materialize.[83]

His argument was patently boosteristic. The AMS, Veblen explained, had been forced since 1923 to publish its journals in *Germany*, owing to the prohibitive cost of publishing them at home.[84] Should the Society not be assured of the financial wherewithal to bring those journals back *home* where they belonged? Once again, Veblen prevailed. The National Academy specifically earmarked almost $5,000 of the GEB money—$1,100 for the *Bulletin*, $2,000 for the *Transactions*, and $1,600 for the *Annals*—for the publication of research-level mathematics.[85] This allowed the printing of both the *Bulletin* and the *Transactions* to return to American shores in 1925 and supplemented the modest coffers of the *Annals*, the latter now a more exclusively research-oriented journal following the end of the MAA's experiment in fostering a greater volume of expository writing there.

Conspicuously absent from this list of supported journals, however, was the fourth and oldest of America's research-level mathematics journals, the *American Journal*. In 1925, when the Academy gave so generously to the other three periodicals, the *American Journal* received no allocation, ostensibly "because of a possible conflict of interest between one of the NAS members on the allocation committee and the J[ohns] H[opkins] U[niversity] Press"

83. See Veblen to Richardson, 3 October, 1924, Box 10, Folder: Richardson, R.G.D. 1924, Veblen Papers.

84. For more on the issue of American dependence on Germany relative to mathematical publishing in the 1920s, see Reinhard Siegmund-Schultze, "The Emancipation of Mathematical Research Publishing in the United States from German Dominance (1878–1945)," *Historia Mathematica* 24 (1997), 135–166, especially pp. 140–148.

85. See Richardson to Veblen, 29 September, 1925, Box 10, Folder: Richardson, R.G.D. 1925, Veblen Papers. The first quotation in the next paragraph is also from this letter. The "real wealth" of $5,000 in 1924 dollars is equivalent to $73,400 in 2018 dollars (https://www.measuringworth.com).

which published it. There had, in fact, been a growing perception, at least in certain circles, that the *American Journal* had been slipping in quality and that its editorial standards had become lax. Unfortunately, this diminution had coincided with the surge in American mathematical output. It seemed unjust that weak papers on outmoded mathematical topics should be published in the *American Journal*, while strong papers on cutting-edge topics had to be turned away or delayed by the *Transactions* and the *Annals*. The AMS Council was concerned. Would it have to found yet another journal—thereby further compromising its financial situation—in order to provide adequate space for quality mathematical output? Or could some means of cooperation be devised between Hopkins and the AMS to restore the *American Journal* to its former luster?[86] These were very delicate questions as the AMS leadership, principally Birkhoff as President and Richardson as Secretary, recognized. Things would have to be handled carefully.

In September 1926, Richardson was charged by the AMS Council to open lines of communication with the powers-that-be at Hopkins, and particularly with Frank Morley, former AMS President and former editor of the *American Journal*. Richardson arranged for AMS founder Thomas Fiske to journey to Baltimore from New York to talk with Morley as well as with Hopkins President Frank Goodnow. In some sense, the timing of this meeting and of these negotiations was auspicious. Hopkins was celebrating its semicentennial in 1926, and Fiske hit upon the face-saving idea of an announcement on that occasion to the effect that "the American Mathematical Society seeks the honor and the privilege of being associated with the University in the maintenance of the *American Journal of Mathematics*."[87] After some two months of intense negotiations, the AMS prevailed. In getting key representation by what the AMS leadership viewed as the *right* kind of mathematicians on the journal's editorial board, the AMS succeeded in "bracing up" the journal

---

86. Richardson to Frank Morley, 12 October, 1926, Box 11, Folder: Richardson, R.G.D. 1926, Veblen Papers as well as the letters exchanged between Veblen and Earle Hedrick in 1926 in Box 5, Folder: Hedrick, E. R. 1924–32, Veblen Papers. Indeed, the *American Journal* had been targeted for possible takeover by the AMS as early as the late 1890s owing to the perception of lax editorial standards. In that instance, the founding of a new journal, the *Transactions of the American Mathematical Society* resulted. Steve Batterson tells this story based on an interesting cache of archives found in the Simon Newcomb Papers at the Library of Congress in *American Mathematics 1890–1913: Catching Up to Europe*, pp. 127–136.

87. Fiske to Richardson, 13 October, 1926, HUG 4231.2, Box 6, Folder: P–R, Birkhoff Papers.

and in "assist[ing] in building up mathematics as a whole."[88] From 1927 on, the *American Journal* had a financial line item on the AMS's books and thus partook in the publication subventions that the AMS received through the National Academy from the GEB. The GEB ultimately extended its various financial commitments to American mathematics at least through 1931.

Although the AMS's financial problems stemmed primarily from its publication of research journals, they also owed—to a somewhat lesser but nevertheless non-negligible degree—to the Society's efforts to publish research-level mathematics books. As early as 1894, the AMS had been instrumental in the publication by Macmillan & Co. of New York of *The Evanston Colloquium Lectures on Mathematics* that Felix Klein had given just a year earlier in connection with the World's Columbian Exposition in Chicago.[89] In 1896, and with this precedent in mind, the AMS had decided to initiate roughly biannual, week-long series of so-called "Colloquium Lectures" that would follow its annual summer meeting and that would consist of six, two-hour talks by an "expert lecturer" on topics that would serve to provide for the continuing education of America's mathematicians.[90] The contents of the earliest of these lectures were duly sketched in brief notices and published in the *Bulletin*, but the AMS made the bold move of securing in book form the publication of the 1903, fourth set of Colloquium Lectures, by convincing the London home branch of Macmillan & Co. to assume the financial responsibility for the venture.[91] After a subvention from Yale University allowed the Society to print in 1910 the 1906, fifth Colloquium Lectures, the

88. The two quotations are in Richardson to Veblen, 29 November, 1926, and Richardson, "Memorandum on Journal Negotiations," (undated but late 1926), respectively, in Box 11, Folder: Richardson, R.G.D. 1926, Veblen Papers.

89. Felix Klein, *The Evanston Colloquium Lectures on Mathematics* (New York: Macmillan & Co., 1894). As discussed in Parshall and Rowe, *Emergence*, pp. 331–354, these lectures represented a key moment in the evolution of a research-level community of mathematicians in the United States.

90. Thomas Fiske, "The Buffalo Colloquium," *BAMS* 3 (1896), 49–59 on p. 49. Archibald gives a brief history of the Colloquium Lectures in *Semicentennial History*, pp. 66–73. I characterize the AMS Colloquium Lectures to 1940 in " 'Increasing the Utility of the Society': The Colloquium Lectures of the American Mathematical Society," *Philosophia Scientiae* 19 (2015), 153–169.

91. Edward Van Vleck, Henry White, and Frederick Woods, *The Boston Colloquium: Lectures on Mathematics*, vol. 1 (London: Macmillan & Co., 1905). Hosted by MIT, these lectures were given by three of Klein's former students: White of Northwestern, Woods of MIT, and Van Vleck of Wesleyan University.

AMS published the next set of lectures using its own resources and, in 1914, finally resolved "henceforth regularly [to] publish the Colloquia."[92] The AMS thus officially went into the business of publishing technical mathematical monographs.

World War I, as well as the many financial issues that consumed the AMS in the first half of the 1920s, disrupted not just the AMS's publication plans but the very staging of the Colloquia themselves. None was given in the four consecutive years from 1921 through 1924. By 1925, however, the Society and its Colloquium Lecture series were both back on track, and even larger plans were in the works.

Veblen, in his capacity as chair of the NRC's Division of Physical Sciences, had chanced to mention the Colloquium Lectures during the course of a meeting in New York with the members of the National Academy's committee on publication funds. That committee would be charged with overseeing the funding that the GEB was then on the verge of making available for the publication of technical scientific materials. Writing to Birkhoff in June 1925, Veblen related that when he "spoke about the Colloquium and the plan for publishing one book a year, this was very favorably received . . . . There was no suggestion that such an enterprize [sic] would be outside the domain of the fund, and I am strongly convinced that it would be worth while [sic] to lay the project before the Committee."[93] Once again, Veblen had read the situation astutely. GEB money, as administered by the National Academy, was also allocated in support of the publication of the AMS's Colloquium Lecture series.[94]

This financial vote of confidence further emboldened the AMS. By the end of 1925, a new Colloquium Editorial Committee had been set up with the sixteenth, seventeenth, and sitting eighteenth AMS Presidents—Bliss, Veblen, and Birkhoff, respectively—as its members. With Veblen as chair, the committee very quickly issued a statement of its enlarged vision for the series. In addition to a continuation of the practice of publishing volumes based on the lectures of the various Colloquium speakers, the series would grow to include "a number of monographs and expositions of new mathematical

92. Frank Cole, "The October Meeting of the Society," BAMS 20 (1914), 169–176 on p. 170.

93. Veblen to Birkhoff, 3 June, 1925, HUG 4213.2, Box 5, Folder: S–V, Birkhoff Papers.

94. From 1925 to 1936, the GEB contributed in excess of $43,000 toward the publication of research-level mathematics in the United States. Archibald, Semicentennial History, p. 34. For more on this support, see chapter four.

developments which may be submitted by their authors on their own initiative without special invitation from the Council."[95] To help assure the
financial solvency of the new venture, the committee also stipulated "that all
proceeds from the sales of old Colloquia shall go to the Colloquium Fund and
be used to defray the expenses of the publication of new Colloquia."

Both of these new ideas were quickly realized. Between 1927 and the
United States' entry into World War II in 1941, an average of two AMS Colloquium volumes appeared each year, with ten of these being monographs
submitted independently of texts written by invited Colloquium lecturers.
Moreover, Veblen's 1916 Colloquium Lectures on *Analysis Situs* had been
published in a print run of 600 in 1922 and had sold out by 1929, thereby earning $1,000 for the AMS. Some of that money was used to defray the costs of
bringing his volume out in a second edition in 1931; the rest went to support
the series as a whole. In general, the Colloquium volumes enjoyed steady—
and, in some cases, even brisk—sales.[96] It was a successful experiment in the
monographic publication of research mathematics made possible in part by
philanthropic dollars from the Rockefeller-funded GEB as filtered through
the National Academy of Sciences.

In the 1920s and into the 1930s, the GEB both bridged the AMS's repeated
gaps in funding for the publication of mathematical research and allowed the
AMS's publications ultimately to become self-sustaining, even as the production of the American mathematical research community continued to
increase.[97] Whether thanks to philanthropic support or to the munificence
of private donors, America's mathematicians had succeeded in the 1920s in
marshalling funds for the publication of their work.

Championing American Achievements

Enabling mathematicians to do and to publish their research was certainly
among the functions of the professional mathematical society as it had

95. George Birkhoff, Gilbert Bliss, and Oswald Veblen, "The Colloquium Publications,"
BAMS 32 (1926), 100. The next quotation is also on this page.

96. See Birkhoff to Veblen, 26 June, 1929, Box 2, Folder: Birkhoff, George D. 1912–47,
Veblen Papers; American Mathematical Society: Study of Finances of Colloquium Publications 1896–1944, Box 6, Folder 17, Whyburn Papers; and Archibald, *Semicentennial History*,
p. 72.

97. Archibald, *Semicentennial History*, pp. 34–36 and compare Feffer, "Oswald Veblen and
the Capitalization of American Mathematics," pp. 482–484.

evolved particularly over the last half of the nineteenth century.[98] Another that had emerged—as not only the sciences and mathematics but also academic disciplines in general continued to professionalize into the twentieth century—was the explicit recognition of specialized work in specific disciplines through the awarding of prizes and the bestowal of other honors and recognition.

Although James McKeen Cattell, editor of *American Men of Science*, had instituted the practice of "starring" those 1,000 scientists, among them mathematicians, deemed by their peers to be the best in their field, perhaps the highest honor that an American mathematician could receive in the 1920s was election to the National Academy of Sciences. Founded by Congress in 1863, the NAS was a highly elite group that, as noted, had served the chemists and physicists particularly well at the time of World War I in their efforts both to direct science during the war and to position themselves and their fields in a postwar world. In the early 1920s, as part of his mathematical activism, Veblen was thus particularly keen to get the nation's most talented and energetic mathematicians—"a number of the live young men" as he put it— into the Academy and to enlarge the NAS's Mathematics Section.[99] In 1924, for example, the Academy consisted of fourteen members in anthropology and psychology and fifteen each in botany, mathematics, and engineering. Astronomy, however, had nineteen members, geology and paleontology twenty-six, chemistry twenty-eight, and physics twenty-nine. In the view of Veblen and others, there was no reason for mathematics to be less well-represented than chemistry and physics, especially since "the scientific qualifications" of "a number of mathematicians not members" "compare favorably with those members of the Mathematical and other Sections of the Academy."[100] It was critically important at this time of forward momentum in mathematics that the field and its best practitioners be recognized at the highest national level.

98. See the comparative discussion in Karen Hunger Parshall, "Mathematics in National Contexts (1875–1900): An International Overview," in *Proceedings of the International Congress of Mathematicians: Zürich*, 2 vols. (Basel: Birkhäuser Verlag, 1995), 2: 1581–1591.

99. Veblen to Leonard Dickson, 17 October, 1924, Box 4, Folder: Dickson, L. E. 1910–33, Veblen Papers.

100. The numbers may be found in Gilbert Bliss, Oswald Veblen, Hans Blichfeldt, Luther Eisenhart, and George Miller to the Members of the Mathematical Section of the NAS, 30 April, 1924, Box 26, Folder: NAS Mathematical Section Officers 1922–25, Veblen Papers. The quotations are also from this letter.

Like academy membership, prizes, too, had long been part of the fabric of science. For example, another academy, the Paris Académie des Sciences, had, since its founding in the seventeenth century, held regular, and often highly publicized, prize competitions for the solution of explicitly set problems, and other scientific societies had followed its example. In 1731, the Royal Society of London had inaugurated its Copley Medal for exceptional achievement in any branch of science, and by 1901, the Nobel Prize had been established in Sweden to honor outstanding contributions internationally in chemistry, physics, and physiology or medicine as well as in literature and in the promotion of peace. Given the notoriety of two-time Nobelist, the French physicist and chemist, Marie Curie, the Nobel Prize, in particular, fairly quickly took on a certain cachet.

In the sciences, Americans Albert Michelson (in physics in 1907), Theodore Richards (in chemistry in 1914), and Robert Millikan and Arthur Compton (in physics in 1925 and 1927, respectively) brought home the United States' first Nobels. These men thus added further luster to chemistry and physics, and, once again, the mathematicians were at a disadvantage. There was no Nobel Prize in mathematics,[101] and, on American shores at least, there were no prizes in mathematics at all. In order to be competitive, the American mathematical community needed at least some means of acknowledging extraordinary attainment and thereby of beginning to establish parity with the other sciences within scientific and broader American culture.

Although not a prize, the AMS's Colloquium Lectures had been established as a way both to provide postdoctoral, continuing education in research-level mathematics and to acknowledge research expertise. By 1925, and following the Colloquium's four-year hiatus, the lectures had taken on a more explicitly honorific as well as an overtly political role. This fact was well appreciated by the members of the Colloquium committee charged with selecting the speakers and the site for the next colloquium. Birkhoff, Veblen, and Edward Van Vleck were joined in the deliberations by Lefschetz, professor of mathematics then at Kansas but en route to Princeton, Arnold Dresden then still at Wisconsin, and Pennsylvania's John Kline. Lefschetz, Dresden, and Van Vleck represented the "West," while the others hailed from the East. Their discussions were frank and wide-ranging as they debated not only the merits—and demerits—of possible individual speakers but also what

101. Michael Barany tells this story in "The Myth and the Medal," *NAMS* 62 (January 2015), 15–20.

mathematical areas might best be featured. Lefschetz, Dresden, and Kline seemed agreed—at least in March 1925—that the topics should be mathematical physics and geometry/topology; they were less in accord as to the speakers.

Two things were clear. First, Lefschetz could not *"stress too much the Colloquium as [a] means of recognizing eminence,"* since "[t]he Society has had practically no other way of" doing that.[102] And, second, the selection of the sites of the AMS's Summer Meeting and of its Colloquium Lectures could have a real impact politically. Again, to Lefschetz's way of thinking, it should be "Madison sans Colloquium in 1926, thus giving the Middle West its chance. Ohio State with Colloquium in 1927. It is East and West, and we would thus give recognition to a young and growing department which really is in need of it." In fact, the next Colloquium Lectures were held in Madison, Wisconsin, in 1927 with E. T. Bell of Caltech lecturing on "Algebraic Arithmetic" and Bryn Mawr's Anna Pell Wheeler speaking on "The Theory of Quadratic Forms in Infinitely Many Variables and Applications."[103] Talent and eminence had been recognized, and political and geographical concerns had been addressed, even if the specifics of the latter were somewhat different from those on Lefschetz's mind in 1925.

The AMS's first *official* prize was authorized in 1918 on the occasion of the death of Maxime Bôcher, Harvard mathematician and one of the leaders of the emergent, turn-of-the-twentieth-century, American mathematical research community. Thanks to over $1,000 in donations raised from members of the Society, the Bôcher Memorial Prize of $100 was first awarded in 1923 to George Birkhoff for "Dynamical Systems of Two Degrees of Freedom," judged by a committee of his peers to have been the best paper published in volumes 18 to 22 of the AMS's *Transactions*.[104] Initially, limited both to papers published in the *Transactions* and to mathematicians under the

---

102. Lefschetz to Dresden and Kline, 10 March, 1925, HUG 4213.2, Box 5, Folder: H–J, Birkhoff Papers (his emphasis). The quotation that follows is also from this letter. Compare also Parshall, " 'Increasing the Utility of the Society,' " p. 164.

103. As noted in the previous chapter, Bell's lectures were ultimately published as *Algebraic Arithmetic*, the seventh volume in the AMS Colloquium Publications. Although an abstract of Pell Wheeler's lectures appeared in the *Bulletin* (see Theophil Hildebrandt, "The Second Madison Colloquium," *BAMS* 33 (1927), 663–665 on pp. 664–665), they never appeared in book form.

104. See Archibald, *Semicentennial History*, p. 37 and Birkhoff, "Dynamical Systems of Two Degrees of Freedom."

age of forty, the Bôcher Prize was shared a year later by Bell for his number-theoretic paper on "Arithmetical Paraphrases" and Lefschetz for research "On Certain Numerical Invariants of Algebraic Varieties with Application to Abelian Varieties."[105]

It had been a difficult decision. Henry White of Vassar and Chicago's Leonard Dickson made up the selection committee, and they were initially split. As White explained in a letter to Birkhoff, the latter in his role as then AMS President, they had sought Francesco Severi's opinion of Lefschetz's paper during the course of the 1924 International Congress of Mathematicians in Toronto and had been assured "emphatically that it was sound and reliable," while Dickson, who had heard Bell lecture in Chicago, had strongly advocated for the latter.[106] In naming two winners, the committee thus set the precedent that the "best" paper in the *Transactions* was not necessarily unique.

By 1933—Princeton's James Alexander had won the third prize in 1928—it had been decided that the award should be made every five years, that it should be limited to those under the age of fifty, and that it should specifically reward "notable research in analysis," Bôcher's area of expertise. Two years later, the stipulation on the place of publication was also altered to include any "recognized journal published in the United States or Canada."[107] In just over a decade, the Bôcher Memorial Prize had evolved into a means for acknowledging achievement in analysis—and not just by the youngest mathematicians—as reflected in the publication community defined by North America.[108]

Achievement explicitly in algebra had been singled out for recognition by the AMS even earlier. On the occasion of his retirement as AMS Secretary and as Editor of its *Bulletin* in December 1920, Columbia's Frank Cole was presented with a sum of money collected from the Society's membership in

105. E. T. Bell, "Arithmetical Paraphrases" and Solomon Lefschetz, "On Certain Numerical Invariants of Algebraic Varieties with Application to Abelian Varieties," both in *TAMS* 22 (1921), 1–30 and 327–406, respectively.

106. White to Birkhoff, 29 November, 1924, HUG 4213.2, Box 4: Correspondence 1924 T–Z, Birkhoff Papers.

107. Archibald, *Semicentennial History*, pp. 37–38.

108. Sloan Despeaux uses the notion of a "mathematical publication community" as an analytic tool in "The Development of a Publication Community: Nineteenth-Century Mathematics in British Scientific Journals" (unpublished doctoral dissertation, University of Virginia, 2002) and "Fit to Print?: Referee Reports on Mathematics for the Nineteenth-century Journals of the Royal Society," *Notes and Records of the Royal Society* 65 (2011), 233–252 on pp. 246–247.

thanks for his years of service. Cole promptly transferred the gift to the AMS to use as it saw fit, and in 1922, the Frank Nelson Cole Prize in Algebra, to be awarded every five years, was created "for the best memoir offered in competition upon some question in the theory of Galois groups, or the theory of numbers, or some other part of algebra."[109] This way of conducting a mathematical prize competition, that is, by specifying in advance specific questions to be answered, differed from the model adopted for the Bôcher Prize and was more reminiscent of the seventeenth- and eighteenth-century prototype of, say, the Paris Académie des Sciences.

The first Cole Prize competition was announced in 1927 for 1928, and among the topics "suggested as suitable" were: the "determination of all division algebras of rank 5"; a "direct proof, without the theory of the rank equation, that an algebra whose units form a group is semi-simple"; the "classification of nilpotent algebras"; the "investigat[ion of] the multiplicative relations between units of S and units of N," given that "every algebra A is the sum of a semi-simple subalgebra S and the maximal nilpotent invariant subalgebra N"; the "develop[ment of] a rational theory of algebras A over any field by starting with the rank function of the general element of A expressed in terms of the rank functions of the simple components of S" (given the previous set-up); and, finally, the extension of "the theory of arithmetics of algebras" by "invent[ing] a theory of ideals for division algebras."[110] These open questions had been suggested by Gaetano Scorza's then-recent 1921 treatise on *Corpi numerici ed algebre* as well as by Leonard Dickson's 1923 *Algebras and Their Arithmetics*. It perhaps came as little surprise to his contemporaries, then, that, in 1928, the first Frank Nelson Cole Prize in Algebra was awarded to Dickson for *Algebren und ihren Zahlentheorie*, his 1927 elaboration on and extension of the results he had presented in English in his earlier tract on the subject.

In 1929, another Cole Prize—this one perhaps somewhat surprisingly in number theory given that field's relative underdevelopment in the American mathematical community—was also created. At this time, too, the terms of the awards of both Cole prizes were amended to parallel those for the Bôcher Prize, namely, winners could be no older than fifty and the papers for consideration should have been published in recognized North American journals.

---

109. Roland Richardson, "The Frank Nelson Cole Prize in Algebra," *BAMS* 29 (1923), 14 and Archibald, *Semicentennial History*, pp. 38–39.

110. Henry White, "The Frank Nelson Cole Prize in Algebra," *BAMS* 21 (1925), 289.

These would also no longer be competitions based on set themes; rather, they would simply recognize exemplary work in algebra and number theory.[111] The first Cole Prize in Number Theory went to Texas's Harry Vandiver for his work toward a proof of Fermat's theorem but, in particular, for his paper entitled "On Fermat's Last Theorem."[112] Citing the absence of papers of sufficiently high quality in both algebra in 1934 and in number theory in 1936, however, the AMS failed to make scheduled awards in those years.[113] This, too, set an interesting precedent, one that reflected the true distinction that the prizes were meant to confer.

In 1925, Lefschetz had recognized and bemoaned the fact that the American mathematical community "had practically no . . . way of recognizing eminence." By the end of the decade, it had at least six: honors—membership in the National Academy of Sciences and selection as a Colloquium Lectures speaker—and prizes—the MAA's Chauvenet Prize, the Bôcher Prize, and the Cole Prizes in both algebra and number theory. While these prizes, in particular, were neither as high-profile nor as well publicized as the Nobel Prizes that Americans in physics and chemistry had begun to bring home, they nevertheless served as key public relations vehicles for legitimizing and advertising a thriving American mathematical endeavor.

———

The 1920s seemed different, at least to America's mathematical leaders. Their community's development seemed to have accelerated significantly; they sensed that significant changes were under way. For one thing, their community was growing. In the fifteen-year period from the end of 1915, when the MAA was formed, to 1930, the MAA had doubled the size of its membership to 2,200, while the AMS had grown even more sharply from 732 to 1,940 members, thanks both to the ongoing membership drive spearheaded particularly by Clara Smith and to the sharp increase in the number of American Ph.D.s in mathematics.[114] There was, moreover, the perception that with

111. Roland Richardson, "The Society's Prizes," *BAMS* 36 (1930), 3–4.

112. Harry Vandiver, "On Fermat's Last Theorem," *TAMS* 31 (1929), 613–642.

113. Archibald, *Semicentennial History*, p. 39.

114. For the membership numbers, compare *American Mathematical Society: List of Officers and Members, January 1916* (New York: American Mathematical Society, 1916), p. 30 and *American Mathematical Society: List of Officers and Members, 1929–1930* (New York: American Mathematical Society, 1930), p. 46, and see "One Hundred Percent Membership," *AMM*

greater numbers came greater strength. For example, writing to Veblen in May 1924, Richardson offered the opinion that "America has in the past quarter-century made great strides in both pure and applied science and it is hoped that by this move . . . new forces will be let loose which will contribute toward putting America in the front rank which should be hers."[115] Hedrick, from his vantage point as editor of the *Bulletin*, saw things similarly. "I imagine," he wrote to Birkhoff a year later in July, "that we are entering upon a somewhat new *era*."[116] What had given Richardson, Hedrick, and others this feeling of being at a historical turning point?

Over the course of the 1920s and into the 1930s, they had participated in what were ultimately successful initiatives in fund-raising for mathematics. They had joined with corporate sponsors in helping to assure the financial solvency of their community. They had witnessed private philanthropies—like the Carnegie Corporation and various Rockefeller philanthropies—come to support their actual research as well as their publication of it. They had watched as their universities began to acknowledge their institutional importance through the creation of research professorships and the construction of state-of-the-art buildings that made manifest their presence. They had been gratified to realize that private patrons, like Mary Hegeler Carus, would come forward to encourage their work. And, they had taken pride in acknowledging, through NAS membership and prizes to the creation of which they had contributed, the accomplishments of their peers. Their community was maturing. It was establishing its place in academic—as well as in American—culture.

Yet, the members of that community also appreciated that these advances would likely not have been made without the active intervention and tireless devotion to the cause of a handful among them. Again, Richardson, the keen observer and mathematical insider, understood the dynamic. "Everyone who knows intimately the history of the Society for the last two years," he stated

---

(1930), 563. The numbers for the AMS given here differ slightly from those found in Archibald, *Semicentennial History*, p. 44. According to Roland Richardson, the United States produced 129 Ph.D.s in mathematics between 1920 and 1924, but 227 between 1925 and 1929. See Richardson, "The Ph.D. Degree and Mathematical Research," p. 203 (or p. 366 in the reprint edition).

115. Richardson to Veblen, 7 May, 1924, Box 10, Folder: Richardson, R.G.D. 1924, Veblen Papers.

116. Hedrick to Birkhoff, undated (but sometime shortly after 13 July, 1925), HUG 4213.2, Box 5, Folder: Correspondence, 1924/1925, H–J, Birkhoff Papers (Hedrick's emphasis).

categorically to Veblen in September 1925, "realizes that you have been the statesman who has guided us through troubled waters to a moderately calm sea."[117] Indeed, the dramatic developments within the American mathematical community of particularly the 1920s owed, in no small measure, to the evolution and efforts of a new breed of which Veblen and his immediate successor as AMS President, Birkhoff, were emblematic, the mathematician as savvy politician.

117. Richardson to Veblen, 29 September, 1925, Veblen Papers.

3

Breaking onto the
International Scene

Mathematical ambassadorship was clearly in evidence in 1926 when AMS President George Birkhoff was asked to engage in a fact-finding mission in Europe for the Rockefeller-funded International Education Board (IEB). The IEB had been founded three years earlier in an effort to promote "science on an international scale" through individual traveling fellowships to Americans and others and, by the end of the 1920s, through grants for the construction and maintenance of strategically located scientific institutes such as Paris's Institut Henri Poincaré.[1] The forty-two-year-old Birkhoff's objective, on this his first trip abroad, was to assess the lay of the European mathematical landscape and to compare it, in so far as possible, with that of the United States. The IEB wanted to make the most informed judgments possible when allocating its money to particular people for particular projects in particular areas of research.

The import of this trip was not lost at home. Birkhoff was perceived as representing abroad the entire American mathematical research endeavor. As Earle Hedrick told him, "[w]e in America are proud of the impression that you are making: it will go far toward convincing Europe that the study of mathematics in America is to be taken very seriously by them."[2] Hedrick was quick to admit, however, that "[i]t is not strange, perhaps, that their views of

1. The quotation is from the title of a memorandum by IEB President Wickliffe Rose dated April 1923. See Siegmund-Schultze, *Rockefeller and the Internationalization of Mathematics*, p. 27.

2. Hedrick to Birkhoff, 17 July, 1926, HUG 4213.2, Box 6, Folder: Correspondence 1926, H–J, Birkhoff Papers. The next quotation is also from this letter.

America are not true, for they cannot sense the development here in the past thirty years."

Hedrick, unlike Birkhoff, had experienced key aspects of mathematical Europe firsthand. After earning his 1901 doctorate in Göttingen, he had spent time in Paris acquainting himself with the mathematicians there. The first President of the MAA and the then-current editor of the AMS's *Bulletin*, he had worked to build the American mathematical community of the 1920s, so he naturally wanted to see its advances more widely appreciated abroad.

The American mathematical community had looked to Europe at least since the 1890s when some of its number, like Hedrick, had gone abroad to pursue their advanced studies.[3] By the 1920s, however, in light of what it saw as its significant progress, it was anxious to assert itself on the international— that is, at least at this moment in time, *European*—mathematical scene. How, though, could the Americans change the European perception from one of apprentice/master to one of mathematical equals? How could Europe, especially Germany but to a lesser extent France, Italy, England, and elsewhere, come fully to "sense," as Hedrick put it, "the development" of mathematics in the United States? If such changes could be effected at all, they would likely involve American and European mathematicians in active dialogue, working shoulder to shoulder in Europe and in the United States, and publishing side by side in journals on both sides of the Atlantic.

Engaging in the International Politics of Mathematics

In 1920, an opportunity at least to attempt to showcase American mathematics arose at the International Congress of Mathematicians (ICM) held in Strasbourg at the end of September. International congresses had been hosted in Zürich in 1897, in Paris in 1900, in Heidelberg in 1904, in Rome in 1908, and in Cambridge, England, in 1912, but World War I had forced the cancellation of the 1916 Congress that was to have been held in Stockholm.[4] Even

3. Interestingly, and symptomatic of this shift, the *Bulletin* essentially ceased after 1920 to provide detailed information on the courses being offered at European universities. In that year, moreover, information was only given for the University of Strasbourg and the Italian universities, not for the French, English, or especially German universities which had long attracted American mathematical travelers. See "Notes," *BAMS* 26 (1920), 425–430 on p. 429 and "Notes," *BAMS* 27 (1920), 39–46 on pp. 39–41, respectively.

4. What has been called the "zero-th" International Congress of Mathematicians was held in Chicago in the fall of 1893 in conjunction with the World's Columbian Exposition. For

before the war's end in 1918, however, plans for the resumption of international scientific relations had been made, but, in retrospect, those plans were tainted from the beginning by hatred engendered by the war.

Meeting in London in October 1918, representatives of the scientific academies of the Allied nations declared that they would "not be able to resume personal relations in scientific matters with their enemies until the Central Powers can be readmitted into the concert of civilized nations."[5] For those representatives, that "readmission" would require a renunciation by the scientists from those countries of the political policies that had led to the war and its many atrocities. Given these strongly negative and unconciliatory sentiments, it is perhaps not surprising that, when an International Research Council (IRC) was formed in Brussels in 1919, the former Central Powers were excluded from membership despite the explicit charge of "coordinat[ing] international efforts in the different branches of science and its applications." As one of its first acts, the IRC drafted a set of statutes that formed an International Mathematical Union (IMU) ostensibly to "promote international cooperation in mathematics"; it would be comprised of sections with members selected by the national academies of each of the adhering countries. The new IMU was thus an organ of the IRC and bound by its strictures.

Reporting on the IMU's formation, the writer for the AMS's *Bulletin* wryly noted in January 1920 that "information of [this] action taken six months ago now reaches American mathematicians for the first time by indirect ways."[6] Moreover, and not without more than a bit of sarcasm, the opinion was offered that "[b]efore the war international mathematical activities were carried on very efficiently by *mathematicians* and it may be hoped that *they* will soon resume charge of *their* affairs." Other American *scientists*, like astrophysicist George Hale, had been present at the birth of the IRC, but no American *mathematicians* had been party to the discussions regarding the formation of the IMU, and none of those actually present

---

that designation, see Donald Albers, Gerald Alexanderson, and Constance Reid, *International Mathematical Congresses: An Illustrated History* (New York: Springer-Verlag, 1987). For more on the gathering itself, see Parshall and Rowe, *Emergence*, pp. 295–330 and Parshall and Rowe, "Embedded in the Culture."

5. Olli Lehto, *Mathematics without Borders: A History of the International Mathematical Union* (New York: Springer-Verlag, 1998), p. 18. For the quotations that follow, see pp. 19 and 25, respectively.

6. "Notes," *BAMS* 26 (1920), 184–188 on p. 184. The quotation that follows is also on this page (with my emphases).

there had thought to notify their American colleagues in a timely way.[7] This lack of communication between scientists in positions of influence and their mathematician colleagues would not prove an isolated event (see chapter eight).

The Americans also soon learned that the IMU—without consulting either them or the mathematicians of Great Britain, that is, the mathematicians of two of its eleven initially adhering countries—had decided to hold an International Congress of Mathematicians in 1920 in politically charged Strasbourg, a city that had been under German control since the Franco-Prussian War and that had then-recently been annexed by France. The mathematicians of Germany, Austria, and Hungary as well as Bulgaria and Turkey would not be invited to participate there.[8] From that moment on, engaging in mathematics on the international stage meant dealing with the politics thrust upon the field by the IMU.

The Strasbourg ICM was ultimately a very small affair. Although only 200 mathematicians participated, eighty of whom were from France, twenty-seven countries were represented. The handful of Americans in attendance were led by the two representatives selected by the American section of the IMU, Chicago algebraist, immediate AMS Past President, and one of the Congress's Vice Presidents, Leonard Dickson, and Princeton differential geometer, Luther Eisenhart.[9] In addition to their official duties, they, like their compatriots, had come to represent American mathematics. In particular, Dickson had been invited to be one of the Congress's five plenary speakers. He thus found himself in the distinguished company of Cambridge University's Joseph Larmor, Niels Nörlund then of the University of Sweden, Louvain's Charles de la Vallée Poussin, and Rome's Vito Volterra. It was a privileged place for an American in 1920, but was it the result of postwar international politics, or actual recognition of American mathematical achievements, or possibly both?

7. Lehto, pp. 16–24 and 310.

8. Archibald, *Semicentennial History*, p. 19.

9. "Notes," *BAMS* 26 (1920), 464–469 on p. 464. The official record of the Strasbourg ICM lists the names of ten Americans in attendance but gives eleven as the total number. See *Comptes rendus du Congrès international des mathématiciens (Strasbourg, 22–30 septembre 1920)*, ed. Henri Villat (Toulouse: Imprimerie et Librairie Éduoard Privat, 1921), and compare pp. ix–xiv and xv. David Eugene Smith of Columbia College and the President of the MAA as well as Norbert Wiener and James (Sturdevant) Taylor both then at MIT were also listed as official members of the American delegation to the Congress (p. viii).

Regardless of the motives behind the invitation, Dickson took the opportunity to highlight results that had come out of research he had done for the 1920 publication of the second volume of his *History of the Theory of Numbers*. In discussing "Some Relations Between the Theory of Numbers and Other Branches of Mathematics," he not only showed how to use geometrical methods to determine all rational solutions of Diophantine equations of the forms $x_1^2 + x_2^2 + x_3^2 = x_4^2$ and $x_1^2 + \cdots + x_5^2 = x_6^2$, but he also employed the theory of integral algebraic numbers and integral quaternions to find the integral solutions for them.[10] It was a solid paper that augured more interesting results to come, yet the shadows of war hung over the purely mathematical proceedings of which Dickson's address was a part.[11]

In his opening welcome, Congress President Émile Picard had reminded the participants that one of their objectives was "to establish personal relations" that had been "ruptured" as a result of "the dreadful torment" which had been World War I.[12] "Rapprochements are necessary between scientists," he acknowledged, but there was a caveat. Such rapprochements would be between scientists "who hold each other in esteem." Picard closed the Congress making it absolutely clear that mathematicians from the former Central Powers were *not* among this group. In his view—and reprising the same exclusionary language that had been employed at the organization of the IRC in 1918—they would have to offer "a sincere repentance" before being permitted to "reenter" into "the concert of civilized nations."[13] It was on this politically charged international stage that American mathematicians made a postwar leadership debut that proved less than successful.

Dickson and Eisenhart, in their official capacity as IMU representatives, extended an invitation to the IMU to hold the next, the 1924, Congress in

10. Leonard Dickson, "Some Relations Between the Theory of Numbers and Other Branches of Mathematics," in *Comptes rendus du Congrès international des mathématiciens (Strasbourg)*, pp. 41–56.

11. As Della Fenster argues, this lecture may well have served as at least a partial inspiration for the fundamental research that Dickson would soon present in his 1923 book, *Algebras and Their Arithmetics*, and in his Cole Prize–winning follow-up text of 1927, *Algebren und ihre Zahlentheorie*. See her "Leonard Eugene Dickson and His Work in the Arithmetics of Algebras" and "American Initiatives toward Internationalization."

12. Émile Picard, "Allocution" (Séance d'Ouverture du Congrès), in *Comptes rendus du Congrès international des mathématiciens (Strasbourg)*, pp. xxvi-xxix on p. xxviii (my translation). The quotations that follow in the next two sentences are also on this page.

13. Émile Picard, "Allocution" (Séance de Cloture du Congrès), in *Comptes rendus du Congrès international des mathématiciens (Strasbourg)*, pp. xxxi-xxxiii on p. xxxiii (my translation).

FIGURE 3.1. Leonard Eugene Dickson (1874–1954). (Photo courtesy of R. L. Moore Legacy
Collection, 1890–1900, 1920–2013. e_math_00172, Archives of American Mathematics,
Dolph Briscoe Center for American History, University of Texas at Austin.)

the United States. For them, that meant a truly international congress, that
is, one, unlike the one they were attending in Strasbourg, in which mathematicians from all countries could and would participate. Rather than raising
this point explicitly at the meeting with their fellow IMU representatives,
and thereby risking the unpleasantness of a politicized confrontation, they
hoped that in four years' time French sentiments would have changed sufficiently to allow for such an all-inclusive congress.[14] That was their first

14. G. H. Hardy to Gösta Mittag-Leffler, 30 September, 1921, as quoted in Elaine McKinnon Riehm and Frances Hoffman, *Turbulent Times in Mathematics: The Life of J. C. Fields and
the History of the Fields Medal* (Providence: American Mathematical Society, 2011), p. 130.

political mistake. Their second was not vetting the invitation with the AMS before extending it.[15]

By 1922, it was clear to the Americans that the French were not going to budge and that it would be impossible to raise the necessary funding if the congress were not open to all. They were obviously in a political bind, but what to do? Some within the American mathematical community favored going ahead regardless of the exclusions imposed by the IMU. Others were adamant that a congress in the United States, closed especially to the community of German mathematicians that had played such a key role in the development of American mathematics, was inconceivable. Many hands were wrung and much ink was spilled before the offer of the Hopkins-trained, Canadian mathematician John Fields to host the congress in Toronto—under the IMU's constraints—got the Americans out of their immediate predicament.[16] In the longer term, however, they would still have to decide if or when they would confront the IMU's exclusionary policies.

Fields worked tirelessly to make the Toronto Congress a success,[17] but the new reality of international mathematical politics was in evidence in the gathering's very name. Whereas each preceding congress had, regardless of its primary language, been termed an "International Congress of Mathematicians," the 1924 affair was called the "International Mathematical Congress." It was an international congress on *mathematics* in which not all *mathematicians* were free to participate. Fields, who was sympathetic to both the French and German positions and who was also committed to fostering international mathematical relations, was walking a very fine line.

Some 444 mathematicians from thirty-three different countries participated in Toronto in August 1924 to take part in the event.[18] Perhaps not surprisingly, the countries with the most attendees were the United States—with 191 or just over 40%—and Canada—with 107 or almost 25%. Great Britain and France were next with fifty-eight (13%) and twenty-four participants (5%), respectively.

15. Dickson did at least try to vet the proposal with the National Research Council, the body responsible for the American section of the IMU. Oswald Veblen to Wilbur Tisdale, 3 January, 1924, Box 14, Folder: Tisdale, W. E. 1923–24, Veblen Papers.

16. For a detailed account of the controversy, see Riehm and Hoffman, pp. 130–137.

17. Riehm and Hoffman paint a vivid picture of Fields's organizational activities on pp. 137–161.

18. For these and the data that follow, see *Proceedings of the International Mathematical Congress Held in Toronto, August 11–16, 1924*, ed. John Fields, 2 vols. (Toronto: University of Toronto Press, 1928), 1: 30–44 and 48.

In contrast to Picard's politically laden words of welcome in Strasbourg four years earlier, Fields struck a more upbeat and apolitical tone. He emphasized the strides—of which the Europeans in attendance might be unaware—that North America had made in research-level mathematics since the founding of the Johns Hopkins University in 1876. Those strides were precisely what had made it not unnatural, at least, for an ICM to convene for the first time on the other side of the Atlantic. Still, Fields could not help but allude to especially the German universities when he acknowledged that "[i]t would not be easy to differentiate between the influence of the Johns Hopkins University and that of immediate European contact on the development of other great American universities."[19] It was a subtle allusion, but one that would not have been lost on his audience, many of whom, like himself, had made mathematical pilgrimages to Germany.

In his response to Fields on behalf of the delegates to and members of the congress, however, de la Vallée Poussin refused, in his role as President of the IMU, to set international politics completely aside. As if his audience needed reminding, he opened by recalling that the previous congress had been held in Strasbourg, a city the choice of which had "had an evident moral significance." Strasbourg, he continued, had been "a symbol and a celebration . . . of the deliverance of the science [that is, mathematics] that sacrilegious hands had for too long subjugated to criminal purposes." Despite Fields's best efforts, past hostilities were still evident in Toronto.

The Toronto Congress did succeed in fostering a different sort of coming together: that of the pure and applied aspects of the discipline. Fields and his co-organizers—and especially his young University of Toronto colleague, the Irish mathematician and physicist John Synge—had striven to bring "together the theoretical man and the applied scientist" in Toronto in an effort to drive home the point "that practice cannot be divorced from theory and that the pure scientist does his share in contributing to the welfare of the community."[20] To this end, the Toronto Congress incorporated the usual sections that covered algebra, the theory of numbers, and analysis; geometry; mechanics and mathematical physics; and history, philosophy, and didactics with distinct sections on engineering as well as statistics, actuarial science, and economics. Moreover, the section on mechanics and mathematical physics

19. Ibid., 1: 54–55 on p. 55. The two quotations in the next paragraph are in 1: 56–58 on p. 56 (my translation). Riehm and Hoffman also highlight these passages in their analysis of the Toronto Congress.

20. Ibid., 1: 54–55 on p. 55.

was divided into two subsections, one on "Mechanics, Physics" and the other on "Astronomy, Geophysics," while that on "Engineering" was subdivided into separate subsections on "Electrical, Mechanical, Civil and Mining Engineering" and "Aeronautics, Naval Architecture, Ballistics, Radiotelegraphy."[21] It was the first time in the history of the ICMs that such careful attention had been paid to such a full range of applied topics. Although this might well have exposed the Americans in attendance to the kind of applied mathematics that some among the AMS leadership had tried to encourage, most of the actual American contributions to the congress were pure in nature.

For example, once again, Dickson was both a Congress Vice President and a plenary speaker. This time, he highlighted the innovative work he had presented in his recently published book, *Algebras and Their Arithmetics*, by giving an outline of that theory to date.[22] In many regards, his 1924 lecture was a natural outgrowth of the results he had presented in Strasbourg for it "sketch[ed] in a broad way the leading features of the origin and development of a new branch of number theory," namely, the arithmetics of algebras, "which furnishes a fundamental generalization of the theory of algebraic numbers." Interestingly, Dickson's new theory had bested results of both the Zürich-based, German mathematician, Adolf Hurwitz, and his Swiss student, Louis-Gustave du Pasquier, a point highlighted prominently in the Toronto talk. Dickson's contributions to the congresses in both Strasbourg and Toronto demonstrated that American mathematics was asserting itself in the international arena. It remained to be seen whether those assertions— as well as Fields's emphasis on more general mathematical developments in North America—would spark the kind of recognition that Earle Hedrick would still lament two years later in his July 1926 letter to a Birkhoff abroad as an IEB scout.

While mathematics—presenting new results of one's own and being exposed to the latest work of others—dominated the Toronto Congress for most of the Americans in attendance, a handful were equally, if not more, concerned with the broader political implications that had resulted in that congress being held in Canada and not in the United States. Those politics

---

21. Ibid., 1: 74.

22. Leonard Dickson, "Outline of the Theory to Date of the Arithmetics of Algebras," ibid., 1: 95–102. The quotation that follows appears on p. 95. In Toronto, Dickson also lectured on "Further Development of the Theory of Arithmetics of Algebras," in which he presented results he had obtained after publishing his 1923 book. Ibid., 1: 173–184.

had, in fact, kept some of the prominent members of the American mathematical community away from the event, notably Veblen, then AMS President, and many of his Princeton colleagues. Yet, some of those boycotting the affair in Toronto were behind a planned political move—on the part of the American section of the IMU—to have the restrictions on future congress attendance officially lifted. If the Americans had their way, there would be political fireworks in Toronto, but not everyone was enthusiastic about those prospects.

Just a week before the congress's start, Richardson, who, along with Bliss, had been designated the AMS's representatives there, wrote agonizingly to Veblen, admitting that "I should enjoy staying away from Toronto. But I feel it my duty to go, just as you feel it incumbent upon you to stay away."[23] As AMS Secretary, Richardson had been at the epicenter of the controversy about whether or not to make good on the American offer to serve as host in 1924, receiving and passing on letters from all possible sides in the debate and just generally trying to mediate the situation. His was a middle-of-the-road position, hence one reason for his presence in Toronto. Veblen took a more radical stance, hence his boycott. Still, Richardson did not feel that his diplomatic skills were up to the task ahead of him. "Can't you come to help swing the thing to the right side?" he pleaded to Veblen. As Richardson saw it, "our differences here in America are largely those of method. I think your attitude is the one which is the more likely to lead to strife: you think mine is." He nevertheless felt compelled to do his best "to keep the peace" and entreated Veblen almost fatalistically to "Pray for us!" Richardson's extreme unease stemmed from the fact that he, together with Arthur Coble then of Hopkins and Virgil Snyder of Cornell, were the official delegates in Toronto of the American section to the IMU.[24] He knew that he would be in the thick of the battle, yet, as he confessed to Veblen, "Coble & Snyder are stronger on your side than am I."

The Americans made their move on 15 August at the meeting of the IMU. They read a prepared statement that had been unanimously endorsed:

23. Richardson to Veblen, 4 August, 1924, Box 10, Folder: Richardson, R.G.D. 1924, Veblen Papers. The quotations that follow in this paragraph are also from this letter.

24. Princeton's Luther Eisenhart, Harvard's Edward Huntington, then the chair of the American section, and the University of Iowa's Henry Rietz, the section's secretary, had also been chosen as section delegates to the IMU meeting in Toronto. They, however, were not in attendance at the Congress. "Notes," *BAMS* 30 (1924), 376–378 on p. 377 and compare the list of attendees in *Proceedings of the International Mathematical Congress Held in Toronto*, 1: 30–44.

"Resolved, that the American Section of the International Mathematical Union requests the International Research Council to consider whether the time is ripe for the removal of the restrictions on membership now imposed by the rules of the Council."[25] They requested, moreover, that their resolution "be transmitted to the International Research Council by the Executive Committee of the International Mathematical Union." The representatives from Denmark, England, Italy, the Netherlands, Norway, and Sweden spontaneously asked that "the resolution be recorded as the *unanimous* views of their delegations also."[26]

It had been a shrewd political maneuver. By requesting that their resolution be "transmitted" rather than voted on by those assembled in Toronto, they had done their best not to insult the French. They had also avoided being ruled out of order by IMU President, the Belgian de la Vallée Poussin, and especially IMU Secretary General, the Frenchman Gabriel Koenigs, a particularly strong supporter of the exclusionary policy. As Richardson described the dynamics, "the French were beaten at every point where they opposed the Americans." At the same time, "the extremists of both ends in our own country were not allowed to have their way." It was a satisfactory outcome, but Richardson nevertheless felt "pretty well fed up." "I have a dark-brown taste in my mouth as I leave the Congress," he confessed to Birkhoff. "But I am convinced that the best thing the Americans could do was to come here and fight it out." They had gotten their position on the record, but the battle had been hard-fought, and the war was far from over.

Three months later in November, the American section was still at work. This time, it unanimously "[r]esolved, that . . . the National Research Council of the United States . . . strive for such a modification of the statutes of the International Research Council as will remove the present restrictions on membership in the Union."[27] This statement was more forceful, as it urged its own national contingent within the IRC to lobby actively and directly for the cause. In the section's view, "no successful international mathematical

25. Edward Huntington to the Division of Physical Sciences of the National Research Council, 3 November, 1924, HUG 4213.2, Box 4, Folder: Correspondence 1924 E–J, Birkhoff Papers. The next quotation is also from this letter.

26. Roland Richardson to George Birkhoff, undated (but at the time of the Toronto Congress), HUG 4213.2, Box 4, Folder: Correspondence 1924 R–S, Birkhoff Papers (Richardson's emphasis). The quotations that follow in the next paragraph are also from this letter.

27. Huntington to the Division of Physical Sciences of the National Research Council, 3 November, 1924. The next quotation is also from this letter.

congress can again be held under the present rules, which exclude the Central European Powers." American mathematicians had taken their first major stand as players on the international stage, and although their actions did not result in immediate changes to the IMU's exclusionary policy, they did reflect a symptom of the disease that led to the IMU's death at the Zürich Congress in 1932.[28]

Americans Abroad

The revival of the ICMs in the 1920s reflected a partial resumption of international mathematical contacts in the aftermath of World War I, but as their lobbying efforts in Toronto suggested, the Americans wanted more. They had a sense of destiny. They saw it playing out at home with the success of their research agendas and of their efforts to strengthen their community's infrastructure, but they also aspired to see it play out much more broadly. Richardson fully captured that spirit when he opined to Veblen that "America has in the past quarter-century made great strides . . . and it is hoped that by this move . . . new forces will be let loose which will contribute toward putting America in the front rank which should be hers."[29] The "front" of that "rank" was clearly measured by an international yardstick. Obviously, producing high-quality mathematical results would be key, but such results had already been and were being produced without garnering, in the Americans' view, the international recognition they merited, Dickson's ICM plenary talks in Strasburg and Toronto aside. The problem was that from the last quarter of the nineteenth century to World War I, American mathematicians had, by and large, been *students* abroad.[30] At that time, they had gone as far as they could in the United States—some earning doctorates there—but they recognized that their best education at home was inferior to what could be obtained abroad.

Things were different in the 1920s. They had come to sense that higher mathematical education in the United States was much more on a par with that in Europe. They thus had to alter the European perception of them as *students*. Their foreign colleagues needed to recognize them as *peers*. One

28. Lehto details the events surrounding the IMU's death in Zürich on pp. 50–60. See also chapter five.

29. Richardson to Veblen, 7 May, 1924, Box 10, Folder: Richardson, R.G.D. 1924, Veblen Papers.

30. See, for example, Parshall and Rowe, *Emergence*, chapter 5 on "The *Wanderlust* Generation."

way to begin to achieve that broader recognition was personally to represent their maturing community by directly sharing the fruits of their mathematical labors abroad. Dickson and others had done exactly that in Strasbourg and Toronto in 1920 and 1924, respectively. Americans capitalized on other opportunities over the course of the 1920s.

In May 1924, for example, the Société mathématique de France (SMF) celebrated—two years late—its fiftieth anniversary. Although the Moscow Mathematical Society had been founded in 1864 with the London Mathematical Society following a year later, these, at least nominally, were local organizations, while the SMF had more national ambitions.[31] By 1924, it had over 400 members from all over France as well as an international contingent that included some two dozen Americans.[32] Notably absent from the SMF roster, however, were the names of German and Austrian mathematicians. They had been deleted as of 14 January, 1920, and would only be reinstated following "a formal request" from the former member that would then be submitted for consideration to the SMF Council. Not surprisingly, given the considerable overlap in their leadership, the SMF and the IMU adopted policies unfavorable to the participation of what the SMF termed "mathematicians from enemy nations."

These policies did not prevent the AMS from sending an official representative to the SMF anniversary in Paris in May, just as they could not stop those American mathematicians who wished to from attending and engaging actively in the Toronto ICM three months later in August. The University of Wisconsin's Göttingen-trained Edward Van Vleck joined with the twenty-seven other foreign guests at a banquet at which toasts were raised to the achievements of the SMF and, in general, to the "progress of the mathematical sciences." The Americans were literally at the European mathematical table.

31. Hélène Gispert details the history of the SMF in *La France mathématique: La Société mathématique de France (1872–1914)* (Paris: Société française d'histoire des sciences et des techniques & Société mathématique de France, 1991); rev. ed., *La France mathématique de la Troisième République avant la Grande Guerre* (Paris: Société mathématique de France, 2016). Adrian Rice and Robin Wilson explore the history of the London Mathematical Society and its more national thrust despite its local name in "From National to International Society: The London Mathematical Society 1867–1900," *Historia Mathematica* 25 (1998), 185–217. Finally, I take a comparative look at the rise of mathematical societies in a variety of national settings in "Mathematics in National Contexts (1875–1900): An International Overview."

32. "Vie de la Société" (Supplément spécial), *Bulletin de la Société mathématique de France* 52 (1924), 1–67. The list of members as of July 1924 is on pp. 2–14. The next two quotations are on p. 2. The final quotation in this paragraph may be found on p. 23 (my translations).

By December 1924, they were present in a very different international venue. The third Pan-American Scientific Congress was held in Lima, Peru from 20 December, 1924, through 6 January, 1925, and brought together scientists of all kinds from throughout the Americas. What had started out in 1898 as a *Latin American* Scientific Congress had expanded by 1908 to include the United States in a *Pan-American* Congress in recognition of the fact that many of the issues under discussion would be of interest to their neighbor to the north.[33] Although most of the science represented at the third Congress involved such immediate concerns as public health, agriculture, mining, and engineering, the United States was represented at the small section on mathematics by Harvard's Edward Huntington, as an official representative of Boston's American Academy of Arts and Sciences, the American Mathematical Society, and the Mathematical Association of America. There were two sessions. At the second, a dozen papers were simply read by title, but there was also a short business meeting in which "a resolution was adopted calling attention to the desirability of courses in the mathematics of finance in the departments of mathematics and economics," a call that would still be heard in the early decades of the twenty-first century.

The first session was more substantive. There, some fifteen people were present to hear three papers, one by Florencio Jaime, the President of the newly formed Argentinian Mathematical Society, on "Conjugate Ordinates and Their Geometric Applications," one by Alejandro Guevara, honorary professor of engineering in Lima, "On the Descriptive Geometry of the Sphere," and one by Huntington on "Elementary Types of Order."[34] Although abstracts of the talks were not part of the official record, Huntington's presentation was likely related to recent postulate-theoretic work that he had presented before the U.S. National Academy of Sciences.[35] In recognition of his mathematical contributions, the Universidad Nacional Mayor de San Marcos in Lima conferred upon him the degree of Honorary Doctor in the Faculty of Sciences.[36] The Latin Americans, if perhaps less

33. *Report of the Delegates of the United States of America to the Third Pan-American Scientific Congress* (Washington: Government Printing Office, 1925), p. 2. The quotation that follows is on p. 34.

34. Edward Huntington, "The Pan-American Scientific Congress," *BAMS* 31 (1925), 290.

35. Compare Edward Huntington, "Sets of Completely Independent Postulates for Cyclic Order," *PNAS* 10 (1924), 74–78.

36. *Report of the Delegates ... to the Third Pan-American Congress*, p. 4. Seven American scientists were so honored, but Huntington was the only mathematician.

so the Europeans, acknowledged American mathematical achievements.[37] Attempts to strengthen mathematical interaction in the Americas, of which Huntington's visit to Peru represented an early indication, would continue into the 1930s and 1940s in direct scientific response to Franklin Roosevelt's "good neighbor" foreign policy. By then, however, they would reflect an agenda broader than mere mathematical dialogue (see chapter eight).

Foreign policy of another kind was manifest in the 1920s in Rockefeller Foundation initiatives and, particularly, in the Rockefeller-funded International Education Board. By taking the lead in promoting international scientific cooperation, those at the helm saw an opportunity simultaneously to lay the groundwork for reconstructing national and international cultures in the aftermath of World War I. Theirs was a broad and idealistic vision.[38]

Although physics, chemistry, and biology were the sciences on which the IEB initially focused, mathematics, thanks to the successful contemporaneous efforts of Veblen and others, was quickly brought into the IEB's philanthropic sphere. As a result, the IEB needed better to understand the state of mathematics in Europe, and it was to this end that it enlisted Birkhoff's help in 1926.[39] Home-based in France, Birkhoff spent the second half—the winter, spring, and summer—of what became an extended year-long sabbatical from Harvard, traveling through Belgium, Denmark, France, Germany, Great Britain, Italy, the Netherlands, Sweden, and Switzerland. He gave lectures on his research. He found time to pursue that research further. And, he made personal contacts with his European mathematical *confrères* that gave him, for the first time, a more balanced impression of the mathematical endeavors on either side of the Atlantic.

37. George Birkhoff and Solomon Lefschetz were at least two exceptions here. Birkhoff won a cash prize in 1918 from the Fondazione Querini Stampalia of Venice for his work on the periodic solutions of differential equations and was made a foreign member of the Danish and Göttingen scientific societies by 1922. He became even more celebrated in Europe after his 1926 trip abroad, but he would not receive his first European honorary degree until 1933 from the University of Poitiers. Lefschetz won the Borodin Prize of the Paris Académie des Sciences in 1919 for his paper "On Certain Numerical Invariants of Algebraic Varieties with Application to Abelian Varieties," which appeared in print in 1921. As noted in the previous chapter, Lefschetz also won the AMS's Bôcher Prize for this paper.

38. Kohler, p. 142.

39. The account that follows draws from Siegmund-Schultze's invaluable documentary study, *Rockefeller and the Internationalization of Mathematics*, especially pp. 46–56. Siegmund-Schultze reproduced in full Birkhoff's report of 8 September, 1926, to Augustus Trowbridge of the IEB (pp. 265–271) as well as other key, archival sources.

Birkhoff concluded that Paris and Göttingen were the leading European centers for mathematics with Rome a clear but more distant third.[40] As for the best European mathematicians, he saw things this way:

> The greatest mathematician of Europe is Hilbert at Göttingen, but he is nearly at the end of his career. Since the War, Hardy of Oxford has perhaps done the most spectacular work. In range and power, Hadamard of Paris seems nearest to Hilbert. The principal leaders of European mathematics are: Volterra and Levi-Civita in Italy; Picard, Hadamard, Lebesgue and Borel in France; Hilbert, Landau, Hecke, Carathéodory in Germany; Brouwer in Holland; Weyl in Switzerland; H. Bohr in Denmark; and Hardy and Whittaker in Great Britain. There are other men of almost the same rank.[41]

It was a sobering list, and Birkhoff had to admit not only that he had "hitherto under-estimated the power of Europe in the scientific direction" but also that "unless the situation in the United States alters substantially, we will occupy a position not so much commensurate with that of Europe as a whole, but rather with that of a single nation of Europe like Great Britain or Germany." As leaders of the American mathematical research community, Birkhoff, Veblen, and others had long been trying "to 'boost' mathematics [in the United States] into the position it deserves to have," and that position, in their view, should be one commensurate with nothing less than the whole of Europe.[42] Now, as a result of these, his first, experiences abroad, Birkhoff at least had a much better idea of just what such "boosting" would entail.

Two years later, Birkhoff was back in Europe, this time to participate with over fifty of his American colleagues in the International Congress of Mathematicians that was held at the University of Bologna in September 1928. He and Veblen had been invited to give two of the congress's seventeen plenary lectures, the only two Americans so honored.[43] Both had accepted and were actually in attendance.

40. Birkhoff to Trowbridge, 8 September, 1926, in ibid., p. 268.

41. Ibid., p. 270. The next two quotations are also on this page.

42. Birkhoff to Veblen, 17 February, 1927, Box 2, Folder: Birkhoff, George D. 1912–47, Veblen Papers.

43. Other than Italy, which, not surprisingly given the venue, had the most plenary speakers at seven, France had three, the United States and Germany had two, and Great Britain, Russia, and Switzerland each had one.

Things had and had not changed in the four years since the close of the politically charged Toronto Congress. The International Research Council (IRC), of which Émile Picard, described as a "recalcitrant ultra," was President, had finally (in 1926) revoked its exclusionary policies relative to the scientists of the former Central Powers.[44] There was, however, little enthusiasm for joining among the countries formerly boycotted. German scientists, in particular, felt aggrieved and wanted little, indeed nothing, to do with the IRC and its various unions. As long as the International Congresses of Mathematicians were entwined with the IMU, then, it seemed clear to some that IRC politics would interfere with mathematicians' efforts to encourage international contact through their ICMs.

In a bold move, Salvatore Pincherle, President of the IMU, organizer of the Bologna ICM, and proponent of inclusion as called for by the American delegation in 1924, sidestepped the IMU and its politics by arranging to have the invitations to the congress extended not by the IMU but by the Rector of the University of Bologna. Fed up with the politics that tainted the IMU, he resigned from its presidency at the Bologna Congress, halfway through his eight-year term.[45] As a result of his initiative, Birkhoff and Veblen were able to share their ideas—Birkhoff on "Some Mathematical Elements of Art" and Veblen on "Differential Invariants and Geometry"—with over 800 mathematicians, among whom were Hilbert and seventy-five of his German colleagues.[46]

Birkhoff's talk, given in French, outlined his evolving ideas on mathematics and aesthetics.[47] In many respects, it was a strange topic to choose on the occasion of an international congress, when he could, for example, have focused on the work on dynamical systems that he had then recently published as a Colloquium volume of the AMS and that would have highlighted major American research achievements. Veblen, on the other hand, adopted the strategy that Dickson had embraced in both Strasbourg and Toronto. He

44. Lehto, pp. 37–44. The description is on p. 40.

45. Lehto, pp. 44–48 and Riehm and Hoffman, pp. 166–174.

46. George Birkhoff, "Quelques éléments mathématiques de l'art" and Oswald Veblen, "Differential Invariants in Geometry," in *Atti del Congresso internazionale dei matematici, Bologna 3–10 settembre 1928*, 6 vols. (Bologna: Nicola Zanichelli Editore, 1929), 1: 315–333 and 181–189, respectively. The attendance data are on p. 63.

47. He would later develop those ideas into the book, *Aesthetic Measure* (Cambridge: Harvard University Press, 1933). Interestingly, Veblen omitted this book in the selected bibliography that he published in his obituary of Birkhoff. Oswald Veblen, "George David Birkhoff (1884–1944)," *Yearbook of the American Philosophical Society* (1946), 279–285.

opted to highlight, against the backdrop of European results, the latest work in differential geometry being done by himself and other Americans such as his student Tracy Thomas and his NRC postdoctoral fellow Joseph Thomas (Penn Ph.D. 1923). In so doing, he was responding directly to the lecture his geometric rival, Élie Cartan, had given in Toronto four years earlier on "La théorie des groupes et les recherches récentes de géométrie différentielle," a presentation that had placed French results at the forefront.[48] In Bologna, key Americans were participating in a more open ICM and, so, on a more fully international stage, but Birkhoff's lecture, unlike Veblen's, may have represented a missed opportunity to show before such an audience just how well American mathematical research stacked up.[49]

That point, however, was increasingly being driven home by younger American mathematicians who were taking advantage of postdoctoral fellowship opportunities both to work abroad and to make international contacts. As noted in the previous chapter, the first NRC fellows in mathematics were named in 1924 thanks to the initiatives of Veblen and others. One of those, the same Joseph Thomas whose work in differential geometry Veblen would later highlight in Bologna, took his fellowship to Harvard and Princeton, coming under Veblen's wing at the latter. Their collaboration promptly produced a joint paper on "Projective Normal Coordinates for the Geometry of Paths" that appeared in the *Proceedings of the National Academy of Sciences* in 1925 and that explicitly identified Thomas as a "National Research Fellow in Mathematics."[50]

The NRC, although committed to supporting Americans' study on home soil, also allowed Thomas to travel to Paris, one of the seats of differential geometry on the continent thanks to the presence there of Cartan and Hadamard. Ultimately, Thomas did not publish jointly with any of the Parisian mathematicians, but traces of his sojourn in the City of Light may be found in a 1928 paper that he presented to the AMS in September at the very moment that Veblen was lauding his earlier work at the Bologna Congress. Thomas returned from France to take up a position at Penn, his

48. Élie Cartan, "La théorie des groupes et les recherches récentes de géométrie différentielle," *Proceedings of the International Mathematical Congress Held in Toronto, August 11–16, 1924*, ed. John Fields, 1: 85–94. Compare, too, Ritter, p. 168.

49. The Bologna Congress was predominantly a (Western and Eastern) European and North American affair. There were eleven mathematicians in attendance from Japan, one from Guatemala, five from India, six from Palestine, and two from Turkey.

50. Oswald Veblen and Joseph Thomas, "Projective Normal Coordinates for the Geometry of Paths," *PNAS* 11 (1925), 204–207 on p. 207.

alma mater, and subsequently published his Parisian findings in the *Annals of Mathematics*.[51] While at Penn, he continued to focus on theorems about the existence of solutions for systems of partial differential equations that Charles Riquier had given in his Poncelet Prize–winning book, *Les systèmes d'équations aux dérivées partielles*, and that Maurice Janet had more recently reprised in a lengthy article in the *Journal de mathématiques pures et appliquées*.[52] Thomas had discovered a way significantly to simplify Janet's treatment of Riquier's work.

If the NRC only rather grudgingly allowed its fellows to take their financial support abroad, the International Education Board, after its founding in 1923, had no such qualms. It not only granted foreign postdoctoral fellowships of its own, but it also cooperated with the NRC by actually funding the fellowships of its awardees—like Joseph Thomas—who wanted to go beyond U.S. borders.[53] In the 1920s, some sixteen NRC/IEB fellows took all or part of their fellowships abroad.[54] Among these, Jesse Douglas, another differential geometer, proceeded for the 1928–1930 academic years to Paris to interact primarily with Cartan and Hadamard after visiting the "big three" of Princeton, Harvard, and Chicago in the United States.

Douglas had earned his Ph.D. under Edward Kasner at Columbia in 1920 and won his NRC fellowship six years later on the basis of a proposal to work toward a solution of Plateau's problem. Posed by Joseph–Louis Lagrange in 1760 and still unsolved in 1926, Plateau's problem was to find, given a contour, a surface of minimal area that spans it. In October 1929, Douglas was writing to Cartan hoping to set up a meeting at the Sorbonne and rather proudly let Cartan know that one of the results he had obtained in the geometry of paths had just been referenced by the University of Hamburg's Wilhelm Blaschke.[55]

51. Joseph Thomas, "Riquier's Existence Theorems," *AM* 30 (1928–1929), 285–310.

52. Charles Riquier, *Les systèmes d'équations aux dérivées partielles* (Paris: Gauthier-Villars, 1910) and Maurice Janet, "Les systèmes d'équations aux dérivées partielles," *Journal de mathématiques pures et appliquées*, 8th ser., 3 (1920), 65–151.

53. Assmus, p. 169. This cooperation was natural given that the money for both the NRC and the IEB came from the same source, the Rockefeller philanthropies.

54. Compare Siegmund-Schultze, *Rockefeller and the Internationalization of Mathematics*, pp. 288–301 (where the number would seem to be fourteen) and *National Research Fellowships 1919–1938*, pp. 22–28 (where the number would seem to be sixteen). Four recipients appear in the latter but not in the former source, while two appear in the former but not in the latter. As Siegmund-Schultze candidly noted, the list that he gave "should be about complete" (see p. 288).

55. Jesse Douglas to Élie Cartan, 10 October, 1929, Carton 5: Correspondences: Jesse Douglas: 6.52, Fond Élie Cartan.

By February 1930, he was telling Veblen that Hadamard was "enormously interested" in his work on Plateau's problem and that Cartan had characterized his "point of view as distinctively new."[56]

A month later, Douglas had more good news to report. He had not only taken the position at MIT that both Veblen and Cartan had helped him to secure, but he was also "in possession of a complete and rigorous solution" of Plateau's problem, having presented his full result in Hadamard's seminar at the Collège de France late in December 1929.[57] Douglas had begun his research on the problem in earnest in Princeton at the beginning of his NRC/IEB fellowship in 1926, and he had completed it for publication in 1930 as he was returning home.[58] At the 1936 International Congress of Mathematicians held in Oslo, he would be awarded one of the first two Fields Medals for this work (see chapter five). Named after the same John Fields who had engineered the Toronto ICM in 1924, the Fields Medal came to represent the "missing" Nobel Prize for mathematics. The conferral of it to Douglas represented a striking "proof of concept" of the NRC and IEB fellowship programs in mathematics.

Yet another philanthropy, the John Simon Guggenheim Foundation, shared the IEB's commitment to the encouragement of foreign study through postdoctoral fellowships. Established in 1925, it also fostered research but in a wider variety of fields than the NRC: the arts, humanities, social sciences, and sciences, including mathematics. Initially (and until 1941), Guggenheim fellowships supported postdoctoral study abroad *only*. The Foundation named its first fifteen fellows in the 1925–1926 academic year, and one of them was a mathematician, the Penn point-set topologist John Kline, who took his award to Göttingen.

56. Douglas to Veblen, undated but in the week before 21 February, 1930, Box 4, Folder: Douglas, Jesse 1926–39, Veblen Papers.

57. Douglas to Cartan, 26 March, 1930, Carton 5: Correspondences: Jesse Douglas: 6.52, Fond Élie Cartan and Douglas to Veblen, 30 March, 1930, Box 4, Folder: Douglas, Jesse 1926–39, Veblen Papers. The quotation is from the latter letter.

58. Jesse Douglas, "Solution of the Problem of Plateau," TAMS 33 (1931), 263–321. Douglas detailed what he called the "successive stages" in his progress on the problem in footnote * on p. 263. As discussed in the next chapter, the problem was solved independently by Ohio State's Tibor Radó. On their respective contributions to the solution, see Jeremy Gray and Mario Micallef, "About the Cover: The Work of Jesse Douglas on Minimal Surfaces," BAMS 45 (2008), 293–302. See also the essays in The Problem of Plateau: A Tribute to Jesse Douglas & Tibor Radó, ed. Themistocles Rassias (Singapore: World Scientific, 1992).

Dominated by a Hilbert fully imbued at this point in mathematical physics, Göttingen was an odd choice for a mathematician of Kline's ilk, a fact he seemed to realize only after he got there. Writing to his mentor, R. L. Moore, back in Austin, Kline related that "practically no one here is interested in our line" of research and that the attitude toward American mathematics in general was "extremely patronizing."[59] His was thus a very different sense of reception than that Douglas would experience three years later in Paris. Regardless, shortly after his return to the United States, Kline shared the purely mathematical insights he had gained abroad with the American mathematical community in an address "by invitation of the Program Committee" before a meeting of the AMS in New York City. There, he carefully traced—and made a point to emphasize—what he saw as the fully intertwined nature of the work of American and European point-set topologists.[60] As he made clear, that field was growing thanks to an international dialogue in which American topologists were, in fact, equal interlocutors (see this chapter's last section). His may not have been an area that greatly interested Hilbert and his Göttingen colleagues, but the latter were not the only mathematicians of note in Europe.

By 1930, eleven Guggenheim fellowships had been awarded to ten mathematicians who, like Kline, capitalized on the opportunity to gain perspective, to make contacts, and to pursue their researches outside the United States. The only woman among them, Dickson's former student, Olive Hazlett, actually received back-to-back awards in 1928 and 1929 which allowed her to work in Switzerland, the home base of both Hurwitz and du Pasquier, as well as in Germany and Italy. In particular, she presented to the mathematicians assembled at the Bologna Congress a natural extension both of Dickson's and of her own earlier research in the arithmetic of algebras that, like Dickson's and her Toronto Congress lectures four years earlier, stressed the superiority of their American approach over that being advocated in Switzerland.[61]

59. Kline to Moore, 6 April, 1926, Box 4RM74, Folder: Kline, John Robert (1925–1928), Moore Papers.

60. John Kline, "Separation Theorems and Their Relation to Recent Developments in Analysis Situs," BAMS 34 (1928), 155–192.

61. Olive Hazlett, "Integers as Matrices," in Atti del Congresso internazionale dei matematici, Bologna, 2: 57–62. On the fellows in mathematics appointed by the Guggenheim Foundation, see http://www.gf.org/fellows. As with the NRC fellowships, news of the Guggenheims was regularly reported in the Bulletin. See, for example, "Notes," BAMS 35 (1929), 740–746 on p. 743. For more on Hazlett's life and career, see Green and LaDuke, especially pp. 196–197.

The cases of Joseph Thomas, Douglas, Kline, Hazlett, and others tes-
tify to the fact that postdoctoral traveling fellowships like those provided
by the NRC, the IEB, and the Guggenheim Foundation allowed American
mathematicians not only to make valuable connections—mathematical and
personal—abroad but also to compete as peers with their European counterparts. Unlike their predecessors in America's *Wanderlust* generation around
the turn of the twentieth century, they were *post*doctoral students, who, on the
basis of their training *in the United States*, were in a position to go to Europe
and to dive right into the mathematical scene there. They had no need to do
the "remedial work" of earning a "real" Ph.D.

Europeans in the United States

Still, unless the Europeans came to the United States to experience the American mathematical research community firsthand, it would be hard for them
fully to assess the extent of that community's maturity. It was one thing to
encounter the crème de la crème of American mathematics—the George
Birkhoffs among the movers and shakers and the Jesse Douglasses among the
younger generation—but it was quite another to be immersed in the mathematical culture and directly to take the measure of its depth and breadth.
Moreover, a generational shift was in process. Felix Klein, to whom Americans had been drawn in the 1880s and 1890s, died in 1925. David Hilbert, at
least in Birkhoff's view in 1926, was "nearly at the end of his career."[62] Mathematicians being trained in the 1920s still looked for mentorship to those, like
Hilbert, Cartan, Hadamard, and others born in the 1860s, but those leaders
were two mathematical generations removed from them. They thus naturally
also sought out those who were making their marks with new, original, and
exciting results in the generation just before theirs, that is, those born in the
1880s. For the first time, Americans—like Veblen and Lefschetz at Princeton
and Birkhoff at Harvard—were among those numbers.

Given this state of affairs, the IEB also needed a better sense of the lay of
the *American* mathematical landscape. For that, it had turned to, among others, Göttingen's Richard Courant, whom IEB head Augustus Trowbridge had
met in the summer of 1926. Courant found his assignment difficult. As he
admitted at the outset, "[o]ur personal contacts with the numerous American

---

62. Birkhoff to Trowbridge, 8 September, 1926, in Siegmund-Schultze, *Rockefeller and the
Internationalization of Mathematics,* p. 270.

mathematicians are not really close enough to enable a really sound judgment on most of them."[63] Still, based on his reading of the journal literature—he had never actually traveled to the United States at this point—Courant offered the opinion that the United States had had its greatest mathematical success in algebraic topology. In his view, "Harvard and, in the first place, Princeton, are the best places in the world today to study this discipline, the representatives of mathematics there would do honor to every European mathematical chair." In sum, these two universities—"perhaps less so Chicago—seem to offer the best possibilities for mathematics." This represented high praise, indeed, for a country in which, "several decades ago," "scientific mathematics . . . was lacking and European assistance required."

As for specific American mathematicians, Courant singled out quite a few. At the "big three," he cited Alexander, Veblen, Lefschetz, Eisenhart, and Wedderburn at Princeton; Birkhoff, Osgood, Coolidge, and Graustein at Harvard; and E. H. Moore, Dickson, Bliss, and Moulton at Chicago. Elsewhere in the Northeast, he mentioned Ritt at Columbia; Richardson and historian of mathematics Raymond Archibald at Brown; and Wiener at MIT. Finally, in the Midwest and West, he noted Miller at Illinois; Blichfeldt of Stanford; and Harry Bateman and Paul Epstein at Caltech. Still, in his view, the "big three" clearly dominated, so it is perhaps not surprising that of the nine (of eighty-five) IEB fellows named in the 1920s who came to the United States, five went to Princeton to work in topology with Veblen, Lefschetz, and/or Veblen's student, Alexander, three were drawn to Harvard by Birkhoff's presence there, and one went to Chicago to engage in algebraic discussions with Dickson.[64] As Veblen had put it to Trowbridge, "[i]t is high time that some of the younger Europeans should get in touch with the mathematical work which is going on in this country."[65] The IEB made that possible.

63. Courant to Trowbridge, 27 April, 1927, in ibid., pp. 272–274 on p. 272 (Siegmund-Schultze's translation). The quotations that follow in this paragraph are also from this page. For the names singled out in the next paragraph, see pp. 272–273.

64. They hailed from England (Maxwell Newman), France (Lucien Féraud), Germany (Heinz Hopf and Wilhelm Maier), Hungary (Belá Kerékjaártó), Japan (Yusuki Hagihara), Romania (Florin Vasilesco and Gheorge Vranceanu), and Russia (Paul Alexandroff). Alexandroff, Hopf, Kerékjaártó, Newman, and Vranceanu were attracted to Princeton; Féraud, Hagihara, and Vasilesco went to Harvard; and Maier traveled to Chicago. Siegmund-Schultze lists the IEB fellows in ibid., pp. 288–301.

65. Veblen to Trowbridge, 14 October, 1926, Box 14, Folder: Trowbridge, Augustus 1923–31, Veblen Papers.

The 1927–1928 academic year was particularly auspicious at Princeton in this regard with Veblen and his colleagues hosting two IEB fellows: Paul Alexandroff (the romanization he wrote under—and which I adopt here—of Pavel Alexandrov) from Russia and Heinz Hopf from Germany. Alexandroff had earned his doctoral degree from Moscow University in 1921 and had ultimately accepted a position there, while Hopf had taken his doctorate at the University of Berlin in 1925 and had spent the 1925–1926 academic year in Göttingen. The two young topologists had met at the latter during Hopf's stay and had quickly become fast friends. It is perhaps not surprising, then, that they both applied for IEB fellowships in the same year, and that, when awarded, they both took them to Princeton.

Their year abroad proved productive. They participated actively in the vibrant topological community that the Princeton mathematicians animated, sitting in on lectures by and talking with Veblen, Lefschetz, and Alexander. They attended an informal topology conference in Princeton that brought mathematicians from Harvard (Marston Morse), the University of Iowa (Edward Chittenden), and the University of Pennsylvania (John Kline) to talk about matters of mutual interest before participating in a "lively" symposium on topology chaired by Veblen at the April meeting of the AMS in New York City.[66] They also met daily with each other to discuss thoroughly "all the freshly absorbed scientific impressions and thought."[67] Both did seminal work: Alexandroff on the homological theory of dimension and Hopf on the homology of manifolds. Both published the early fruits of their American mathematical labors in American journals in 1928: Alexandroff in the *Annals of Mathematics* and Hopf in the *Proceedings of the National Academy of Sciences*. The references in their papers to the work of their American

---

66. "Notes," *BAMS* 34 (1928), 385–394 on pp. 386–387 and Arnold Dresden, "The April Meeting of the Society," *BAMS* 34 (1928), 419–432 on pp. 419–420 (the quotation is on p. 420). Chittenden had earned his Ph.D. at the University of Chicago in 1912 for a thesis on "Infinite Developments and the Composition Property $(K_{12}B_1)$ in General Analysis" (*Rendiconti del Circolo matematico di Palermo* 39 (1915), 81–108) under the direction of E. H. Moore. After six years as an instructor at the University of Illinois, he moved to Iowa where he climbed the academic ladder, becoming a professor there in 1925. He studied under Moore during the latter's "general analysis" period and was interested in topology's applications to analysis, thus in the more analytic, point-set variety.

67. From a manuscript account of Hopf's IEB year in Princeton, as quoted in Günther Frei and Urs Stammbach, "Heinz Hopf," in *History of Topology*, ed. Ioan James, pp. 991–1008 on p. 997 (their translation from the original German).

mentors clearly reflected the influences on their research of their year in Princeton.[68]

In particular, Hopf was inspired by Lefschetz's then-recent work on fixed points of continuous maps of manifolds and almost immediately found a new proof of the Princeton mathematician's main theorem. He sketched his ideas in print before leaving the United States, but he demonstrated them in detail only in 1929 after his return to Germany.[69] His year in Princeton also found Hopf thinking about the problem of constructing maps from the three-dimensional sphere into the two-dimensional sphere, and he made the unexpected discovery of the existence of such a map that was *not* homotopic to a constant map. Although he sketched a proof of this fact in an April 1928 letter that he sent from Princeton to his student, Hans Freudenthal, back in Berlin, he only published its full proof in 1931.[70] Likely owing to the quality of all of this work as well as to his abilities in English, Hopf was offered a position in Princeton's Mathematics Department in 1929, although he opted to stay in Berlin.

As for Alexandroff, he returned to Moscow to take up a university teaching position after his IEB year in Princeton, but it in no way slowed down his research. By December 1930, Veblen and his colleagues were inviting him back to teach during the spring semester of 1931. In addition to providing instruction for the "considerable number" of new graduate students "who are interested in analysis situs," they wanted him to give "an account of [his] own recent work in some supplementary lectures."[71]

68. Paul Alexandroff, "Untersuchungen über Gestalt und Lage abgeschlossener Mengen beliebiger Dimension," *AM* 30 (1928), 101–187, which was received by the editors of the *Annals* (among whom was Lefschetz) on 16 April, 1928; Heinz Hopf, "A New Proof of the Lefschetz Theorem on Invariant Points," *PNAS* 14 (1928), 149–153, which was communicated on 9 January, 1928; and Heinz Hopf, "On Some Properties of One-valued Transformations of Manifolds," op. cit. 14 (1928), 206–214, which was communicated on 11 February, 1928.

69. See the previous note as well as Heinz Hopf, "Über die algebraische Anzahl von Fixpunkten," *Mathematische Zeitschrift* 29 (1929), 493–524. Robert Brown discusses this work of Lefschetz and Hopf in his article, "Fixed Point Theory," pp. 275–280.

70. Willem van Est, "Hans Freudenthal, 17 September 1905 – 13 October 1990," in *History of Topology*, ed. Ioan James, pp. 1009–1019 on p. 1011 (note 6) and Heinz Hopf, "Über die Abbildungen der dreidimensionalen Sphäre auf die Kugelfläche," *Mathematische Annalen* 104 (1931), 637–665.

71. Veblen to Alexandroff, 15 January, 1931, Box 2, Folder: Alexandroff, Paul 1927–38, Veblen Papers.

Hopf and Alexandroff maintained the ties they had made in the late 1920s with the mathematicians in Princeton, in particular, and with the American mathematical community, in general, for the rest of their careers. Hopf returned for extended visits during the 1946–1947 and 1955–1956 academic years; Alexandroff participated with his former American colleagues, Lefschetz and Alexander, in the first international congress on topology that was held in Moscow in 1935 (see chapter six).[72] They represented the new generation of European mathematicians that, in seeking out American mentorship, viewed American mathematics, at least in some areas, as better than or on a par with mathematics in Europe. As Americans in the 1890s had returned home edified by their (primarily) German mathematical mentors and impressed by what they had experienced abroad, so young foreigners in the 1920s were in a position to do the same relative to their American mentors and their mathematical experiences in the United States. New perceptions emerged.

At the same time that these younger mathematicians were coming to the United States from abroad on their IEB postdoctoral fellowships, the AMS, under Birkhoff's leadership as outgoing President, launched a new program in December 1926 to expose more established, indeed "distinguished," foreign mathematicians to the United States.[73] While individual universities had, on occasion, hosted a European mathematician for an extended stay—for example, Klein at Northwestern in 1893 and Hadamard at Yale in 1920—the AMS had never before taken on this kind of leadership role.[74] While, for financial reasons, it planned to piggyback on invitations from particular universities to

72. Hassler Whitney, "Moscow 1935: Topology Moving toward America," in *A Century of Mathematics in America*, ed. Peter Duren et al., 1: 97–117. Whitney, who had earned his Ph.D. under Birkhoff at Harvard in 1932, was another American invitee.

73. George Birkhoff, Gilbert Bliss, and Earle Hedrick, "The Visiting Lectureship of the American Mathematical Society," *BAMS* 34 (1928), 22. Birkhoff had conceived of the idea of the Visiting Lectureship at least as early as February 1927 (Birkhoff to Gilbert Bliss and Earle Hedrick, 10 February, 1927, HUG 4213.2, Box 8: Correspondence 1927 A–Z, Birkhoff Papers). Initially, he had thought that this might also involve American mathematicians, since, as he put it in his letter to Bliss and Hedrick, "the post would be an honor so notable as to fix the standing of any young man selected and thus encourage him greatly." In the end, it was only implemented with visiting *foreign* mathematicians.

74. As is well-known, Klein gave the Evanston Colloquium Lectures while at Northwestern (Parshall and Rowe, *Emergence*, pp. 331–361). During his stay in the United States, Hadamard delivered not only the Silliman Memorial Lectures at Yale but also lectures at the Rice Institute in Houston. See Jacques Hadamard, *Lectures on Cauchy's Problem in Linear Partial Differential Equations* (New Haven: Yale University Press, 1922) and "The Early Scientific Work of Henri

particular mathematicians, it would capitalize on the presence of one such mathematician per year by designating him (or, theoretically, her) an AMS Visiting Lecturer and coordinating a nationwide lecture tour. In so doing, it "hoped that knowledge of recent and important mathematical advances [could] be brought more quickly to the various centers, and that . . . informal contacts . . . of scientific value" could be established. To kick off its new program, the AMS tapped the University of Munich's Constantin Carathéodory, one of the European mathematicians whom Birkhoff had deemed the "best" in his 1926 IEB report. Perhaps not surprisingly, Harvard had invited him for the spring semester of the 1927–1928 academic year to fill the gap left by a Birkhoff once again in Europe.

The fifty-four-year-old Carathéodory, who had started out as an engineer and who, in earning his Ph.D. in mathematics in 1904, was actually in the mathematical generation of those born in the 1880s, had made his reputation in the theories of real- and complex-valued functions, the calculus of variations, and measure theory over the course of an academic career spent largely in Germany. His American sojourn found him lecturing in January on his work at the Universities of Pennsylvania, Iowa, and Michigan as well as at Ohio State, Cornell, and Adelbert College (or what is now Case Western Reserve University) in Ohio before settling down at Harvard for the rest of the spring semester. There, he taught both lower- and upper-level courses, with his function-theoretic interests particularly complementing and updating those of Osgood, a mathematician trained, as noted, by the earlier generation of Klein at Göttingen and Max Noether at Erlangen. Carathéodory also took the opportunity to participate—with Alexandroff and Hopf—in the symposium on analysis situs that formed part of the AMS's April meeting in New York City. With the end of the semester on the East Coast, he next journeyed—with an academic stop along the way at the University of Texas, Austin—to the University of California, Berkeley for the summer. There, he joined the Rice Institute's Griffith Evans and MIT's Henry Phillips as a special summer lecturer.[75]

Poincaré," *Rice Institute Pamphlet–Rice University Studies* 9 (3) (1922), 111–184. Hadamard returned to the United States and to Rice in 1925, continuing his earlier lectures on Poincaré's work. The 1925 lectures were only published in 1933 as Jacques Hadamard, "The Later Scientific Work of Henri Poincaré," *Rice Institute Pamphlet–Rice University Studies* 20 (1) (1933), 1–86.

75. Parshall and Rowe discuss Osgood's training in *Emergence*, pp. 229–234. On Carathéodory's peregrinations, see "Notes," *BAMS* 34 (1928), 121–124 on p. 122; Dresden,

That Carathéodory's visit was a success is confirmed by the facts that he was seriously considered for a professorship at Harvard and received an actual offer from Stanford. In granting him leave to visit the United States for a semester, the German Foreign Ministry had hoped to work to counteract the negative propaganda that had persisted in the aftermath of the war by sending forth one of its best people. It apparently succeeded in its choice of Carathéodory as a cultural ambassador.[76] In a sense, then, both Germany and the United States profited as a result of the AMS's new initiative.

Hermann Weyl of the Eigenössische Technische Hochschule in Zürich followed Carathéodory as the AMS Visiting Lecturer in the 1928–1929 academic year. Weyl had accepted—at least for one initial year—Princeton's offer of the Thomas D. Jones Research Professorship in Mathematical Physics.[77] This position, made possible by the continuing beneficence of the same Gwethalyn Jones, who, together with her uncle, had endowed Veblen's Fine Research Professorship as well as Princeton's Fine Hall, was regarded as "the best chair in America" at the time.[78] From his Princeton home base, Weyl had already traveled late in the fall to the Universities of Iowa and Michigan, two of the universities that had hosted Carathéodory just months earlier, as well as to Columbia. While in New York, he, like Carathéodory, participated in a special symposium, this one on Fourier series and associated with the AMS's March meeting.

Another intellectually exciting event, the Fourier series symposium was chaired by Birkhoff and showcased the work of Weyl and G. H. Hardy.[79] Hardy, who was also visiting Princeton (see the next section), lectured on "Modern Work in the Theory of Ordinary Trigonometric Series," while Weyl spoke on "Fourier Series and Almost Periodic Functions from the Standpoint of the Theory of Groups." Their presentations then served as the point of departure for discussions, in the case of Hardy's lecture, by Einar

---

"The April Meeting of the Society," p. 419; and "Notes," *BAMS* 34 (1928), 533–538 on p. 536.

76. Maria Georgiadou provides more details on his trip to the United States in *Constantin Carathéodory: Mathematics and Politics in Turbulent Times* (Berlin: Springer-Verlag, 2004), pp. 220–233.

77. Weyl ultimately decided not to stay beyond the first year. As is well-known, he returned to Princeton as a member of the Institute for Advanced Study in 1933. See the next chapter.

78. Veblen to Weyl, 15 June, 1929, Box 16, Folder: Weyl, Hermann 1928–34, Veblen Papers.

79. Arnold Dresden, "The March Meeting in New York," *BAMS* 35 (1929), 433–446 on pp. 433–434.

Hille then of Princeton, the University of Cincinnati's Charles Moore, and Columbia's Thomas Gronwall, and, in that of Weyl, by Norbert Wiener and Philip Franklin of MIT, Harvard's Marshall Stone, and Michigan's Theophil Hildebrandt. It was a mathematical give-and-take in which the American-based mathematicians shared the stage with two more on Birkhoff's list of the "best" Europeans.

Again, like Carathéodory the year before, Weyl followed his academic year in Princeton with a cross-country trip—making academic stops, in his case, at the Rice Institute and elsewhere—before spending a summer lecturing in Berkeley. He also gave a talk "On the Foundations of General Infinitesimal Geometry" at the June meeting in Berkeley of the AMS that he then published as a paper in the AMS's *Bulletin*.[80] That talk and that paper provided an interesting comparative perspective on American and French work.

Although he may have already been introduced to the Princeton group's work through the lecture Veblen had just given at the Bologna Congress, Weyl said that the topic of his Berkeley talk had been suggested to him by the seminar on differential geometry in which he had actually participated in Princeton. As he explained, "it seemed desirable" to him "to clarify the relations between the work of" what he regarded as "the Princeton school and that of Cartan."

Cartan had developed an approach to differential geometry that, like Klein's *Erlanger Programm* of 1872, exploited the notion of a group of transformations to understand and interpret a geometry's structural features.[81] Unlike Klein's approach, however, in which a group G of transformations was applied to an m-dimensional manifold M itself, Cartan applied it to the tangent plane T_P associated with a point P of M. If P' is a point "near" P, then there is an equivariant isomorphism between the representation spaces of T_P and T'_P, called the displacement, in terms of which the concept of curvature can be defined. Supposing that the tangent plane T_P of P is displaced "along a curve L on M which leads back to P, the tangent plane returns in a new position or orientation. The final position is obtained from the original by a certain isomorphic representation of T_P onto itself" called the curvature along L.

80. Benjamin Bernstein, "The June Meeting in Berkeley," *BAMS* 35 (1929), 593–606 on p. 603 and Hermann Weyl, "On the Foundations of General Infinitesimal Geometry," *BAMS* 35 (1929), 716–725 on p. 716. The quotations that follow in the next two paragraphs are on p. 717.

81. Élie Cartan, "Sur les variétés à connexion affine et la théorie de la relativité généralisée," *Annales de l'École normale supérieure* 40 (1923), 325–412.

Cartan was then able to give analytic expressions of the displacement of P to all neighboring points P' as well as of the associated curvature and to embed, in a particular way, any given tangent plane locally (around a point in T_P that covers P on M) into the manifold. His embedding, however, worked nicely—that is, the tangent plane T_P was uniquely determined by the nature of the underlying manifold M—only if the group G was affine, and not if it was "a more extensive group."[82] Weyl saw this as "a deficiency of the theory" that "the infinitesimal-geometric researches of Eisenhart, Veblen, T. Y. Thomas, and others in Princeton have remedied ... for projective and conformal geometry." He then proceeded to show just how, liberally and appreciatively citing the work of the Princetonians. In a companion paper co-authored with then-Caltech but soon-to-be Princeton mathematical physicist Bob Robertson, moreover, some of the explicitly group-theoretic ramifications of the set-up were developed further.[83] Weyl's assessment and his collaboration must have been particularly gratifying for an American mathematical community seeking both to have its accomplishments recognized and to be acknowledged as a worthy mathematical competitor.

Other AMS Visiting Lecturers followed—such as Enrico Bompiani of the University of Rome for the 1929–1930 academic year and Hamburg's Wilhelm Blaschke for 1930–1931—and the profiles of their visits were strikingly similar to those of Carathéodory and Weyl.[84] The universities on the lecture tours—Iowa, Michigan, Texas, Rice, Berkeley, and others—reflected the evolving landscape of American mathematics as institutions of higher education beyond the "big three" of Chicago, Harvard, and Princeton aspired to move more aggressively into research (see the next chapter).[85] And,

82. Weyl, "On the Foundations of General Infinitesimal Geometry," p. 719. The quotations that follow are also on this page.

83. Howard Robertson and Hermann Weyl, "On a Problem in the Theory of Groups Arising in the Foundations of Infinitesimal Geometry," *BAMS* 35 (1929), 686–690.

84. Archibald lists the AMS Visiting Lecturers to 1937, the program's last year before the outbreak of World War II, in *Semicentennial History*, p. 21.

85. Birkhoff gave a strongly positive assessment of the first three years of the Visiting Lecturer program, noting that "the number of institutions which have aided mathematical lecture[r]s of this type ... has been very large," including "Brown, Bryn Mawr, California Institute of Technology, Columbia, Cornell, Dartmouth, Harvard, Iowa State College, Massachusetts Institute of Technology, Ohio State, Pomona, Princeton, Rice Institute, Yale, Universities of California at Berkeley and Los Angeles, of Chicago, Cincinnati, Illinois, Indiana, Iowa, Michigan, Minnesota, Pennsylvania, North Carolina, Texas, Wisconsin." Birkhoff to "Colleagues," 23 April, 1930, Box 2, Folder: Birkhoff, George D. 1912–47, Veblen Papers.

the lecturers themselves, deemed among the most "distinguished" of European mathematicians, interacted while in the United States as peers. They attended seminars. They participated in symposia. And, at least in Weyl's case, they published in American journals—singly and in collaboration—explicitly highlighting American research accomplishments against a European backdrop. These Europeans experienced and appreciated a mathematical community in the United States different in many respects from their own—vast, diverse, decentralized—but comparable in many others—talented, committed, driven. In taking its measure, they took it seriously.

An Englishman in America and an American in England: The Hardy-Veblen Exchange

Another intriguing indication of the perception of a maturing American mathematical research community was the physical exchange that transplanted G. H. Hardy to Veblen's position at Princeton and Veblen to Hardy's at Oxford during the 1928–1929 academic year.[86] The two mathematicians, who had been corresponding at least since 1924, began discussing such a possibility in earnest in the fall of 1927. Both had started to lay the necessary groundwork at their respective home institutions.

For Hardy's part, it was "rather an elaborate affair" to get all of the i's dotted and t's crossed in Oxford. He had to clear the idea with the other named professors, the Vice-Chancellor, the mathematics faculty, the Faculty of Natural Science, and the so-called Visitatorial Board. He also had to see about securing Veblen's admission as a temporary member of the Senior Common Room in New College, the college to which his chair, the Savilian Professorship of Geometry, was attached.[87]

Things were somewhat easier for Veblen. He had little trouble convincing his colleagues in the Mathematics Department of the merits of the idea, but he also had to arrange for an audience with Henry Fine, the Dean of the Faculty. "[I]f he is favorable," Veblen explained to Hardy, "I don't think there will be any difficulty about the red tape."[88]

86. Harvard geometer, Julian Coolidge, had been appointed an "exchange professor" at the Sorbonne in 1927, but his was a one-way "exchange." "Notes," *BAMS* 33 (1927), 499–508 on p. 506. Coolidge had strong French connections based on his wartime service in Paris and had, in fact, already been named a chevalier of the Légion d'Honneur in recognition of those efforts.

87. Hardy to Veblen, 14 October, 1927, Box 6, Folder: Hardy, G. H. 1924–47, Veblen Papers. The quotation is from this letter.

88. Veblen to Hardy, 27 October, 1927, Box 6, Folder: Hardy, G. H. 1924–47, Veblen Papers.

All of the approvals had been secured on both sides of the Atlantic by January 1928. It only remained to decide on the courses that each man would offer. For that, each had to understand better the environment in which he would find himself.

As far as Veblen's teaching at Oxford was concerned, Hardy explained that to satisfy the statutory requirements of the Savilian chair, Veblen would need to offer just one course of two lectures a week, even though he himself tended to do much more than that. He had always felt obliged to teach at least some geometry, given that the very word was in his chair's title, but unless he also lectured on his own "pet subjects," especially analytic number theory, there was "no one else who could make any show" of them in Oxford.[89] Presumably, though, the geometry part of the position would present no problem at all for Veblen. "It is quite understood by everybody," Hardy told him, "that you will be there as a real 'geometer' & will confine yourself to the things in which you are a real expert." Hardy had, in fact, played up this point in arguing for the exchange. As he saw it, "it w[oul]d be a heaven sent chance for Oxford really to learn something about the pure mathematics of relativity," and he thought that Veblen would actually "find a real demand for it." Still, Veblen also needed to understand the audience he would encounter in Oxford. According to Hardy, he would "find the people quite keen and intelligent, but with a restricted range of knowledge—on the other hand with very good technique." Hardy thus counseled him to begin "at the beginning." If he did, he "would find it possible to go quite a long way."

The situation in Princeton was, not surprisingly, different. There, as Veblen explained, the "usual lecture course calls for three hours a week," but there was a certain amount of flexibility.[90] Hardy could, for example, "divide it up into two sections, say two hours of elementary and intermediate and one hour of advanced." This was how Lefschetz taught his course in analysis situs, thereby allowing Veblen "and other 'highbrows' in this line" to attend "the advanced sessions," while leaving the "elementary ones, unembarrassed to the elementary students." Veblen was sure that those same "highbrows," as well as mathematicians in the Princeton environs, would attend a one-hour advanced session per week, but "they probably would not have time to do . . . more." The main thing, he counseled Hardy, would be "for you to follow whatever

89. Hardy to Veblen, undated but probably January 1928, Box 6, Folder: Hardy, G. H. 1924–47, Veblen Papers. The quotations that follow in this paragraph are also from this letter.

90. Veblen to Hardy, 9 January, 1928, Box 6, Folder: Hardy, G. H. 1924–47, Veblen Papers. The quotations that follow in this paragraph are also from this letter.

program gives you the most pleasure and serves best as propaganda for your brand of analysis."

In the end, Hardy decided to offer "Chapters in the Theory of Functions." Its suitably vague title would give him the opportunity to take the measure of his audience and to adjust accordingly, but what he had in mind was "a half systematic account of those parts of the theory of functions which lead up to the theory of Fourier series & power series on the circle of convergence."[91] As he freely admitted, "this . . . covers a great deal both of classical & modern stuff," in particular, work of Edmund Landau, Ludwig Bieberbach, and Henri Lebesgue. For his part, Veblen opted to lecture on "Tensors and Other Differential Invariants," exactly in line with Hardy's suggestion that he give a course on the differential geometry that he and his Princeton colleagues had been developing for the theory of relativity.

Both men would be filling a gap in the mathematical expertise of their respective host institutions. As Hardy had lamented to Veblen early in 1927, "there is in England no real 'Einsteinian' geometer," while although Princeton had the young Hille in analysis, Hardy ultimately found that his own "sort of math[ematic]s simply doesn't exist" in Princeton.[92] This, in fact, had also been the basis for the argument that Veblen had used to convince the Rockefeller-funded IEB to make a small grant to both him and Hardy so that they would not be out of pocket for their travel expenses.[93] The exchange was thus recognized as an occasion for real mathematical cross-fertilization, and both men took advantage of it.

Veblen arrived in Oxford having just published his Cambridge Tract on *Invariants of Quadratic Differential Forms* the year before. This gave his audience a natural place to start in preparing for his course, and Veblen had made just that recommendation when Hardy sought his advice on the matter. One of the students who wanted to take advantage of Veblen's presence in England was J.H.C. "Henry" Whitehead, nephew of Alfred North Whitehead and an honors student in mathematics at Oxford's Balliol College before

91. Hardy to Veblen, undated but probably January 1928. The next quotation is also from this letter.

92. Hardy to Veblen, undated but shortly after his arrival in Princeton in the fall of 1928, Box 6, Folder: Hardy, G. H. 1924–47, Veblen Papers. As Hardy put it in this letter, Princeton may have had Hille, but he "isn't very inspiring."

93. Veblen to Hardy, 24 January, 1928 and 4 May, 1928, Box 6, Folder: Hardy, G. H. 1924–47, Veblen Papers as well as Siegmund-Schultze, *Rockefeller and the Internationalization of Mathematics*, pp. 73–74.

embarking on a career in finance. Whitehead had decided to return to mathematics, and Hardy thought that Veblen "might be able to put him in the way of doing something later on," especially if Whitehead could bone up over the summer before Veblen's arrival.[94] For Whitehead's reading list, Veblen suggested, in addition to his Cambridge Tract, Eisenhart's *Riemannian and Non-Riemannian Geometry*, as well as Cartan's *La géométrie des espaces de Riemann*.[95] It was a lot to tackle, but Whitehead apparently rose to the challenge, for one of the outcomes of Veblen's stay in Oxford was his acquisition of Whitehead as a new graduate student at Princeton beginning in the summer of 1929.

Three years later, Whitehead had earned his Ph.D. under Veblen for a thesis on "The Representation of Projective Spaces"; he and Veblen had published their now-classic *The Foundations of Differential Geometry*; and he had begun studies, thanks to Lefschetz's influence, that would shape his subsequent mathematical career in topology.[96] Whitehead returned to England to take up a fellowship at Balliol in 1933 and to establish himself in English mathematics. As Hardy quipped to Veblen, "you or God or both have really worked wonders with him!"[97] Whitehead was thus an early example of a foreign, American-trained mathematician who quickly made good, and his transformation was directly linked to Hardy and Veblen's 1928 exchange.

In addition to working with students in Oxford, Veblen participated in the London Mathematical Society. On 14 February, 1929, for example, he was in the capital city to give a lecture before the Society on "Generalized Projective Geometry." There, he endeavored to expose a broader English audience than the one he had in Oxford to work being done on what Hardy had called "Einsteinian geometry," that is, the research of Weyl, Cartan, and others in Europe as well as himself, Eisenhart, Douglas, and Joseph and Tracy Thomas

---

94. Hardy to Veblen, undated but from the spring of 1928, Box 6, Folder: Hardy, G. H. 1924–47, Veblen Papers.

95. For the references to Eisenhart's book, recall chapter one, and see Élie Cartan, *La géométrie des espaces de Riemann* (Paris: Gauthiers-Villars, 1925).

96. J.H.C. Whitehead, "The Representation of Projective Spaces," *AM* 32 (1931), 327–360; for the reference to Veblen and Whitehead, recall chapter one; Solomon Lefschetz and J.H.C. Whitehead, "On Analytical Complexes," *TAMS* 35 (1933), 510–517. On Whitehead's life and work, see Maxwell Newman, "John Henry Constantine Whitehead," *Biographical Memoirs of Fellows of the Royal Society* 7 (1961), 349–363.

97. Hardy to Veblen, undated but 1931, Box 6, Folder: Hardy, G. H. 1924–47, Veblen Papers.

in the United States. As at the Bologna ICM, he took the opportunity in London to show how fully intertwined American work was with that of the Europeans and how central it was to developments in the field.[98]

Hardy's stay in the United States was similarly eventful, after a settling-in period that found him, on the one hand, trying to adjust to such American ways as "the long evenings after so early a dinner" but, on the other, glorying in American baseball.[99] He also had to get to know his new colleagues. "Alexander & Lefschetz seem obviously attractive people," he wrote of his first impressions to Veblen, but his Princeton sojourn also coincided with Weyl's year as Jones Professor of Mathematical Physics, and Hardy confessed that he doubted if he would "like him very much . . .: I may, but I don't feel sure." Hardy was also initially a bit discouraged about his lectures . He found that "there is . . . no real kick in the analysis here, and no one is happy until he can put at least 3 subscripts under every letter, or with less variables than 6."[100] As he and his audience adjusted to one another, though, things improved in this regard.

Hardy's presence on the East Coast was also the occasion for his participation in the activities of the AMS in nearby New York City. Although a brief illness prevented him from personally delivering the sixth Gibbs Lecture on "An Introduction to the Theory of Numbers," it was given in his stead in December 1928 before a joint session of the AMS and the American Association for the Advancement of Science.[101] Recovered by

---

98. Oswald Veblen, "Generalized Projective Geometry," *Journal of the London Mathematical Society* 4 (1929), 140–160. On the work of Douglas and the two Thomases, see, for example, Jesse Douglas, "The General Geometry of Paths," *AM* 29 (1927–1928), 143–168; Tracy Thomas, "On the Projective and Equi-projective Geometries of Paths" and "On Conformal Geometry," *PNAS* 11 (1925), 199–203 and 12 (1926), 352–359, respectively; and Joseph Thomas, "Conformal Invariants," *PNAS* 12 (1926), 389–393 and (with Oswald Veblen) "Projective Invariants of Affine Geometry of Paths," *AM* 27 (1926), 279–296. Although he did not, Veblen might also have mentioned the work of another Princeton colleague, the mathematical physicist, Howard Robertson. See Howard Robertson, "Dynamical Space-Times Which Contain a Conformal Euclidean-Space," *TAMS* 29 (1927), 481–496 and his joint paper with Hermann Weyl discussed in the previous section.

99. Hardy to Veblen, undated but shortly after his arrival in Princeton in the fall of 1928. The quotations in the next sentence are also from this letter.

100. Hardy to Veblen, 10 December, [1928], Box 6, Folder: Hardy, G. H. 1924–47, Veblen Papers.

101. G. H. Hardy, "An Introduction to the Theory of Numbers," *BAMS* 35 (1929), 778–818.

March, however, Hardy was in New York to participate with Weyl in the special symposium on Fourier series. Birkhoff, in fact, had even more visible AMS involvement on his mind on first learning from Veblen that he and Hardy would be trading places: "Would he not make an admirable Visiting Lecturer?" Birkhoff ventured.[102] As Veblen knew, though, Robert Millikan at Caltech had already contacted Hardy about an extended stay on the West Coast, so he was unavailable. This made the choice of Weyl as the AMS's second Visiting Lecturer all the easier. The AMS would not have to decide between two mathematicians of the caliber of Hardy and Weyl! In addition to his six-week stay in Pasadena, Hardy also gave invited lectures on a variety of topics at Lehigh in Pennsylvania, Ohio State, the University of Chicago, UCLA, and Berkeley, much as if he had been the AMS's official Visiting Lecturer.[103]

If, as Birkhoff had concluded in his 1926 report to the IEB, Hardy was among the Europeans doing "the most spectacular" work of the immediately postwar period, then Veblen, an American, was literally taking his place. One of America's best was thus deemed interchangeable, in a sense, with one of the best in Europe. This represented strong symbolism to an American mathematical research community seeking international validation and appreciation.

Even more telling, had they known of it, was Hardy's comparative assessment of research-level mathematics in the United States and England. On a scrap of paper and employing a calculus as arcane as the scoring of his beloved cricket (see Fig. 3.2), Hardy, excluding himself and Veblen, listed and ranked the American mathematicians he viewed as the ten best—Birkhoff, Lefschetz, Dickson, Alexander, E. H. Moore, Wiener, Morse, Osgood, Eisenhart, and Bliss—against the ten best English mathematicians—John Littlewood, William Young, Ernest Hobson, Abram Besicovitch, Edward Titchmarsh, Edmund Whittaker, George Watson, Alfred Young, Percy Daniell, and Augustus Love. His conclusion, boldly circled? "U.S. wins by 9 runs to 7." At least as Hardy saw it, the United States had surpassed England on the international mathematical playing field by the close of the 1920s.

102. Birkhoff to Veblen, 9 January, 1928, Box 2, Folder: Birkhoff, G. D. 1912–47, Veblen Papers.

103. For his itinerary and the titles of his talks, see "Notes," *BAMS* 35 (1929), 279–283 on p. 280.

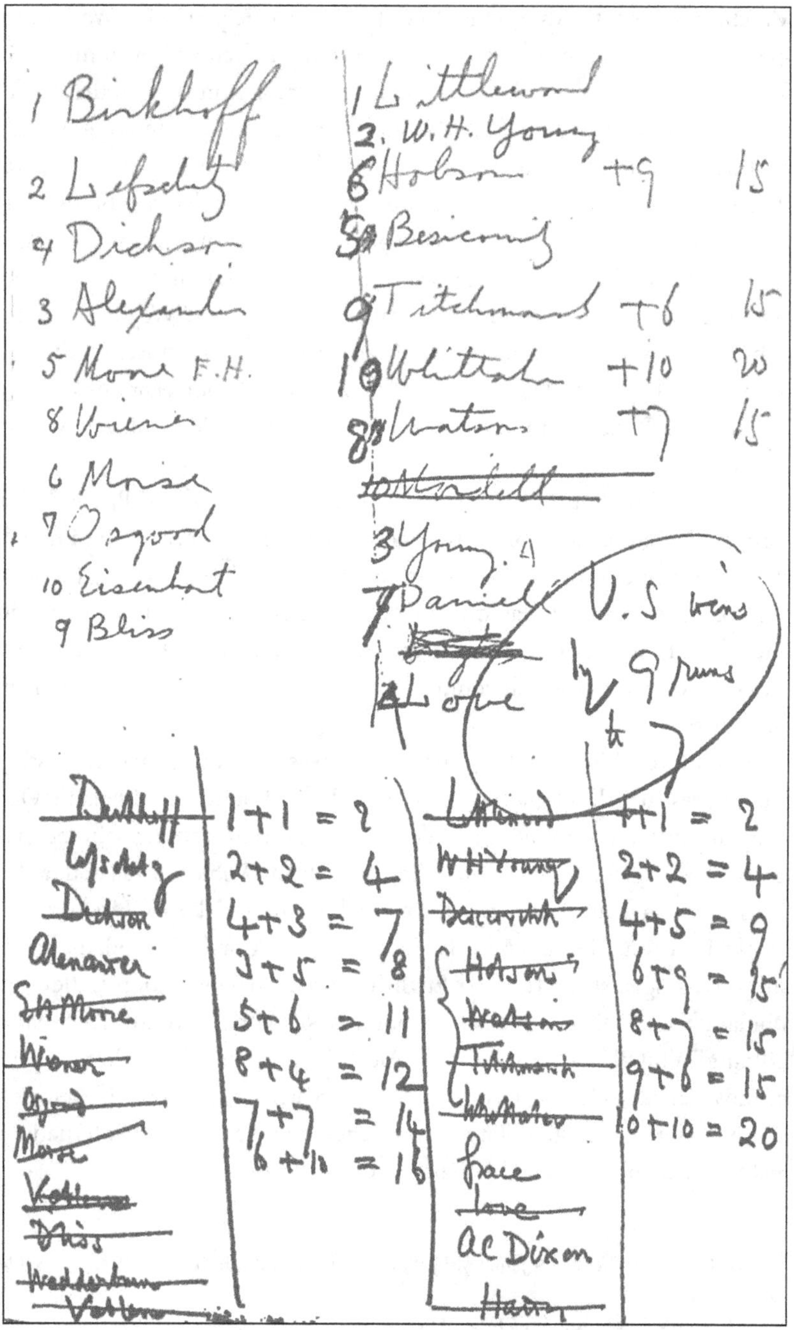

FIGURE 3.2. G. H. Hardy's "calculus" and "score card" for ranking mathematicians in the United States and England. (Photo courtesy of Veblen Papers, Library of Congress.)

International Competition and Collaboration: The Case of Point-Set Topology in Poland and the United States

While Veblen had lectured in Oxford on the differential geometry that in Weyl's words constituted part of a "Princeton school," he and his colleagues were equally renowned for their closely related research in combinatorial, that is, algebraic topology. The latter work had elicited Courant's strong kudos in 1927 and had prompted the mathematical pilgrimages of Alexandroff and Hopf. Courant had failed, however, to single out for honorable mention—or, indeed, for mention at all (!)—the other brand of American topology, that is, the point-set topology that R. L. Moore and his students had established as a significant American mathematical research specialty beginning in the 1920s.

The existence of two flavors of topology was not, however, limited to the United States. In particular, as the American school of point-set topology had grown, so too had a roughly contemporaneous and initially independent Polish school of just the same stripe. Although the political situations of Poland and the United States could not have been more different, both countries sought, especially in the decades after the First World War, to assert themselves on the international mathematical stage. In the case of Poland, Wacław Sierpiński, his doctoral student, Stefan Mazurkiewicz, and the Paris-trained student of Henri Lebesgue, Zygmunt Janiszewski, all found themselves on the faculty of the University of Warsaw and all began to discuss mathematical ideas of mutual interest, especially Cantorian set theory and its applications to topology. Soon, Mazurkiewicz's students, Casimir (also rendered as Kazimierz) Kuratowski and Bronisław Knaster, had also been won over to this sort of mathematical research, and a new research journal, *Fundamenta Mathematicae*, had been founded (in 1920) specifically to support their work.[104]

It had been Janiszewski, in fact, who, as early as 1917, had conceived the idea of starting a journal that would, by targeting specific branches of mathematics, stimulate mathematical research in Poland. Moreover, such a journal "would accept articles in each of the four languages considered in

---

104. Roman Duda discusses the founding of this journal in *"Fundamenta Mathematicae* and the Warsaw School of Mathematics," in *L'Europe mathématique/Mathematical Europe,* ed. Catherine Goldstein, Jeremy Gray, and Jim Ritter (Paris: Éditions de la Maison des sciences de l'homme, 1996), pp. 481–498. The quotations that follow in the next paragraph are on p. 484. I have also made the point in the next paragraph in Karen Hunger Parshall, "Mathematics and the Politics of Race: The Case of William Claytor (Ph.D., University of Pennsylvania, 1933)," *AMM* 123 (2016), 214–240 on pp. 220–221.

mathematics as international," namely, English, French, German, and Italian, and "would contain, aside from original contributions, . . . translations of those articles which are valuable and were not published in 'international' languages, above all in Polish." In so doing, it would "become indispensable for everyone working in that branch of mathematics," "find readers everywhere," and "attract serious collaborators from abroad." In order to realize this vision, *Fundamenta Mathematicae* focused on research in set theory, topology, mathematical logic, and the theory of functions of a real variable. It thus helped define what would ultimately be recognized as a Polish school of point-set topology[105] at the same time that it provided a critical line of communication between Polish and especially American topologists.

American point-set topologists—R. L. Moore and his student Kline— appeared in print in *Fundamenta* as early as the journal's third volume in 1922, and it was clear from their contributions that they were reading and actively engaging with the research issuing from Poland. In his short paper, Kline explored further an example that Knaster and Kuratowski had published (in French) in *Fundamenta* the year before. They had constructed a connected point set M (we would call it a connected *topological space* today) containing a point P such that $M - P$ is totally disconnected, and Kline showed that "no connected point set can have more than one point such that when it is removed, the remainder is totally disconnected."[106]

Moore's contribution was more substantive. He responded to a result that Sierpiński had published (also in French) in *Fundamenta*'s first volume. There, the Polish topologist had proved that a closed and connected set of points M is a continuous curve if and only if "for every positive number ϵ, the connected point-set M should be the sum of a finite number of closed and connected point-sets each of diameter less than ϵ."[107] Thus, as Moore noted, applied to point sets that are closed, bounded, and connected, Sierpiński's necessary and sufficient condition was equivalent to the notion

105. For this designation, see Krzysztof Ciesielski and Zdzislaw Pogoda, "The Beginning of Polish Topology," *MI* 18 (1996), 32–39 on p. 32.

106. Bronisław Knaster and Casimir Kuratowski, "Sur les ensembles connexes," *Fundamenta Mathematicae* 2 (1921), 206–255 (the example is on p. 241) and John Kline, "A Theorem Concerning Connected Point Sets," *Fundamenta Mathematicae* 3 (1922), 238–239 (the quotation is on p. 238).

107. Wacław Sierpiński, "Sur une condition pour qu'un continu soit une courbe jordanienne," *Fundamenta Mathematicae* 1 (1920), 44–60 and R. L. Moore, "Concerning Connectedness im kleinen and a Related Property," *Fundamenta Mathematicae* 3 (1922), 232–237 on p. 232. The quotations in this and the next paragraph are also on this page.

termed connectedness *im kleinen*.[108] In a move that had already become characteristic of Moore's mathematical style, he rethought the conditions underlying Sierpiński's result and asked if they were optimal for getting the most broadly applicable results. His answer? No.

Moore defined a point set M "to have the property S if and only if, for every positive number ϵ, M is the sum [that is, the union] of a finite number of connected point-sets M_{1n}, M_{2n}, ..., $M_{(n-1)n}$ each of diameter less than ϵ." As he was quick to note, "[t]he only difference between property S and Sierpiński's is that Sierpiński stipulates that the connected sets" M_{in} must be continuous curves, hence closed, while Moore's property S does not assume this. It was vintage Moore, and by making this adjustment, he was able to demonstrate numerous ways in which his property S was superior to that of connectivity *im kleinen*, thereby improving on Sierpiński's result.

But just as the American point-set topologists were reading the work of the Poles, so the Poles were responding to research of the Americans. For instance, in 1923, the year after the Americans' publication debut in *Fundamenta*, Kuratowski was citing examples there that Moore had published in both the *Bulletin* and the *Transactions* of the AMS as well as results of Veblen and Wiener.[109] Four years later in 1927, he and Knaster took their results directly to the AMS's *Bulletin*, extending work that R. L. Moore had published in that same outlet a year earlier. Moore had shown that there exist point sets connected and connected *im kleinen* that contain no arc, and they had succeeded in proving that there exists such a point set that contains no perfect subset, that is, that contains no subset A such that A equals the set of all its limit points.[110] By 1929, Kuratowski's eye had also fallen on the work of one of Moore's students, Gordon Whyburn, and he was building— in a paper explicitly entitled "Quelques applications d'éléments cycliques de M. Whyburn"—on a concept that Whyburn had defined.[111]

108. A point set M is connected *im kleinen* (or is regular) if, for every point p and every $\epsilon > 0$, there exists a $\delta > 0$ such that if x is any point in M at a distance less than δ from p, then both x and p lie in some connected subset of M of diameter less than ϵ.

109. Casimir Kuratowski, "Sur la méthode d'inversion," *Fundamenta Mathematicae* 4 (1923), 151–163 on pp. 151 and 161.

110. R. L. Moore, "A Connected and Regular Point Set Which Contains No Arc," *BAMS* 32 (1926), 331–332 and Bronisław Knaster and Casimir Kuratowski, "A Connected and Connected im kleinen Point Set Which Contains No Perfect Subset," *BAMS* 33 (1927), 106–109.

111. Gordon Whyburn, "Cyclically Connected Continuous Curves," *PNAS* 13 (1927), 31–38 and Casimir Kuratowski, "Quelques applications d'éléments cycliques de M. Whyburn,"

Given a peanian space, that is, a metric space that is the continuous image of a closed interval, a cyclic element is either a point p that cuts the space or, if not, it is the set of all x such that no point cuts the space between x and p. As Kuratowski explained, "[t]he utility of this notion stems from the fact that a large number of properties of the (peanian) space depend on properties analogous to cyclic elements." In his paper, he demonstrated one of them, namely, that the dimension of the space is, with one exception, the maximum of the dimensions of the cyclic elements.

Nor were these mathematicians merely interacting with one another in print. Another of Moore's students, William Ayres, was awarded an NRC/IEB postdoctoral fellowship for the 1928–1929 academic year, and he decided to take it to Vienna to work principally with Hans Hahn and his student, Karl Menger. While in Vienna, though, he was in contact with Kuratowski, who had enlisted his help from Lwów in proofreading articles in English to be published in *Fundamenta*. As he acknowledged in a letter to his mathematical "brother" Whyburn back in Austin, "[t]he first paper I got was one of yours for volume 13," namely, the fourth of what would be many more of Whyburn's *Fundamenta* papers. The real purpose of Ayres's letter, however, was to offer Whyburn counsel on his possibilities for a postdoctoral year abroad.

Whyburn was debating whether or not to apply for a Guggenheim, and Ayres advocated actually applying for both that and an NRC fellowship, just in case the Guggenheim should fall through. As Ayres saw it, "if you could spend part of the year here and part in Warsaw that ought to satisfy you fairly well."[112] In fact, Whyburn got the Guggenheim and took it, as Ayres had suggested, to both Vienna and Warsaw, but Whyburn also met and corresponded with Kuratowski who was, in Ayres's words, "worth going a long way to meet and talk to."[113] In fact, Kuratowski and Whyburn actually collaborated, while the latter was in Lwów, on a paper that drew from the prior work of both men on cyclic elements.[114]

By the end of the 1920s, it is fair to say that the Polish and American schools of point-set topology were fully entwined. Their respective members had published side by side in *Fundamenta*—with contributions in English

Fundamenta Mathematicae 14 (1929), 138–144. For the quotation and the definitions that follow in the next paragraph, see p. 138 (note 3) (my translation).

112. Ayres to Whyburn, 28 December, 1928, Box 1, Folder 9, Whyburn Papers.

113. Ayres to Whyburn, 16 April, 1929, Box 1, Folder 9, Whyburn Papers.

114. Casimir Kuratowski and Gordon Whyburn, "Sur les éléments cycliques et leurs applications," *Fundamenta Mathematicae* 16 (1930), 305–331.

comprising as much as a quarter of the journal's articles by the decade's close—as well as in the AMS's *Bulletin* in the United States. Their respective mathematical results were informed by and in response to work done in both the New and the Old Worlds. They were engaged in a lively mathematical dialogue and a friendly competition that resulted in what has been called the "golden age"—the 1920s to the 1960s—of point-set topology.[115] American mathematical research was thus highly visible on at least one part of the international mathematical stage. In fact, it was increasingly visible throughout Europe as Americans also published in the venerable *Mathematische Annalen*, among other journals.

———

The American mathematical research community had striven to raise its visibility internationally over the course of the 1920s. It had participated actively both in the scientific programs of the International Congresses of Mathematicians and in the geopolitics that had marred them. It had sent senior emissaries abroad to represent it physically and, in so doing, to emphasize its presence on the mathematical map. It had encouraged many of the most gifted of its youngest generation to work side by side with their European counterparts, and it had welcomed those counterparts—both junior and already distinguished—to share in its expertise and to join in its activities. It had witnessed the active incorporation of its results into, and the direct stimulation of its ideas on, the international body of mathematics. As the decade closed, its members had the strong sense of being part of what Richardson unabashedly characterized as "the fast-growing School of Mathematics in America," a national community competing increasingly successfully in an international arena.[116] No less a mathematician than Hardy apparently concurred.

---

115. For the quotation, see Teun Koetsier and Jan van Mill, " 'By their fruits ye shall know them': Some Remarks on the Introduction of General Topology and Other Areas of Mathematics," in *History of Topology*, ed. Ioan James, pp. 199–239 on p. 199. For an interesting case study of the interconnections between this American school of topology and its Polish counterpart, consult Thomas Bartlow and David Zitarelli, "Who Was Miss Mullikin?" *AMM* 116 (2009), 99–114, which details the work of R. L. Moore's third student, the 1922 University of Pennsylvania Ph.D. Anna Mullikin, and traces its impact in the literature.

116. Richardson to Frederick P. Keppel, President of the Carnegie Corporation, 27 January, 1928, Box 11, Folder: Richardson, R.G.D. 1928, Veblen Papers.

PART II

1929–1941: "A generation ago we were in need of direct stimulation . . . now we could well interchange."

—Griffith Evans, 16 January, 1934

4

Sustaining the Momentum?

The strides made in the 1920s had been impressive. American mathematicians had produced new mathematical research in a wide range of areas from academic homes across the country. They had reinforced a professional community that strongly supported their needs and interests both as researchers and as teachers. They had begun seriously to establish themselves in the international mathematical arena. By the end of the decade, they had the strong sense of being "on the verge of important steps forward."[1]

And then the stock market crashed. The worldwide economic depression that followed the October 1929 financial debacle in the United States lasted well into the 1930s and affected all aspects of American life. There was much uncertainty, and, like the population at large, the nation's mathematicians questioned prospects for the future. Would America's colleges, but especially its universities, maintain what had appeared to be their growing appreciation of original research—in addition to teaching—as an essential role of a modern faculty, in general, and of a modern mathematics faculty, in particular? Would the philanthropies continue to underwrite the mathematical research endeavor, thereby helping to assure the further honing of what was perceived as an increasingly competitive edge? Would American mathematicians be able to deepen the professional and personal inroads they had made on the international mathematical scene? Would, in other words, it be possible, in such straitened times, to sustain into the 1930s the momentum that the American mathematical community had managed to build in the 1920s? Only time would tell.

1. Roland Richardson to "Professors G. A. Bliss, L. P. Eisenhart, E. V. Huntington, Dunham Jackson, and R. L. Moore," undated (but apparently 1926), Box 4RM75, Folder: Richardson, Roland George Dwight (1922–1947), Moore Papers.

Trying to Do Mathematical Research in the Early 1930s

After the financial dust began to settle into the opening years of the 1930s, mathematicians came fairly quickly to recognize the challenges of trying to carry on the "business" of mathematics "as usual." Some had it better than others. Mathematicians at Chicago, for example, moved into brand-new, state-of-the-art Eckhart Hall in 1930, followed a year later by their *confrères* in Princeton's new Fine Hall. These capital improvements had both been made possible by money that had been raised and construction that had begun well before the stock market crashed (recall chapter two). Still, by May 1932, Chicago's Bliss was ruefully noting to Princeton's Veblen that "[a]ll of the departments [here] seem to be struggling with their budgets."[2] After all, the endowments at private schools had not been immune to the effects of the crash, nor had donors' investment portfolios.

Mathematicians at other schools were ostensibly less fortunate. Topologist Raymond Wilder had left Ohio State in 1926 for what he hoped would be the greener pastures of an assistant professorship at Michigan. Three years later, things were looking good; he had been promoted to associate professor. In the years following the stock market crash, however, the state's budget became ever more precarious. By April 1933, Wilder was writing despairingly to Gordon Whyburn, then on the Hopkins faculty, about his department's immediate financial outlook. After inquiring whether privately endowed Hopkins was undergoing cuts in either faculty numbers or in salary, Wilder lamented that, at Michigan, "[w]e've been cut both ways, and it looks like we are due for a beautiful cut next year. Beginning with our April salary, we are going to get only half pay for the rest of the year (this means for four of our ten months' salary), the rest to be paid maybe, when, and if the state is willing."[3] But, the future actually looked even bleaker. "The proposal in the legislature, now taking up the appropriation for next year," Wilder recounted, "is to cut us 50%. [I]n addition there is a joker in the bill making our allowances payable only on prop[o]rtion as the taxes are paid. If only 50% of the taxes are taken in that would mean a 75% cut." "Well," he concluded, "enough of such mournful stuff."

2. Bliss to Veblen, 24 May, 1932, Box 3, Folder: G. A. Bliss 1923–40, Veblen Papers.

3. Wilder to Whyburn, 18 April, 1933, Box 86-36/14, Folder 5: General Correspondence Whyburn, G. T., Wilder Papers (also quoted in Parshall, " 'A New Era,' " p. 285). The two quotations that follow are also from this letter.

The situation at Michigan *was* bad. From 1930 to 1934, the budget of its Department of Mathematics was cut by just over 25%, although Wilder's salary over the same period decreased by 12.5%.[4] State schools, in general, were hard hit as legislatures were forced to make difficult choices. It was reported that "[m]ost college and university faculties took salary cuts of 10 to 15 percent around 1932 or 1933."[5] Wilder's cut fell in middle of that range.

As salaries were being slashed, positions, as at Michigan, were also going unfilled. Trained mathematicians were out of work. In 1932, Roland Richardson estimated that just over 10%—some 200 of 1883—of the AMS membership was unemployed.[6] By the winter of 1934, the MAA had appointed a commission to study the unemployment situation specifically for mathematics Ph.D.s, and it had circulated a questionnaire to fifty "leading universities" to try to get a handle on the number of recent Ph.D.s who were actively seeking positions for the 1934–1935 academic year.[7] A subsequent questionnaire circulated in October 1934 determined that of the 180 Ph.D.s on the job market, 112 or two-thirds of them "had obtained positions of a nature more or less satisfactory to them," while the other third had not. Jobs deemed "satisfactory" included those in the government and/or business for which "mathematical training was a direct asset" as well as in teaching, but of the eighty-eight employed in the latter, only twenty-one (or slightly more than

4. See Mathematics Department, Total Budget from 30 June, 1923, to 30 June, 1934, and Budget Sheet (listing individual salaries for the 1929–1930 through the 1932–1933 academic years), Box 1, Folder: Budget 1934–1935, University of Michigan, Department of Mathematics Records: 1913–1981.

5. Ivan Niven provides a firsthand account of this state of affairs relative specifically to mathematics in "The Threadbare Thirties," especially pp. 211–212. For the quotation, see p. 211.

6. Tomlinson Fort gives the membership figure in "The Annual Meeting in New Orleans," *BAMS* 38 (1932), 145–154 on p. 148. Richardson's estimate is given in Nathan Reingold, "Refugee Mathematicians in the United States of America, 1933–1941: Reception and Reaction," *Annals of Science* 38 (1981), 313–338; reprinted in *A Century of Mathematics in America*, ed. Peter Duren, et al., 1: 175–200 on p. 179.

7. Elton Moulton, "The Unemployment for Ph.D.'s in Mathematics," *AMM* 42 (1935), 143–144 on p. 143. The quotations that follow in the next two sentences are also on this page. The Depression had similar effects in physics. See Weart, pp. 309–310. Data on shrinking faculty sizes in the various regions of the country between 1930 and 1933 may be found in "Academic Unemployment," *Bulletin of the American Association of University Professors (1915–1955)* 19 (October 1933), 354–355. In all, faculty sizes shrank, on average, by some 7.5%.

10% of those on the market) had landed a university job, fifty-three (or 30%) had found some sort of college post, and fourteen (or just under 8%) had taken positions in either a normal school, a junior college, or a secondary school. A year later in the summer of 1935, Wilder had the (exaggerated) impression that there were "over a hundred applicants for any position that opens up."[8] In his view, it was "like a lottery operation." Regardless of the actual numbers, the situation seemed bad to those on the ground, even if it certainly could have been worse.

Nor were the mathematical societies spared fiscal hardship. Although the MAA supported the *Monthly* almost completely through membership dues and had the philanthropic support of Mary Hegeler Carus for its monograph series, it was feeling the pinch by 1932.[9] When he handed over the *Monthly's* editorial reins in January of that year, the University of Minnesota's William Bussey noted ruefully that "[t]he *depression* seems to have hit the advertising business."[10] One company, the publisher Ginn & Co., had cut the number of paid advertisements that it ran in the *Monthly* in half; another, McGraw-Hill, had taken itself off the list of "*regular* advertisers," although it had not precluded doing some advertising on an ad hoc basis.[11] The absence of ads was causing the insertion of "more than the usual amount of '*filler material*,'" and this situation had grown even worse just two months later as the April issue was in preparation.[12] By September, the decision had been made to decrease advertising charges somewhat in an effort to assure continuing ad dollars.[13] Chicago's Herbert Slaught, the *Monthly's* managing editor, stoically took a long view of the situation. "[W]e are not going to worry too much about the

8. Wilder to Gordon Whyburn, 14 August, 1935, Box 2, Folder 96: Wilder, Raymond L., Whyburn Papers for this and the next quotation.

9. In the early 1930s, however, the Carus family was in arrears relative to its financial obligations to the MAA, a situation that had largely resolved itself by 1933 (William Cairns to Raymond Archibald, 4 July, 1933, Box 86-14/65, Folder 5, MAA Records). Even so, only two Carus monographs were printed in the lean 1930s: John Young's *Projective Geometry* and David Smith and Jekuthiel Ginsburg's *A History of Mathematics in America before 1900* (see the bibliography for the full references). The series also limped through the war-torn 1940s, although a number of the earlier volumes were reprinted. Compare chapter eight.

10. Bussey to Herbert Slaught, 27 January, 1932, Box 86-14/66, Folder 5, MAA Records (his emphasis).

11. Bussey to Slaught, 18 March, 1932, Box 86-14/66, Folder 5, MAA Records (his emphasis).

12. Bussey to Slaught, 27 January, 1932 (his emphasis).

13. Bussey to Slaught, 2 September, 1932, Box 86-14/66, Folder 5, MAA Records.

effects of the depression on the advertising business," he wrote to Bussey, "but rather we will forge ahead as much as possible and wait for better times to come."[14]

As for the AMS, after successfully putting its financial house in order in the 1920s, a "crisis . . . had arisen" by 1934 that proved difficult to address.[15] Not surprisingly, the problem centered on publications. As noted in chapter two, the American mathematical community, in general, and the AMS, in particular, had significantly increased its publication capacity over the course of the 1920s in direct response to the growing number of Ph.D.s and the requirement that their dissertation research be publishable. Journal capacity thus represented a key component of the community's growth and development. The AMS had also gone into the business of monographic publication with the indirect financial help of various of the Rockefeller philanthropies, in particular General Education Board funds funneled to the National Academy of Sciences. From the beginning, though, that money was understood to be temporary. It was thus no great surprise when the GEB let it be known that its funding would come to an end in 1934. The timing, in the depths of the Depression, could not have been worse. Recognizing that, the Rockefeller Foundation agreed to extend the arrangements until 1936 in order to give the AMS the opportunity "to stabilize its finances."[16]

AMS President in 1933 and 1934, the algebraic geometer (at that point at Illinois) Arthur Coble got immediately to work, forming in October 1933 a committee on financial policy chaired by Harvard differential geometer William Graustein.[17] In an open letter sent out some seven months later to the

14. Slaught to Bussey, 4 August, 1932, Box 86-14/66, Folder 5, MAA Records. Banta Company, the printer of the *Monthly* as well as of the AMS's *Transactions* and *Bulletin*, had, after an informal meeting with representatives of both societies, agreed to reduce its charges for all three journals. Excluding postage, charges for the *Monthly* and *Transactions* were lowered by 10% and for the *Bulletin* by 6% (Earle Hedrick to the Editors and Officers of the American Mathematical Society and the Mathematical Association of America, 18 July, 1932, Box 86-14/68, Folder 5, MAA Records). By Slaught's calculation, the savings to the *Monthly* would essentially offset the lost advertising revenues.

15. Mark Ingraham to Members of the American Mathematical Society, 15 June, 1935, Box 8, Folder 5: AMS–Miscellaneous Correspondence RE the AMS (1935–1959), Whyburn Papers.

16. Archibald, *Semicentennial History*, p. 34.

17. That committee's work was based on detailed statistical studies of the American mathematical publication scene that included such data as the number of pages of research published in the various American journals by institution. The top ten institutions in terms of number

AMS membership, Coble laid out the problem and sketched a course of action to address it. He called for two things: the establishment of contributing individual memberships—the latter at a rate of $15.00 per year or higher, "the amount being determined by the members themselves"—and a 33% increase in the price of the AMS's *Transactions* from $9.00 to $12.00 per year with no concomitant increase in the number of published pages.[18] More importantly, he outlined a strategy by which the nation's colleges and universities would be asked "to become institutional [AMS] members with dues varying in proportion to the amount of publication space which has been used by these institutions in recent years."[19]

The AMS leadership was making a bold move, but one that had already been made by the American physics community.[20] It was officially acknowledging and justly seeking to capitalize on the new reality that had evolved in academe especially since the end of World War I. As Coble put it, "[t]he universities which are constantly encouraging a larger and larger output of research and are thereby gaining more and more prestige might well be expected to contribute" to the costs of publication.[21] "It is recognized that to the universities goes the chief credit for having stimulated research in America," he continued. "This is certainly a great achievement, but it has brought problems which can best be solved through the cooperation of these same institutions." It was proposed that that institutional "cooperation" be determined based on pages published in the AMS's *Bulletin* and *Transactions* as well as in the *American Journal* and the *Annals* according to the following formula: "$2.75 a page for the average number of pages published from that institution in the four above-mentioned journals during the seven years 1927–1933,"

---

of pages published in the *American Journal, Annals of Mathematics, Bulletin, Transactions,* and the Colloquium publications over the seven-year period from 1926 through 1932 were: Princeton, Harvard, Yale, the University of Texas, the University of Illinois, the University of Chicago, Columbia, Caltech, Rice, and the Johns Hopkins. See Roland Richardson, "Report of Activities of the American Mathematical Society during the Years 1925–1932 and Statistics of Publication by Universities, 1926–1932," both in Minutes of the Board of Trustees, 1923–1960, AMS Records, Ms.75.1, AMS Papers.

18. Ingraham to "Members of the American Mathematical Society," 15 June, 1935. See also Mark Ingraham, "The Trip of Associate Secretary M. H. Ingraham in the Interests of the Society," *BAMS* 40 (1934), 641–643. Recall from chapter two that the AMS had already created the category of contributing *institutional* membership in the 1920s.

19. Ingraham to "Members of the American Mathematical Society," 15 June, 1935.

20. Archibald, *Semicentennial History,* p. 35.

21. Ibid. The quotations in the next two sentences may also be found here.

where "[t]he charge of $2.75 per page represents, roughly, one third of the cost of printing in these periodicals."[22] If institutions of higher education were going to place increasing emphasis on research in hiring, in professional advancement, and in the establishment of comparative rankings, then they should bear at least some fair share of the financial burden that the publication of that research entailed. Theirs should become a more symbiotic relationship with the professional societies that worked to set and maintain the publication standards on which they relied.

In an effort to get the colleges and universities on board, the AMS, with Rockefeller Foundation support, sent its Associate Secretary, the University of Wisconsin algebraist Mark Ingraham, on a yearlong, nationwide tour beginning in the fall of 1934. His goal was to convince as many schools as possible of the justness of this plan. By June 1935, he was able to report that he had visited eighty-five campuses, had asked seventy-seven schools to consider institutional membership, and thus far had gotten the green light from fifty of them. Moreover, he had talked personally with about half (!) of the AMS's members, describing the contributing membership scheme. Some ninety-nine had subsequently signed up for dues varying between the $15.00 minimum and $50.00. Also gratifying was the fact that the subscription rate increase for the *Transactions* had not resulted in a decrease in the overall number of subscribers. The drive was not yet over, but $7,630 of the $8,500 goal had already been raised by June 1935 thanks to Ingraham's efforts.

Even during the Depression, then, the AMS was managing successfully to make its financial case both to its members and to the broader academic community. The philanthropies were no longer able to shore up its publications, but it had been able creatively to bridge the funding gap. A minority of the AMS membership, however, had advocated a solution radically different from the direct solicitation of ever more funds. Theirs harkened back to the controversy that had surrounded the initial delineation of the MAA from the AMS back in 1915.

Earle Hedrick—who had been an editor of the *Monthly* (from 1913 to 1915), the first President of the MAA (in the calendar year of 1916), the

---

22. Ingraham to "Members of the American Mathematical Society," 15 June, 1935. The account in the next paragraph also draws from this letter. Ingraham gave an even more extensive account of his activities in his report to the trustees of the AMS dated 27 December, 1935. See Minutes of the Board of Trustees, 1923–1960, AMS Records, Ms.75.1, AMS Papers.

twentieth President of the AMS (over the calendar years of 1929 and 1930), and the editor-in-chief (since 1921) of the AMS's *Bulletin*—was remarkably well placed in both societies to make an informed judgment on their then-current financial woes. His solution was radical. As he saw it, "[a] still greater degree of economy might be got by a thoroughgoing junction of the *Society* and the *Association* . . . . It seems absurd to many to have the double over-head, and the double publication."[23] Still, he was a realist, recognizing that "[a]t the beginning the Association authorities would have favored a combi-nation. Now, I do not feel sure that this would be the case, but it would be possible to investigate it." He actually floated his idea with MAA stalwart, Her-bert Slaught, but it gained no traction. Even though the issue of MAA-AMS relations seemed to raise its head in times of financial trouble,[24] by the 1930s, it was clear to most that both societies were there to stay and that publication capacity would simply have to increase if the members of the American math-ematical community were actually going to be able to sustain into the 1930s the momentum they had built over the course of the preceding decade.

In fact, the financial situation did not seem adversely to affect employed mathematicians' ability to pursue their research. Although belts may have been tightened at those universities like Chicago, Harvard, and Hopkins, where the research ethos had been in place the longest, mathematicians there continued to maintain their research productivity. Princeton, which under Eisenhart, Veblen, and Lefschetz had vaulted into the top three in the 1920s, also continued strong into the 1930s. In fact, its position was further cemented after 1933 with the opening of the Institute for Advanced Study also in Princeton (see this chapter's penultimate section). Among other universi-ties that had begun to appear conspicuously on the research horizon by the end of the 1920s, Penn and MIT in the East, Ohio State, Wisconsin, and Illi-nois in the Midwest, Stanford and Caltech in the West, and the Rice Institute and the University of Texas, Austin in the South all continued to produce new mathematical research as well as fresh Ph.D.s. The research ethos had become firmly embedded in American mathematical and academic culture. (Chap-ter six gives a technical overview of the new work generated by the American community in the 1930s.)

23. Hedrick to William Longley (of Yale), 13 November, 1933, Box 86-14/68, Folder 5, MAA Records. The next quotation is also from this letter.

24. Recall chapter two and the discussions that took place in 1924 surrounding Veblen's fund-raising initiatives for the AMS.

FIGURE 4.1. Earle Raymond Hedrick (1876–1943). (Photo courtesy of Mathematical Association of America Records, 1916-present, e_math_00970, Archives of American Mathematics, Dolph Briscoe Center for American History, University of Texas at Austin.)

In the East, the mathematics programs at Penn and MIT represented interesting cases. Both lived in the shadow of strong departments, Princeton being just forty-five miles from Philadelphia and Cambridge serving as home to Harvard as well as MIT. At Penn, where the faculty was viewed as burdened "by a heavy schedule of teaching" and working in "a not particularly inspiring environment," topologist John Kline, together with his colleagues Howard Mitchell in group theory and James Shohat in analysis, steadily generated an average of two Ph.D.s a year—by contrast Princeton produced double that number—throughout the 1930s with 1933 seeing a graduating class of five.[25] Among the 1933 crop, Kline's student, William Claytor, was the third African-American to earn a doctoral degree in mathematics in the United States (see fig. 6.2). He had been preceded by Elbert Cox, a 1925 Cornell Ph.D., and Dudley Woodard, another Kline student who finished at Penn three years later in 1928.[26] Both Cox and Woodard took positions at Howard University in Washington, D.C., arguably the best of the historically black institutions of higher education in the United States but a school which, in the 1930s, only supported a Master's program in mathematics. Claytor, who had been a product of that education, nevertheless aspired to a career as a research mathematician. The realities of the 1930s—the Depression combined with societal attitudes and the fact that the research universities were an almost exclusively white-male preserve—made his an uphill battle (see chapter six).

MIT's Mathematics Department was larger than that at Penn, with Philip Franklin working in the broad view of analysis situs that he had learned from his adviser, Veblen; Dirk Struik in differential geometry as well as tensor and vector calculus; Norbert Wiener in applied areas as well as in a panoply of mostly analytic subfields but primarily potential theory; George Rutledge in group theory; and Frederick Woods in geometry. Together they produced some thirty new Ph.D.s in the 1930s, a decade that, in Wiener's view, witnessed "MIT's new impetus of life" under the enlightened leadership of the school's thirteenth President, physicist Karl Compton.[27] Still, Wiener, always

25. Interestingly, it was Veblen at neighboring Princeton who characterized Penn in such negative terms in Veblen to Henry Moe, 7 May, 1925, Box 5, Folder: Guggenheim Foundation 1924–30, Veblen Papers. Veblen had written in strong support of Kline's successful Guggenheim application. Recall the previous chapter. Information here and below on the Ph.D.s produced in mathematics has been culled from issues of the *Bulletin of the American Mathematical Society*, which annually recorded the names, schools, and dissertation titles of Ph.D. recipients. The Mathematics Genealogy Project's website http://www.genealogy.ams.org also contains much valuable information. Compare, too, Richardson, "The Ph.D. Degree and Mathematical Research," p. 366 (in the reprint edition), which gives composite data through 1934.

26. On Cox, see James Donaldson and Richard Fleming, "Elbert F. Cox: An Early Pioneer," *AMM* 107 (2000), 105–128. On Claytor, see Parshall, "Mathematics and the Politics of Race." Woodard is discussed in both articles.

27. Compton assumed the MIT presidency in 1930 and held it until 1948. See Wiener, *I Am a Mathematician*, p. 141 for the quotation. Dirk Struik discusses the situation at MIT in "The MIT Department of Mathematics During Its First Seventy-Five Years: Some Recollections," in *A Century of Mathematics in America*, ed. Peter Duren et al., 3: 163–177.

touchy, felt particularly undervalued and, indeed, persecuted by the mathematicians at nearby Harvard and especially by Birkhoff. From his perspective, not only did Birkhoff exert "continued hostile pressure," but he also actively "saw to it that many academic offers which otherwise might have opened up to [Wiener] were diverted elsewhere."[28] It was obviously hard for at least some at MIT to live and work in such close proximity to a department that not only boasted a number of the country's best mathematicians but also produced some two-and-a-half times more Ph.D.s in mathematics in the ten years from 1930 through 1939.

The situation in the Midwest was different. Northwestern was near the nation's third mathematical powerhouse, Chicago, but it only began to mount a serious graduate program at the very end of the decade (see chapter seven). The midwestern dynamic may be better characterized by on-going efforts at a number of state-supported and land-grant schools.

At Ohio State, for example, Harry Kuhn, a 1901 Cornell Ph.D. under George Miller in the theory of groups, had joined the Mathematics Department in 1901 and became its chair in 1926. One of his first administrative acts was the institution of a full-fledged graduate program.[29] Ohio's land-grant institution, Ohio State had been founded in 1870 with the purely pragmatic objectives of providing training in agriculture and the so-called "mechanical arts" of engineering, but had fairly quickly embraced a more comprehensive and classical sense of its mission. Graduate training and research, while engaged in by some, had nevertheless not been a strong component of academic life in Columbus.

Relative to mathematics, Kuhn sought to change that. Already in 1924, he had successfully argued that Ohio State hire Raymond Wilder in topology and Cyrus MacDuffee in algebra.[30] Although Wilder left for Michigan

28. Wiener, *I Am a Mathematician*, p. 116.

29. See https://math.osu.edu/about-us/history. Ohio State had technically conferred its first mathematics Ph.D. in 1909 in the absence of any real graduate program, with Kuhn supervising the doctoral work of Grace M. Bareis on imprimitive substitution groups of degree sixteen.

30. The state-supported University of Cincinnati also began to move in a more research-oriented direction in the 1920s. In 1924, Harris Hancock, senior member of Cincinnati's mathematics faculty, wrote to Veblen to tell him that "[t]he administration wants a fine univ[ersity] here" and to ask if he could recommend any recent Princeton Ph.D.s for the two new positions that were being created (Hancock to Veblen, 30 November, 1924, Box 6, Folder: Hancock, Harris 1924–29, Veblen Papers).

in 1926, MacDuffee remained at Ohio State until 1935 (when he moved to the University of Wisconsin). He had been joined in 1930 by the Hungarian analyst, Tibor Radó, who had studied at the University of Szeged under Alfréd Haar and Frigyes Riesz, had won the fellowship from the International Education Board that had allowed him to further his studies in Munich and Leipzig, and had spent the 1929–1930 academic year in the United States as a visiting lecturer at both Harvard and the Rice Institute.[31] The Hungarian's appointment at Ohio State served to reinforce the research mission Kuhn promoted, especially after Radó's solution, independent of Jesse Douglas's, of Plateau's problem.[32]

Ohio State's local goal of moving into the research ranks soon also became well-known within the broader mathematical community.[33] Over the course of the 1930s, MacDuffee and Radó produced half of the roughly thirty mathematicians who earned their doctoral degrees at Ohio State and went on to positions elsewhere. Their colleagues—function-theorist Henry Blumberg, and Frederic Bamforth and Lincoln La Paz, the latter both Bliss students in the calculus of variations—accounted for the rest of Ohio State's graduate training in mathematics.

The Universities of Wisconsin and Illinois had larger programs than Ohio State in terms of both the number of faculty members actively training graduate students and Ph.D. output. Each of these programs, however, had begun to move into research in the opening decade of the century, significantly earlier than Ohio State. As a result, both already appeared among the thirteen mathematics programs judged to be the country's best in Raymond Hughes's 1924 survey (recall chapter one). Illinois ranked fourth; Wisconsin eighth.[34]

31. See Jaques Cattell, ed., *American Men of Science*, 7th ed. and Siegmund-Schultze, *Rockefeller and the Internationalization of Mathematics*, p. 297 as well as Erwin Kreyszig, "Remarks on the Mathematical Work of Tibor Radó," in *The Problem of Plateau: A Tribute to Jesse Douglas & Tibor Radó*, ed. Themistocles Rassias, pp. 13–32 on pp. 19–20.

32. Recall the previous chapter and see Tibor Radó, "The Problem of Least Area and the Problem of Plateau," *Mathematische Zeitschrift* 32 (1930), 763–796 and *On the Problem of Plateau*, Ergebnisse der Mathematik und ihrer Grenzgebiete, vol. 2, no. 2 (Berlin: Springer-Verlag, 1933).

33. See, for example, Raymond Wilder to Oswald Veblen, 9 December, 1935, Box 17, Folder: Wilder, Raymond L. 1935–47, Veblen Papers. Wilder and Veblen were trying to secure positions at Ohio State for Leo Zippin and Karl Menger. As Wilder put it to Veblen, "I know Kuhn wants to build up a strong department, but sometimes he goes about the matter in a peculiar way."

34. Penn came in at twelfth, while MIT did not appear on the list.

At Wisconsin, Klein's student, Edward Van Vleck, helped to build a graduate program until his retirement in 1929, while at Illinois, Edgar Townsend, who had earned his doctorate under Hilbert at Göttingen, embraced the same research-supportive agenda, hiring Miller away from Cornell in 1906. In the 1930s, each of these programs produced over fifty new members of the American Ph.D.-holding ranks. Wisconsin's graduates tended to focus in analysis and algebra thanks to the presence there of Rudolph Langer and Mark Ingraham, respectively, while Illinois's pursued original research in analysis, geometry, and algebra, owing to the efforts of analyst and Birkhoff student Robert Carmichael, geometer Arthur Coble, and, after Miller's retirement in 1931, algebraists Henry Brahana and Olive Hazlett.[35]

Continuing the sweep across the country all the way to California on the West Coast, Stanford and Caltech—neither of which was rated in Hughes's survey but both of which were already on Courant's radar screen in 1927—continued into the 1930s to build on the gains they had made in the 1920s. Both were privately endowed. Stanford had been founded in Palo Alto in 1891 with money from the coffers of robber baron Leland Stanford, while Throop University had opened in Pasadena as a vocational school in 1891, becoming the California Institute of Technology in 1921 under physicist Robert Millikan. Both institutions viewed themselves as rivals of the state's land-grant university in Berkeley, which had come in at eleventh, just ahead of Penn, in Hughes's poll.

Stanford had begun to make its move in the 1920s, particularly after the group and number theorist Hans Blichfeldt became department head in 1927.[36] Although he initiated a series of summer lectureships in mathematics that attracted notable American and European mathematicians to the West Coast, the department and graduate program remained small in the 1930s: Blichfeldt himself generated no new doctorates; William Manning produced at least two in group theory; and the Russian-born James Uspensky oversaw the number-theoretic work of some four more. Mathematical prospects at Stanford improved dramatically starting in 1938 when, on Blichfeldt's retirement, it hired the Hungarian émigré Gábor Szegő to succeed him (see the next chapter).

35. Edna Stanford provides more details in her 1940 Illinois Master's thesis in mathematics, entitled "The History of the Department of Mathematics at the University of Illinois."

36. On the program at Stanford, see Halsey Royden, "A History of Mathematics at Stanford," in A Century of Mathematics in America, ed. Peter Duren et al., 2: 237–277.

The Caltech program, while also small, was more robust thanks particularly to the efforts of number theorist E. T. Bell, and, after his move from Ohio State, Aristotle Michal.[37] Between them, these two pure mathematicians trained more than a dozen graduate students in the 1930s, but their colleague Theodore van Kármán, in Caltech's Guggenheim Aeronautical Laboratory, single-handedly almost doubled their output in his applied areas of mathematical focus.[38]

Although the West had been represented in Hughes's survey by one school, Berkeley, the South had failed to appear in it at all. Indeed, it had been a matter of some concern—at least as early as the 1920s—that mathematicians in the South be more fully brought into the mathematical fold.[39] In 1924 and as part of his AMS fund and membership drives, for instance, Veblen had acknowledged to Richardson that representation in the South was perhaps not what it could or should have been. He proposed that the University of North Carolina's Archibald Henderson be enlisted "in gathering in members," since Henderson had already volunteered to help get more Southerners to attend the AMS's 1924 annual meeting to be held—in some sense, just north of the South—in Washington, D.C.[40] Veblen even floated the idea of sending "out some sort of a circular to people in that part of the country when we send out other campaign material." The South apparently required

37. Bell's student Morgan Ward and applied mathematician Harry Bateman were also on the Caltech faculty as was Harry Van Buskirk, although they were not as actively a part of graduate training there in the 1930s.

38. For more on mathematics at Caltech, see Angus Taylor, "A Life in Mathematics Remembered," *AMM* 91 (1984), 605–618 on pp. 608–611; Judith Goodstein and Donald Babbitt, "E. T. Bell and Mathematics at Caltech between the Wars," *NAMS* 60 (2013), 686–698; and John Greenberg and Judith Goodstein, "Theodore van Kármán and Applied Mathematics in America," *Science* 222 (1984), 1300–1304; reprinted in *A Century of Mathematics in America*, ed. Peter Duren et al., 2:467–477. Judith Goodstein gives a history of the university as a whole in *Millikan's School: A History of The California Institute of Technology* (New York: W. W. Norton & Company, 1991).

39. This had also been a concern more broadly. The General Education Board had made grants to both the University of North Carolina and to the University of Virginia in the 1920s as "tentative experiments in aiding science in regions where academic high culture had not flourished for many decades." Kohler, p. 201.

40. Veblen to Richardson, 25 July, 1924, Box 10, Folder: Richardson, R.G.D. 1924, Veblen Papers. The quotation that follows is also from this letter. The MAA had already targeted collegiate mathematics teachers in the South, founding its Southeastern Section—for the states of Alabama, Florida, Georgia, North Carolina, South Carolina, and ultimately Tennessee—in 1922. A Louisiana-Mississippi Section followed in 1930.

special targeting, and to that end the AMS, the MAA, and the American Association for the Advancement of Science convened in Nashville, Tennessee, in December 1927 for the first truly southern meeting of the New York City–based AMS.[41] When the three societies returned to the region four years later in 1931, however, the turnout proved disappointing despite their best efforts and what should have been the draw of the New Orleans venue.[42]

Although those elsewhere may have had the general impression of mathematical inactivity in the South, at least two schools there—the University of Texas at Austin and Houston's Rice Institute—had already begun to make important strides in the 1920s. State-supported Texas had opened in Austin in 1883, but an initial "false start" and "a faculty of one" characterized its program in mathematics prior to the "new beginnings" marked in the 1920s by the hiring of topologist R. L. Moore.[43] Moore, although a native Texan, had been reluctant to leave his position at the University of Pennsylvania when he got the call to Texas in 1920. Penn, after all, was on an upward trajectory after World War I, and Moore had played a key role in that upswing. Moore's move back to Austin—he had been an undergraduate there before doing his graduate work at Chicago—at least did not deprive him of the opportunity to train graduate students. Over the course of the 1920s, four young men earned their Ph.D.s under his supervision, and two of those, his first Texas student, Wilder, and his third, Whyburn, went on to animate programs of their own at Michigan and Virginia, respectively (see the next section). Moore continued to put Texas on the mathematical map in the 1930s through his instruction at the graduate level, through the publication in 1932 of his 1929 Colloquium

41. Arnold Dresden, "The Thirty-fourth Annual Meeting of the American Mathematical Society," *BAMS* (1928), 129–154 on p. 130 and compare Archibald, *Semicentennial History*, pp. 83–84.

42. William Cairns, "The Sixteenth Annual Meeting of the Association," *AMM* 39 (1932), 123–134 on pp. 124–125. There had, indeed, been a certain amount of discussion about how best to organize a meeting in the South. For instance, Cairns, who was MAA Secretary-Treasurer at the time, joined Herbert Slaught in advocating Rice University's Griffith Evans as program chair, since Evans "has better ideas than any other one in the South as to the character of papers we should have at that time and we think he knows the Southern situation well enough to know what topics should be presented and how strong papers they can stand!" (William Cairns to E. T. Bell, 6 April, 1931, Box 86-14/66, Folder 1, MAA Records). Ultimately, not Evans but the University of Kentucky's Claiborne Latimer served as program chair.

43. Albert Lewis, "The Building of the University of Texas Mathematics Faculty, 1883–1938," in *A Century of Mathematics in America*, ed. Peter Duren et al., 3: 205–239.

Lectures on *The Foundations of Point Set Theory*, and through his presidency in 1937 and 1938 of the AMS.[44]

Some 150 miles away in Houston, mathematical strides were also being made at the Rice Institute, a private school that had opened in 1912. Under the leadership of its first President, mathematician Edgar Lovett, Rice supported a lecture series that brought the best scholars from the United States and abroad to experience directly Texas's new educational experiment.[45] During Rice's inaugural year, the distinguished measure theorist Émile Borel journeyed from Paris and his position at the École normale supérieure to deliver a series of lectures that was later published as a *Rice Institute Pamphlet*.[46] Borel was followed on the Rice dais in 1920 and 1925 by his Parisian colleague Jacques Hadamard, in 1924 by Charles de la Vallée-Poussin, then in North America following the 1924 Toronto ICM, and in 1929 by Hermann Weyl during his Princeton sojourn.[47] By 1932, Birkhoff, Veblen, and R. L. Moore had all also journeyed to Rice to deliver their thoughts on "A Mathematical Theory of Aesthetics and Its Application to Poetry and Music," "Certain Aspects of Modern Geometry," and "Fundamental Theorems Concerning Point Sets," respectively.[48] A regular stop, too, for the AMS's Visiting Lecturers, Rice had

44. Moore did not do this singlehandedly. Hyman Ettlinger had served as an instructor at Texas beginning in 1913 after earning his Harvard M.A. in 1911 and before taking his Ph.D. there under Birkhoff in 1920. He trained a number of students in his brand of analysis in the 1920s and 1930s. In number theory, Harry Vandiver, although he did not have even a high school diploma, was hired in 1925 and won the first Cole Prize in Number Theory six years later (recall chapter two). An idiosyncratic mathematician, Vandiver trained no Texas doctoral students until the 1940s and 1950s.

45. Lovett had earned his doctoral degree under Sophus Lie at the University of Leipzig in 1896. (Parshall and Rowe mistakenly gave this date as 1898 in *Emergence*, p. 237.) After holding a temporary post at Hopkins, he joined the Princeton faculty in 1897 and had moved up the ranks to professor by 1905. He left Princeton to assume the presidency of the Rice Institute in 1908.

46. See Émile Borel, "Molecular Theories and Mathematics," *Rice Institute Pamphlets–Rice University Studies* 1 (1915), 163–193 and Émile Borel, "Aggregates of Zero Measure" and "Monogenic Uniform Non-analytic Functions," *Rice Institute Pamphlets–Rice University Studies* 4 (1917), 1–21 and 22–52.

47. For the specific references to these and the following lecture series, see https://scholarship.rice.edu/handle/1911/8328

48. Yale's mathematical astronomer and the AMS's thirteenth President, Ernest Brown, preceded them by one year, lecturing on "Elements of the Theory of Resonances Illustrated by the Motion of a Pendulum" in 1931. See *Rice Institute Pamphlets–Rice University Studies* 19 (1932), 1–60.

become a sort of mathematical destination, but it had also—thanks to the presence on the faculty there (until 1934) of Harvard-trained analyst Griffith Evans—come to support a small but strong graduate program that produced some twenty Ph.D.s in the 1920s and 1930s.

As these snapshots evince, mathematical research and the training of graduate students continued largely unabated into the 1930s at schools that had chosen to focus on such endeavors in the 1920s or earlier. In a 1935 report to the Commission on the Training and Utilization of Advanced Students of Mathematics, in fact, Richardson documented a more than 50% increase in American Ph.D. production in mathematics from the five-year period 1925–1929 to that of 1930–1934, although he also rather ruefully noted that while "America seems in recent years to be adding to the *quantity* of personnel," it seems "not [to be] improving the average *quality* as judged by the number of papers published." [49] Moreover, he estimated that "[a]bout 60 (or 5%) of the doctors are responsible for half of the published pages of research" issuing from the American mathematical research community as a whole. That others in that community also seemed to sense this problem is reflected in the pointed comment that E. T. Bell made to Harry Vandiver in November 1933. "I don't blame you for getting away from the damned students," Bell wrote. "The more I see of them, the more I am convinced that trying to *train* people to do research is a waste of time. What few ideas a trainer has left after ten years of it are too precious to be thrown away. A man who is worth a damn will train himself." [50]

The early graduate ideal—at, for example, Hopkins in the 1880s and Chicago in the 1890s—had been the fostering of original research and the training of future researchers. Had, by the 1930s, the Ph.D. evolved into more of a credential and metric for comparing programs and less of a stepping stone to an actual career as an active researcher? Perhaps it was only natural that a stratification would take place, that, the shared credential of a Ph.D. aside, those with research in their blood would be separated from the rest of the cohort. [51] After all, the majority of mathematics Ph.D.s were employed at

49. Richardson, "The Ph.D. Degree and Mathematical Research," p. 363 (for Ph.D. production) and p. 374 (for the quotation; my emphasis). For the quotation that follows, see p. 373.

50. Bell to Vandiver, 1 November, 1933, Box 4RM155, Folder: Correspondence: Beckenbach, Beeger and Bell, 1929–1956, Vandiver Papers (Bell's emphasis).

51. A committee of the MAA actually proposed the creation of two Ph.D.s, one for those who would become active researchers and one for those who would mainly concentrate on

junior colleges, four-year colleges, and normal schools where teaching loads were heavy and where they were largely unrewarded for actively pursuing research. Regardless, select mathematicians and their supportive administrators joined forces in the 1930s, as the Depression dragged on, further to develop the American mathematical landscape at the research level.

The Targeted Building of Programs in the 1930s

The South, perceived in the 1920s as lagging behind in mathematical developments, finally began to make significant gains in the 1930s. For example, Thomas Jefferson's University of Virginia—which had been founded in 1819 as a new educational experiment in which students could study in one or more of the separate schools of ancient languages, chemistry, law, mathematics, medicine, modern languages, natural philosophy, and moral philosophy— had struggled in the decades between the end of the American Civil War and World War I.[52] When its first President, Edwin Alderman, died in 1931 (Jefferson had stipulated that the University be run not by a President but by a Rector and Board of Visitors), second President John Newcomb continued the modernization that Alderman had begun especially in the 1920s. Relative to mathematics, that took the form of hiring "a man who would assume leadership in reorganizing" the department, and represented "an important professorship" both for Virginia and for the South.[53]

In the early 1930s, the small mathematics group was dominated by two men. William Echols had been trained as a civil engineer in the early 1880s and had joined the faculty in 1891; James Page had earned a doctorate under Sophus Lie at Leipzig in 1887 and had come to Virginia in 1896. Although Page had tried to continue research in his special area of transformation

college teaching (Elton Moulton, "Report on the Training of Teachers of Mathematics," *AMM* 42 (1935), 263–277, especially on 267–270). David Roberts treats this notion of stratification in a mathematical context in "Albert Harry Wheeler (1873–1950): A Case Study in the Stratification of American Mathematical Activity," *Historia Mathematica* 23 (1996), 269–287.

52. This period is covered in Philip Bruce, *History of the University of Virginia: 1819–1919*, 5 vols. (New York: Macmillian, 1920–1922), vols. 3–5.

53. For the first quotation, see Gordon Whyburn to John Luck, 27 April, 1934, Box 10, Folder 47: Department of Mathematics, Hiring a Professor of Mathematics (1933–1934), Whyburn Papers. The characterization in the second is from Roland Richardson to Percey Smith, 19 March, 1934, Box 11: Correspondence 1931–1934 'R-W,' Folder: Smith, Percey Franklin, Richardson Papers.

groups through the 1890s, he and Echols had never animated a true graduate program during their long tenures at Virginia.[54] It was the announcement of Page's retirement effective at the end of the 1933–1934 academic year that prompted Newcomb to look for a new mathematician to begin in the fall of 1934. He wanted someone fashioned in the research-oriented mold.

The seventy-four-year-old Echols served as one of Newcomb's principal advisors. Luckily for Virginia mathematics, Echol's son Robert, also a mathematician, just happened to be at the start of a two-year stint at the newly opened Institute for Advanced Study in Princeton in the fall of 1933 (see the next section on the Institute's founding). At his father's request, the younger Echols sought Veblen's advice about the situation at Virginia, and Veblen counseled the university to "get a fairly young man, who has already attained a fair degree of prominence."[55] In particular, he suggested Richard Courant as "an extremely prominent man and a good organizer," but Echols recognized that the fact that Courant was Jewish would be a definite "drawback to his getting the job" at the Southern school.[56] As a result of his own casting around in Princeton, Echols *fils* suggested topologist Raymond Wilder, a fellow visitor at the Institute that year, as well as differential geometer Tracy Thomas and logician Alonzo Church, both former Veblen students. Also under consideration were the Chicago differential geometer Ernest Lane, the displaced Hungarian analyst Gábor Szegő, and Hopkins topologist Gordon Whyburn.

By 27 April, 1934, and after bringing in a number of candidates among them Wilder, Newcomb had given the nod to Whyburn, but the terms the President had proposed had left the mathematician cold. Newcomb had offered a professorship of mathematics at the "minimum salary—$3825 net,"

54. In 1897, for example, Page published *Ordinary Differential Equations: An Elementary Text-Book with an Introduction to Lie's Theory of the Group of One Parameter* (London and New York: Macmillan & Co., 1897). Between 1880 and 1934, Virginia did award some eleven Ph.D.s in mathematics (Richardson, "The Ph.D. Degree and Mathematical Research," p. 366).

55. Robert Echols to William Echols, 5 December, 1933, Box 10, Folder 47: Department of Mathematics, Hiring a Professor of Mathematics (1933–1934), Whyburn Papers. The quotations that follow in the next sentence are also from this letter.

56. As is well known, the University of Virginia had hired the prominent Anglo-Jewish mathematician, James Joseph Sylvester, in 1841, but he left after only four-and-a-half months. His religion was one of the reasons, although not the only one, for his ill-fated tenure in Charlottesville (Parshall, *James Joseph Sylvester: Jewish Mathematician in a Victorian World*, pp. 49–80).

a sum that Whyburn could "not possibly accept."[57] That was less than his reduced, Depression-Era Hopkins salary, a salary that was promised soon to be restored to its pre-Depression level. Moreover, Whyburn felt "quite definitely that a person could not be effective" in meeting the challenge of substantially rebuilding a department "unless he commands the maximum salary for professors in the University." Fortunately, this initial bump did not ultimately derail negotiations. By May, Veblen was writing to the Director of Virginia's Leander McCormick Observatory that the university "had done extremely well in adding Whyburn to its faculty. . . . I think he is one of the best mathematicians of his generation in this country."[58]

With Newcomb's support, Whyburn not only "completely revised" the curriculum but also decided on what "general fields of specialization and research" the department should pursue and provided "new direction . . . to both faculty and student scholarly efforts and activities."[59] Whyburn made two key appointments in the 1930s: Edward J. "Jimmy" McShane in the calculus of variations and Gustav (Arnold) Hedlund in topological dynamics.[60] Together these three men had built a small but vibrant graduate program by the end of the 1930s. (Duke University also made a major move in mathematical research in that decade thanks largely to the founding there of the *Duke Mathematical Journal* in 1935 (see this chapter's final section).)

Whyburn's departure from Hopkins in 1934 had consequences not just for Virginia and the South. The Maryland school immediately launched a search to replace him and netted Whyburn's academic "brother" at Texas and unsuccessful competitor at Virginia, Wilder. There were, however, conditions. Wilder would come not in the fall of 1934 but rather for the 1935–1936 academic year. In the meantime, he would finish up a year at the Institute and return to Michigan, where the university had declared itself unable

---

57. Whyburn to Luck, 27 April, 1934. The next two quotations are also from this letter. The "real wealth" of $3,825 in 1934 dollars is equivalent to $76,300 in 2021 dollars (https://www.measuringworth.com).

58. Veblen to Samuel Mitchell, 24 May, 1934, Box 10, Folder 47: "Department of Mathematics, Hiring a Professor of Mathematics (1933–1934)," Whyburn Papers.

59. Gordon Whyburn, "School of Mathematics," typescript, Box 10, Folder 53: Report on the Department of Mathematics to the Scientific Advisory Committee of the University of Virginia 1952, Whyburn Papers. Compare Parshall, " 'A New Era,' " pp. 288–289.

60. McShane went by the nickname Jimmy but generally wrote under E. J. McShane; Hedlund wrote under Gustav A. Hedlund and G. A. Hedlund but, at least later in life, went by his middle name, Arnold.

to counteroffer a year early. Wilder's acceptance of the Hopkins overture thus came "with the understanding that [he would be] free to consider any other offers (including Michigan) in the meantime," but this left him in a quandary.[61] "Offhand," he mused to R. L. Moore, "it looks like a good situation for me there, since outside of [Abraham] Cohen (who will retire in a few years), there would be only [Francis] Murnaghan as full professor, and [Oscar] Zariski and myself as assoc. profs. Here I have eight full professors over me, most of whom will not be retiring for over ten years."[62] On the other hand, Wilder, ultimately a father of four, found Ann Arbor "a nice family town," and, as he saw it, "although there are many disadvantages in the dept., I have finally succeeded, I think, after much work and opposition, in getting courses lined up fairly satisfactorily."

One of those "disadvantages" was what Wilder viewed as the disproportionate ratio at Michigan between non-research-oriented full professors and the newer blood, like himself, who had fully embraced the research ethos. The latter, in fact, had formed an informal and secret club as early as 1927 actively to foster research. The so-called "Small C," which met weekly, set itself up in reaction to the larger "Department Club," which met monthly and "was not accomplishing very much in the development of interests in research."[63] When 1895 Harvard Ph.D. James Glover decided to step down as department chair in 1934, these two factions clashed in deliberations over who Glover's successor should be.

Glover, although not a "Small C" member, was nevertheless the man responsible for hiring Wilder in topology—and others like George Rainich in the theory of relativity and Theophil Hildebrandt in functional analysis—and thus for beginning the department's orientation toward research. As Wilder explained to his friend and fellow topologist Leo Zippin (then at the Institute

---

61. Wilder to Moore, 3 September, 1934, Box 86-36/8: Folder 6: General Correspondence R. L. Moore, Wilder Papers. The quotations that follow in this paragraph are also from this letter.

62. Francis Murnaghan, an Irish-born and -trained applied mathematician, was chair of the Hopkins department. His colleague, analyst Abraham Cohen, had earned his Hopkins Ph.D. in 1894 and had worked his way through the ranks at his alma mater. He actually retired only in 1940. Russian-born topologist, Oscar Zariski, had begun his association with the department in 1927 and had become an associate professor there in 1932. For more on Zariski, see chapters six and ten.

63. Raymond Wilder, "Reminiscences of Mathematics at Michigan," in *A Century of Mathematics in America*, ed. Peter Duren et al., 3: 191–204 on p. 193.

for Advanced Study), the research faction tried "to sell the idea of having a mathematician for chairman, but many, especially the eight full professors, insist on an *administrator*—of course the chairman will be a member of the mathematics dept., but you know what I mean. It does not seem to count with the full professors that it is vastly more important to have a real mathematician who will do the dept. some good—they act as if they were electing the president of a country club."[64] Given the uncertainty of the political situation, Wilder added that "it is nice at a time like this to have the J[ohns] H[opkins] position tucked up one's sleeve."

In the end, the functional analyst and "Small C" member Hildebrandt succeeded Glover as chair. He had joined the department in 1909, the year before finishing his Ph.D. under E. H. Moore at Chicago, and had clearly proven himself to be one of the "research men" with his work in the 1920s on, among other topics, uniform boundedness for Banach spaces (recall the discussion in chapter one). Apparently, at this key transition point, the die had been cast in favor of research by the powers-that-be at Michigan in making Hildebrandt's appointment. As Wilder reported to Zippin, "[t]he dope is that he is supposed to build up the research side of the dept."[65] And, although Hildebrandt was unable to match Hopkins's offer to Wilder at the end of the 1934–1935 academic year when the latter had to decide whether to leave for Hopkins or to stay at Michigan, he did succeed in getting Wilder promoted to full professor. That was apparently enough, for the topologist stayed and joined forces with Hildebrandt and Rainich to raise Michigan's profile as a research department into the 1940s. It, in fact, had already made a certain reputation for itself in research, appearing at number ten in Hughes's 1924 survey.[66]

Just behind Michigan in those rankings, Berkeley also made a major move relative to research in the 1930s under the guidance of the school's eleventh President, Robert Sproul. Founded in 1868 as California's land-grant institution, Berkeley had tried to build its Mathematics Department as early as the 1890s by hiring Hopkins-trained Irving Stringham and Klein's student Mellen Haskell, but it had failed to maintain it "at the proper standard" in the opening decades of the twentieth century.[67] Even Haskell, as department chair in

---

64. Wilder to Zippin, 15 October, 1934, Box 86-36/14, Folder 5: General Correspondence Zippin, L., Wilder Papers (his emphasis). The next quotation is also from this letter.

65. Ibid.

66. Compare Parshall, " 'A New Era,' " pp. 286–287.

67. Parshall and Rowe, *Emergence*, pp. 191–211. For the quotation, see Robin Rider, "An Opportune Time: Griffith C. Evans and Mathematics at Berkeley," in *A Century of Mathematics*

the 1920s, had to admit that "[w]e have an excellent staff of teachers, but they have never been strong in research."[68]

With urging from the departments of chemistry and physics as well as from the School of Engineering—sectors of the university that had already benefitted from the post–World War I boost in support for the sciences—Sproul turned his attention to mathematics.[69] The argument he heard from his scientists was not unlike the one Veblen had used to such great success in the 1920s to secure Rockefeller support for mathematics: the strongest science programs are underpinned by a strong program in mathematics. Sproul appointed a blue-ribbon committee to canvass the nation in search of a mathematician who could replace Haskell on his retirement in 1933 and who could carry out the sort of major mathematical overhaul envisioned. It came back with the unanimous recommendation to hire Griffith Evans.

Evans, as a mathematician of broad scope, had the capacity to bridge Berkeley's disparate pure and applied concerns. He had published extensively on the theory of integral equations in the 1910s influenced particularly by Vito Volterra as a result of a study trip to Rome. By 1920, however, his interests had shifted to potential theory, where he pioneered the use of measure and integration theory in tackling classical problems in the field such as the so-called Dirichlet Problem (recall chapter one) at the same time that he had developed the deep interests in mathematical economics that marked him as a "pioneer" in that field.[70] According to his biographer, "[a]t a time when most economists in this country disdained to consider mathematical treatments of economic questions," Evans "boldly formulated several mathematical models of the total economy in terms of a few variables and drew conclusions about these variables." By 1930, he had written his *Mathematical Introduction to Economics*, a text described by statistician Harold Hotelling (then at Stanford but

in America, ed. Peter Duren et al., 2: 283–302 on p. 286. There, she cited a letter from Vice President and Provost, Monroe Deutsch, to Sproul, 1 September, 1932. Compare also Parshall, " 'A New Era,' " pp. 287–288.

68. Rider, p. 297, quoting a letter from Haskell to Griffith Evans, 25 October, 1928.

69. Verne Stadman treats Sproul and his presidency in *The University of California, 1868–1968* (New York: McGraw Hill Book Company, 1970). Rider details Sproul's quest for a mathematician on pp. 286–289 of her article.

70. The characterization and many more details on Evans's scientific work may be found in Charles Morrey, "Griffith Conrad Evans," *Biographical Memoirs*, vol. 54 (Washington, D.C.: National Academy of Sciences, 1983), pp. 126–155 on p. 140. The quotation that follows is also on this page.

at Columbia by the fall of 1931) as one that "helps to lay a groundwork upon which future contributions to political economy of first-rate importance may be expected to be based."[71] It was thus a mathematician with rather unique credentials upon whom Sproul and his advisers had settled, but what would it take to uproot him from Houston and transplant him to Berkeley?

Even as the State of California and the university struggled financially, Sproul's budget committee essentially gave him carte blanche relative to the situation in mathematics. "The department is in vital need of leadership, which . . . cannot be found in the present membership," the committee reported. "[T]he availability of an exceptionally strong leader at this moment is an opportunity which . . . should not be allowed to pass, whatever the conditions may be."[72] Ultimately, Evans was offered a princely annual salary, especially for a state school, of $9,000, although that figure would be reduced by whatever percentage salary cut the university might implement for all faculty salaries.[73] Salary aside, the uncertainty surrounding the possible reduction gave Evans pause. He declined the offer. By May 1933 and after both a flood of correspondence from the scientists at Berkeley urging him to change his mind and a firming up of the offer, Evans had accepted. Staying on at Rice for one more year to settle his affairs, he relocated to the West Coast in time for the start of the 1934–1935 academic year.

In the 1930s alone, Evans oversaw the appointments at Berkeley of specialists in, among other fields, the theory of functions of several complex variables, the calculus of variations, mathematical logic, and statistics.[74] Moreover, the graduate program, which had produced some eleven new Ph.D.s between 1930 and Evans's arrival in 1934, generated almost double that number between 1935 and the end of the decade under his encouragement. Evans's accomplishments both mathematical and organizational were recognized when he became the twenty-fifth President of the AMS in 1939, succeeding R. L. Moore. As an indication that both the South and the West

71. Griffith Evans, *Mathematical Introduction to Economics* (New York: McGraw Hill Book Company, Inc., 1930) and Harold Hotelling, Review of *Mathematical Introduction to Economics*, *Journal of Political Economy* 39 (1931), 107–109 on p. 109.

72. Rider, p. 287, quoting the budget committee's report to Sproul, 20 March, 1933.

73. The "real wealth" of $9,000 in 1933 dollars is equivalent to $185,500 in 2021 dollars (https://www.measuringworth.com).

74. Rider, pp. 301–302, lists the appointments made at Berkeley from the academic year 1930–1931 through 1949–1950. For the impact of mathematical emigration on the program at Berkeley, see the next chapter.

Coast had finally begun to take their place in the American mathematical research community in the 1930s, Moore at Texas and Evans at Berkeley were the first AMS Presidents from the South and the West, respectively. The American mathematical research community was becoming more truly national.[75]

A New Experiment: The Institute for Advanced Study

Just as, paradoxically, the Depression Era of the 1930s witnessed the targeted development of new, strong, research-oriented programs in mathematics like those at Virginia, Michigan, and Berkeley, so, too, it saw the development of an innovative experiment in the encouragement of properly postdoctoral research, the Institute for Advanced Study (IAS) in Princeton.[76] Veblen had floated the idea of an institute in support of mathematical research at least as early as 1924 in connection with his successful lobbying for the NRC's extension of its postdoctoral fellowship program from physics and chemistry to mathematics. As he saw it, a mathematics institute would take the NRC concept one step further. Postdoctoral fellowships would give talented, fresh Ph.D.s the opportunity to establish a program of research before embarking on careers in academe that would quickly find them juggling administration, teaching, and research. A mathematics institute would provide respite from administration and teaching as well as provide a block of uninterrupted time for original work.[77] Although Veblen did not succeed in convincing the Rockefeller-funded philanthropies to underwrite his vision of a mathematics institute in 1924, by 1930 things had changed in unexpected ways.[78]

75. The MAA had already had three Presidents from the West by the end of the 1930s: Florian Cajori of Colorado College was President in 1917; E. T. Bell of Caltech in 1931–1932; and Aubrey Kempner of the University of Colorado in 1937–1938. It would only have its first President from the South in 1953–1954 when Virginia's E. J. McShane assumed the post.

76. The story of the Institute's creation is well-known. The account here draws principally from Nathan Reingold and Ida Reingold, ed., *Science in America: A Documentary History 1900– 1939* (Chicago: University of Chicago Press, 1981), pp. 433–470; Armand Borel, "The School of Mathematics at the Institute for Advanced Study," in *A Century of Mathematics in America*, ed. Peter Duren et al., 3: 119–147; and Steve Batterson, *Pursuit of Genius: Flexner, Einstein, and the Early Faculty at the Institute for Advanced Study* (Wellesley: A K Peters, Ltd., 2006).

77. Feffer, pp. 488–489.

78. In the meantime, Veblen had persistently pushed the idea to any and all who would listen. Writing to Birkhoff on 13 March, 1926, for example, he remarked that "I have recently had two opportunities to explain the [institute] idea to the heads of other institutions, which

Abraham Flexner, a man with strong Rockefeller connections and a scholar of higher education in general and of medical education in particular, had been approached at the close of 1929 by representatives of Newark-based department store magnate Louis Bamberger and his sister, Caroline Bamberger Fuld.[79] In what had proved to be a shrewd business move, the siblings had sold out to New York City–based Macy's in June 1929. Once the deal was finalized, brother and sister found themselves in possession of a $15 million stock interest in Macy's and $11 million in cash proceeds.[80] Although their stock holdings took a hit when the stock market crashed at the end of October, their cash was firmly in hand, and they sought to give back to the citizens of New Jersey by founding a medical school in Newark. To realize their idea, however, they needed the guidance of an expert in medical education. Flexner was perfect for this role.[81]

The brother of the same Simon Flexner whom Veblen had had to convince of the merits of mathematics for Rockefeller support, Abraham had earned his B.A. in classics from Hopkins in the mid-1880s, a decade after the trend-setting university had opened under the enlightened leadership of its first President, Daniel Gilman. In setting up the new school, Gilman had sought to find and fill a gap in American higher education.[82] What was lacking, in his view, was a European- and particularly German-style university in which the training of future researchers and, concomitantly, the active promotion of

---

opportunities I did not let slip. While nothing may come of it in any special case, I think it worth while to make it propaganda. At present I say it would cost 2 million and should include 3 leading positions at $15,000 each. In one case we even discussed the personnel" (HUG 4213.2, Box 6, Folder: Correspondence 1926, S-Z, Birkhoff Papers). The "real wealth" of $2M in 1926 dollars is equivalent to $30.2M in 2021 dollars, while that of $15,000 is $213,000 (https://www.measuringworth.com).

79. Abraham Flexner, *I Remember: The Autobiography of Abraham Flexner* (New York: Simon and Schuster, 1940) and Thomas Bonner, *Iconoclast: Abraham Flexner and a Life in Learning* (Baltimore: Johns Hopkins University Press, 2002) provide more on Flexner's life and place in American higher education.

80. In 2021 dollars, $11 million represents $171.5M in "real wealth" (https://www.measuringworth.com).

81. Batterson, *Pursuit of Genius*, pp. 35–36.

82. Much has been written on the creation of the Johns Hopkins University and on its place in the history of American higher education. Hugh Hawkins, *Pioneer: A History of the Johns Hopkins University 1874–1889* (Ithaca: Cornell University Press, 1960) and Lawrence Veysey, *The Emergence of the American University* (Chicago: University of Chicago Press, 1965) are standard references.

original research among faculty and students alike was paramount. Gilman's Hopkins thus opened doing what no other American institution of higher education really did in the 1870s.

It started small. Gilman focused first on attracting the best mathematician, the best classicist, and the best physicist or chemist he could find. The laboratory sciences were expensive; mathematics and classics were not. He built from this central core.[83] To support his faculty's research mission and to facilitate the communication of its results, Hopkins under his leadership also underwrote the publication of research journals. Finally, a strong undergraduate component generated a pipeline of students capable of moving into the research ranks.

Flexner had directly experienced only the undergraduate side of Gilman's experiment, but, as a scholar of higher education, he had learned from and was deeply impressed by Gilman's example. Still, he had come to feel that, over time, Gilman's graduate ideal had been compromised by undergraduate concerns. In responding to the Bamberger family's approach, then, Flexner employed Gilman's find-and-fill-a-gap strategy to steer them away from their original idea and toward his own. The United States had many medical schools, Flexner argued, but it still did not have a *true* university, one in which *graduate* education and research were the *sole* missions. By endowing such an institution, the Bambergers would take Gilman's experiment to the next level and thereby make an indelible mark on American higher education.[84]

Flexner's powers of persuasion proved strong. On 8 June, 1930, the *New York Times* carried the front-page story "Bamberger Gives $5,000,000 for . . . Foundation in Newark to Aid Advanced Learning." Abraham Flexner was at that institution's helm.[85]

---

83. Francesco Cordasco laid out Gilman's plan for the Johns Hopkins University in *Daniel Coit Gilman and the Protean Ph.D.: The Shaping of American Graduate Education* (Leiden: E. J. Brill, 1960), pp. 65–67.

84. Batterson, *Pursuit of Genius*, pp. 38–41. Clark University in Worcester, Massachusetts, had been founded in 1889 as a true university in Flexner's sense, but it had had to reorient itself after 1892 when William Harper, the President of the then-newly-forming University of Chicago, lured away almost all of its faculty. This story may be found in many places, but Parshall and Rowe give the mathematical perspective in *Emergence*, pp. 271–275 and 286–289.

85. The full headline read: "Bamberger Gives $5,000,000 for Study: He and Sister, Mrs. Felix Fuld, Set Up Foundation in Newark to Aid Advanced Learning: Dr. Flexner the Director: Institution, First of Kind in This Country, Provides for Research and Training Post-Graduates." *New York Times*, 8 June, 1930, p. 1. The "real wealth" of $5M in 1930 dollars is equivalent to $79.9M in 2021 dollars (https://www.measuringworth.com).

Veblen read this story with great interest. He and Flexner had then recently been in correspondence on a similar topic, sparked by another *New York Times* story. At the end of December 1929, Veblen had spoken at the AMS's annual meeting at Lehigh University and had called for what the *Times* reporter termed "a Seat of Learning" that focused more exclusively on original research.[86] Veblen, recently returned from England and his faculty exchange with Hardy, had been impressed by what he had experienced. He was reported to have averred that "the work in [the United States] exceeds that abroad in *quantity*, but has not yet approached it in *quality*," and Flexner, who was in complete sympathy with both of these positions, wrote to Veblen to tell him so.[87] In reply, Veblen elaborated on his ideas further, stating in particular that his "mathematical institute which has not yet found favor may turn out to be one of the next steps. Anyhow it seems to me to fit in with the concept of a seat of learning."[88] Could the new institution under Flexner's direction, Veblen wondered, not fulfill that lofty goal? And, moreover, could it not be expansive enough to encompass a mathematics institute?

Writing to congratulate Flexner on the news of the Bamberger-Fuld endowment, Veblen ventured to suggest that Flexner consider locating his new venture in or near Princeton "so that you could use some of the facilities of the University and we could have the benefit of your presence."[89] It was a bold proposition, and one counter to the donors' desire to set up shop in Newark. In the end, not only did this suggestion for a site prevail, but it was also decided that the new institute's first school would be one of mathematics.[90] Veblen's broader vision for mathematics at Princeton—and for the United States as a whole—was being realized thanks to Flexner's expertise and philanthropic dollars.

In a final Princeton coup, it was decided by the fall of 1932 that Veblen and Albert Einstein would be the first two members of the Institute's faculty;

86. "Says America Lacks a Seat of Learning: Professor Veblen Tells Mathematicians at Lehigh We Need Such an Institution as Cambridge," *New York Times*, 29 December, 1929, p. 3. The next quotation is also from this article with my emphases.

87. Flexner to Veblen, 21 January, 1930, Archives of the Institute for Advanced Study, as referenced in Batterson, *Pursuit of Genius*, p. 69.

88. Veblen to Flexner, 24 January, 1930, Archives of the Institute for Advanced Study, as quoted in Armand Borel, p. 119. Compare Batterson, *Pursuit of Genius*, p. 69.

89. Veblen to Flexner, 10 June, 1930, Archives of the Institute for Advanced Study, as quoted in Armand Borel, p. 121. Compare Batterson, *Pursuit of Genius*, p. 69.

90. Batterson lays out the many details in *Pursuit of Genius* on pp. 69–79.

Flexner's negotiations with his first choice for a mathematician, Birkhoff at Harvard, had fallen through that spring. Once again, the news made the front page in New York City. Although Einstein would not come until the Institute's official opening in the fall of 1933, it was reported that "Professor Veblen has already assumed his new duties and during the coming year will assist Dr. Flexner in preparing for the opening of the Institute."[91] Indeed, during that unofficial first year, Veblen conducted a seminar on the "Modern Differential Geometry" to which he and his Princeton colleagues had contributed so fundamentally.

Much planning also needed to be done prior to the official opening. In particular, decisions had to be made as to just who would be eligible for appointment. Originally, Flexner thought that a limited number of talented students would actually come to the Institute to earn a Ph.D., working under a member of the permanent faculty, but he was fairly quickly won over to Veblen's idea of purely postdoctoral appointees. After all, if the Institute were going to be different, it would not duplicate the model in place at, for example, Flexner's paradigmatic Johns Hopkins.

Veblen tested the postdoctoral idea out on various of his contacts within the American mathematical community and received strong support. His friend Bliss, for example, highly approved of the "plan to include short term temporary positions. It seems to me," he told Veblen, "that you can have a fine influence on the development of mathematics in this country if you can give opportunities for the completion of research in progress to men holding university positions. . . . Such opportunities will have an excellent effect in the encouragement of men to pursue aggressively their research interests."[92]

How, though, should the financing of such a scheme be worked out relative to a potential visitor's home institution? Bliss and Veblen were also in discussion about that. Adrian Albert, then an assistant professor at Chicago, had been tapped as one of the visitors during the Institute's inaugural year, and Bliss, as department chair, was naturally concerned about the pecuniary implications of such a visit for a then financially strapped Chicago. As Veblen explained, "the sort of plan which would apply is what Flexner proposes to call a 'grant in aid.' This would mean that if the University gave

91. "Einstein Will Head, School, Opening Scholastic Centre," *New York Times*, 11 October, 1931, p. 1.

92. Bliss to Veblen, 13 December, 1932, Box 3, Folder: G. A. Bliss 1923–40, Veblen Papers.

[Albert] sabbatical leave at half pay the Institute would pay the other half of his salary."[93] That might have seemed like a fine plan in the abstract, but it meant that, in the depths of the Depression, Chicago would have to pay half of Albert's salary while it would have none of his service.

From Flexner's point of view, that was perfectly appropriate. Schools like Chicago, Harvard, and others were, after all, "rich institutions, thoroughly able to give their men additional opportunities, of which they themselves will reap the benefits."[94] Why should "[t]he institutions and the men . . . between them [not] make the sacrifice." Although Bliss was well aware of those "benefits"—Albert would have the opportunity to devote his full energies to his research and that would be of paramount importance not only for his subsequent career but also for his department's reputation—those holding Chicago's purse strings were less convinced. As Bliss confessed to Veblen, "I have had some difficulty in securing agreement from the various administrative officers concerned in view of the serious financial situation. . . . Because of financial difficulties, such as I suppose almost every institution is having, they did not feel justified in letting [Albert] go without service in return for the payments which will be made to him."[95] In the straitened 1930s, the notion of a sabbatical was not viable. The deal was finally sealed when Albert agreed both to work full-time at Chicago in the summer before his departure and to take on "some extra responsibilities" upon his return in the summer of 1934. The research ethos had ultimately trumped finances, but a quid pro quo had been exacted.[96]

Not all institutions were ultimately as accommodating as Chicago, however. At Harvard, Morse's 1930 Ph.D. student, Albert Currier, was serving as an instructor during the 1932–1933 academic year and seemed like another prime candidate for the Institute's inaugural year. Morse had certainly pushed for his candidacy, but financial considerations—like those that Chicago and other schools had agreed to—proved intractable at Harvard. "I am sorry," Veblen wrote with resignation to Morse, "that we could not have had one of the best Harvard men in the first year of the Institute. Perhaps after a year or

---

93. Veblen to Bliss, 16 January, 1933, Box 3, Folder: G. A. Bliss 1923–40, Veblen Papers.

94. Flexner to Veblen, 23 January, 1933, in Reingold and Reingold, p. 444. The next quotation is also on this page.

95. Bliss to Veblen, 27 January, 1933, Box 3, Folder: G. A. Bliss 1923–40, Veblen Papers. The quotation that follows is also from this letter.

96. The story of another inaugural year participant, Raymond Wilder from the University of Michigan, was similar. Compare Parshall, " 'A New Era,' " p. 286.

two, the aims and conditions of the Institute will be better understood and cooperation will be easier. I have had a lot of correspondence and negotiations about these junior appointments during the last month or two and this has determined a set of conditions that seem necessary and also acceptable by the other universities."[97] The Institute was a work in progress. Policies were evolving and, with them, not only the sabbatical concept but also the Institute's place within the landscape of higher education.

When it finally opened in the fall of 1933 in space in Fine Hall shared with the Princeton Mathematics Department, the Institute's School of Mathematics had a faculty of four: Einstein, Veblen, and Princeton's James Alexander as well as its recently hired Hungarian émigré John von Neumann. After much discussion and negotiation, the rest of the school's faculty was finally in place by its second year of operation, with the addition of Göttingen's Hermann Weyl starting in January 1934 and Marston Morse who left Harvard in January 1935.[98] These six men constituted the school's permanent faculty for the next decade.

Twenty-three so-called "workers" participated in the IAS experiment in its first official year of operation, 1933–1934. Four were NRC fellows. Three were fellows funded by Rockefeller's International Education Board, two from the United Kingdom and one from Denmark. Two others—the Austrian Kurt Gödel and the Romanian Isaac Schoenberg—also came from abroad. Most were recent Ph.D.s, having taken their degrees between 1930 and 1933, but four had earned their doctorates before 1930, among the latter Albert and Wilder. Two, Mabel Schmeiser (later Barnes) and Anna Stafford (later Henriques), were women. Schmeiser had done her doctoral work in analysis under Henry Blumberg at Ohio State in 1931, while Stafford had taken a Chicago Ph.D. under Mayme Logsdon.[99] This cohort well reflected the

97. Veblen to Morse, 26 January, 1933, Box 8, Folder: Morse, Marston, Veblen Papers.

98. Batterson provides a very detailed account of how the faculty came to be in *Pursuit of Genius*, pp. 81–157. Weyl had moved from the ETH in Zürich to Göttingen in 1930.

99. For the list of members, see *Institute for Advanced Study Bulletin No. 3* (Princeton: Princeton University Press, 1934), pp. xiii-xiv (the quotations that follow in this paragraph are on p. iv); compare Janet Mitchell, ed., *A Community of Scholars: The Institute for Advanced Study: Faculty and Members 1930–1980* (Princeton: The Institute for Advanced Study, 1980). On Schmeiser and Stafford, see https://www.agnesscott.edu/lriddle/women/barnes.htm and https://www.agnesscott.edu/lriddle/women/stafford.htm, and Georgia Whidden, "Anna Stafford Henriques," *Attributions: Newsletter from the Development Office of the Institute for Advanced Study* 1 (2001).

FIGURE 4.2. John von Neumann (1903–1957). (Photographer unknown. Photo courtesy of Shelby White and Leon Levy Archives Center, Institute for Advanced Study, Princeton, NJ.)

ideals stated and reiterated in each of the Institute's early published *Bulletins*, namely, that "[i]t is fundamental in our purpose, and our express desire, that in the appointments to the staff and faculty, as well as in the admission of workers . . . , no account shall be taken, directly or indirectly, of race, religion, or sex." For Flexner and his colleagues, this was ostensibly a non-negotiable condition for what they termed "the pursuit of higher learning." Yet, the fact that they shared space with the Mathematics Department of a white-males-only university in a city in which Jim Crow laws prevailed made such an idealization almost impossible to realize in the 1930s and beyond.[100]

In its first year, that pursuit took several forms. There was, of course, independent study and informal conversation, but there was also a weekly "mathematical club" run jointly with the Princeton Mathematics Department at which a paper was presented and discussed. At least three of the visitors ran seminars, one each in algebra, algebraic logic, and topology, while the faculty offered additional courses and seminars. Alexander and Lefschetz collaborated on a "seminar on topology, and particularly on group theoretical problems connected with it," while Veblen and von Neumann did the same in a seminar on differential geometry and quantum theory that focused primarily on "the theory of spinors and the Dirac equation, projective relativity, and conformal geometry."[101] Von Neumann gave a course on the theory of functional operators, covering the "theory of integration, [the] general theory of Hilbert space, and bounded operators." And, following his arrival in January, Weyl added to the offerings with a course on the structure and representations of continuous groups. Finally, the Institute entered into an agreement with the university to publish the *Annals of Mathematics* jointly, under the co-editorship of von Neumann representing the former and Lefschetz the latter. Like Gilman, Flexner recognized the desirability of supporting strong research by being associated with a publication outlet for it.

The next year, 1934–1935, witnessed a sharp increase—to thirty-five—in the number of visiting "workers," owing at least in part to the ever-worsening political situation in Germany (see the next chapter). Alexander and Lefschetz as well as Veblen and von Neumann continued their collaborative seminars, with von Neumann extending his lectures on functional operators to the theory of unbounded operators and the algebra and analysis of operators, and with Weyl both lecturing and conducting a seminar on continuous and infinitesimal groups and their representations. Emmy Noether, who had left Germany to take a post at Bryn Mawr, and Carl Ludwig Siegel, who was visiting Princeton from Göttingen, augmented the IAS's offerings

100. When William Claytor, for example, had the opportunity to go to the Institute in 1939 following its move out of Fine Hall and into its own Fuld Hall, he declined. "There's never been a black at Princeton," he was reported to have said, "and I'm not going to be a guinea pig." See Raymond Wilder to M. Solveig Espelie, 7 December, 1979, Box 86-36/2, Folder 4, Wilder Papers. The African-American probabilist David Blackwell would be the first to break the color barrier at the Institute in the 1941–1942 academic year. See chapter ten.

101. *Institute for Advanced Study Bulletin No. 3*, p. 6. The quotation that follows is on p. 7.

with seminars on class field theory and the analytic theory of quadratic forms, respectively.[102] These numbers and this level of activity in mathematics would remain roughly steady through the 1930s, although the IAS itself would expand dramatically in the 1935–1936 academic year with the openings of both a School of Economics and Politics and one of Humanistic Studies.

Still, in this first decade, it was in mathematics that the IAS made its greatest impact. Writing to Flexner in 1938 from his privileged view as AMS Secretary, Richardson stated categorically that "[t]he Institute has had a very considerable share in the building up of . . . mathematics to its present level. . . . Not only has [it] given ideal conditions for work to a large number of men, but it has influenced profoundly the attitude of other universities."[103] Research had become ever more firmly embedded in American mathematics, in particular, and in American higher education, in general.

By 1939, the seventy-two-year-old Flexner had retired as Director in the face of mounting dissatisfaction with, among other things, his management style, and the IAS had moved into its own space, Fuld Hall, on a parcel of land not far from the university.[104] The IAS faculty numbered sixteen strong over the three schools with some nine assistants or associates supporting them. The annual number of "members" (happily, the terminology had changed in 1936 from "workers") had settled in the forties.[105] By all measures, Flexner's experiment—bankrolled by Bamberger-Fuld capital—had succeeded.

Journals in an Evolving Research Community

One aspect of that success had been the IAS's partnering with Princeton in the publication of the *Annals of Mathematics*. A journal with a somewhat checkered past, the *Annals*, founded in 1884, had changed hands from Virginia to Harvard in 1899 before moving to Princeton in 1911. There, it had struggled with its identity, accepting the subvention from the MAA in the late 1910s and early 1920s that allowed for more printed pages devoted to expository articles (recall chapter two). When that initiative was finally abandoned later

---

102. *Institute for Advanced Study Bulletin No. 4* (Princeton: Princeton University Press, 1935), pp. 6–7.

103. As quoted in Armand Borel, pp. 127–128.

104. Batterson tells the story with all its many intrigues in *Pursuit of Genius*, pp. 215–237.

105. *Institute for Advanced Study Bulletin No. 9* (Princeton: Princeton University Press, 1940), pp. viii-xiii.

in the 1920s, the *Annals* firmly embraced the publication *solely* of high-level research. That decision apparently paid off.

In 1932, for example, then Princeton analyst and, with Lefschetz, *Annals* co-editor Einar Hille had reported to Veblen that none other than the Director of Sweden's Mittag-Leffler Institute, Fritz Carlson, had "offered his opinion that the *Annals* was now the best mathematical periodical *in English*."[106] Hille was quick to add that this was by no means faint praise, for Carlson "has severe standards and a wide range of interests. . . . Both he and M[arcel] Riesz were actually reading papers in the *Annals* when I called on them, and it was obviously not staged (I know of such cases!)." It was a sure sign that, as Veblen and others had long hoped would be the case, American mathematics was being taken seriously in Europe. The IAS partnership would only further strengthen that perception.

Indeed, as Gilman had already demonstrated at Hopkins, the relationship between a journal and its host institution could be symbiotic. That lesson was not lost on other developing institutions. In Durham, North Carolina, Duke University debuted the *Duke Mathematical Journal* in 1935, giving testament to mathematical developments in the South.[107]

In 1924, James B. Duke, a hugely successful tobacco and electric power industrialist, had established the Duke Endowment with a $40 million trust fund that benefitted, among other concerns, higher education.[108] One of the institutions he singled out was Trinity College, a school that had been the target of his father's financial generosity at the end of the nineteenth century. In honor of Duke's father and in recognition of the beneficence of the

---

106. Hille to Veblen, 24 October, 1932, Box 6, Folder: Hille, Einar 1926–32, Veblen Papers (my emphasis). The quotation that follows is also from this letter. Hille moved to a professorship at Yale in 1933.

107. As noted in chapter two, MIT brought out the first volume of its *Journal of Mathematics and Physics* in 1922 as a publication outlet for those on its staff. By the mid-1930s, however, it had "relaxed its rigid Institute policy" and begun to accept "some outside papers of high caliber." See "Massachusetts Institute of Technology, Department of Mathematics, Report to the Visiting Committee," March 13, 1935, Box 12: Correspondence, 1935–1939, Folder: Massachusetts Institute of Technology, Richardson Papers.

108. For more on Duke and the founding of his eponymous university, see Robert Durden, *Bold Entrepreneur: A Life of James B. Duke* (Durham: Carolina Academic Press, 2003) and *The Launching of Duke University* (Durham: Duke University Press, 1993). The "real wealth" of $40M in 1924 dollars is equivalent to $624.7M in 2021 dollars (https://www.measuringworth.com).

Duke family, Trinity College was renamed Duke University and embarked on a major expansion of both its physical plant and its faculty.

From the beginning, Duke's strategy in mathematics had been to put itself on the map by establishing a new research periodical. By April 1927, it had offered the University of Illinois's Robert Carmichael a professorship and the founding editorship of its planned journal.[109] Carmichael was an interesting choice. Birkhoff's first doctoral student, he had earned his Ph.D. at Princeton in 1911, the year before Birkhoff left for the position at Harvard that he would hold for the rest of his career. After four years at the University of Indiana, Carmichael moved to Illinois, where, by 1927, he had already directed the dissertation research of over a dozen students in Birkhoff's approach to asymptotic expansions of ordinary linear differential equations and boundary-value problems. His was a prime manifestation of America's westward expansion of quality research-level graduate programs in mathematics. And, even more was true. Carmichael had also served for one year as editor-in-chief of the *American Mathematical Monthly*. Thus, in addition to being a proven researcher and graduate adviser, he had been at a journal's helm. There was just one hitch. Carmichael was unsure whether a new research-level journal was actually needed. A month-long deliberation and consultation with various colleagues, among whom was his advisor, Birkhoff, ultimately ended in the decision that the time was *not* right to start a new journal, and he declined the offer.[110]

The matter did not end there, however. In September 1927, the AMS met for its thirty-third summer meeting in Madison, Wisconsin, and some twenty-eight mathematicians—among them, oddly, Carmichael, given his decision in the spring—joined in signing a letter in strong support of Duke University's journal initiative. In their view, "there exists an insistent need for an additional mathematical periodical devoted to mathematical research and of the standards and scope of those already established."[111] Moreover, they believed that "Duke University is in a strategic position to serve the cause of mathematics

---

109. Carmichael to R. L. Moore, 21 April, 1927, Box 4RM73, Folder: R. D. Carmichael (University of Illinois at Urbana), Moore Papers.

110. Carmichael to Birkhoff, 21 April, 1927, HUG 4213.2, Box 8, Folder: Correspondence 1927 A–Z, Birkhoff Papers as well as R. L. Moore to Carmichael, 14 May, 1927, and Carmichael to Moore, 24 May, 1927, Box 4RM73, Folder: R. D. Carmichael (University of Illinois at Urbana), Moore Papers.

111. This and the next quotation may be found in the letter dated 6 September, 1927, Box AMM-MNR/1, Folder: Duke Mathematical Journal, Duke Mathematical Journal Records, 1927–1934.

by embarking on such a project." Even with this strong vote of confidence, the initiative came to naught in 1927.

Four years later in 1931, the issue was once again under discussion, this time among the members of the Council of the AMS. The output of the American mathematical research community had only continued to increase in the intervening four years (see fig. 4.3). Richardson laid out the problem this way:

> Each year there is a demand for additional space for printing mathematical research in the country, and this in spite of gradually rising standards. This increase in the *Annals, Bulletin, Journal,* and *Transactions* has averaged 200 pages a year for 9 years, there being now 3600 pages as against 1800 in 1922, and the demand for the increase seems not to abate. Before 1932 begins, there will be on hand accepted material to fill more than 2500 pages and this is considerably more than in any recent year. To make the situation worse, mathematics is forced by the financial depression to curtail printing next year by several hundred pages (probably by a minimum of 400 in total).[112]

In light of these increasingly dire conditions, the AMS Council voted unanimously in favor of the establishment of "a new mathematical journal of high grade by Duke University."[113] Yet, as in 1927, not everyone in the American mathematical community agreed that a new journal was the answer.

In his retiring presidential address before the MAA in September 1931, Dartmouth's John Young, for one, raised a number of questions for his audience to consider. The first concerned publications and their financial drain on the AMS. More and more papers were being submitted for consideration, and the reaction to that reality had been to continue to increase the number of journals and so journal pages. "If our editorial policy remains the same," Young noted, "these publications will have to continue their expansion."[114] "Is this policy sound," he wondered, "or has the time come to raise materially our editorial standards? Are we publishing too much?" Food for

112. Richardson to the members of the AMS Council, 18 November, 1931, Box AAM-MNR/1, Folder: Duke Mathematical Journal, Duke Mathematical Journal Records, 1927–1934.

113. Report by Roland Richardson dated 2 December, 1931, Box AAM-MNR/1, Folder: Duke Mathematical Journal, Duke Mathematical Journal Records, 1927–1934.

114. John Young, "Functions of the Mathematical Association of America," *AMM* 39 (1932), 6–15 on p. 6. The quotations that follow in this paragraph are also on this page.

FIGURE 4.3. Pages published per year per journal, 1920–1939.[a]

	1920	1921	1922	1923	1924	1925	1926	1927	1928	1929	1930	1931	1932	1933	1934	1935	1936	1937	1938	1939
AJM	286	290	316	314	287	301	297	614	636	660	922	936	802	707	663	942	880	1004	948	1008
AM	333	303	331	395	357	321	747	571	540	313	726	838	791	878	914	992	936	957	944	948
BAMS	491	367	503	504	599	600	760	832	831	928	936	936	935	1016	920	936	944	888	888	952
AMM	498	501	438	469	516	538	544	554	582	559	574	608	624	630	654	646	664	688	714	674
TAMS	458	543	755	608	494	600	786	848	855	931	944	999	938	972	893	1101	1001	995	1073	975
DMJ	—	—	—	—	—	—	—	—	—	—	—	—	—	—	—	555	750	755	800	962
JSL	—	—	—	—	—	—	—	—	—	—	—	—	—	—	—	—	219	188	212	194

[a]The numbers for the AMS's *Bulletin* and *Transactions* differ slightly from those given in Archibald, *Semicentennial History*, pp. 55 and 64, respectively. He was apparently also counting title pages, etc. The jump in the *AJM*'s printed pages from 1926 to 1927 came in the wake of the official involvement in that journal of the AMS (Archibald, *Semicentennial History*, pp. 62–63). The fluctuation in the *AM*'s printed pages between 1926 and 1930 coincided with a reorganization following an infusion of funds from the National Research Council in 1925–1926 (ibid., p. 16). The journal had been publishing volumes but then with undifferentiated numbers), but moved in 1930 to publishing volumes in calendar years. Issues became undifferentiated with volume 28 in 1926. The numbers given for the years 1926, 1927, 1928, and 1929 were obtained by simply dividing the number of pages published in the complete volumes by two and allocating them over the two years. The *TAMS* printed "double" issues between 1935 and 1939. The present table does not include numbers for MIT's *Journal of Mathematics and Physics* since it was primarily an in-house journal during this period.

thought. Young clearly did not advocate that the AMS "solve its financial problems by retrenchment," but he did think that editorial standards should be further tightened and that, in particular, generalization for generalization's sake should be questioned when evaluating a given paper's suitability for publication. More papers could and should be turned away.

Although he was not alone in his advocacy of ever higher editorial standards, Young had to admit that his was a minority position. Given the absence of "any more valid canons of evaluation than we now possess," he told his audience, "the overwhelming majority" of mathematicians were of the opinion that "cutting down on the volume of our publication would be a calamity."[115] Still, he argued, even if standards of content did not change, "editors should insist on more careful methods of exposition" so that "not only could space in our journals be saved but also the value of the papers enhanced."

Countercurrents like the one Young's views represented ultimately proved insufficient to stop the publication juggernaut. It may have taken some time for the project of a new Duke-underwritten journal to begin actually to get off the ground, but that likely owed more to the fact that the country as a whole was in the Great Depression's depths than to massive resistance. The majority favored expansion.

In January 1934, Joseph Thomas wrote to Duke's President to revive the journal idea. Thomas, an ambitious and talented differential geometer, had come to Duke in 1930 with strong credentials. After earning his Penn Ph.D., he had worked with some big guns: Veblen in Princeton, Birkhoff at Harvard, and Hadamard and Cartan in Paris thanks (as noted in the previous chapter) to NRC traveling fellowships. His was a research-centered point of view, and he argued emphatically for *"the need for immediate action,"* asserting that "[f]rom every standpoint, except the financial, the present moment is propitious for founding the journal at Duke: the field is clear of competitors, the necessary outside cooperation and support are assured, the need is urgent, and the service to science is the greater and the more appreciated if rendered in difficult times."[116] As for why Duke University, in particular, should undertake the project, the advantages, in Thomas's view, were equally clear: "The scientific prestige which will accrue to the University will be large. The journal

---

115. Ibid., p. 8. The quotations that follow in this paragraph are on p. 9.

116. Thomas to Robert Flowers, 5 January, 1934, Box AAM-MNR/1, Folder: Duke Mathematical Journal, Duke Mathematical Journal Records, 1927–1934 (his emphasis). The quotation that follows is also from this letter.

will also provide an easy outlet for the research of the Mathematics Department and will have a stimulating effect on its members, as was the case at Johns Hopkins and Princeton." Thomas's argument carried the day, even given the extremely lean financial times. By December 1934, he was announcing that "Duke University is about to start publishing a new mathematical journal designed to relieve pressure on the existing journals," as he sought to line up an editorial board.[117]

The first number of the *Duke Mathematical Journal* appeared in the spring of 1935 and was made up of eleven research articles comprising over 100 pages. By the end of its first year, it had published over forty papers running to more than 550 pages authored by mathematicians from the Northeast, the Midwest, and the South, including two papers by Leonard Carlitz, a Penn-trained algebraist and number theorist who had been hired at Duke just three years earlier. It had been an impressive inaugural year, which the Council of the AMS acknowledged by officially expressing "its grateful appreciation of the service rendered by the Journal to mathematical science" and offering, not without some hyperbole, "the congratulations of the Society on the distinguished place which it has assumed from the beginning among the significant mathematical periodicals of the world."[118] The interwar years thus witnessed not only a 25% increase in research-level journal support but also an announcement by mathematicians in the South that they should be recognized as a vital part of the American mathematical research community.

The *Duke Mathematical Journal* represented a fifth, *general* research-level journal, that is, it, like the *American Journal*, the *Annals*, the *Bulletin*, and the *Transactions*, published new research results regardless of mathematical field. Papers on analysis, algebra, geometry, topology, and other areas appeared side by side and completely interspersed in all five of these journals. They each, in principle, reflected mathematics broadly construed.

In 1936, however, a completely new experiment in American mathematical publication was launched. The *Journal of Symbolic Logic* of the new Association for Symbolic Logic (ASL) was a *specialized* quarterly, co-edited in the 1930s by Princeton's Alonzo Church and Cooper Langford of the University of Michigan, that aimed to support what was, in the 1930s, a growing field within—but also somewhat contiguous to—the American mathematical

---

117. Thomas to Gordon Whyburn, 10 December, 1934, Box 8, Folder 12: Duke Mathematical Journal (1934–1969), Whyburn Papers.

118. Thomas to Whyburn, 16 January, 1936, Box 8, Folder 12: Duke Mathematical Journal (1934–1969), Whyburn Papers.

community.[119] Before 1936, American symbolic logicians like Church and Langford as well as Benjamin Bernstein at Berkeley, Cornell's Barkley Rosser, Wisconsin's Stephen Kleene, Emil Post at the City College of New York, and Haskell Curry at Penn State had received their training often, but not exclusively, in mathematics departments at Berkeley, Princeton, and elsewhere and had published their work in the by then five available mathematical research publications. Yet, there, as at meetings of the AMS, their papers were often islands—that drew the serious attention of few—in a sea of "pure" mathematics. Indeed, Cyrus MacDuffee, an editor of the *Transactions* in the late 1930s and early 1940s, summed up the situation well in response to a paper that Rosser submitted to his journal in 1939. "Very few persons in this country," he stated unequivocally, "are both expert in analytic number theory and interested in foundations."[120]

That was precisely why, three years earlier in 1936, what they deemed a critical mass of symbolic logicians had resolved to delineate that field "from pure mathematics on the one hand and pure philosophy on the other."[121] They sought to fulfill the call that Veblen had made some twelve years earlier in his 1924 AMS retiring presidential address, namely, "that formal logic has to be taken over by the mathematicians," since "there does not exist an adequate logic at the present time, and unless the mathematicians create one, no one is likely to do so."[122] In the process, they were intent on separating themselves as "symbolic logicians" from "mathematicians" as well as from "philosophers," even though a mathematician no less influential than Weyl opposed such a move, viewing the subject a part of mathematics.[123]

119. "A Statement of Policy," *JSL* 1 (1936), 1. The Association for Symbolic Logic had an initial membership of over 160. See "List of Officers and Members of the Association for Symbolic Logic," *JSL* 1 (1936), 106–109.

120. MacDuffee to Rosser, 3 June, 1939, Box 1, Folder: Correspondence Erdős 1938–1939, J. Barkley Rosser Papers.

121. Alonzo Church, "A Bibliography of Symbolic Logic," *JSL* 1 (1936), 121–123 on p. 121. The quotations that follow in the next paragraph are on pp. 121–122 (with his emphasis).

122. Veblen, "Remarks on the Foundations of Geometry," p. 141 also quoted in Mac Lane, "Topology and Logic at Princeton," p. 219. Indeed, Veblen had tried to do his part at Princeton as chronicled by William Aspray in "Oswald Veblen and the Origins of Mathematical Logic at Princeton," in *Perspectives on the History of Mathematical Logic*, ed. Thomas Drucker (Boston: Birkhäuser, 1991), 346–366.

123. Ivor Grattan-Guinness, *The Search for Mathematical Roots 1870–1940: Logics, Set Theories and the Foundations of Mathematics from Cantor through Russell to Gödel* (Princeton: Princeton University Press, 2000), pp. 568–569.

In a free-standing reviews section, they adopted a set of criteria for inclusion—a definition of symbolic logic—that followed the lead of the "Bibliography of Symbolic Logic" that Church had compiled. Even Church had to admit, though, that the "line is . . . difficult to draw on both sides. . . . By symbolic logic is understood the formal structure of propositions and of deductive reasoning investigated by the symbolic method." In practice, this meant the exclusion of "[w]orks on traditional, or Aristotelian, logic . . . unless they contain[ed] application of, or appear[ed] to have some especial bearing on, those parts of formal logic which are more usually described as symbolic" as well as of "[w]orks on postulate theory" and of "mathematical treatises in which use of notations of symbolic logic is merely an abbreviation, or occurs only incidentally." "The test," Church explained, was "substantial use, explicit or implicit, of formal criteria of inference, so that the symbolism becomes a calculus rather than a mere shorthand." This then implied the inclusion of "treatises on foundational questions connected with the axiom of choice, on the paradoxes of which Burali-Forti's and Russell's are typical, and on Intuitionism and related topics" in addition to "[w]orks on Boolean algebra and *closely* related topics" and "discussions of logical diagrams, such as those of Venn and Peirce, . . . when there is thought to be some relevance to symbolic logic." By following these criteria, the *Journal of Symbolic Logic* served to carve out its niche in an otherwise ill-defined borderland between mathematics and philosophy and actually to define the new field of symbolic logic.

Like the MAA and its *Monthly*, the ASL financed its new publication venture primarily through membership dues, but it also managed to secure institutional support from a number of colleges and universities, among them, Brown, Columbia, Harvard, Michigan, Princeton, the Institute, the University of Rochester, and two of the women's colleges, Bryn Mawr and Smith. Its inaugural volume contained eleven original research papers, over fifty substantive reviews, and Church's 100-page bibliography. Indeed, through the 1930s and into the 1940s, one-third to as much as one-half of the journal was devoted to reviews of contemporaneous work both in the United States and in Europe. It aimed to keep its readers abreast of the latest work in the field as well as to provide careful critique of that work in order to promote the highest standards of scholarship.[124]

124. Herbert Enderton, "Alonzo Church and the Reviews," *BSL* 4 (1998), 172–180 on p. 175.

As the cases of the *Duke Mathematical Journal* and the *Journal of Symbolic Logic* attest, journal formation goes hand in hand with key developments in the evolution of a mathematical research community.[125] In the case of the former, it signaled the arrival on the scene of an area of the country that had previously been considered a mathematical backwater. In that of the latter, it underscored the disciplinary delineation that can occur as mathematical communities grow in size and as like-minded practitioners come to seek common cause. The American mathematical community's continuing maturation through the 1940s and into the 1950s, moreover, brought with it more journals in these molds: the geographically oriented *Pacific Journal of Mathematics* founded in 1951 and the specialized *Quarterly of Applied Mathematics* and *Mathematical Tables and Other Aids to Computation* both begun in 1943 (see chapters eight, nine, and ten).[126] Interestingly, the American mathematical community first experienced both of these evolutionary changes over the course of a decade, the 1930s, that had opened with an extreme sense of foreboding.

———

If there had been questions—following the stock market's crash and the precipitous slide into the Great Depression—about the sustainability of the advances that the American mathematical research community had made over the course of the preceding decade, it was clear by the mid-1930s that

125. For more on this idea in the American context, see Karen Hunger Parshall, "Journals in the Evolution of a National Research Community: The Case of Mathematics in the United States (1776–1940)," to appear.

126. Another example of an attempt at subdisciplinary delineation through the establishment of journals involved the history of mathematics. The journal *Scripta Mathematica* was founded by Jekuthiel Ginsburg at Yeshiva University in 1932 and ran until 1973. In addition, in March 1939, Brown University proposed, in conjunction with the MAA, to launch *Eudemus*, an international journal devoted to the history of mathematics and astronomy to be co-edited by Raymond Archibald and the then-recent émigré Otto Neugebauer, both at Brown. Ultimately, only one volume of *Eudemus* was published, in 1941 (Archibald to William Cairns, 23 March, 1939, Box 86-14/65, Folder 5, MAA Records). Even as early as 1920, the MAA had aimed to support the history of mathematics by revivifying *Bibliotheca Mathematica*, a journal that had been edited by the Swedish historian of mathematics Gustav Eneström and that had ceased publication in 1915. Failing to secure the necessary financial support for the venture in 1920, it had tried—and failed—again in 1928 ("Bibliotheca Mathematica," *Science* 68 (1928), 474–475).

that community had matured sufficiently to maintain itself under duress. Despite financial stringencies that had prevented some from pursuing careers in academe and that had rolled back, at least temporarily, the salaries of many with academic jobs, American mathematicians swelled the pages of their journals with new results and successfully financed their publication in the wake, particularly, of disappearing Rockefeller philanthropic support for technical publication. The lesson? While such support unquestionably facilitated and enhanced research and while it was most certainly welcome and appreciated, mathematical production would and could go on even if it was curtailed. Institutions like Princeton in the case of the *Annals of Mathematics* or Duke in that of the *Duke Mathematical Journal* had come to acknowledge—as had Daniel Gilman at Hopkins in the 1870s—that providing publication outlets for research served to enhance academic reputation. Professional societies—the research-oriented AMS, the teaching-oriented MAA, the even more specialized ASL—also recognized it as their duty to support publication, even if they sometimes had to scrabble for funds.

Nor, paradoxically, did the Depression prevent quite a number of universities from coast to coast and north to south—Penn, MIT, Ohio State, Wisconsin, Illinois, Michigan, Stanford, Caltech, Berkeley, Texas, Rice, Virginia, Duke, among others—from building their departments of mathematics and embracing the production of original research as a key criterion for success within the academy more broadly. This sense of mission was further reinforced by the Institute for Advanced Study, a new experiment in American higher education that unabashedly privileged research at the purely postdoctoral level. The Institute's existence, and the policies it developed, forced the universities to assess the depth of their commitment to research by confronting them with the notion of the paid research sabbatical and by providing a mechanism for sharing in its cost.

These represented important changes in the American mathematical landscape of the 1930s. In interesting and unexpected ways, they were associated with the decade's financial instability as mathematicians, like so many others, did their best to follow what little money there was. Other key changes in the 1930s, however, reflected a geopolitical upheaval that ultimately resulted in the outbreak of another world war.

5

Adapting to Geopolitical Changes

As American mathematicians managed not only to sustain their research but also further to develop their research community throughout the financially fraught 1930s, so, too, did they sustain the international presence that they had so carefully cultivated in the 1920s. In particular, they attended and were politically and mathematically active in the two International Congresses of Mathematicians that took place in the 1930s, one in Zürich in 1932 and one four years later in Oslo. They also won their bid at the latter to host the 1940 ICM in Cambridge, Massachusetts, and thereby, in some sense, officially to assert their role as an international mathematical leader.

International participation became increasingly problematic after April 1933, however, when Nazi policies compelled the expulsion of so-called non-Aryans from German universities. That act set in motion a flight of Jewish and other mathematicians—first from Germany and then from other European countries—as Hitler gained power and began the rampage across Europe that would result in the outbreak in 1939 of the Second World War. Some eighty of those German-speaking mathematical émigrés ultimately found refuge in the United States thanks largely to the efforts of a handful of well-placed and influential humanitarian and mathematical activists. At least another forty came from other countries, for more than 120 in all.[1] The relationships that

1. Exact numbers are hard to determine, but in *Mathematicians Fleeing from Nazi Germany*, Siegmund-Schultze gives the dates of the first and final (until 1945) place of immigration for the 145 *German-speaking* refugee mathematicians (to all countries, including the United States) in his sample space (see Appendix 1.1, pp. 343–357). In his article "Refugee Mathematicians in the United States of America, 1933–1941," Nathan Reingold estimates that "the total migration was somewhere between 120 and 150" (p. 176 in the reprinted edition). "More

developed represented very different international ties, *international* ties that became *national* associations within the context of the American mathematical community.

Although forced to leave Europe and its well-established mathematical networks behind, these émigrés nevertheless found themselves transplanted into a mathematical environment on the other side of the Atlantic that by the 1930s supported a robust research and publication endeavor centered on the AMS and, to a lesser extent, on the MAA. The émigrés did not make the American mathematical research community. It was already fully formed at the time of their arrival.

Yet, if developments within American mathematics, especially in the 1920s, rendered the émigrés' *mathematical* transition relatively easy, their *social* transition proved less so. Arriving in the 1930s when colleges and universities were still in the throes of the Depression, they were perceived by some as depriving deserving American mathematicians of positions. It also seemed that they were being provided with more research-oriented— as opposed to time-intensive teaching—positions because of their perhaps imperfect command of English and their lack of familiarity with the American undergraduate. The majority of them, moreover, were Jewish at a time when anti-Semitism was far from unknown in academic and other American social circles. Could they be absorbed and, if so, in ways that strengthened rather than undermined the American mathematical community? Activists like Veblen and Richardson hoped that they could but recognized equally that success would hinge on the ability of the transplanted Europeans and the home-grown Americans mutually to adapt.

International Business as Usual?

After the dramatic renunciation by Salvatore Pincherle and others of the International Mathematical Union, and, by association, its umbrella organization, the International Research Council, at the 1928 ICM in Bologna (recall

---

than 120" would thus seem to be a safe characterization of the total number. See also Arnold Dresden, "The Migration of Mathematics," *AMM* 49 (1942), 415–429 for data that Reingold characterizes as "both incomplete and rather peculiar in some specifics" (p. 176, note 4).

To give a sense of scale, in 1945, the AMS had 2,828 members. Even if none of the émigrés had joined the AMS, they would have represented some 4% of the American mathematical research community, at least as defined by the AMS, in 1945. For the 1945 membership figures, see Richard Bruck, "The Annual Meeting of the Society," *BAMS* 52 (1946), 35–47 on p. 39.

chapter three), the first ICM of the 1930s, held in Zürich in 1932, represented something of a new beginning. It, like the Bologna Congress, was independent of the IMU and its politics, and, so, as had Bologna, Zürich welcomed mathematicians independently of international science politics and regardless of their nationality. Of the 667 mathematicians who attended, some 140 or 22% were from Switzerland, but 111 or 17% hailed from previously banned Germany with North American and French participation at 68 or 10% each. In all, thirty-five countries were represented.[2]

Over the course of the congress's six working days, the men and women in attendance had the opportunity to hear twenty-one, hour-long, invited addresses in addition to some 200 fifteen-minute presentations divided among eight mathematical sections. Two Americans—Morse and Alexander—gave invited talks, and they found themselves in eminent mathematical company.[3] Morse, who spoke on "The Calculus of Variations in the Large," shared the spotlight on Wednesday, 7 September, with Paris's Élie Cartan, Ludwig Bieberbach of the University of Berlin, Göttingen's Emmy Noether, and Harald Bohr of the University of Copenhagen, while Alexander took the podium the following day to lecture on "Some Problems in Topology" alongside the University of Rome's Francesco Severi, Rolf Nevanlinna of the University of Helsinki, the University of Geneva's Rolin Wavre, and Frigyes Riesz of the University of Szeged in Hungary. It was not without more than a little pride and American boosterism that Richardson described the talks of his two fellow countrymen as "among the very best in material and form" of all the invited addresses given.[4]

In his lecture, Morse reprised for his Zürich audience the topic to which he had devoted his AMS Colloquium Lectures a year earlier in Minneapolis. As he explained, his calculus of variations in the large—loosely

---

2. Roland Richardson, "International Congress of Mathematicians, Zurich, 1932," *BAMS* 38 (1932), 769–774 gives the number of countries as thirty-seven (p. 769), but the congress proceedings lists thirty-five. See Walter Saxer, ed. *Verhandlungen des Internationalen Mathematiker-Kongresses Zürich 1932*, 1: 36. Italy and Great Britain had the next largest numbers of attendees at 63 (just under 10%) and 38 (5%), respectively.

3. The Germans gave four of the hour addresses; the French and Swiss, three each; the Americans and Russians, two each; and one each for the Austrians, Poles, Hungarians, Finns, Italians, Danes, and Swedes. See Saxer, ed., 1: table of contents. Wolfgang Pauli, who by this time taught at the University of Zürich, is counted as Swiss but was omitted from Saxer's list, presumably because he did not submit his paper for publication.

4. Richardson, "International Congress of Mathematicians, Zurich, 1932," p. 769.

speaking, understanding maxima and minima (critical points) using nonlinear techniques—involved carefully adapting, to the setting of functionals,
the "three principal steps" he had taken in the context of functions f in n variables. In that case, he had first considered "the topological characteristics of
the domain R of definition of" f, namely, in this case, the well-known topology of a subspace of real n-space.[5] He had next assigned what he called "type
numbers" to the critical points of f, that is, those points at which all of the
first partial derivatives of f vanish. Third, and finally, he determined "the relations between the topological characteristics of R and the type numbers of the
critical sets of f." In other words, type numbers allowed him to see through
to the underlying topology. A key problem in effecting the extension of these
three steps to the theory of functionals, however, was that, whereas in the case
of functions "an adequate topological theory is already at hand," in that of
functionals the topological theory had to be developed from scratch. In his
lecture, he sketched just how the type numbers, duly adapted, allowed him to
do that.

Instead of a function in n variables, then, take a functional, that is, a point in
function space. The context becomes a domain in function space as opposed
to an n-dimensional region in Euclidean n-space. Extremals of the functional
replace the critical points of the function. Morse thus had to find topological invariants for domains in function space, in terms of which he could
define their connectivities or, in other words, their underlying topological
structure.[6] Since he clearly could not give all of the details in an hour talk,
however, he referred his audience to the AMS Colloquium volume that he
was in the process of completing but that would only appear in 1934 (see the
next chapter).[7]

Alexander took a different tack in his lecture. While Morse gave a technical
talk aimed at a specialist audience, Alexander opted for a more free-wheeling,
more broadly accessible presentation. He opened by distinguishing between
point-set and combinatorial topology. These were precisely the two camps
into which topology, as pursued in the United States and elsewhere, had

5. Marston Morse, "The Calculus of Variations in the Large," *Verhandlungen der Internationalen Mathematiker-Kongresses Zürich 1932*, ed. Walter Saxer, 1: 173–188 on p. 173. The
quotations that follow in this paragraph are also on this page.

6. Arnold Dresden, Review of "*The Calculus of Variations in the Large*. By Marston Morse."
BAMS 42 (1936), 607–612 on p. 613.

7. Marston Morse, *The Calculus of Variations in the Large*, AMS Colloquium Publications,
vol. 18 (New York: American Mathematical Society, 1934).

divided by the 1930s, and it was clear that Alexander, a topologist of the combinatorial persuasion, was making a pitch on the international stage for his own way of thinking. In his view, "[w]henever we attack a topological problem by analytic methods," that is, whenever we think about a problem from a point-set topological point of view, "it almost invariably happens that to the intrinsic difficulties of the problem, which we can hardly hope to avoid, there are added certain extraneous difficulties in no way connected with the problem itself, but apparently associated with the particular type of machinery used in dealing with it."[8] Thus, in his view and not without a tinge of false modesty, "it is often better to avoid the artificial subtleties of analysis by using a more simple minded type of machinery," namely, that of combinatorial analysis or the brand of topology that he and his Princeton colleagues had been hard at work developing through the 1920s and into the 1930s. Alexander also used his international platform to pose a series of open questions about three-dimensional manifolds and, in particular, to link that theory to the theory of knots, a topic that had occupied him since the mid-1920s.

If Americans accounted for 10% of the invited addresses in Zürich, they were responsible for somewhat more—over 13%—of the short papers presented there. Although no Americans contributed to the section on the "Technical Mathematical Sciences and Astronomy" and although that on "Algebra and Number Theory" was represented only by Yale's Oystein Ore, the sections on "Analysis" and "Geometry" reflected the robustness of American production in those areas.[9] Among the speakers in the geometry section, moreover, Vassar College's Louise Cummings underscored—as had Olive Hazlett in Toronto and Bologna—the presence of women in the American mathematical research community.[10]

8. Alexander, "Some Problems in Topology," p. 249. For the next quotation, see p. 250.

9. Oystein Ore, "Theory of Non-Commutative Polynomials," *Verhandlungen der Internationalen Mathematiker-Kongresses Zürich 1932*, ed. Walter Saxer, 2: 19–20.

10. Cummings had joined the Vassar faculty as an instructor in 1902 and had worked mathematically with Henry White, her colleague there, before officially earning her Ph.D. in 1914 under Charlotte Scott at Bryn Mawr. She had risen to the rank of full professor by 1927.

In Zürich, Emmy Noether was the only woman who gave an hour-long, invited address, but some three dozen women participated in the congress's mathematical, as opposed to social, side. See Saxer, ed., 1: 27–36. Indeed, after Noether, the next woman to give an ICM plenary address was Karen Uhlenbeck in 1990.

The purely mathematical agenda at the ICM continued to be accompanied, however, by a disruptive political agenda. The International Mathematical Union, ostensibly the organizer and overseer of the congresses, had been defied in Bologna and Zürich. To make matters worse for it, its statutes had actually expired in 1931. Despite all this, a meeting was held at the Zürich ICM to consider, in Richardson's words, "what steps, if any, should be taken to perpetuate the organization."[11] Along with Veblen, Cornell's Virgil Snyder, and Cincinnati's Charles Moore, Richardson participated in a discussion that came to the damning conclusion that "a permanent international organization had no problems important enough to warrant its existence." Although this point would be debated once again in 1934 in advance of the 1936 Oslo Congress,[12] the IMU was effectively put out of its misery in Zürich in 1932.[13] The Americans, as well as mathematicians internationally, overwhelmingly concurred that there was no need for some sort of a new IMU—within a body like the International Research Council—to oversee them.

In light of these developments, some within the American mathematical community had begun to think seriously—in contradistinction to the ill-considered invitation made in 1920—about hosting an ICM in the United States. Although the next congress was already slated to take place in Oslo in July 1936, it would be there that any invitation to host the 1940 ICM would have to be extended and discussed. Veblen, for one, was already at work and had sounded out his friend, Birkhoff, on the issue. He knew that, as early as 1930, Birkhoff had been looking ahead to 1936 or 1940 for a possible American bid for the ICM and had even begun to explore the possibility of support from the Rockefeller Foundation for such a venture.[14] He was thus likely unsurprised to learn that Birkhoff was not only "strongly in favor of staging a modest international mathematical congress" in the United States in 1940, but that he was also of the opinion that "[i]f we do not hold an American congress soon the effect would be very unfortunate" relative to the larger agenda of both showcasing American mathematical accomplishments and

11. Richardson, "International Congress of Mathematicians, Zurich, 1932," p. 773. The quotation that follows is also on this page.

12. See, for example, the correspondence in Box 24, Folder: International Mathematical Union 1934, Veblen Papers.

13. Lehto details the tortured early history of the IMU on pp. 15–71.

14. Birkhoff to Veblen, 17 March, 1930, Box 2, Folder: Birkhoff, George D. 1912–47, Veblen Papers.

FIGURE 5.1. Richard Courant, G. H. Hardy, and Oswald Veblen
photographed in Zürich in 1932. (Photo in the collection of the
Mathematisches Forschungsinstitut Oberwolfach.)

playing a more important and visible international leadership role.[15] Hosting
an international congress could well be a way, as Birkhoff had put it in 1927,
finally "to activate this situation."

Just days after Veblen received Birkhoff's letter in April 1936, Lefschetz,
then the AMS President, formed an eight-person committee—including,
not surprisingly, both Veblen and Birkhoff but also Eisenhart and Morse—
to represent the Society at the Oslo Congress that summer.[16] By the time
they converged on the Scandinavian city in July, $15,000 had already been
secured in support of an ICM on American shores—$7,500 each from the

15. Birkhoff to Veblen, 6 April, 1936, Box 2, Folder: Birkhoff, George D. 1912–47, Veblen
Papers. For the quotation that follows, see Birkhoff to Gilbert Bliss and Earle Hedrick, 10
February, 1927, HUG 4213.2, Box 8, Folder: Correspondence 1927 A–Z, Birkhoff Papers.

16. John Kline, "The April Meeting in New York," *BAMS* (1936), 449–455 on p. 451.
The other delegates were Lefschetz himself, together with Hans Blichfeldt, Virgil Snyder, and
Norbert Wiener.

Carnegie Corporation and the Rockefeller Foundation—and the American delegation, chaired by Eisenhart, had been authorized to extend an American invitation for 1940.[17] As Morse later reported, that invitation "was accepted with thanks" at the Congress's closing meeting on 18 July.[18] It would be time for the Americans to get to work (see chapter seven), but not before they savored the fact that at Oslo "two young men *on the staffs* of *American* universities"—the Finn Lars Ahlfors then at Harvard and the American Jesse Douglas then at MIT—had won the first two Fields Medals.[19] The selection committee—chaired by Francesco Severi and rounded out by Birkhoff, Carathéodory, Cartan, and Teiji Takagi—had singled out Ahlfors's early work in complex analysis and Douglas's proof of Plateau's Problem (recall chapter three) for special recognition on the international stage.

Also at Oslo, effects of the worsening European political situation were in evidence. Four years earlier, Hermann Weyl and his wife had made the trip from their home in Göttingen to Zürich. The Oslo ICM, however, found Weyl giving a fifteen-minute lecture on "Riemannian Matrices and Factor

17. Archibald, *Semicentennial History*, p. 20 and Roland Richardson, "Minutes of Meeting of Organizing Committee: International Congress of Mathematicians," 29 December, 1937, Box 93-372/3, Folder: International Congress of Mathematicians 1940, Price Papers. In terms of "real wealth," $15,000 in 1936 dollars is equivalent to $289,000 in 2021 dollars (https://www.measuringworth.com).

On early Carnegie support for mathematics, consult Fenster, "Funds for Mathematics"; on Rockefeller support, see Siegmund-Schultze, *Rockefeller and the Internationalization of Mathematics*.

18. Marston Morse, "The International Congress in Oslo," *BAMS* 42 (1936), 777–781 on p. 781. For the quotation that follows, see p. 777 (with my emphasis). What Morse does not recount is that the Americans were not alone in bidding for the 1940 ICM. The fact that they already had money in hand gave them the edge over the Greeks, who were also vying to be congress hosts. See Christopher Hollings and Reinhard Siegmund-Schultze, *Meeting under the Integral Sign?: The Oslo Congress of Mathematicians on the Eve of the Second World War*, HMATH, vol. 44 (Providence: American Mathematical Society, 2020), p. 128.

19. Ahlfors returned to Finland in 1938, where he took up a professorship at the University of Helsinki, but by March 1945, he had left Finland for a safer position at the Eigenössische Technische Hochscule in Zürich. Finding Switzerland less than amenable, he accepted Harvard's call in 1946 and remained on the faculty there until his retirement in 1977. Douglas left MIT in 1936 for a visiting position at the Institute for Advanced Study. After holding Guggenheim fellowships in 1941 and 1942, he taught at Brooklyn College and Columbia University, ultimately taking a position at the City College of New York in 1955.

Systems" as a member not of the faculty of Göttingen University in Germany but of the Institute for Advanced Study in Princeton.[20] He and his family had emigrated to the United States near the close of 1933. Richard Courant, too, had ultimately left Germany for the United States—but in 1934 after a year's stay at Cambridge University in England—and had established himself at New York University in New York City. In Oslo, his fifteen-minute talk "On Plateau's Problem" extended Douglas's methods and was given under his new credentials, not in association with his former directorship of the Rockefeller-funded Mathematisches Institut in Göttingen.[21] In the immediate aftermath of the Austrian Anschluss in March 1938, at least two other Oslo speakers, Karl Menger and his student at the University of Vienna, the econometrician and future computer scientist Franz Alt, had sought and found refuge in the United States. Another Oslo speaker, historian of mathematics Otto Neugebauer (at Göttingen until 1933 and then at the University of Copenhagen), followed in January 1939 (see chapter seven). The political situation in Europe was directly affecting the American mathematical community.[22]

Embracing Foreign Mathematicians

The American mathematical research community had actually welcomed foreign mathematicians from its very inception. One of them, James Joseph Sylvester, had arguably animated the first research-level program in the field in the United States at the Johns Hopkins University after his transplantation from London to Baltimore in 1876.[23] Subsequent decades had brought additional talent from abroad. As noted, in the 1890s, the Germans Oskar Bolza

20. Hermann Weyl, "Riemannsche Matrizen und Faktorsysteme," *Comptes rendus du Congrès international des mathématiciens Oslo 1936,* 2 vols. (Oslo: A. W. Brøggers Boktykkeri A/S, 1937), 2: 3.

21. Richard Courant, "Über das Problem von Plateau," op. cit., 2: 143. See also the account of this paper in Constance Reid, *Courant in Göttingen and New York: The Story of an Improbable Mathematician* (New York: Springer-Verlag, 1976), p. 181. Siegmund-Schultze discusses the institute's founding in *Rockefeller and the Internationalization of Mathematics,* pp. 144–155.

22. Hollings and Siegmund-Schultze discuss the participation of refugees mathematicians in Oslo on pp. 117–119.

23. Parshall, "America's First School of Mathematical Research"; Parshall and Rowe, *Emergence,* pp. 53–146; and Parshall, *James Joseph Sylvester: Jewish Mathematician in a Victorian World,* pp. 225–277.

and Heinrich Maschke formed two-thirds of the first mathematics department at the University of Chicago from its opening in 1892.[24] Canadian
Roland Richardson followed his 1906 Yale Ph.D. with the position at Brown
University that he would hold for the rest of his career. Scots algebraist Joseph
Wedderburn returned home following his Carnegie year at Chicago only to
be lured back to the United States and Princeton where he remained until his
retirement in 1945. These émigrés, like others who followed, were of various
religious persuasions—Sylvester was Jewish—and had a variety of reasons for
leaving their homelands.

The 1920s witnessed continued mathematical migration to the United
States, but as the direct result of geopolitical shifts such as the Russian Revolu-
tion in 1917 and the turbulent, post-revolutionary political situation there.[25]
For example, the Jewish mathematical physicist George Rainich left Kazan
for Baltimore via Istanbul in 1923. After three years on the Hopkins faculty, he
moved to Michigan, where he quickly rose through the ranks at the same time
that he—together with Raymond Wilder—was instrumental in orienting
the Michigan Mathematics Department toward research. Algebraic geometer
Oscar Zariski joined the Hopkins faculty the year after Rainich's departure for
Ann Arbor. Zariski had left Russia in 1920 to earn his doctoral degree in Rome
and, with Lefschetz's encouragement, had come to the United States to make
a mathematical career that ultimately found him at Hopkins until 1946, at
Illinois for a year, and then at Harvard until his retirement in 1969.[26] Russian-
born analysts James Shohat and Jacob Tamarkin settled down to careers at
Penn (in 1930) and at Brown (in 1927), respectively, while their countryman, probabilist James Uspensky went to Stanford in 1929 to be joined seven
years later by applied mathematician Stephen Timoshenko, the latter having

24. Parshall, "E. H. Moore and the Founding of a Mathematical Community in America,"
and Parshall and Rowe, *Emergence*, pp. 279–294.

25. Lipman Bers, "The Migration of European Mathematicians to America," in *A Century of
Mathematics in America*, ed. Peter Duren et al., 1: 231–243 on pp. 232–233. At least one noted
Russian-born mathematician, Solomon Lefschetz, had come to the United States earlier. The
son of Turkish citizens, Lefschetz, whose father was an importer, was educated as an engineer
in France and came to the United States in 1905 to pursue a career in that field. He only moved
into mathematics in 1910 after an accident in the laboratory resulted in the loss of both of his
hands.

26. For more on Zariski's life and work, see Carol Parikh, *The Unreal Life of Oscar Zariski*
(Boston: Academic Press, Inc., 1991).

held positions in the 1920s at Westinghouse and, like Shohat, at Michigan. Yet another Russian applied mathematician, Ivan Sokolnikoff, made his way to the United States via China in 1922, earned a bachelor's degree in electrical engineering at the University of Idaho in 1926 and a Ph.D. in mathematics at Wisconsin in 1930, and served on the faculty at Wisconsin until his move to UCLA in 1946. All of these early political émigrés quickly established themselves and represented their fields in the American mathematical research community.

The efforts of the International Education Board and of individual institutions also resulted in the importation of mathematical talent from abroad in the 1920s. Following their IEB fellowships, the Dutch differential geometer Dirk Struik and the Hungarian analyst Tibor Radó took positions at MIT in 1926 and at the Ohio State University in 1930, respectively. Yale and Caltech, however, like Hopkins before them, sought European talent directly in the 1920s, with Yale hiring Norwegian number theorist Oystein Ore in 1927 and Caltech converting its half-year arrangement with the Hungarian-born engineer and applied mathematician Theodore von Kármán into the full-time directorship of its Guggenheim Aeronautical Laboratory in 1930.[27]

These transatlantic crossings continued into the 1930s, even before the political turn of events in Germany in 1933. The Leipzig-trained Hungarian analyst Aurel Wintner and the Belgian-born topologist Egbert van Kampen joined Zariski at Hopkins in 1930 and 1931, respectively, while another Hungarian, John von Neumann, took the lectureship at Princeton in 1930 that became a professorship there a year later.[28] As noted, von Neumann was tapped in 1933 as a member of the first faculty of the Institute, remaining in that post until his death in 1957 and making fundamental contributions to ergodic theory as well as to game theory and computer science.

27. Siegmund-Schultze provides more examples in *Mathematicians Fleeing from Nazi Germany*, pp. 30–58.

28. According to Stanislaw Ulam, "[w]ith his typically rational approach, Johnny computed that the expected number of professorial appointments [in Germany] within three years was three, the number of Dozents was 40! He also felt that the coming political events would make intellectual work very difficult." See Stanislaw Ulam, "John von Neumann 1903-1957," *BAMS* 64 (1958), 1–49 on p. 3. On the lives of Wintner and van Kampen, see Philip Hartman, "Aurel Wintner," *Journal of the London Mathematical Society* 37 (1962), 483–503 and Robbert Fokkink, "A Forgotten Mathematician," *European Mathematical Society Newsletter* 52 (2004), 9–14, respectively.

The story of the Austrian-born but also German-trained ergodic theorist Eberhard Hopf was different. A Rockefeller Foundation, National Sciences Division fellowship brought him to Harvard in 1930 to work with Birkhoff in mathematics and Harlow Shapley in astronomy, and he accepted an assistant professorship at MIT in 1932.[29] Increasingly dissatisfied with his Depression-Era salary, however, Hopf returned to Germany when the University of Leipzig offered him a full professorship in 1936.[30] Working there throughout the war, he returned to the United States in 1947 to take a visiting position at New York University. Two years later, he became a U.S. citizen and assumed the professorship at Indiana that he would hold until his retirement in 1972.[31]

If Hopf, an Aryan, went back in the 1930s to a Germany under Hitler, many Jewish and politically dissenting Germans actively sought refuge in the United States following the dismissals of German academics over the course of the three weeks between 13 April and 4 May, 1933. This state of affairs was brought dramatically to light by the *Manchester Guardian* on 19 May in a widely read, full-page article entitled "Nazi 'Purge' of the Universities: A Long List of Dismissals."[32] The register of 196 names with the dates and places of dismissal "read like a Who's Who of scholarship" and sent shock waves through the academic world. By the end of May 1933, two organizations, the Academic Assistance Council in England and the Emergency Committee in Aid of Displaced German Scholars (EC) in the United States, had been formed to try to address the situation.[33]

29. Siegmund-Schultze, *Rockefeller and the Internationalization of Mathematics*, pp. 103–105 and 292.

30. Wiener, *I Am a Mathematician*, pp. 209–211.

31. "In Memoriam Eberhard Hopf: 1902–1983," *Indiana University Mathematics Journal* 32 (1983), i–ii. See also M. Decker, "Eberhard Hopf 04-17-1902 to 07-24-1983," *Jahresbericht der Deutschen Mathematiker-Vereinigung* 92 (1990), 47–57.

32. The article is reproduced in Charles Weiner, "A New Site for the Seminar: The Refugees and American Physics in the Thirties," in *The Intellectual Migration: Europe and America, 1930–1960*, ed. Donald Fleming and Bernard Bailyn, *Perspectives in American History*, vol. 2 (Cambridge: Charles Warren Center for Studies in American History of Harvard University, 1968), pp. 190–234 on p. 234. The quotation that follows is on p. 204.

33. Stephen Duggan and Betty Drury, *The Rescue of Science and Learning: The Story of the Emergency Committee in Aid of Displaced Foreign Scholars* (New York: The Macmillan Company, 1948), p. 5. France, the Netherlands, Belgium, Switzerland, and the Scandinavian countries also contributed to the effort, but their effectiveness was much shorter-lived, given the subsequent Nazi advance on Europe.

Mathematicians and the Emergency Committee in Aid of Displaced German Scholars

The Emergency Committee grew out of the Institute of International Education that had been founded in 1919 by 1912 Nobel Peace Prize laureate Elihu Root, Columbia University President Nicholas Murray Butler, and Stephen Duggan, a historian and educator widely known in his day as "the apostle of internationalism."[34] The Institute had the idealistic goal of serving "as an instrument to develop understanding and good will between the people of the United States and the peoples of other countries."[35] In May 1933, it was approached by several individuals, well-connected in philanthropic circles, to find a way "whereby assistance might be extended to scholars." Duggan and his colleagues set to work immediately, forming an Executive Committee comprised of an elite group of some two dozen college and university presidents as well as other academics. Among this number was one mathematician, Oswald Veblen.[36]

The broader American mathematical community had also felt the need to act. After all, American mathematicians had come to know their European counterparts personally as a result of their direct participation in the mathematical scene abroad. Some of the Europeans had also visited the United States as research fellows and visiting lecturers. They were thus not just names on book spines and in the tables of contents of mathematics journals. They were friends and compatriots in a shared endeavor. In May, Veblen and Richardson, together with Bliss and Dartmouth's Louis Silverman, in their capacity as "persons interested in doing something," met in New York and decided that it would be prudent for America's mathematicians to "tie up with the Emergency Committee" in their efforts to address the plight of their German colleagues.[37] Visible mathematicians like Richardson and Veblen were increasingly receiving letters from abroad detailing the situation and urging

34. E. C. Condon, "Duggan, Stephen Pierce," in *Biographical Dictionary of American Educators*, ed. John Ohles, 3 vols. (Westport: Greenwood Publishing Group, 1978), 1: 402–403. The characterization appears on p. 402.

35. Duggan and Drury, p. 6. The quotation that follows in the next sentence is also on this page.

36. Duggan and Drury give the full list of members on p. 177.

37. Richardson to Hans Blichfeldt, Henry Rietz, and Oswald Veblen, 25 July, 1933, Box 23, Folder: Emergency Committee in Aid of Displaced German Scholars, Veblen Papers.

aid for specific, high-profile colleagues, among them, Richard Courant, Otto Neugebauer, and Emmy Noether in Göttingen, Issai Schur in Berlin, and Kurt Reidemeister in Königsberg.[38]

The AMS's summer meeting in Chicago in June was also abuzz with news from abroad. Held in conjunction with the "Century of Progress" World's Fair (see the next chapter), it benefitted from the presence of a number of European mathematicians—particularly Tullio Levi-Civita from Rome and Lipót Fejér from Budapest—who had been brought over to join in the celebration of the city of Chicago's hundredth anniversary.[39] Celebrating a "Century of Progress" in the depths of the Depression and in the aftermath of the Nazi purge? The irony could not have failed to strike at least some of those present in Chicago, but the fair's organizers actually hoped that their event would serve as an antidote to the financial and political worries of the day by instilling confidence in the future. In the mathematical sciences, the fair and the AMS's meeting also attracted the Danish Nobel laureate in physics, Niels Bohr, who, together with his mathematician brother, Harald, had quickly become an information conduit between Europe, England, and the United States on the status of Germany's mathematicians and physicists.[40] These and other Europeans were barraged for information on the unfolding situation on the other side of the Atlantic.[41]

Based on input from all of these sources, the AMS Council appointed an "informal committee" comprised of Stanford's Hans Blichfeldt, the University

38. See, for example, Heinz Hopf to George Polyá, 25 May, 1933, forwarded with Polyá to Veblen, 25 June, 1933; John von Neumann to Veblen, 19 June, 1933; and Harald Bohr to Richardson, 30 May, 1933, all in Box 23, Folder: Emergency Committee in Aid of Displaced German Scholars, May–September 1933 (hereinafter EC, May–Sept. 1933), Veblen Papers. On Courant, Neugebauer, Noether, and Schur, see below. Reidemeister was dismissed on purely political grounds from his position at Königsberg in 1933 but was named to a chair in Marburg in 1934 following Wilhelm Blaschke's intervention on his behalf. Reidemeister remained in Marburg throughout the 1930s as well as World War II.

39. Mark Ingraham, "The Summer Meeting in Chicago," *BAMS* 39 (1933), 633–640.

40. For more on the efforts of the brothers Bohr as well as on Denmark as an intermediate destination for particularly German refugees, see Henrik Sørensen, "Confluences of Agendas: Emigrant Mathematicians in Transit in Denmark, 1933–1945," *Historia Mathematica* 41 (2016), 157–187. Among the mathematicians who came to the United States via Denmark were Vilim (later William) Feller, Otto Neugebauer, and Max Dehn (see chapter seven).

41. Bohr, in fact, spoke specifically "on the situation of mathematicians and physicists in Central Europe" at the meeting's closing dinner for the assembled members of the AMS and the MAA. See Ingraham, "The Summer Meeting in Chicago," p. 635.

of Iowa's Henry Rietz, and Veblen as chair "to represent the mathematicians in cooperating with other committees concerning the professors who have been ousted in Germany for racial or political reasons."[42] How, though, could a three-person, "informal," AMS committee hope actually "to help our persecuted colleagues"? It had no funding and no real power. It represented more of a political statement than a concerted call to action. Still, of its three members, at least Veblen, through his participation in the Emergency Committee, was actually in a position to fulfil the charge to "represent the mathematicians" in the relief efforts.

By the end of June, the Emergency Committee had raised $35,000 primarily from wealthy Jewish donors and had decided to make $2,000 available to each of fifteen institutions—Berkeley, Brown, Bryn Mawr, Chicago, Columbia, Cornell, Harvard, Hopkins, MIT, Michigan, Minnesota, Ohio State, Princeton, Stanford, and Yale—"for the partial support of one German professor" should they wish to serve as a temporary host.[43] The EC's support would help cover costs for one year, but the schools were urged to make two-year appointments and to raise the rest of the money required from other sources. They were specifically "advised to apply to the Rockefeller Foundation," one philanthropic concern that the EC had already persuaded to help.

Theirs had not been a hard sell. The Foundation had made significant investments in European science in the 1920s through its International Education Board, even though it was shifting its focus away from the natural sciences and toward the social sciences in the 1930s. Relative to mathematics in particular, the IEB had invested impressively in Göttingen's Mathematisches Institut, a group gutted by the Nazi proclamation.[44] The EC's leadership felt certain that the Foundation would want to salvage its investments, and it was right. In the 1930s, it shifted its European philanthropic focus from developing institutions abroad to aiding foreign scholars threatened by Nazi policies.

42. Richardson to Blichfeldt et al. The quotation that follows is also from this letter.

43. Edward R. Murrow (in his role as Assistant Secretary of the EC), Minutes of the Executive Committee Meeting, 29 June, 1933, Box 23, Folder: EC, May–Sept. 1933, Veblen Papers. The quotation that follows is also from these minutes. See also Stephen Norwood, *The Third Reich in the Ivory Tower: Complicity and Conflict on American Campuses* (New York: Cambridge University Press, 2009), pp. 31–32. The "real wealth" of $35,000 in 1933 dollars is $721,200 in 2021 dollars (https://www.measuringworth.com).

44. Raymond Fosdick, *The Story of the Rockefeller Foundation* (New York: Harper & Brothers, 1952), pp. 277–278.

In partnership with the Foundation, the EC adopted the strategy that each of the universities tapped would choose a "displaced scholar of such eminence . . . that there would be no thought of competition with young American scholars."[45] It would then contact the EC for support. In this way, the EC would work directly with universities and not with individual refugee scholars. In theory at least, proven talent, not mere promise and not the poignancy of personal appeals, would serve as their selection criterion. Those chosen would, moreover, be, except in rare cases, over the age of thirty and invited into an *honorary* position for a *limited* number of years so that it would "be clearly understood that any commitment on the part of the university itself or of the committee should stop at the close of that period." These conditions were explicitly viewed as "safeguards" against the possible "resentment" of American scholars who were coping with the lean Depression-Era academic job market as well as against an anti-Semitic backlash, given that most of the displaced scholars were Jewish.

Worries on these scores were not idle. Meeting late in 1933 and explicitly discussing the émigré issue and the EC's role in it, the presidents of the state universities, for example, were almost unanimously of "the opinion . . . that there was too much dynamite in this [émigré] situation for State Universities."[46] In their view, "in spite of the fact that [the EC's] money came from special and private sources, . . . the public in each state could not be made to see that fact and would be outraged by seeing foreigners called to positions which, from their point of view might better be filled by well-trained Americans who were also out of positions." (See chapter seven for an example of precisely this attitude as exhibited by at least one citizen of the state of Idaho.)

Moreover, the quarter-century between the closes of the two World Wars marked what has been deemed "the worst period" of anti-Semitism in American history,[47] with prominent Americans agitating against what was viewed as the ever-increasing financial, political, and cultural influences of Jews

45. Duggan and Drury, p. 186. For the quotations that follow in this paragraph, see pp. 174–175, 176, and 186, respectively.

46. Roy Flickinger, Professor of Classics at the State University of Iowa, to Murrow, 23 December, 1933, Series II: Educational and Research Institutions, Box 149, Folder 16: New York University (1937–1938), Emergency Committee Papers. The quotation that follows is also from this letter.

47. Leonard Dinnerstein, "Antisemitism in Crisis Times in the United States: The 1920s and 1930s," in *Anti-Semitism in Times of Crisis*, ed. Sander Gilman and Steven Katz (New York: New York University Press, 1991), pp. 212–226 on p. 212.

in America. To take just two examples, Harvard President A. Lawrence Lowell played a leading role in the Immigration Restriction League in the 1920s, while the *Dearborn Independent*, bankrolled by automaker Henry Ford, strongly reflected Ford's vehemently anti-Semitic stance. Lowell, in fact, refused to cooperate with the EC as, initially, did his successor as Harvard President, James Conant. As the *New York Times* reported on 30 January, 1934, Harvard "would not make a place on its faculty for any man because he was an emigré, or as a protest to the Nazi removal of educators from German universities."[48] The EC clearly needed to tread carefully.

By July 1933, arrangements for the appointments of eight German scholars had already been made at American institutions thanks to Emergency Committee funds. Three of them—the Germans Felix Bernstein and Emmy Noether from Göttingen and the Hungarian Otto Szász from Frankfurt—were Jewish mathematicians. Indeed, some twenty émigrés had already found places in the United States, in England, and in the Netherlands as a result of the active cooperation between the EC and the English Academic Assistance Council.[49]

Connections in these early days were critical. The fifty-five-year-old Bernstein found refuge at Columbia owing to the intervention there of his friend, anthropologist Franz Boas. A biomedical statistician and director of Göttingen's Institute for Mathematical Statistics, Bernstein was perhaps better known in the American mathematical community for the set-theoretic result establishing the notion of cardinality that he had proved at the age of eighteen while on vacation with his father's friend and fellow Halle resident Georg Cantor.[50] Fifty-one-year-old Noether, one of the founders of modern algebra, received an invitation from Bryn Mawr, the only women's college in the

48. *New York Times*, 30 January, 1934, as quoted in Norwood, p. 33. For more on Harvard and the so-called "Jewish question," see Morton and Phyllis Keller, *Making Harvard Modern: The Rise of America's University* (New York: Oxford University Press, 2001), especially pp. 47–51.

49. Murrow to Veblen, 24 July, 1933, Box 23, Folder: EC, May-Sept. 1933, Veblen Papers.

50. The so-called Schröder-Bernstein Theorem states that if each of two sets is equivalent to a subset of the other, then the two sets are equivalent. On Bernstein's life and career, see M. Frewer, "Felix Bernstein," *Jahresbericht der Deutschen Mathematiker-Vereinigung* 83 (1981), 84–95; Reinhard Siegmund-Schultze, "Rockefeller Support for Mathematicians Fleeing from the Nazi Purge," in *The "Unacceptables": American Foundations and Refugee Scholars between the Two World Wars and After*, ed. Giuliana Gemelli (Brussels: P.I.E.–Peter Lang, 2000), pp. 83–106 on pp. 101–102; and Siegmund-Schultze, *Rockefeller and the Internationalization of Mathematics*, pp. 116–117 and 204.

United States with a graduate program in mathematics, at the suggestion of a Solomon Lefschetz unprepared to advocate for a woman on the faculty at Princeton.[51] MIT hosted the forty-nine-year-old harmonic analyst Szász on Wiener's recommendation even though Wiener viewed him as "not in the first rank but . . . a very useful mathematician of about the second rank."[52]

Although all of these appointments were initially temporary, Bernstein had had strong indications that his would become permanent. Unfortunately, he did not fit in well at Columbia, found a new home beginning in the 1936–1937 academic year at New York University teaching biometry, and ultimately made little impression on the American mathematical scene.[53] As is well-known, Noether died unexpectedly in 1935 of complications from abdominal surgery just as the Bryn Mawr administration was trying to secure funding to keep her on beyond her second year. Szász held temporary posts at MIT and jointly at MIT and Brown before finally moving to the University of Cincinnati in 1936, where he spent the rest of his career.

Like Columbia, Bryn Mawr, and MIT, Brown University also acted quickly but with somewhat more deliberation. Richardson, the chair of its

51. In Lefschetz's view, Noether was "the holder of a front rank seat in every sense of the word" in the mathematical world. "In fact," he continued, "it is no exaggeration to say that without exception, all the better young German mathematicians are her pupils. Were it not for her race, her sex, and her liberal political opinions (they are mild), she would have held a first rate professorship in Germany and we would have no occasion to concern ourselves with her." See Lefschetz to Jacob Billikopf, the Honorary Consultant of the Resettlement Division of the National Coordinating Committee for Aid to Refugees and Emigrants Coming from Germany, 31 December, 1934, Series V: General Correspondence, Box 177, Folder 10: Dresden, Arnold (1934, 1940–1944), Emergency Committee Papers. See also the case study on Noether in Siegmund-Schultze, *Mathematicians Fleeing from Nazi Germany*, pp. 214–217. Noether's life story has been well documented in, for example, Auguste Dick, *Emmy Noether 1882–1935* (Boston: Birkhäuser Verlag, 1981).

52. Karl Compton to Veblen, 31 May, 1933, Box 23, Folder: EC, May-Sept. 1933, Veblen Papers. On Szász's life, see Gábor Szegő, "Otto Szász," BAMS 60 (1954), 261–263.

53. The correspondence in Series I: Grantees, Box 2, Folder 13: Bernstein, Felix, Emergency Committee Papers documents the protracted saga of Bernstein's complaints about salary and Columbia's efforts to find him a job elsewhere. As early as April 1934, Murrow had formed the opinion that Bernstein "does not appear to be a particularly adaptable individual." Murrow to Leslie Dunn, Professor of Zoology at Columbia, 24 April, 1934 in this same folder. See also Siegmund-Schultze, *Rockefeller and the Internationalization of Mathematics*, p. 204 as well as the case study on Bernstein in Siegmund-Schultze, *Mathematicians Fleeing from Nazi Germany*, pp. 262–266. Bernstein returned to Germany in 1948 at the age of seventy and was made professor emeritus at Göttingen.

mathematics department, was, as AMS Secretary, a natural clearinghouse for information on mathematicians and so a natural channel to Veblen and the Emergency Committee. On 24 July, 1933, he wrote to Veblen not only to pass on recently acquired news on how to reach various Europeans but also to fill him in on Brown's discussions about its allotted EC funding. The Swiss-born, Göttingen logician and set theorist Paul "Bernays has been considered by our group here," Richardson recounted, "but would not fit into our situation so well as Lewy, Bochner, or Fenchel. We have not decided yet which department is to call a man, but we are considering the three departments mathematics, theoretical physics, and economics."[54] Six weeks later, the decision had been made. Of the four Jewish mathematicians considered, Hans Lewy, a twenty-nine-year-old analyst, one of Bernays's colleagues at Göttingen, and Courant's former assistant there, was invited to Brown for a two-year associateship.[55]

Although he did not find a place at Brown, analyst Salomon Bochner was also among the earliest European mathematicians to find professional refuge in the United States. The thirty-four-year-old Bochner had earned his doctorate at the University of Berlin in 1921 and had held an IEB fellowship from 1924 to 1926, first in Copenhagen, where he had worked with Harald Bohr, and then at Oxford and Cambridge attracted by Hardy and Littlewood, respectively. On returning to Germany, he had taken a lectureship at the University of Munich, where his colleagues Carathéodory and Oskar Perron had quickly argued for a permanent position for him on the basis of what they deemed his exemplary research output. That he was not German—he had been born in Cracow, a city that had passed from Austria-Hungary to Poland in the aftermath of World War I—ultimately proved too controversial within university circles to secure that result. Since he was also an observant Jew, the Nazi dictates of 1933 effectively ended any employment prospects he may have had not just in Munich but in all of Germany.

In 1933, Bochner went to Princeton as a temporary "associate," not on Emergency Committee funds, but on money Eisenhart, in his role as Dean of the Faculty, had secured after the university administration opted to use its

---

54. Richardson to Veblen, 24 July, 1933, Box 23, Folder: EC, May-Sept. 1933, Veblen Papers. On Salomon Bochner, see immediately below. Werner Fenchel, a geometer, earned his doctorate at the University of Berlin under Ludwig Bieberbach in 1928 before moving to Göttingen as Edmund Landau's assistant. An IEB fellowship allowed him to work with Levi-Civita in Rome and with Harald Bohr in Copenhagen.

55. Murrow to Veblen, 8 September, 1933, Box 23, Folder: EC, May-Sept. 1933, Veblen Papers.

Emergency Committee allocation for a professor of Romance languages.[56] Moving into an assistant professorship the next year, Bochner successively worked his way up through the ranks. In the process, he trained some forty graduate students in a variety of areas, ultimately holding the Henry Burchard Fine Professorship of Mathematics from 1959 until his retirement in 1968.[57]

Whereas mathematicians at some of the fifteen schools initially privileged with funding by the EC acted quickly, a couple, notably, did not. At Harvard, Birkhoff agreed with the stance taken by Lowell and Conant, feeling "a sense of increased duty toward our own promising American mathematicians" in the matter of jobs for "newcomers."[58] The Harvard mathematicians thus had no incentive to make an early push to have an émigré mathematician among them. At Chicago, views were more open, but the school's Depression-Era financial situation precluded it from doing much early on, despite Bliss's early interest in working with Veblen on the émigré cause.[59] (See chapter seven for more on Harvard and Chicago relative to the émigré problem.)

The first placements had thus been hit-or-miss. By 5 September, 1933, the Emergency Committee had compiled an extensive roster of displaced scholars in all fields in an effort to effect more systematic and informed decision-making.[60] Some sixty mathematicians ranging in age from twenty-four to sixty-five were identified there, but even with such a list in hand, thorny questions remained. Were the EC's initial selection criteria the "right" ones? Should older, proven mathematicians necessarily take precedence over younger, emerging ones? And, what did "older" mean, anyway? If a mathematician—regardless of his or her attainments—was close to retirement age, how could a pension possibly be afforded? If, on the other hand, established, middle-aged mathematicians should not be privileged, how could convincing arguments be made for employing younger, less proven

56. Richardson to Veblen, 24 July, 1933, Box 23, Folder: EC, May-Sept. 1933, Veblen Papers. Finally, the scholar originally selected did not go to Princeton, but an economist from Berlin did. See "Dismissed German Scholars Placed in American Institutions, 28 September, 1933, Box 23, Folder: Emergency Committee in Aid of Displaced German Scholars, May-September 1933, Veblen Papers.

57. Bochner retired from Princeton, but he immediately accepted a chaired professorship at Rice University, spending the rest of his long and productive career in Houston. On Bochner's life, see Robert Gunning, "Bochner, Salomon," Dictionary of Scientific Biography, ed. Charles C. Gillispie, 18 vols. (New York: Charles Scribner's Sons, 1970–1990), 17: 88–90.

58. Birkhoff, "Fifty Years of American Mathematics," p. 277.

59. Norwood, p. 32.

60. See Box 23, Folder: EC, May-Sept. 1933, Veblen Papers.

German mathematicians when younger, less proven American mathematicians were hard-pressed for jobs? Was it actually possible to establish some sort of rank-ordering of mathematicians from the "best" to "lesser" talents? Answers were unclear, especially as specific cases—actual faces associated with names on a list—presented themselves. Even harder to answer was the question of just how many foreign mathematicians the United States could realistically absorb.

By the end of September 1933, five Jewish mathematicians on the Emergency Committee's list had been placed: Bernstein, Bochner, Lewy, Noether, and Szász.[61] By year's end, it had also placed one more. Richard Brauer, a thirty-two-year-old algebraist recently dismissed from the University of Königsberg, was offered a post at the University of Kentucky thanks to EC support, to the generous donations of the Lexington Jewish community, and to the intervention of Leon Cohen, who was then in Kentucky's mathematics department (see below).[62] Einstein, moreover, had personally arranged for the forty-five-year-old Austrian Jewish mathematician Walther Mayer to join him at the Institute for Advanced Study not as a scientific equal but as his assistant. The Institute had also hired the forty-eight-year-old, acknowledged mathematical world leader, Hermann Weyl, into one of its coveted professorships. Although Weyl was not Jewish, his wife Hella was. This made precarious the position he had assumed in the summer of 1933 as director of the Mathematisches Institut in Göttingen after founder-director Courant, a Jew, had found himself placed on indefinite leave of absence at the end of April.

An abstract set of criteria had evidently been difficult to follow, and although the number of mathematicians at risk was large, not even a dozen had yet been placed in the United States, either through the Emergency Committee or otherwise.[63] Still, Veblen was already worried in November that "we

61. Murrow to Veblen, 8 September, 1933, Box 23, Folder: EC, May-Sept. 1933, Veblen Papers.

62. On bringing Brauer to Kentucky, see Series I: Grantees, Box 4, Folder 6: Brauer Richard, Emergency Committee Papers. See also Walter Feit, "Richard D. Brauer," *BAMS*, n.s., 1 (1979), 1–20 on p. 4. Cohen had been a student of Raymond Wilder in topology at Michigan and had followed an NRC fellowship at Princeton with an assistant professorship at Kentucky in 1931. For Cohen's recollections, see the interview with him at https://findingaids.princeton.edu/collections/AC057/c48, the website of the oral history project entitled "The Princeton Mathematics Community in the 1930s."

63. Again, exact numbers are hard to determine. Recall footnote 1 in this chapter.

are approaching the point where we shall have invited as many of the displaced Germans as we safely can."[64] Like his fellow EC members, Veblen recognized "the danger of stirring up anti-semitism" when "so many of our own young men are without employment."[65] He was also well aware that many American colleges and universities had actual admission quotas for Jewish students and were loath to appoint too many—or, in some cases, *any*—Jews to their faculties. These numbers were already so small as to put American Jewish scholars at a distinct disadvantage, much less foreign Jewish academics. As early as the closing months of 1933, then, not even a year after the Nazi dismissals, Veblen feared that while "there are still *certain other fields* in which the *saturation point* has not been reached, . . . we are very close to it in mathematics."[66]

In light of the fact that the United States ultimately absorbed more than 120 mathematical émigrés, Veblen would seem to have been overly pessimistic in 1933. Perhaps his vantage point in Princeton, with both the university's Department of Mathematics and the Institute's co-located School of Mathematics, colored his early perception. Lefschetz, in the department, was Jewish as was Flexner, the Institute's Director. Between the two interconnected groups, five foreign scholars—Einstein, Mayer, von Neumann, and Weyl in the Institute and Bochner in the Department—four of whom were Jewish, had already gotten permanent or soon-to-be permanent positions by the end of 1933. While the Institute was independent of the EC, Veblen was not. As Veblen saw it, the EC's stated principle of temporary appointments seemed not to be at work in Princeton; some of the country's most prized mathematical jobs *were* going to foreign Jews. Moreover, as Flexner explained to Einstein's wife, Elsa, in November 1933, "[i]t is perfectly possible to create an anti-Semitic feeling in the United States," but "such a feeling would [only] be created by the Jews themselves," particularly unassimilated Eastern

64. Veblen to John Kline, 15 November, 1933, Box 23, Folder: EC, May-Sept. 1933, Veblen Papers.

65. Veblen to Bernard Flexner, 13 December, 1933, Box 23, Folder: EC, May-Sept. 1933, Veblen Papers.

66. Ibid. (my emphasis). See also Veblen to Stephen Duggan, 17 November, 1933, Series I: Grantees, Box 5: Folder 13: Courant, Richard, Emergency Committee Papers where Veblen makes this same argument, but one month earlier. On admissions quotas for Jewish students, see Jerome Karabel, *The Chosen: The Hidden History of Admission and Exclusion at Harvard, Yale, and Princeton* (New York: Houghton Mifflin and Company, 2005). On attitudes towards Jews, specifically among mathematicians, in the 1930s, see Ralph Phillips, "Reminiscences about the 1930s," *MI* 16 (3) (1994), 6–8.

European Jews, even though the Institute's Jewish members did not fall into that category.[67] As Flexner saw it, the Institute was nevertheless running the risk, through its high-profile hires, of anti-Semitic sentiment that could jeopardize the reputation he was so hard at work to establish. Indeed, Veblen may have been right that the saturation point in Princeton *had* already been reached, even if the Institute, with its program of short-term visitors would continue to offer temporary accommodation to mathematicians—Jewish and otherwise—throughout the 1930s and into the 1940s.

A First Wave of Mathematical Refugees Hits the Northeast

The initial placements that ensued after the Nazi purge in the spring of 1933 were followed by a first wave of mathematical refugees that broke on American shores through the end of a tumultuous 1938.[68] That year witnessed the partial realization of Hitler's goal of reunifying German-speaking Europe: first, the so-called Anschluss in March that brought Austria under Nazi control; second, the Munich Agreement of September that allowed for Hitler's annexation of the German-speaking people of Czechoslovakia's Sudetenland; and, third, the Kristallnacht pogrom in November against Jews in Germany. Not surprisingly, the Northeast absorbed the largest number of academic émigrés in this first wave, given its relatively high concentration of colleges and universities and given that New York City was the principal port of entry for those coming from Europe. In mathematics, several northeastern programs, besides the Institute, recognized and seized the opportunity for program-building afforded by the geopolitical crisis. New York University (NYU) represents a particularly interesting example of this strategy.

NYU had been founded in 1831 by a group of influential New York City businessmen, among whom Albert Gallatin had served as the fourth Secretary of the Treasury under both Thomas Jefferson and James Madison and was

67. Abraham Flexner to Elsa Einstein, 15 November, 1933, Einstein Papers, Institute for Advanced Study, reproduced in full in Reingold and Reingold, ed., pp. 451–453 on p. 452.

68. As noted in the preface, the aim here and elsewhere in this book is to be representative as opposed to exhaustive. Relative to the émigrés, Siegmund-Schultze gives the fullest treatment to date of those *German-speaking* émigrés who came in the wake of events in Europe in the 1930s in *Mathematicians Fleeing from Nazi Germany*. In *Illustrious Immigrants: The Intellectual Migration from Europe 1930/41* (Chicago: University of Chicago Press, 1971), Laura Fermi, an immigrant herself and the wife of noted physicist Enrico Fermi, gives a broad overview of immigrants in all areas, including a short section on mathematicians (pp. 283–295).

the university's first President. By the early twentieth century, and following decades of immigration to the United States, NYU had become one of the largest universities in the country as new arrivals, and especially their children, swelled its undergraduate classrooms.[69] It was, however, primarily an undergraduate institution. In mathematics, it had awarded a very small number of Ph.D.s by the early 1930s but had no real graduate program in the field.[70] Some hoped that might change.

In 1929, University of Pennsylvania Ph.D. Donald Flanders followed two years as an NRC Fellow at Princeton with an instructorship at NYU that turned into an assistant professorship there in 1931. Having written his 1927 postulate-theoretic dissertation on "Double Elliptic Geometry in terms of Point, Order, and Congruence" under the direction of John Kline, he had quite naturally sought postdoctoral guidance from the mathematicians at Princeton, given their strength in geometry and topology. By 1933, Flanders, who had imbued the research ethos under Kline but whose own research program had faltered, sought out Veblen's "advice about improving the mathematical situation in New York University."[71] As Veblen reported, Flanders was "very anxious" for NYU "to call in some real mathematicians," even though that "would mean calling people in who outrank[ed] him in every respect." Veblen recommended Richard Courant.

Courant was a well-known quantity in the United States. Many American mathematicians had encountered him as they passed through the IEB-funded Mathematisches Institut he had founded and directed in Göttingen, and many more had met him personally on the coast-to-coast lecture tour that he had made in the summer of 1932 and that had taken him to many of the country's major mathematics departments. His plight had thus been of much concern since the rash of dismissals in April 1933. At various moments, Cornell, Berkeley, and the Institute had approached the Emergency Committee about hiring him, and Berkeley, his main West Coast stop of the previous summer,

69. Paul Mattingly, "New York University," in *The Encyclopedia of New York City*, ed. Kenneth Jackson (New Haven: Yale University Press, 1995), pp. 848–849.

70. In "The Ph.D. Degree and Mathematical Research" (p. 366), Richardson found that NYU had produced one Ph.D. over the period from 1915 to 1919 and four between 1925 and 1934. A search using the key phrase "New York University" of the site http://www.genealogy.ams.org (on 15 June, 2018) yields six NYU Ph.D.s over that time period, although one of those would appear to have been in physics.

71. Veblen to Richard Courant, as quoted in Reid, *Courant*, p. 158. The quotations that follow in this paragraph may also be found here.

had actually extended an offer of a one-semester lectureship to him for the spring semester of 1934.[72]

Relative to the EC and its policies, however, Courant was on shaky ground. He had been placed on leave of absence from Göttingen, but he had not been fired. There was thus a chance that he might be allowed to return to his post, especially in light of the fact that he had served in World War I and so, by the Nazis' own regulations, should have been exempt from the purge altogether. He fought hard to be reinstated.[73]

Under these circumstances, the EC could not see its way clear to allocate funds to bring him to the United States. It had resolved to adhere "very closely to the policy of making grants only for the support of those scholars who are *definitely* without the possibility of continuing their work in Germany."[74] Finally, it was Cambridge University, and not an American institution, that provided Courant with a position for the 1933–1934 academic year, thanks to a subsidy from the Rockefeller Foundation. Grateful for the opportunity, Courant nevertheless felt that while he would "pass" in England, "if serious prospects develop in America, they will have to be given preference for the long term."[75]

Efforts on his behalf did continue in the United States. By December 1933, the EC had been convinced by the Institute's Abraham Flexner, the Rockefeller Foundation's Warren Weaver, and others that Courant had "practically no possibility" of being able to continue in his position at Göttingen.[76] Two months later in mid-February 1934, he had been offered and had accepted a two-year, visiting professorship of mathematics at NYU at a salary of $4,000 a year, half coming from the EC and half from the Rockefeller Foundation.[77]

72. Brittany Shields, "A Mathematical Life: Richard Courant, New York University, and Scientific Diplomacy in Twentieth-Century America" (unpublished doctoral dissertation, University of Pennsylvania, 2015), pp. 32–35.

73. Reid, *Courant*, pp. 147–152.

74. Stephen Duggan to Harry Chase, Chancellor of New York University, 16 November, 1933, Series I: "Grantees," Box 5, Folder 13, Emergency Committee Papers (my emphasis). See also Shields, p. 36.

75. Courant to James Franck, as quoted in Reid, *Courant*, p. 156.

76. Murrow to Harry Chase, 18 December, 1933, Series I: Grantees, Box 5, Folder 13, Emergency Committee Papers, as quoted in Shields, p. 41.

77. This salary was roughly a third of what he had been earning as a professor in Göttingen. See Reid, *Courant*, p. 158 and Shields, p. 43. Deep salary cuts, initially lesser positions, and a loss of prestige all affected academic refugees coming to the United States in the 1930s. In 2021 dollars, $4,000 represents $79,800 in "real wealth" (https://www.measuringworth.com).

Courant, his wife, Nina, and their four children arrived in New York City at the end of August, the future unclear but the present secure.

The Courants found their footing quickly. A commodious house and garden in nearby New Rochelle, a steady stream of émigré visitors, congenial school placements for the children, and the resumption of musical soirées with Courant on the piano and his wife on the violin, all helped smooth their cultural transition. Mathematically, though, Courant was at sea. He had long been trying to finish the second volume of *Methoden der mathematischen Physik*, a work inspired by Hilbert's Göttingen lectures on mathematical physics and ostensibly, although not effectively, written with Hilbert's collaboration.[78] The first volume had appeared in 1924, but the building of the institute in Göttingen in the second half of the 1920s, followed in the 1930s by his dismissal and relocations first to England and then to the United States, had profoundly affected his ability to concentrate on the project. Instead, during that first year at NYU, he focused on perfecting his English, teaching his courses, and thinking about ways to improve mathematics at his new academic home.[79]

One of his strategies was to capitalize on Princeton's proximity to New York City—as well as on his connections with Flexner—by arranging for visitors at the Institute to supplement through single lectures or even mini-courses on their work what he and his colleagues were able to teach. Another was to invite mathematicians from other universities as colloquium speakers. In the spring of his first year, for example, his former assistant at Göttingen, the analytic number theorist Carl Siegel, traveled from Germany to spend a semester at the Institute and accepted Courant's invitation to give a lecture at NYU. MIT's Jesse Douglas, still exploring ramifications of his solution of Plateau's problem, was another mathematical guest that spring.[80] It had been a promising first year, and the university was pleased.

Harry Chase, NYU's Chancellor, praised Courant's efforts to tie the mathematics program at NYU with that at the Institute and let him know that the university was prepared to make him a permanent offer at the end of his two-year contract.[81] As Chase saw it, "we have a real opportunity to develop here at New York University a strong department of mathematics at the graduate

78. Richard Courant and David Hilbert, *Methoden der Mathematischen Physik*, 2 vols. (Berlin: Julius Springer, 1924 and 1937).

79. Reid provides more details in *Courant*, pp. 169–176.

80. Ibid., pp. 173–174.

81. Harry Chase to Courant, 4 June, 1935, as quoted in Shields, p. 65. The next quotation is also from this letter.

level." That was all the encouragement that a proven mathematical structure-builder like Courant needed. By December 1935, he, together with two of his colleagues, Flanders and Robert Putnam, had drawn up and submitted for Chase's consideration their vision of a graduate program in mathematics and a detailed plan for effecting it.[82]

They argued that such a program should center on courses, seminars, and research in pure mathematics at the same time that it interacted and cooperated with NYU's graduate programs in physics, engineering, biology, economics, and philosophy, among others. In the latter spirit, it should develop a school of applied mathematics that would create ties with the various independent laboratories and actuarial concerns in and around New York City. But, higher-level mathematics is built on a foundation of strong undergraduate and high school teaching, so it should also provide for teacher training at those levels. In order to carry out these interrelated missions, moreover, it would need its own building—equipped with a well-stocked library, work and study spaces, lecture and seminar rooms, offices, a support staff—in addition to a "permanent teaching and research staff" and funding for visiting professorships, colloquium speakers, and graduate student fellowships.[83] The vision that Courant and his colleagues had for graduate mathematics at NYU, then, had much in common with that of Courant and his mentor, Klein, for a Mathematisches Institut in Göttingen. Such a vision would take time and money to realize in the United States, just at it had in Germany, but Courant and his colleagues were eager to get the go-ahead to pursue it.

Chase, for one, was behind the plan, even though the country and his university were still feeling the debilitating effects of the Depression. His argument was pragmatic and, by this point, not unfamiliar in the American context. He recognized that NYU could not afford to elevate all of its departments. Mathematics, however, was central to the sciences as well as to a number of "professional fields which require its use"; it did not need elaborate equipment or laboratories and so was relatively inexpensive; it had the cooperation of the nearby Institute for Advanced Study; and it had Courant "to head up our work."[84] With Chase's moral backing, Courant began fund-raising.

---

82. Richard Courant, Donald Flanders, and Robert Putnam, "Memorandum Concerning a Graduate Department of Mathematics at New York University," 23 December, 1935, as discussed in Shields, pp. 65–69.

83. Ibid. See Shields, p. 68 for the quotation.

84. Harry Chase to David Sarnoff of RCA, 6 May, 1936, as discussed in Shields, p. 74. This argument was again reminiscent of the one that Gilman had made in the 1870s for starting

By the beginning of the 1936–1937 academic year, the university had provided the graduate mathematics program with three rooms; Courant had raised $5,000 a year for two years for graduate fellowships and scholarships; and the mathematics library had gotten a modest start.[85] The enrollments in Courant's two graduate courses had also doubled in the two years since he had joined the faculty. Numbering some forty students each, about half were doing graduate work part-time on top of their jobs as high school teachers, engineers, etc., while about half of the full-time students were either in physics or preparing for teaching positions. The 1936–1937 academic year also saw NYU's mathematical offerings augmented by over two-dozen guest lectures each by two IAS visitors, algebraist Reinhold Baer and geometer Herbert Busemann, both former Göttingen Ph.D.s. The program was beginning to take shape, but it still lacked an adequate number of first-rate, *permanent* faculty members. That changed over the course of the next academic year when Courant succeeded in hiring Kurt Friedrichs, his former graduate student and assistant, and James Stoker, then an assistant professor at the Carnegie Institute of Technology in Pittsburgh.

Friedrichs, a professor at the University of Braunschweig who was not Jewish, emigrated to the United States in 1937 in order to escape Nazism's racist strictures. He had fallen in love with a Jewish woman in 1933 and wanted to be able to marry her.[86] As early as 1935, while in New York to work with Courant on the then still unfinished second volume of *Methoden der mathematischen Physik*, Friedrichs had sounded Courant out about the possibility of a job at NYU, and Courant had done the ultimately fruitful spadework for such a move with the EC and the Rockefeller Foundation. Not only was Friedrichs a former student and long-time friend, but he also combined just the sort of mathematical talents that Courant sought for his new program. "Friedrichs, through his rather unique combination of pure mathematics and applications," Courant explained to the EC's John Whyte in March 1937, "might be an invaluable addition to science over here." Applied mathematicians Theodore von Kármán at Caltech and Stephen Timoshenko at Stanford both agreed.[87]

---

the Johns Hopkins University with a program in mathematics. See, for example, Parshall, "America's First School of Mathematical Research," pp. 163–165.

85. Shields details the events of this year on pp. 76–81.

86. Reid treats Friedrichs's emigration to the United States in *Courant*, pp. 195–197.

87. See Courant to Whyte, 31 March, 1937; von Kármán to Courant, 16 March, 1937; and Timoshenko to Courant, 29 April, 1937, all in Series I: Grantees, Box 10, Folder 9:

Stoker also combined pure and applied mathematics. Born in Pennsylvania, he had started his professional life as a mining engineer but had decided to leave fieldwork in the 1930s to pursue a doctoral degree in mechanics at the Eigenössische Technische Hochschule in Zürich.[88] There, however, he was quickly captivated by the pure mathematical work of Heinz Hopf and wrote a dissertation instead on differential geometry co-directed by Hopf and George Polyá. Hopf—who had known Courant since the summer of 1926 when, together with Alexandroff, he had been a mathematical visitor in Göttingen—had alerted him to Stoker's unusual combination of talents. Courant's interest was piqued. He invited Stoker to give a colloquium at NYU on his next trip to New York City. That proved to be in December 1936, and by the fall of 1937, Stoker had been hired in the Mathematics Department of NYU's Engineering School, thus creating precisely the sort of tie that Courant and his colleagues had envisioned between their graduate program and other schools.[89] Courant, Friedrichs, and Stoker—two émigrés and one American—thus formed the nucleus of a new experiment in graduate mathematics education at NYU that entwined the pure and the applied aspects of the field.[90]

Elsewhere in the Northeast, the IAS, which had dramatically hired some of Europe's best into its permanent positions, also hosted a steady stream of short-term, European visitors classified as "non-Aryan" by the Nazi regime. Among them were algebraists Richard Brauer for the academic year 1934–35 and Reinhold Baer from 1935 to 1937 and geometer Herbert Busemann from 1936 to 1939.[91] Brauer, following his stints at first the University of

---

Friedrichs, Kurt, Emergency Committee Papers. Shields also discusses Friedrichs as a member of Courant's team at NYU on pp. 81–86.

On Friedrichs's mathematical work, see Cathleen Synge Morawetz, "Kurt Otto Friedrichs 1901–1983," *Biographical Memoirs*, vol. 40 (Washington, D.C.: National Academy of Sciences, 1969), pp. 69–90. The second volume of *Methoden der mathematischen Physik* finally appeared, with Friedrichs's help, in 1937, although his contributions were not acknowledged for fear that such mention would have negative repercussions for his family back in Germany.

88. Wolfgang Saxon, "James Stoker Jr., 87, Ex-Director of N.Y.U. Mathematical Institute," *New York Times*, 22 October, 1992.

89. Reid, *Courant*, pp. 186 and 194–195.

90. It is interesting that, despite his presence at NYU, Felix Bernstein was never part of the Courant circle.

91. Mitchell, ed., *A Community of Scholars*. The highly idiosyncratic, peripatetic Hungarian mathematician, Paul Erdős, also found his first footing in the United States at the Institute from 1938 to 1940.

Kentucky and then as Weyl's assistant at the Institute, moved on to positions at the University of Toronto (1935–1948), the University of Michigan (1948–1952), and finally Harvard (1952–1971); Baer followed his Institute sojourn with an assistant professorship at the University of North Carolina (UNC) for the 1937–1938 academic year but left it for the position at Illinois that he would hold until his return to Germany in 1956; Busemann left the Institute for a year-long instructorship at Swarthmore and proceeded to a string of temporary appointments before settling at the University of Southern California in 1947.[92] The Institute's structure of permanent professors, supplemented annually by visitors in residence, had given a number of émigré mathematicians just the toehold they needed to find long-term employment elsewhere in North America, although some of them, like Baer at Illinois, found the amount of lower-level teaching at American institutions disagreeable.[93]

The situation at Brown and Penn was different. Both had departments of mathematics that had been working to build graduate programs prior to 1933. At Brown, Richardson's addition of Jewish analysts Lewy for the two academic years 1933–1935 and Szász half-time (shared with MIT) in 1934–1935 helped lay the foundation for the later development of a program in applied mathematics beginning in 1941 (see chapter eight).[94] Similarly, at Penn, John Kline had sought to sustain a graduate program in mathematics after the 1920 departure of his advisor, R. L. Moore. The addition there in 1934 of Hans Rademacher, the Göttingen-trained analyst-turned-number-theorist, fundamentally supplemented the department's offerings, especially after the EC- and Rockefeller-supported Rademacher was offered an assistant professorship in 1935.[95] Forty-three years old, Rademacher had been both a full professor at the University of Breslau and actively engaged in peace and human rights initiatives. His dismissal on the latter political grounds in 1934 had forced him to seek asylum and—like so many others—to take what he was offered. Rademacher remained on the Penn faculty, retiring as Thomas

92. See Feit and Jaques Catell, ed., *American Men of Science: A Biographical Dictionary*, 9th ed., vol.1, "Physical Sciences" (Lancaster: The Science Press and New York: R. R. Bowker Company, 1955).

93. Siegmund-Schultze, *Mathematicians Fleeing from Nazi Germany*, p. 243.

94. Jaques Catell, ed., *American Men of Science*, 7th ed.

95. On bringing Rademacher to Penn, see Series I: Grantees, Box 27, Folders 14–15: Rademacher, Hans, Emergency Committee Papers.

A. Scott Professor of Mathematics in 1962 and producing some seventeen new Ph.D.s there.[96]

The First Wave of Mathematical Refugees Spreads over the Rest of the Country

Displaced mathematicians also took up posts in the Midwest, on the West Coast, and, to a lesser extent, in the South.[97] For example, when Szász's EC funding ran out in 1935 and MIT could not offer him a permanent position, he spent the fall of 1935 as an EC-funded itinerant lecturer and the spring as a visiting professor back at MIT. As luck had it, however, one of the universities in the Midwest that had invited him for a series of lectures must have liked what it heard. As noted, the University of Cincinnati appointed him to a research lectureship in its graduate school for the 1936–1937 academic year.[98] Joining forces with fellow analyst Charles Moore, Szász remained in this position until his promotion to a full professorship in 1947. In all, he trained at least eight Cincinnati graduate students, while producing some sixty papers on methods of summability, theorems of Tauberian type, and related topics.[99]

Gábor Szegő, another Jewish, Hungarian analyst, took two years of EC and Rockefeller Foundation support to Washington University in St. Louis in 1934 and remained there until 1938. His position had initially been arranged

96. Bruce Berndt chronicles Rademacher's life and work in "Hans Rademacher (1892–1969)," *Acta Arithmetica* 61 (1992), 209–231. On what has been termed "The Mathematical Family Tree of Hans Rademacher," see Solomon Golomb, Theodore Harris, and Jennifer Seberry, "Albert Leon Whiteman (1915–1995)," *NAMS* 44 (1997), 217–219 on p. 218 and Siegmund-Schultze, *Mathematicians Fleeing from Nazi Germany*, pp. 285–286. Rademacher was a founding editor of *Acta Arithmetica* in 1935, the year he joined the Penn faculty. A publication under the auspices of the Institute of Mathematics of the Polish Academy of Sciences, *Acta Arithmetica* had a troubled start, producing its first and second volumes in 1935 and 1936, but its third in 1939 and its fourth only in 1958.

97. Duggan and Drury provide the numbers and geographical distribution of those placed by the Emergency Committee alone on pp. 68–72.

98. See Series I: Grantees, Box 33, Folder 1: Szász, Otto, Emergency Committee Papers, especially Murrow to Miss Waite, 24 June, 1935, and "Statement RE Professor Szász," 18 September, 1935.

99. Szegő, "Otto Szász," p. 262. For his graduate students, see http://www.genealogy.ams.org. The online database of mathematical publications, MathSciNet, gives a sense of his singly-authored papers as well as those with co-authors from 1936 on.

thanks to Richardson's interventions and to the strong recommendation of Szegő's friend and Richardson's colleague, Jacob Tamarkin.[100] Szegő had studied at both Berlin and Göttingen but had earned his Ph.D. at Vienna while in the Austro-Hungarian Army in 1918. Living in Budapest in 1919–1920, he served as mathematics tutor to the young von Neumann before moving to the University of Berlin to work toward his *Habilitation*. At the time of the Nazi purge in 1933, he was a professor in Königsberg, but because of his wartime service, he was initially exempt from dismissal. It soon became clear, however—at least to his friends Polyá in Switzerland and Tamarkin in the United States—that his situation was becoming increasingly precarious.

With his move to St. Louis, Szegő found an environment supportive of his work. The university's chancellor, George Throop, was intent on both fund-raising and on faculty building, and the Harvard-trained applied mathematician, William Roever, had Throop's ear.[101] When Lefschetz dropped in on the program during a trip to St. Louis in January 1936 as part of his AMS presidential travels, he was impressed. "[N]o refugee mathematician whatsoever is doing more useful work than [Szegő]," he told EC chair Stephen Duggan. "In a region where there is a scarcity of mathematicians, to say the least, it is a fine and most useful thing to have as good a mathematician as Szegő stationed."[102] In addition to overseeing the doctoral research of four students during his four-year stay in the Midwest, Szegő offered, during the 1935–1936 academic year, the course on orthogonal polynomials that served as the basis for the book by the same name that he completed with financial support from the university. Published in 1939 in the AMS's Colloquium Publication series, *Orthogonal Polynomials* appeared only after Szegő's move to

100. See Series I: Grantees, Box 33, Folder 2: Szegő, Gabriel, Emergency Committee Papers for correspondence between Richardson and the EC's Murrow about placing Szegő in St. Louis. Richard Askey and Paul Nevai provide an overview of Szegő's life in "Gábor Szegő: 1895–1985," *The Mathematical Intelligencer* 18 (3) (1996), 10–22. See also Siegmund-Schultze, *Mathematicians Fleeing from Nazi Germany*, pp. 72, 249, and 299.

101. Roever had earned an undergraduate degree in mechanical engineering at Washington University in 1897 before moving to Harvard for his 1906 Ph.D. under Maxime Bôcher. After teaching at MIT for three years from 1905 to 1908, he returned to Washington University, where he was promoted through the ranks, becoming head of the Department of Mathematics and Astronomy in 1932.

102. Lefschetz to Duggan, 10 January, 1936, Series I: Grantees, Box 33, Folder 2: Szegő, Gabriel, Emergency Committee Papers.

Stanford on the West Coast (see below).[103] Polyá, who decided that he, too, needed "to get out before it [was] too late," joined Szegő in Palo Alto in 1940 and remained there for the rest of his life.[104]

Also in the Midwest, the University of Notre Dame, a Catholic institution in South Bend, Indiana, had adopted the development of graduate programs as one of its two main priorities—the other being the building of more residence halls—under its new (as of 1934) president, Father John O'Hara.[105] In 1937, O'Hara hired Karl Menger as a full professor—a rank uncharacteristic for émigrés at this time—specifically to develop the graduate program in mathematics. Born in Vienna in 1902 to a Catholic father and a Jewish mother, Menger was a prominent member of the so-called Vienna Circle, the group of philosophers and scientists that developed logical positivism during the course of their meetings from 1924 to 1936 and that included, in addition to Menger, Rudolf Carnap, Kurt Gödel, and Richard von Mises, among others. Menger was profoundly shaken when the Circle's founder, philosopher and physicist Moritz Schlick, was assassinated in June 1936 by a pro-Nazi, former student.[106] Just weeks later, while lecturing at and serving as one of the Vice Presidents of the International Congress of Mathematicians in Oslo, Menger

103. Gábor Szegő, *Orthogonal Polynomials*, AMS Colloquium Publications, vol. 23 (New York: American Mathematical Society, 1939). See p. vii for his acknowledgments to Washington University. Throop replaced Szegő in 1939 with another émigré mathematician, Stefan Warschawski, who had been dismissed from a position at Göttingen in 1933 during the purge. After short-term appointments in the U.S. at Columbia, the University of Rochester, and Brown, he landed the post at Washington University. He followed wartime service at Brown with a position at Minnesota, where he remained on the faculty until moving to found the Mathematics Department at the University of California, San Diego in 1963. On Warschawski's life, see Frank Lesley, "Biography of S. E. Warschawski," *Complex Variables, Theory and Application* 5 (1986), 95–109.

104. Gábor Szegő to Frank Aydelotte, 24 July, 1940, Box 33, Folder: Refugees, Polyá, George 1940–41, Veblen Papers.

105. Arthur Hope, *Notre Dame: One Hundred Years* (Notre Dame: University of Notre Dame Press, 1943), p. 451.

106. Brief accounts of Menger's life may be found in Seymour Kass, "Karl Menger," *NAMS* 43 (1996), 558–561 and Louise Golland and Karl Sigmund, "Exact Thought in a Demented Time: Karl Menger and His Viennese Mathematical Colloquium," *MI* 22 (2000), 34–45. On Menger's associations in Vienna, see Karl Menger, *Reminiscences of the Vienna Circle and the Mathematical Colloquium*, ed. Louise Golland, Brian McGuinness, and Abe Sklar (Dordrecht: Kluwer Academic Publishers, 1994). Finally, Karl Sigmund discusses the Vienna Circle more broadly in *Exact Thinking in Demented Times: The Vienna Circle and the Epic Quest for the Foundations of Science* (New York: Basic Books, 2017).

described the ever-worsening political situation in Vienna to various friends and colleagues. An offer from Notre Dame followed not long thereafter, with Menger and his family moving to South Bend in the fall of 1937. It was only after the Austrian Anschluss the following March, however, that he officially resigned his professorship in Vienna.

Menger was joined in that first year by another émigré, the Vienna-born, Hamburg-based algebraist Emil Artin. Forced into early retirement in the spring of 1937 because his wife, Natasha, was half Jewish, Artin owed his job at Notre Dame—financed by the university and not by the EC or the Rockefeller Foundation—to Lefschetz's strong lobbying. Like the rest of the mathematical community, Lefschetz recognized not only Artin's profound mathematical talents but also the impact he could have on an emergent program at Notre Dame.[107] Menger and Artin, together with instructors Arthur Milgram and Paul Pepper, launched Notre Dame's graduate program in mathematics that first year. Notably, Artin delivered the series of lectures that would become his classic book, *Galois Theory*, published in the series of Notre Dame Mathematical Lectures that Menger founded.[108] Artin left his temporary position in South Bend after only one year, however, to take up a

107. On Artin's life and works, see Richard Brauer, "Emil Artin," *BAMS* 73 (1967), 27–43 and Hans Zassenhaus, "Emil Artin, His Life and His Work," *Notre Dame Journal of Formal Logic* 5 (1964), 1–9. Della Dumbaugh and Joachim Schwermer specifically focus on Artin's career in the United States in "Creating a Life: Emil Artin in America," *BAMS* 50 (2013), 321–330. On the history of the mathematics program at Notre Dame, see https://math.nd.edu/about/history/.

108. Emil Artin, *Galois Theory*, Notre Dame Mathematical Lectures, vol. 2 (Notre Dame: Notre Dame University Press, 1942). The first volume in the series, Abraham Wald's *On the Principles of Statistical Inference*, Notre Dame Mathematical Lectures, vol. 1 (Notre Dame: University of Notre Dame Press, 1942), was based on a series of lectures he delivered at Notre Dame in 1941. Wald, who had been a student of Menger's in Vienna, was another Jewish émigré. He came to the United States in 1938 under the aegis of the Cowles Commission for Research in Economics.

In addition to the Notre Dame Mathematical Lecture series, Menger also ran the Notre Dame Mathematics Colloquium and oversaw a publication based on it. The latter were the replications on American soil of the Mathematisches Kolloquium and its associated *Ergebnisse* that Menger had animated in Vienna. Golland and Sigmund discuss the Kolloquium and its participants, among them, Gödel, Wald, von Neumann, Franz Alt, and Olga Taussky, and, during their respective stays in Vienna, Americans Gordon Whyburn, Norbert Wiener, and Marshall Stone on pp. 38–41. For a sense of the Mathematics Colloquium, see Frederick Ficken, "*Reports of a Mathematical Colloquium*. Series 2, no. 1. Edited by Karl Menger. Notre Dame University Press, 1939, 64 pp.," *BAMS* 45 (1939), 813–814.

permanent post at Indiana University some two hundred miles to the south in Bloomington. There, he taught a combination of undergraduate and graduate courses, worked to regain his research footing, and directed the doctoral research of two students before accepting Lefschetz's call to Princeton in 1946.[109]

On the other side of the country, the University of California in Berkeley provided a permanent position for Hans Lewy in 1935 on the expiration of his temporary, two-year job at Brown. As noted, Berkeley had begun to invest in mathematics with the appointment of Griffith Evans to head its department in 1933. Fully aware of this, Richardson approached Evans on Lewy's behalf, when changes in the Brown curriculum made it impossible for him to retain the young émigré.[110] In a letter dated 13 March, 1935, Richardson argued that Lewy would be able to "occupy a niche" "in graduate and upperclass work" at Berkeley "that would otherwise be difficult to fill."[111] This, Richardson knew, would fit in well with Evans's broad, program-building plans, even though Evans had earlier expressed his opposition to providing permanent posts to foreign mathematicians at a time when American mathematicians faced such a grim job market. Evans offered to try Lewy out in a temporary position, provided funding could be secured through the Emergency Committee and Rockefeller Foundation. Both agreed to extend their support for an additional two years. Before the end of Lewy's first year at Berkeley, however, Evans and his colleagues had decided to offer him an assistant professorship beginning in 1937.[112] Lewy fairly quickly began to train graduate students in the theory of minimal surfaces as well as in aspects of the theory of partial differential equations. By 1946, he had become a full professor.[113]

Berkeley also decided, with Evans's strong support, to make a move in statistics. In 1938, it hired Jerzy Neyman, a Polish statistician who had been

109. Dumbaugh and Schwermer, pp. 324–328. That Artin's mathematical transition to the United States was not seamless may be reflected in the fact that he published no papers between 1932 and 1940. Zassenhaus gives a complete list of his publications on pp. 6–8.

110. Richardson to Murrow, 29 March, 1935, and Monroe Deutsch, Provost of the University of California, Berkeley, to Stephen Duggan, 15 April, 1935, both in Series I: Grantees, Box 21, Folder 5: Lewy Hans, Emergency Committee Papers.

111. Richardson to Evans, 13 March, 1935, as quoted in Rider, p. 291.

112. Rider, pp. 290–292.

113. Lewy was fired from Berkeley in 1950 for refusing to sign a loyalty oath. Reinstated in 1953, he retired from Berkeley in 1972 and died there in 1988. On his life and work, see David Kinderlehrer, "Hans Lewy: A Brief Biographical Sketch," in Hans Lewy Selecta, ed. David Kinderlehrer, 2 vols. (Boston: Birkhäuser Verlag, 2002), 2: xv-xx.

home-based in London since 1934. Neyman not only successfully bridged Berkeley's pure and applied scientific concerns but also proceeded to build a major program in statistics—the so-called Statistical Laboratory within the Department of Mathematics—on the Berkeley campus (see chapter ten).[114]

Like Berkeley, Stanford sought to take advantage of a key personnel transition to shift the course of its department of mathematics. Under Blichfeldt from 1927 until his retirement in 1938, the department had particularly emphasized his special areas of group theory and number theory. William Manning trained students in the former, while James Uspensky prepared them in the latter, but Stephen Timoshenko augmented their pure with his applied mathematical offerings after taking up a position in Stanford's School of Engineering in 1936.[115] Still, by the late 1930s, again as at Berkeley, Stanford's physical scientists and engineers were lobbying for more mathematical support, and Szegő, who had been doing such good work at Washington University, seemed just the man for the job.[116] Within just a few years, he had begun to turn what has been described as "the provincial mathematics department that Stanford had been under Blichfeldt and Uspensky . . . into one of the leading departments of the country."[117]

Perhaps the most provincial region of the country mathematically in the 1930s, however, remained the South. While the University of Texas at Austin and Rice University in Houston had begun to develop graduate programs in the 1920s, in topology under R. L. Moore at the former and in analysis under Griffith Evans at the latter, and while Virginia, Duke, and several other schools had followed in the early 1930s, the South still lagged mathematically behind the Northeast, Midwest, and Far West. One Southern institution, the University of Kentucky in Lexington, did provide academic homes for two early mathematical émigrés: Brauer, who, as noted, had left after a year, and the more applied mathematician Fritz John, who arrived in 1935 and remained until 1946.

114. Rider, pp. 292–294. Constance Reid details Neyman's life in *Neyman–From Life* (New York: Springer-Verlag, 1982). It should be noted that Neyman was of Roman Catholic heritage and, so, in a situation very different from Lewy's.

115. On Timoshenko's life and work, see C. Richard Soderberg, "Stephen P. Timoshenko 1878–1972," *Biographical Memoirs*, vol. 53 (Washington, D.C.: National Academy of Sciences, 1982), pp. 323–350.

116. Royden, p. 249.

117. Peter Lax, "The Old Days," in *A Century of Mathematical Meetings*, ed. Bettye Anne Case (Providence: American Mathematical Society, 1996), pp. 281–283 on p. 282.

FIGURE 5.2. Fritz John (1910–1994) and his wife, Charlotte, in 1953.
(Photo courtesy of Marion Walter Photograph Collection, 1952–1980s,
undated, e_math_01112, Archives of American Mathematics, Dolph
Briscoe Center for American History, University of Texas at Austin.)

A student of Courant's at Göttingen at precisely the moment Courant had
lost his position there and had left his family behind for Cambridge, John,
who was half Jewish, came under Courant's protective wing. When Courant
returned to Cambridge following a trip back to Germany for Christmas 1933,
he brought not only his wife and eldest son but also John.[118] Courant had
managed to obtain a small fellowship for his young protégé, which was then
supplemented in 1934–1935 by England's counterpart of the EC, the Academic Assistance Council. This had given John and his wife, who had joined

118. Reid, *Courant*, p. 157.

him in England in the spring of 1934, enough money to live on, just barely, for the 1934–1935 academic year.

In the meantime, Courant had taken his new position at NYU. Worried about the young couple's welfare back in England, he mobilized an impressive list of references to write to the EC on John's behalf.[119] After negotiations via the EC to bring Lewy to Kentucky ended in Lewy's acceptance of the offer from Berkeley, the university opted for John, who had ultimately received grants from both the EC and the Rockefeller Foundation. Both had had initial hesitations, owing to John's youth—he was only twenty-four—and relatively untried mathematical skills.[120]

At the end of his initial two-year contract in 1937, Kentucky, which had been "interested in developing the work in mathematics" and which "[f]or the past half dozen years" had been offering "an increasing amount of gradu-ate work ... in that field," hired John as an assistant professor.[121] Two years later, he had been promoted to an associate professorship. The university had explicitly seen John's placement there as "a very wise move for the advance-ment of mathematics in the south," and others like Veblen concurred.[122] John stayed at Kentucky—although he served at the Aberdeen Proving Ground in Maryland during World War II—until 1946 when he was hired by Courant at NYU. He remained in New York for the rest of his career.[123]

The case of Fritz John again drives home the difficulties faced by the Emer-gency Committee in its efforts to relocate displaced scholars in the United States. John was neither an older scholar nor one proven in his field, yet these had been the EC's stated criteria in their efforts to avoid competition between young Americans and Europeans in the tight Depression-Era academic job market and so to forestall anti-Semitic reprisals. Faced, however, with letters

119. Ibid., p. 178 and Shields, pp. 46–49.

120. Series I: Grantees, Box 16, Folder 8: John, Fritz, Emergency Committee Papers documents John's journey to Kentucky.

121. Frank McVey, President of the University of Kentucky, to Frank Hansen of the Rock-efeller Foundation, 10 May, 1935, Series I: Grantees, Box 16, Folder 8: John, Fritz, Emergency Committee Papers.

122. Paul Boyd, Dean of the College of Arts and Sciences at Kentucky, to Edward R. Mur-row, 10 July, 1935, Series I: Grantees, Box 16, Folder 8: John, Fritz, and Veblen to Murrow, 15 June, 1934, Series I: Grantees, Box 4, Folder 6: Brauer, Richard, both in the Emergency Committee Papers.

123. Jürgen Moser, "Fritz John, 1910–1994," NAMS 42 (1995), 256–257.

of support from the likes of Courant, Weyl, Morse, Max Born, and others, the EC was hard-pressed to say "no."[124] Strong lobbying had paid off.

Other cases, even with hard lobbying, were more problematic. The Russian-born Jew Issai Schur had earned his doctoral degree in 1901 under Georg Frobenius at the University of Berlin and was awarded a chair there in 1919, two years after his advisor's death. Schur, who was fifty-eight when placed on leave of absence by the Nazis in 1933, had trained, among others, Richard Brauer and his brother, Alfred, and was known as one of the leading algebraists of the early twentieth century. It is perhaps not surprising, then, that the Stanford algebraist, Blichfeldt, confided to Veblen in a letter dated 14 August, 1933, that he was "very much more interested in the fate of Schur than in any of the others."[125] Stanford's President, however, was unwilling to commit university funds then or in the immediate future to "outside men," viewing it as "unfair to the faculty, under the present salary cut." Still, Blichfeldt hoped that Stanford might "invite Schur for a year or two, assuming the expenses would come from outside sources," even though he wondered "[w]hether this would be a real help to a man of his age."

Indeed, age *was* a problem, since older German émigrés would—owing to the Nazi laws—necessarily come to the United States without the pensions they had earned. American institutions could not afford, especially in the depths of the Depression, to provide, in addition to jobs, pensions for foreign scholars. As Blichfeldt put it to Veblen, since Stanford's retirement age was sixty-five, "the authorities here are definitely opposed to the permanent appointment of a man over 55 (even if we had the funds!)."[126] These considerations did not, however, prevent the University of Wisconsin from approaching the EC for funding for Schur or from offering him the position that he nevertheless declined, feeling unable to uproot himself and to begin lecturing in a new language.[127] Schur was finally fired outright from his post

124. See Series I: Grantees, Box 16, Folder 8: John, Fritz, Emergency Committee Papers and compare Shields, p. 46.

125. Blichfeldt to Veblen, 14 August, 1933, Box 23, Folder: EC, May–Sept. 1933, Veblen Papers. The quotations in the next two sentences are also from this letter.

126. Blichfeldt to Veblen, 3 August, 1933, Box 23, Folder: EC, May–Sept. 1933, Veblen Papers.

127. Alfred Brauer, "Gedankrede auf Issai Schur," in Issai Schur, *Gesammelte Abhandlungen*, ed. Alfred Brauer and Hans Rohrbach, 3 vols. (Berlin: Springer-Verlag, 1973), 1: v-xiii on p. vi. See also Siegmund-Schultze, *Mathematicians Fleeing from Nazi Germany*, pp. 153–154.

in Berlin in 1935 and, in 1939, left Germany for Palestine, where he died two years later a broken man.

———

Mathematicians had entered the decade of the 1930s on an international stage largely liberated from the post-World War I political dissension that had marred its international congresses in the 1920s. Almost immediately, however, they had found themselves impelled by a geopolitical upheaval that, by the end of the decade, had resulted in the outbreak of World War II. The five years from the Nazi purge in April 1933 to the Austrian Anschluss, the Munich Agreement, and the Kristallnacht pogrom, all in 1938, witnessed a first wave of mathematical emigration from Europe to the United States and elsewhere that resulted in the formation of new professional networks as well as in the opening of the American mathematical community to a not insignificant number of foreign scholars over a relatively short period of time and in the depths of the Depression.[128]

Institutions like the Emergency Committee in Aid of Displaced German Scholars and the Rockefeller Foundation worked together and with individual colleges and universities to provide safe havens for refugee scholars, the majority of whom were Jewish, at a time when talented Americans were out of jobs and xenophobia, anti-Semitism, and racism were far from unknown. For some, these efforts represented purely humanitarian concerns. For others, humanitarian concerns were coupled with self-interest. Departments of mathematics that had been working to improve their research profiles in the 1920s and into the 1930s—like Princeton—could hope to snare established or clearly up-and-coming foreign researchers and, in so doing, give an added boost to their efforts. Other departments that had had less institutional support—like NYU, Penn, Notre Dame, and Kentucky—could, by taking advantage of mathematicians' flight from Europe, hope to become more competitive. Veblen captured this motivation well when he offered in a letter to Kline at Penn that "I can see where you might use this means of improving your local situation in a substantial way."[129] Of course, the new Institute for

128. On other countries that made places for refugee mathematicians, see chapter six, "Alternative (Non-American) Host Countries," in Siegmund-Schultze, *Mathematicians Fleeing from Nazi Germany*, pp. 102–148.

129. Veblen to Kline, 15 November, 1933, Box 23, Folder: EC, May–Sept. 1933, Veblen Papers.

Advanced Study of which Veblen was a part had done precisely that in hiring Hermann Weyl.

By the end of 1938, several dozen European mathematicians had found positions across the United States and more seemed likely to come. It was still too soon to tell what the overall effect of this emigration would be on the American mathematical community. Would more institutions, like Washington University during Szegő's four-year stay there and Notre Dame thanks to Emil Artin and especially Karl Menger, develop graduate programs in mathematics where, for all intents and purposes, none had existed before? Or would the example of Fritz John at the University of Kentucky—with a heavy teaching load despite institutional aspirations for fostering research and graduate training—be more typical? It was also too soon to tell how effectively the émigré mathematicians would acculturate to their new environment. Would they, like Richard Courant, succeed in creating new and productive personal and mathematical lives for themselves, or would they, like Courant's NYU colleague Felix Bernstein, ultimately fail to make the transition? Who knew? One thing *was* clear. The American mathematical community's "saturation point" was much greater than Veblen had assumed at the close of 1933.

6

Taking Stock in a Changing World

European mathematical émigrés arrived in the United States as another generational shift was taking place within American mathematics. The year 1932 alone had witnessed the passing of Eliakim Hastings Moore, animator of the research program at Chicago, the Hilbert-trained Harvard analyst Oliver Kellogg, and the geometers Ernest Wilczynski of Chicago and John Wesley Young at Dartmouth. As UCLA's Earle Hedrick wrote to Chicago's Herbert Slaught following Wilczynski's death, this "has added to the many shocks I have experienced this year, in the passing of so many of our very good men, and my own good friends."[1] Five years later, Slaught, who had been so instrumental in the founding of the MAA, was also dead. The American mathematical research community was saying good-bye to the members of its first generation. Those of its second—Veblen, Birkhoff, Dickson, among many others—were "éminences grises." A third generation was already coming into its own.

The 1930s also marked a major mathematical milestone. In 1938, the American Mathematical Society celebrated its golden anniversary, and Columbia University hosted a four-day gala that had been ten years in the planning. The Society's leaders were proud of what had been accomplished in those fifty years. In particular, they had a clear sense of their community's "rise" on the international stage, especially in the two decades following the

---

1. Hedrick to Slaught, 28 September, 1932, Box 86-14/68, Folder 5, MAA Records.

close of the First World War.[2] They thus hoped that their eminent semicentennial speakers would both "reveal what has been accomplished in America since the founding of the Society" and "acquaint mathematicians with current problems in research in many fields."[3] These speakers would not so much look ahead as assess and reflect, although it would be important to lay out where mathematical developments then stood and, in so doing, to hint, at least, at where they might go.

In many regards, the late 1930s should have been a difficult time to try to crystallize a sense of American mathematical accomplishments. A snapshot of the 1920s (as given in chapter one) captures work done in an immediately postwar period of relative stability in which money was not tight and mathematicians optimistically pursued their research. But, a snapshot of the 1930s with its financial crisis, geopolitical storms, and changing demographics? New jobs were in short supply and salaries cut, yet record numbers of new American Ph.D.s in mathematics, as well as established mathematicians fleeing Nazi Germany and elsewhere, sought to enter the American academic ranks. Changing personnel brought the rise of new research emphases, but it also brought the decline of former research agendas as well as new results in old directions. The political situation in Europe rapidly deteriorated, yet the American mathematical community was considering scales both international and national as it planned to host the first International Congress of Mathematicians on its home turf in 1940. These were just some of the contradictions that American mathematicians confronted as they tried to take stock in what one of their number explicitly characterized as the rapidly "changing world" of the 1930s.[4]

Conflicting Perceptions of the Mathematical Endeavor

At the end of May 1933, the Chicago World's Fair, dubbed the "Century of Progress," opened not only to celebrate the hundredth anniversary of the city's

2. Roland Richardson et al., "The Semicentennial Celebration: September 6–9, 1938," *BAMS* 45 (1939), 1–30 on p. 4. The characterization is from the opening remarks of then AMS President R. L. Moore.

3. The Committee on Publication, "Preface," *Semicentennial Addresses of the American Mathematical Society* (New York: American Mathematical Society, 1938), p. i.

4. Typescript of a radio address delivered on 29 November, 1933, by then MAA President Arnold Dresden, entitled "Mathematics in a Changing World" in Box 86-14/67, Folder 4, MAA Records. This lecture was also published as Arnold Dresden, "Mathematics in a Changing World," *The Scientific Monthly* 38 (1934), 568–570.

FIGURE 6.1. Panorama of the Century of Progress World's Fair, Chicago 1933–1934.
(Photograph by Harry Koss. Library of Congress Prints & Photographs
Online Catalog LC-DIG-ds-05672.)

founding but also to give hope for the future—hope through science—for a
nation mired in the Great Depression and nervously watching events unfold
in Germany. It was in a spirit of "scientific idealism," that is, with the "deifica-
tion of the scientific method and [the] glorification of anticipated scientific
solutions to social problems" in mind, that "Chicago . . . asked the world to
join her in celebrating a century of the growth of science, and the dependence
of industry on scientific research."[5]

The almost fifty million visitors who ultimately accepted this invitation
encountered a spectacle that was meant to dazzle with color, light, architectural design, entertainment, and, most importantly, "the ways in which
the discoveries of the basic sciences are utilized" in the modern world.[6] As
these visitors made their ways from the fairground's north entrance down
the impressive Avenue of Flags, their attention was naturally drawn to the
fair's focal point, the massive and futuristic Hall of Science with its exhibits
aimed at demonstrating "*how* these basic sciences—physics, chemistry, biology, geology, mathematics, astronomy—have made it all possible." And, the
first display they encountered once inside? "Mathematics, 'Queen of the
Sciences.'" The title had not been chosen at random.

5. The first two quotations are in Robert Rydell, *World of Fairs: The Century-of-Progress
Expositions* (Chicago: University of Chicago Press, 1993), p. 96; the third is in *Official Guide
Book of the Fair 1933: With 1934 Supplement* (Chicago: The Cuneo Press, Inc., 1934), p. 11.
Rydell discusses the impetus for and the motivations behind the fair's science theme on
pp. 92–98.

6. *Official Guide Book of the Fair 1933*, p. 12. The quotation that follows in this paragraph is
also on this page (emphasis in the original).

FIGURE 6.1. Continued.

Early in 1931, then MAA President, Caltech's E. T. Bell, had been asked by the fair's organizers to contribute a manuscript on mathematics to a "Century of Progress" series. Conceived as short and affordable books for the general public, series volumes would convey "the essential features of those fundamental sciences which are the foundation stones of modern industry."[7] Within two months, Bell had written a 30,000-word typescript entitled *The Queen of the Sciences* after a quotation attributed to one of mathematics' greats, Carl Friedrich Gauss.[8] Using examples from the history of mathematics, Bell illustrated the content of particular mathematical subfields—geometry, algebra, number theory, and analysis—deeper mathematical notions—invariance, the infinite, and existence—and some of mathematics' fundamental applications to physics and other areas. Visitors entering the Hall of Science's mathematics exhibit in Chicago were exposed not just to these ideas but also glimpsed mathematics' recreational value and its even broader applications to such areas as navigation, chronometry, radio communication, and aerodynamics.[9] All of this aimed to make evident "[t]he service

7. Paul Linehan, "Review of *The Queen of the Sciences* by E. T. Bell," *AMM* 39 (1932), 296–297 on p. 296, quoting the publication announcement by the publisher, the Williams and Wilkins Company.

8. E. T. Bell, *The Queen of the Sciences* (Baltimore: Williams and Wilkins Company, 1931). Constance Reid details the circumstances surrounding Bell's writing of this book in *The Search for E. T. Bell Also Known as John Taine* (Washington, D.C.: The Mathematical Association of America, 1993), pp. 245–258, especially on p. 251.

9. *Official Guide Book of the Fair 1933*, pp. 30–33. The quotations that follow in the next sentence are on p. 33. Bell wrote *The Handmaiden of the Sciences* (Baltimore: Williams and Wilkins Company, 1937) as a sequel to *The Queen of the Sciences* in an effort to address, again for a broader audience, precisely this issue of applications. His historically questionable

to mankind of mathematics" as well as "its progress as this service is being performed."

In June, members of the AMS and the MAA met with the American Association for the Advancement of Science in Chicago to capitalize further on the fair and its imagery. In addition to the usual talks associated with the societies' meetings, hour-long addresses, aimed at reporting on the state of four particular fields of mathematics, were given by Leopold Fejér of the University of Budapest and Rome's Tullio Levi-Civita as well as Americans Birkhoff and Dickson.[10] In organizing this program, the AMS recognized that it was vital that the public come, rightly in its view, to understand mathematics—side by side with the other sciences—as fundamental to the progress of society in a changing world and to see American mathematicians taking a leading role. This, after all, was a general appreciation that members of the mathematical community, like Veblen, had actively tried to effect in the 1920s. And, this had been at least part of what had motivated mathematicians to associate themselves with an earlier world's fair in Chicago, the World's Columbian Exposition of 1893.[11]

These same ideas were reinforced in 1933 by Swarthmore's Arnold Dresden, the then President of the MAA, a stalwart member of the AMS, and one of those who had been present two years earlier in Chicago. In a talk produced by Science Service, broadcast on CBS radio on 29 November, and entitled "Mathematics in a Changing World," Dresden sought to convey nothing less than "the importance of mathematics for the maintenance of human existence in such a world."[12] To do that, he needed to go "beyond the superficial aspects of its elements as taught in the schools" to reveal mathematics' "fundamental characteristics." The public needed to appreciate mathematics not in its details but in its overarching concepts. An example was in order.

---

but extremely entertaining *Men of Mathematics* (New York: Simon and Schuster, 1937) also targeted a broader audience, conveying mathematics as a human endeavor.

10. For their talks, see Leopold Fejér, "On the Infinite Sequences Arising in the Theories of Harmonic Analysis, of Interpolation, and of Mechanical Quadratures," *BAMS* 39 (1933), 521–534; Tullio Levi-Civita, "Some Mathematical Aspects of the New Mechanics," *BAMS* 39 (1933), 535–563; George Birkhoff, "Quantum Mechanics and Asymptotic Series," *BAMS* 39 (1933), 681–700; and Leonard Dickson, "Recent Progress on Waring's Theorem and Its Generalizations," *BAMS* 39 (1933), 701–727.

11. David Rowe and I discuss mathematics at that earlier fair in *Emergence*, pp. 295–330 as well as in "Embedded in the Culture."

12. Dresden, typescript of "Mathematics in a Changing World," p. 1 (or p. 568 in the printed version). The quotations that follow in this and the next paragraph are on p. 1 (or pp. 568–569).

The ancient Greeks had conceived of and suitably mathematized a geocentric universe, yet their mathematization required "elaborate systems of epicycles" "to introduce some order into [the planets'] lawless behavior." Copernicus, Kepler, and Newton interpreted things differently in the sixteenth and seventeenth centuries. For them, the Sun was at the center. As a result, their mathematization effected "a great simplification." Could this suggest a "possible direction in which progress may be made in the understanding of our human problems"? Dresden believed it could. While those problems had "been studied preponderantly from national and even sectional points of views," mathematics' "world point of view" revealed the fact "that the apparently capricious way in which our world behaves" is "due to the simultaneous operation of a number of independent forces . . . each capable of exact understanding."

Four things allowed mathematics to make sense of the "apparently capricious": "its powers of analysis, its concern with functional relations, its abstract nature and its interest in invariants under transformations."[13] As Dresden acknowledged, these "are all abstract in character," but "[t]hey have been turned to fruitful account probably because they reveal fundamental qualities of the human mind." Still, and this is an equally great strength, their "*applications* belong to the sciences, natural *and* social."

It is unclear just how this erudite argument may have been received. Americans sitting beside their radios that Wednesday also heard news of the imminent repeal of the Eighteenth Amendment that had, for thirteen years, banned the consumption of alcoholic beverages in the United States. Moreover, the rhetoric and imagery of the "Century of Progress" exhibits aside, mathematics, at least as part of the secondary school curriculum, had been increasingly under fire in the United States since the late 1910s, with the assault intensifying during the Depression as immediate utility and applicability became paramount.

At issue was the usefulness of especially algebra, but to a lesser extent geometry and trigonometry, for the vast majority of high school students. These subjects had long been required in what had come to be viewed as an elitist, undemocratic curriculum at odds with Progressive-Era ideals of utility and social efficiency. By the 1930s, the Depression was forcing educators to rethink that curriculum as part of a "vocationalization" aimed narrowly at

---

13. Ibid., p. 3 (or p. 570). The quotations that follow in this paragraph are also on this page (with my emphases). As noted, Bell had already touched on some of these notions, although in somewhat different ways, in *The Queen of the Sciences*.

giving American students precisely the tools they would need for the types of jobs they would likely have and for the sorts of lives they would likely lead.[14] This movement had resulted in the cutting of mathematics requirements in school districts across the country and in the open questioning of what was termed "academic mathematics."[15] What mathematics was really necessary? Were the mathematicians and teachers of mathematics acting in their own self-interest or truly in the interest of an American society under stress? When Dresden again addressed a large audience, this time one comprised of his fellow mathematicians present in Pittsburgh on the occasion of his retiring MAA presidential address in December 1934, he had these attacks foremost in mind.

Dresden's faith in the power of mathematics remained unshaken, despite the ever-worsening world situation and the still-deepening Depression. "[G]iven a proper share in the education and training of our people," he argued, "mathematics can wield a powerful influence in that reshaping of our world of which the present witnesses perhaps the first significant stirrings."[16] Mathematics, after all, is "an occupation . . . devoted . . . to the highest aspirations of mankind." It was thus imperative that the "mathematical profession" play a "more active" role in bringing its case before the public, and Dresden proposed a three-part plan to guide its actions. First, mathematicians had "[t]o make clear the specific aspects through which mathematics contributes to these high purposes." They then needed "to determine the relation of these aspects to the rest of the subject matter," for this would allow them, finally, "to devise ways in which they can be made effective in the teaching of the subject at its various levels."

Dresden had already addressed the plan's first part in his radio talk a year earlier, but geopolitics had come even more to the fore by the end of 1934. Underscoring mathematics' "cosmopolitanism," he noted that "[i]t is perhaps this same quality . . . which is at least partly responsible for the attitude which the Hitler regime in Germany has taken towards the subject—mathematics does not contribute sufficiently to the national purposes to find favor with

14. Herbert Kliebard and Barry Franklin, "The Ascendance of Practical and Vocational Mathematics, 1893–1945: Academic Mathematics Under Siege," in *A History of School Mathematics*, ed. George Stanic and Jeremy Kilpatrick, 2 vols. (Reston: National Council of Teachers of Mathematics, 2003), 1: 399–440 on p. 420.

15. Ibid., p. 426.

16. Arnold Dresden, "A Program for Mathematics," *AMM* 42 (1935), 198–208 on p. 200. The quotations that follow in this paragraph are also on this page.

the advocates of unrestricted nationalism."[17] In fact, Dresden contended, it is precisely mathematics' cosmopolitanism as well as its universality that can offer hope in such troubled times. "[I]f there is to be true progress," he argued, ". . . the advance must be in the direction of enlarged application of general principles, of the discovery of new principles of wider scope, and of a world wide envisagement of the important problems." It was "[i]n the cultivation of such ability" that "the abstract formulation of problems should play an increasingly important role." It was thus critical to society that mathematics be fostered.

Since the plan's second part was "the very essence of the subject of mathematics, . . . the source of its power," Dresden felt justified in proceeding directly to its third part, educational reform. A child, he contended, needs to see "that the sums in arithmetic with which he is made to wrestle have significance in a great diversity of conditions, in other times and places besides those in which he happens to find himself."[18] For this reason, teachers at all levels should "select and . . . arrange experiences from what is available" so "that the abstract concepts of mathematics can be extracted from them."[19] Yet, mathematics is deductive as well as abstract, and this combination is another of its strengths that teachers should incorporate into their "instruction . . . from the earliest stages to the more advanced ones."[20] Finally, since mathematics deals with existence theorems, concepts that Dresden, like Bell, saw as critical in the modern world, an appreciation of their import was also vital for students. "Insistence upon at least a consideration, if not a solution of the existence problem in connection with social and economic policies would," Dresden maintained, "exercise a very significant, and I think wholesome influence upon the development of human life."[21]

In summary, Dresden quoted no less an authority than the renowned educational reformer and philosopher John Dewey. "The obligations incumbent upon science," Dewey asserted, "cannot be met until its representatives . . . devote even more energy . . . to seeing to it that the sciences which are taught are themselves more concerned about creating a certain mental attitude than they are about purveying a fixed body of information or about preparing a small number of persons for the further specialized pursuit of some particular

17. Ibid., p. 201. The quotations that follow in this paragraph are also on this page.
18. Ibid., p. 202.
19. Ibid., p. 203.
20. Ibid., p. 205.
21. Ibid., p. 206.

science."[22] Both Dewey and Dresden acknowledged that the teaching of science and mathematics needed to change, that the goals of their instruction required revision. Both also agreed that since mathematics was a critical element in our efforts to comprehend and control our world, such instruction was a critical element of a secondary curriculum under pressure to expand in other directions.

Seven months later, another mathematician, this time Earle Hedrick, explicitly isolated at least one source of that pressure. Invited to address the students participating in Columbia's 1935 summer session, Hedrick boldly maintained that "[m]athematics is in danger of being eliminated from required school curricula to make room for propaganda disguised as 'social studies.' "[23] He read the situation as Dresden had. In order to counter the attack, "teachers of science and mathematics [needed] to abandon 'stock methods' for less traditional teaching programs more closely related to everyday living" and, in so doing, demonstrate that "mathematics, when properly presented, has as much 'social' value as history." As cases in point, Hedrick cited the critical importance of the mathematical notion of compound interest as well as the "far-reaching social effects" of life insurance.

This argument apparently convinced at least one member of Hedrick's audience. A writer for the *New York Sun* opined that Hedrick had made a "vigorous defense of mathematics against the inroads of propaganda in favor of this and that quack social doctrine" and had "argued sanely enough that mathematics can be improved, that mathematics can be distinctly social in all its bearings."[24] With tongue in cheek, this same commentator also remarked that "[t]hose school pupils who must race their brains over an algebraic problem will not thank Professor Earle R. Hedrick"!

The motivations of the organizers of the "Century of Progress" World's Fair as well as the elucidations and defenses of mathematics of Bell, Dresden, Hedrick, and others underscored a widespread belief in the 1930s of the social value of mathematics.[25] Yet, these optimistic visions contrasted markedly

22. Ibid., p. 207, quoting John Dewey, "The Supreme Intellectual Obligation," *Science* 49 (16 March, 1934), 240–243 on pp. 241–242.

23. "Mathematics Seen As Social Science: Prof. Hedrick Fears Study May Be Eliminated," *New York Sun*, 13 July, 1935, clipping in Box 86-14/68, Folder 5, MAA Papers. The quotations that follow in this paragraph are also from this article.

24. "Useful Old 'Math,' " *New York Sun*, 16 July, 1935, clipping in Box 86-14/68, Folder 5, MAA Papers. The quotation that follows is also from this article.

25. Michael Barany has argued that this "mathematics awareness" originated during and after World War II. The examples presented in this section demonstrate, however, that such

with the perception of Depression-Era school officials and some educators focused on efficiency, immediate utility, and vocationalization. Caught in the middle, members of the mathematical community—defined by the dual mission of research, championed by the AMS, and of teaching, promoted by the MAA—made common cause in delivering a unified message for their field.

That unification was natural. Many mathematicians, like Bell, Dresden, and Hedrick, were members of and active in both the AMS and the MAA, and some, like Hedrick, even continued to wonder whether two societies were really needed, especially given the financially straitened times. Others, like Bell, had no doubt that each society had its own critical role to play. He considered "legion" those members of the MAA who would find "too advanced" much of the then modern mathematical research that engaged many members of the AMS.[26] This difference of opinion represented not so much a conflicting perception among mathematicians of the mathematical endeavor in the 1930s as it did a divergent sense of the distance between teaching and research. Bell, Dresden, and Hedrick—each committed to teaching—also shared a deep-seated commitment to the research ideal that they and many others made manifest through their work in the 1930s.

Developments in Topology

In topology, that commitment was reflected in the further growth and strengthening of the two rival groups that had emerged in the 1920s. Lefschetz's combinatorial (or algebraic) topological school at Princeton vied with R. L. Moore's point-set topological school at Texas. Both approaches had received codifications early in the 1930s in the form of AMS Colloquium publications, with Lefschetz's *Topology* appearing in 1930 and Moore publishing his *Foundations of Point Set Theory* two years later. If anything, the differences and divisions between these topological camps had become even more pronounced by the 1930s as adherents of both jockeyed for influence within the American mathematical research community.

In a sense, the combinatorial topologists had the edge, for, in 1927, Lefschetz assumed the co-editorship (with Wedderburn until 1930) of what was increasingly perceived as the nation's strongest research journal, the *Annals of Mathematics*. An active editor and a strong personality, Lefschetz had a clear

---

"awareness" was also present throughout the 1930s. See Michael Barany, "The World War II Origins of Mathematics Awareness," *NAMS* 64 (2017), 363–367.

26. Bell to William Cairns, 18 February, 1934, Box 86-14/66, Folder 1, MAA Records.

sense of what was and what was not publishable, and—for the most part, although not exclusively—point-set topology seemed to be in the latter category in the 1930s. As early as 1931, Lefschetz and Whyburn had clashed over this point when the Master's thesis of Whyburn's second Ph.D. student at Hopkins, Beatrice Aitchison, was rejected for publication.

Aitchison's work treated the point-set-theoretic notion of "regular accessibility": a point p is "regularly accessible from a point set M provided all points of M sufficiently near p can be joined to p by arbitrarily small arcs lying in M except for p."[27] In particular, she provided "a systematic development of a large number of the previously known results" about this concept as well as "substantial generalizations of many of" them. Indeed, one such generalization was of a theorem that Whyburn himself had proved in 1928.[28] When it rejected Aitchison's paper, then, the Annals passed a negative judgment explicitly on her work and tacitly on that of her advisor.

Lefschetz's then co-editor and Princeton colleague, Einar Hille, delivered the bad news, explaining the editorial "lack of enthusiasm" to Whyburn this way: "The printing situation of the Annals of Mathematics is unfortunately such that only a very good paper has any chances of rapid publication, meaning less than nine months, and that the ordinary publishable paper usually cannot be accepted for publication owing to lack of space."[29] Hille's justification thus reflected the publication realities of the early 1930s. Mathematical output continued to rise, while financial support for publication tightened (recall fig. 4.3). Aitchison's effort—presumably representative, in the views of the Annals's editorial staff, of the "ordinary publishable paper"—was thus not up to the higher standards that fiscal constraints—among, likely, other motivations—were forcing the Annals to impose.

Not surprisingly, Whyburn was content with neither the explanation nor the assessment. "Why not send" the paper, he asked Lefschetz by way of reply, "to someone *thoroughly familiar* with and active in *our field* such as Moore, Kline, Wilder, Ayres and ask them to referee it before passing final judgement?"[30] In Whyburn's mind, this rejection was clearly political not

---

27. This and the quotations that follow in the next sentence are in Whyburn's review of Beatrice Aitchison, "Concerning Regular Accessibility," *Fundamenta Mathematicae* 20 (1933), 117–125 in *Zentrallblatt für Mathematik und ihrer Grenzgebiete* 7 (1933), 82–83.

28. Gordon Whyburn, "Concerning Accessibility in the Plane and Regular Accessibility in *n* Dimensions," *BAMS* 34 (1928), 504–510. See, in particular, theorem 3 on p. 509.

29. Hille to Whyburn, 10 August, 1931, Box 1, Folder 79: Hille, Einar, Whyburn Papers.

30. Whyburn to Lefschetz, 12 August, 1931, Box 1, Folder 79: Hille, Einar, Whyburn Papers (my emphases).

fiscal. For him, it was a matter of "them against us," that is, the combinatorial vs. the point-set topologists, and a paper from someone in the latter camp would only get a fair assessment from a well-informed, like-minded reader. In retrospect, Aitchson's paper likely ran afoul of both politics and Depression-Era finances. When it finally appeared in print in 1933, it was in the point-set-friendly, foreign journal, *Fundamenta Mathematicae*.[31]

Although Whyburn had published his first paper in the *Annals* in the second volume that Lefschetz had co-edited with Wedderburn, he did not publish in it again until 1939 and, indeed, never thereafter, preferring instead as American outlets for his work the *American Journal of Mathematics*, the journals of the American Mathematical Society, and, after its founding in 1935, the *Duke Mathematical Journal*, of which he was an editorial board member.[32] For the most part, others in the point-set-theoretic camp also opted to publish elsewhere, although Raymond Wilder and Leo Zippin—the former, as noted, a student of Moore's at Texas and the latter one of Kline's at Penn—both published point-set-oriented work there in 1933 before their respective moves into more algebraic topological realms.[33] There was, however, one notable exception. Zippin's mathematical "brother," William Claytor, published two papers in the *Annals* in the 1930s that caused something of a stir in point-set-theoretic circles.

Claytor, the third African-American to earn an American Ph.D. in mathematics, successfully defended his doctoral thesis in 1933 and saw it in print as

31. Beatrice Aitchison, "Concerning Regular Accessibility," *Fundamenta Mathematicae* 20 (1933), 117–125. Aitchison went on to earn her Ph.D. at Hopkins under Whyburn in 1933. After teaching mathematics, statistics, and economics, the latter a field in which she earned an M.A. from the University of Oregon in 1937, she effectively moved in 1942 from academe into the U.S. government, eventually holding high-level positions in the Department of Commerce and the U.S. Postal Service. See Judy Green and Jeanne LaDuke, *Supplementary Material for Pioneering Women in American Mathematics: The Pre-1940 PhD's*, Aitchison, pp. 1–5 (for the URL, see the References).

32. Gordon Whyburn, "Concerning the Complementary Domains of Continua," *AM* 29 (1927–1928), 399–411; "Non-alternating Interior Retracting Transformations," *AM* 40 (1939), 914–921; and Edwin Floyd and F. Burton Jones, "Gordon T. Whyburn (1904–1969)," *BAMS* 77 (1971), 57–72 on pp. 67–72. From 1940 to 1945, however, Whyburn was associated with the *Annals* as one of the two AMS representatives on its editorial board. The MAA also had two editorial board representatives. This practice was discontinued for the MAA in 1945 and for the AMS a year later.

33. Raymond Wilder "On the Linking of Jordan Continua in E_n by $(n-2)$-Cycles" and Leo Zippin, "Independent Arcs of a Continuous Curve," *AM* 34 (1933), 441–449 and 95–113, respectively.

"Topological Immersion of Peanian Continua in a Spherical Surface" in the *Annals* a year later from his position at historically black West Virginia State College.[34] By Lefschetz's lights, then, this work, unlike Aitchison's Master's thesis, must have been a "very good" as opposed to an "ordinary publishable paper." In it, Claytor generalized results of, among others, Kuratowski on the peanian continuum, that is, on the notion of a continuous image of a closed interval. Whereas Kuratowski had determined necessary and sufficient conditions for such a continuum (with the additional assumption that it contain only a finite number of simple closed curves) to be homeomorphic to a subset of the plane, Claytor was interested in the analogous problem in the next higher dimension, that is, in characterizing those continua (with no restrictions imposed) homeomorphic to a Euclidean 2-sphere. In a result that Kline described to R. L. Moore as "the best that I have ever had done under my direction," Claytor found necessary and sufficient conditions for this homeomorphism.[35] A year after the work's publication, Wilder highlighted Claytor's methods in an address before the MAA in Ann Arbor.[36]

By December 1936, during a self-funded research year with Wilder at Michigan, Claytor was presenting new results in a session on topology held in Durham, North Carolina, as part of the joint annual meeting of the AMS and the MAA. Over the summer, he had met and talked with Kuratowski during the Polish mathematician's transatlantic visit to the East Coast, and the two topologists had specifically discussed ideas for how to handle the case of peanian continua that were *not* imbeddable in a Euclidean 2-sphere. Presiding over Claytor's session in Durham, Lefschetz took special note of the young mathematician's latest result, judging it "the best of the session" and reflective of "a noteworthy line of investigation with good chances of continued success."[37] Perhaps not surprisingly then, he snapped up the written

34. William Claytor, "Topological Immersion of Peanian Continua in a Spherical Surface," *AM* 35 (1934), 809–835.

35. Kline to Moore, 24 October, 1933, Box 4RM74, Folder: Kline, John Robert (1929–1954), Moore Papers. I give more on Claytor's life and this particular work in "Mathematics and the Politics of Race," especially pp. 221–223.

36. Raymond Wilder, "Some Unsolved Problems of Topology," *AMM* 44 (1937), 61–70 on pp. 66–67.

37. Lefschetz to the Julius Rosenwald Fund, as excerpted in the file on Claytor's application in the 1936–1937 fellowship round. See Box 402, Folder 6: William W. S. Claytor, Rosenwald Archives. The quotation from Wilder below may also be found here.

FIGURE 6.2. William Schieffelin Claytor (1908–1967) in 1937. (Photo courtesy of
Raymond Louis Wilder Papers, 1914–1982, e_math_02076, Archives of American
Mathematics, Dolph Briscoe Center for American History,
University of Texas at Austin.)

version for the *Annals* and saw to its "rapid publication."[38] Wilder continued
to be impressed, too, proclaiming Claytor "one of the most promising young
men in the field today" based on his two *Annals* papers.

38. William Claytor, "On Peanian Continua Not Imbeddable in a Spherical Surface," *AM*
38 (1937), 631–646. I discuss Claytor's result in more detail in "Mathematics and the Politics
of Race," pp. 227–229.

The *Annals* may largely have failed—Claytor's work aside—to serve the publication needs of the point-set topologists, but the same was not true of those in the combinatorial topological camp.[39] In addition to papers by Lefschetz and his students, Paul Smith, William Flexner, Norman Steenrod, and Albert Tucker, among others, the *Annals* published papers in the 1930s by James Alexander and Samuel Eilenberg in the United States and Paul Alexandroff and Eduard Čech from abroad. In fact, it was precisely research in this vein on which Lefschetz chose to focus in the retiring AMS presidential address he gave—also in Durham in December 1936—on "The Role of Algebra in Topology."[40] There, he acknowledged that "[t]he assertion is often made of late that all mathematics is composed of algebra and topology." Yet, he argued, "[i]t is not so widely realized that the two subjects interpenetrate so that we have an algebraic topology as well as topological algebra." His talk thus aimed to show how "a reasonable blend of the algebraic and topological points of view is possible" using some of the ideas that he had laid out in his 1930 Colloquium volume. This allowed him to characterize "many of the recent very interesting results in combinatorial topology," for example, results of Alexander and Tucker on complexes (recall the definition in chapter one).[41]

If Lefschetz and Whyburn serve to exemplify the rivalry between the two American topological camps, Wilder actively sought a middle ground between them. Taking advantage of the invitation to give a symposium lecture before the AMS at its spring meeting in Chicago in 1932, Wilder considered what he termed "a unified analysis situs."[42] There had been, he noted, extended lectures before the AMS on algebraic topology by Veblen and Lefschetz and on point-set topology by Moore and Kline, but no one had as yet made "clear the relations between these two schools..., why there are two schools and what is the difference between them," and why "the lines of

---

39. Another counterexample to this "rule" is Karl Menger, a topologist primarily of the point-set variety, who published in the *Annals* both before and after his emigration from Austria to the United States.

40. Solomon Lefschetz, "The Role of Algebra in Topology," *BAMS* 43 (1937), 345–359. The quotations that follow in this paragraph are on p. 345.

41. For their work on complexes, see Alexander, "Combinatorial Analysis Situs" and "A Combinatorial Theory of Complexes," *AM* 31 (1930), 292–320 and Albert Tucker, "An Abstract Approach to Manifolds," *AM* 34 (1933), 191–234.

42. Raymond Wilder, "Point Sets in Three and Higher Dimensions and Their Investigation by Means of a Unified Analysis Situs," *BAMS* 38 (1932), 649–692 on p. 649. The quotations that follow in this paragraph are also on this page.

demarcation between these 'branches' " are beginning "to disappear." These would be Wilder's appointed tasks.

The two schools—Wilder viewed them as "two *methods* within topology having the same aim" rather than as two rival *schools*—arose from the approaches to geometrical questions of Poincaré on the one (combinatorial) hand and Cantor on the other (point-set-theoretic) hand.[43] Both were interested in understanding the "*properties of point sets that are invariant under topological transformations,*" but the combinatorialists tended to treat properties of sets in the large, while the point-set theorists focused on their properties locally. Wilder saw limitations in the methods of both groups.

In higher dimensions, he explained, properties in the large "become so complicated that the set-theoretic method has not been successful, by itself," while "the combinatorial method . . . is restricted to closed sets of points, or to compact spaces, and there is no evidence, as yet, that it can be expanded so as to take care of the topology of non-compact sets, excepting, of course, certain particular cases such as the complements of closed sets in" Euclidean n-space. In his paper, Wilder showed, among other things, how "many of the theorems of the set-theoretic topology concerning the separation of space are only special cases of" combinatorial topological methods.

Two different methods may have emerged in topology, but they could reach the same ends. For that reason, Wilder "consider[ed] the present situation of a division into two 'schools' as the greatest menace to the future development of topology, a division which fosters a lack of true perspective, unscientific animosity, and an unnecessary delay in the progress of our investigations of the structure of space." That Wilder's conciliatory efforts largely failed to succeed in the short run was borne out by the fact that in his account of "Fifty Years of American Mathematics" for the AMS semicentennial six years later in 1938, Birkhoff still delineated American topology into two camps.[44]

Such a division was apparently not as sharp abroad as it was in the United States. In planning for what has been deemed "the first truly international

43. The quotations that follow in this and the next two paragraphs are in ibid., p. 650, p. 653, p. 663, p. 664, p. 692, and p. 692, respectively (with his emphases). On the notion of a Moore "school," see Lewis, "The Beginnings of the R. L. Moore School of Topology," and on the notion of "school" in mathematics more generally, see Parshall, "Defining a Mathematical Research School."

44. Birkhoff, "Fifty Years of American Mathematics," pp. 311–313.

conference in a specialized part of mathematics, on a broad scale," topologists at Moscow State University as well as at the newly created Steklov Institute aimed to bring together in Moscow in September 1935 Soviet and non-Soviet topologists of all stripes, about half of the latter to be invited from the United States.[45] Lefschetz was charged with arranging American participation, and although the conference was ultimately a largely combinatorial topological affair (the presence of Polish point-set theorists Kuratowski, Stefan Mazurkiewicz, and Wacław Sierpiński, notwithstanding), Lefschetz had followed the wishes of the Soviet hosts and sought representation from the American point-set topology camp. When Whyburn rather cooly declined his invitation—there was clearly no love lost between Whyburn and Lefschetz— Lefschetz pointedly replied that "[i]t is only fair to say that the decision to invite you did not belong to me alone, but was arrived at jointly with my Moscow colleagues. This is one more proof," he added not perhaps without a certain measure of disingenuousness, "of the fact that your work is appreciated considerably farther east than the Atlantic coast."[46]

Almost forty mathematicians took part in the Moscow meeting.[47] Ten— Alexander, Lefschetz, Stone, Tucker, von Neumann, Zariski, Garrett (son of George) Birkhoff, Paul Smith, Egbert van Kampen, and Hassler Whitney— journeyed from the United States. Of these, only Stone, Birkhoff, and van Kampen had no direct connection to the Princeton group. Whitney, like Stone a Harvard Ph.D., had, however, taken his NRC fellowship to Princeton in 1931–1932, while Zariski had spent the 1934–1935 academic year at the Institute thanks to Lefschetz's endorsement. It was hardly a group representative of American topology, but, as Whitney noted of the conference as a whole, it brought "[m]ost of the world leaders, that is, in the combinatorial direction" into a conversation that had far-reaching repercussions.[48]

45. Whitney, "Moscow 1935," p. 97. A similarly spirited conference in tensor analysis had actually been hosted in Moscow in May 1935 (Solomon Lefschetz to Gordon Whyburn, 20 March, 1935, Box 1, Folder 108: Lefschetz, Solomon, Whyburn Papers).

46. Lefschetz to Whyburn, 27 March, 1935, Box 1, Folder 108: Lefschetz, Solomon, Whyburn Papers.

47. Albert Tucker provides a list in "The Topological Congress in Moscow," p. 764, but for an even more complete account of the meeting and its participants, see Darya Apushkinskaya, Alexander Nazarov, and Galina Sinkevich, "In Search of Shadows: The First Topological Conference, Moscow 1935," *MI* 41 (2019), 37–42.

48. Whitney, "Moscow 1935," p. 108. Papers given by the participants were collected and published in *Matematicheskiĭ Sbornik* 43 (5) (1936).

In particular, after presenting his ideas on a new multiplication for complexes that could be applied to more general spaces, Andrei Kolmogoroff (later Kolmogorov, but I will use this transliteration) learned from Alexander that the American had actually made the same discovery. In effect, these two men had simultaneously and independently discovered cohomology, that is, what is now understood as the association of a sequence of abelian groups to a topological space.[49] Whitney described the atmosphere this way. "From the reputation of these mathematicians," those present sensed that "there must be something real going on; but it was hard to see what it might be" at the time.[50] Almost immediately thereafter, Whitney and the Czech mathematician, Eduard Čech, who was also in the audience in Moscow, provided modifications of the group multiplication that Kolmogoroff and Alexander had defined, thereby providing "an essential improvement."[51] Cohomology developed quickly thereafter at the hands of Americans Steenrod and Whitney, American émigré Samuel Eilenberg, and Čech, among others.[52]

Fifteen months after the gathering in Moscow, another topology conference, this one an American affair in Durham, consciously sought to unite topologists of the point-set and combinatorial persuasions. The idea, which had originated in the spring of 1936 with Richardson in discussion with Lefschetz and Kline, was to associate a free-standing, day-long meeting on some specialized topic with the annual AMS meeting. If this model proved successful, then the AMS could, Richardson hoped, successfully approach one of the foundations for fuller support that would allow it to host meetings,

---

49. See Andrei Kolmogoroff, "Über die Dualität im Aufbau der kombinatorischen Topologie," *Matematicheskiĭ Sbornik* 43 (1) (1936), 97–102 and "Homologiering des Komplexes und des lokal-bikompakten Raumes," *Matematicheskiĭ Sbornik* 43 (5) (1936), 701–706 as well as James Alexander, "On the Chains of a Complex and Their Duals" and "On the Ring of a Compact Metric Space," *PNAS* 21 (1935), 509–511 and 511–512. Dieudonné elaborates on this early history in *A History of Algebraic and Differential Topology*, pp. 78–81.

50. Whitney, "Moscow 1935," p. 110.

51. Eduard Čech, "Multiplications on a Complex," *AM* 37 (1936), 681–697. Whitney presented his multiplication in a letter to Leo Zippin according to James Alexander, "On the Connectivity Ring of an Abstract Space," *AM* 37 (1936), 698–708 on p. 699 (footnote 7). For the quotation, see ibid, p. 699. Whitney ultimately published his results in "On Products in a Complex," *AM* 39 (1938), 397–432.

52. William Massey relates this story in "A History of Cohomology Theory," in *History of Topology*, ed. Ioan James, pp. 579–603. Massey specifically addressed the import of the Moscow conference on these developments on pp. 583–587 and of Čech and Whitney's results on pp. 587–589.

more international in nature, along the lines of the Moscow conference.[53] As a test case, Lefschetz suggested a meeting on topology with Whyburn as the organizer.[54]

Although wary at first, Whyburn fairly quickly satisfied himself that topologists in both camps would cooperate to make such a meeting a success. As early as May 1936, he sketched "a very tentative list of topics which might be covered" provided, as he explained to Kline, "suitable men can be obtained to speak on them and lead the discussion afterwards."[55] Key among those "suitable men" was his advisor, R. L. Moore, but Moore had been cagey when Whyburn sounded him out, asking questions such as "who originated the idea, what were the exact words by which it was referred to (point set theory, topology, or what?), whether it was to include combinatorial topology, and so on." Moore, the American progenitor of the point-set approach, viewed Lefschetz in particular as a not-so-friendly rival in the combinatorial camp, and Whyburn, given his run-in with Lefschetz over Aitchison's thesis, was by no means insensitive to this view. "As to the inclusion of combinatorial topology," Whyburn explained to Moore, "my idea would be to include some of the more modern features of the subject which are most closely related to *our* work and are proving useful *to us* at present."[56] If the Moscow conference had centered on combinatorial topology, the Durham meeting would have a point-set focus. In the end, it not only brought Moore and Lefschetz together on the same stage but also provided a fruitful venue for the exchange of topological ideas.[57] Two days after the stand-alone meeting, the topologists again

53. The hope for outside funding for regular, international symposia went unrealized, although the concept of the topic-focused, auxiliary session proved to have legs. See, for example, the section on strides in Algebraic Research.

54. Kline to Whyburn, 1 July, 1936, Box 5, Folder 18: Lectures and Meetings–1936: Topology Conference, Duke University, Whyburn Papers.

55. Whyburn to Kline, 22 June, 1936, Box 5, Folder 18: Lectures and Meetings–1936: Topology Conference, Duke University, Whyburn Papers. The next quotation is also from this letter.

56. Whyburn to Moore, undated (but shortly after 9 July, 1936), Box 5, Folder 18: Lectures and Meetings–1936: Topology Conference, Duke University, Whyburn Papers (my emphases).

57. Unfortunately, I have not found a copy of the actual program. Whyburn was supposed to have written a report on the conference for the AMS's *Bulletin* but no such account appeared and no notes or drafts remain in his *Nachlass*. Durham's newspaper, the *Herald-Sun*, reported on 27 December, 1936, however, that "Prof. E. W. Chittenden, University of Iowa; Prof. W. Hurewicz, Poland; Prof. S. Lefschetz, Princeton; Prof. R. L. Moore, University of

convened, this time for the session on topology that was part of the AMS's regular programming and in which, despite its location in the Jim Crow South, William Claytor presented his latest results on peanian continua to Lefschetz and others.

Topology, both combinatorial and point-set, thus continued to be a major focus of American mathematical research in the 1930s, but whereas a decade earlier its main players had been Veblen and R. L. Moore, the leadership was changing. Moore was still a force in point-set circles, but he had been joined by members of the next generation like Whyburn. Veblen, whose interests had already shifted in the 1920s, had been replaced by Lefschetz as the leading figure in an algebraic topology that was also attracting such talent as the émigré Samuel Eilenberg (see the next chapter). Wilder led a one-man crusade to unite the two approaches and to create one community of American topologists.[58] Regardless of their topological persuasion, these mathematicians and others participated actively at home and abroad in the further development of their subject. Theirs was a vibrant—if sometimes testy—give-and-take within and across national borders that placed them among the world's leaders in their area.

Geometrical Shifts

If topology in its two variants continued strong into the 1930s, geometry was more in flux. The less abstract, more calculationally oriented, nineteenth-century style of algebraic geometry practiced in the 1920s—already outmoded then—persisted in the work, among others, of Snyder, Coble, and their numerous students. Its waning, however, is ironically suggested by the 1934 NRC-sponsored literature review, *Selected Topics in Algebraic*

Texas; Prof. Oswald Veblen, Institute for Advanced Study; and Prof. Hassler Whitney, Harvard" were all slated to "participate in the conference." See Box 86-14/81, Folder 5, MAA Records.

58. In the summer of 1940, for example, Michigan hosted a "Conference in Topology" that brought together point-set and combinatorial topologists as well as newcomers to the American mathematical scene like Polish-born mathematicians Samuel Eilenberg, who was hired at Michigan in 1940 (see the next chapter), and Witold Hurewicz, who had taken a position at the University of North Carolina following a two-year stint at the Institute from 1936 to 1938. For the program, see Box 5, Folder 19 "Lectures and Meetings–1940: Topology Conference, University of Michigan," Whyburn Papers. The proceedings of this conference were published as Raymond Wilder and William Ayres, ed., *Lectures on Topology* (Ann Arbor: University of Michigan Press, 1941).

Geometry–II, by Snyder, Amos Black, and Leaman Dye that supplemented the 1928 collaborative effort of Snyder, Lefschetz, Coble, Arnold Emch, Francis Sharpe, and Charles Sisam (recall chapter one).[59]

In the 1934 installment, Snyder ruefully acknowledged that, while the two recent Cornell Ph.D.s, Black and Dye, had contributed one chapter each to the new work, "the other chapters have been largely the work of one member," namely, Snyder himself. He may have had "the assistance and support of the" other committee members, but it had unfortunately been the case that "circumstances" had "prevent[ed] a more intensive cooperation." In particular, the "new" algebraic geometry that Lefschetz had covered in the earlier volume was not represented in the 1934 effort, given, most likely, Lefschetz's focus on his duties as editor of the *Annals* and on his own active research program. Coble's eulogist captured well the sense of the slow death of the "old" algebraic geometry in his characterization of Coble's 1929 AMS Colloquium volume, *Algebraic Geometry and Theta Functions*. In his view, "[t]he book as a whole is a difficult mixture of algebra and analysis, as represented by classical invariant theory and theta functions, with intricate geometric reasoning of a type few can follow today. The calculations are formidable; let them serve to our present-day algebraic geometers, dwelling as they do in their Arcadias of abstraction, as a reminder of what awaits those who dare to ask specific questions about particular varieties."[60]

If the "old" algebraic geometry was passing from the scene, a "new" one was issuing from Princeton, Johns Hopkins, and elsewhere to take its place. For his part, Lefschetz continued to push the work in algebraic topology that he had begun in the 1920s, further illuminating in the process not only the strength of adapting topological concepts in algebraic geometric settings but also the blurring of the boundary between geometry and topology. At least initially, Oscar Zariski was one mathematician who followed him along this path.

59. Virgil Snyder, Amos Black, and Leaman Dye, *Selected Topics in Algebraic Geometry–II*, Bulletin of the National Research Council, no. 96 (Washington, D.C.: National Research Council of the National Academy of Sciences, 1934). The quotations that follow in the next two sentences are on p. iii. Snyder also expounded on "Some Recent Contributions to Algebraic Geometry" in a symposium lecture he was invited to give before the AMS at its New York City meeting in March 1934. See *BAMS* 40 (1934), 673–687.

60. Arthur Mattuck, "Arthur Byron Coble," *BAMS* 76 (1970), 693–699 on p. 696, characterizing Arthur Coble, *Algebraic Geometry and Theta Functions*, AMS Colloquium Publications, vol. 10 (New York: American Mathematical Society, 1929).

Zariski had been trained in and was an enthusiast of the Italian algebraic geometric tradition embodied in the 1920s by his advisor, Guido Castelnuovo, and Castelnuovo's colleagues, Federigo Enriques and Francesco Severi. Their approach, however, had come under increasing criticism for what was seen as its over-reliance on intuition and its nonchalance relative to mathematical rigor. In fact, Castelnuovo himself had declared that "[t]he methods of the Italian School have reached a dead end and are inadequate for further progress in the field of algebraic geometry."[61] It was for that reason that he first counseled Zariski to delve into Lefschetz's work and then approached Lefschetz directly on his student's behalf. At least one Italian— Castelnuovo—recognized an American—Lefschetz—as a key mover and shaker in the field internationally.

Thanks to the Princeton mathematician's support, Zariski landed the Johnston Scholarship that allowed him both to spend the 1927–1928 academic year at Johns Hopkins and to participate actively in the American mathematical scene. As luck would have it, that was the same year that topology at Princeton was further energized by the presence of two IEB fellows, Alexandroff and Hopf. Lefschetz saw to it that Zariski was invited to Princeton, introduced to Veblen, Alexandroff, Hopf, and others, and involved in various mathematical gatherings on the East Coast.[62] Indeed, Zariski fell "under the influence of Lefschetz" that year, embracing his topological approach to algebraic geometry and developing it in the work he did over the next decade from what became his Hopkins home base.

Perhaps the most influential and far-reaching research that issued from this phase of Zariski's career was the book, *Algebraic Surfaces*, that he published in 1935 in Springer Verlag's then three-year-old monograph series, Ergebnisse der Mathematik und ihrer Grenzgebiete.[63] There, Zariski took on the challenge of giving "a systematic exposition of the theory of algebraic surfaces emphasizing the interrelations between the various aspects of the theory: algebro-geometric, topological, and transcendental." The algebro-geometric approach was that, on the algebraic side, of Leopold Kronecker,

61. Parikh, p. 36. For more on this school, see Aldo Brigaglia and Ciro Ciliberto, *Italian Algebraic Geometry between the Two World Wars*, Queen's Papers in Pure and Applied Mathematics, vol. 100 (Kingston: Queen's University Press, 1995).

62. Parikh, pp. 43–58. The quotation that follows in the next sentence is on p. 60.

63. Oscar Zariski, *Algebraic Surfaces*, Ergebnisse der Mathematik und ihrer Grenzgebiete, vol. 3, no. 5 (Berlin: Springer-Verlag, 1935). The quotation that follows is in the preface.

Heinrich Weber, and Richard Dedekind and, on the geometric side, of Max Noether and Alexander von Brill as well as Zariski's Italian mentors; the topological point of view had been developed by Lefschetz and others; and the assault by transcendental methods had been spearheaded by Émile Picard and hinged on the analysis of various integrals definable on algebraic surfaces. The problem was that over the course of the late nineteenth and early twentieth centuries, these approaches had tended not only "to diverge up to the threshold of mutual incomprehension" but also, in some instances, to employ techniques that were insufficiently justified.[64] Zariski thus strove to isolate "the typical *methods* of proof used in the theory" of algebraic surfaces as well as "their underlying ideas," since "the *methods* employed are at least as important as the results."[65] At the same time, "due to the exigencies of simplicity and rigor," the proofs he provided "differ[ed], to a greater or less[er] extent, from the proofs given in the original papers." In short, he aimed to put the theory on a more solid foundation.[66]

Among the topics he tackled were the theory and reduction of singularities. The fundamental object of study in classical algebraic geometry—at least in the affine case—was the algebraic variety, that is, the set of solutions of a system of polynomial equations over the field of complex numbers. An algebraic variety of dimension one over its base field is called an algebraic curve, while one of dimension two is termed an algebraic surface. Quite naturally, algebraic geometers first tried to understand algebraic curves—for example, how they behave under birational transformation from one space to another—and then to generalize their results to algebraic surfaces.[67] In particular, they proved this two-part "reduction theorem": 1) "[a]n algebraic curve can always be birationally transformed into a curve" in complex projective three-space "free from singularities and then, by projection, into a plane curve with ordinary double points only," and 2) "[a] plane algebraic curve can always be transformed by a Cremona transformation into a plane curve

64. Dieudonné, *History of Algebraic Geometry*, p. 27. Dieudonné discusses the different approaches on pp. 27–58.

65. Zariski, *Algebraic Surfaces*, preface (my emphases). The quotations that follow in the next sentence are also here.

66. Silke Slembek discusses this work in historical context in "On the Arithmetization of Algebraic Geometry," in *Episodes in the History of Modern Algebra (1800–1950)*, ed. Jeremy Gray and Karen Hunger Parshall, pp. 285–300, especially on pp. 289–292.

67. A birational transformation is one in which coordinates in two spaces are expressed rationally in terms of those in another.

with ordinary multiple points only."[68] In their efforts to generalize this to algebraic surfaces, they only succeeded in giving arguments that *supported* the truth of the analog of the result. As Zariski cautioned, "[i]t is important . . . to bear in mind that in the theory of singularities the details of the proofs acquire a special importance and make all the difference between theorems which are rigorously proved and those which are only rendered highly plausible." In his book, he analyzed the analog's purported proofs—by Pasquale del Pezzo; Beppo Levi; his former teacher, Severi; Giacomo Albanese; and Oscar Chisini—found all of them wanting, and indicated their specific shortcomings.

The result, which had long been accepted, did not remain in limbo for long. His manuscript in page proofs in 1933, Zariski became aware of a paper then just out in the *Annals*. Robert Walker, one of Lefschetz's students, had succeeded in providing "a complete function-theoretic proof of the reduction theorem for algebraic surfaces" that "stands the most critical examination and settles the validity of the theorem beyond any doubt."[69] Still, Walker's proof was function-theoretic. Could a more intrinsic proof be found? And, if so, what would "more intrinsic" mean in an area that had developed from so many different points of view?

If *Algebraic Surfaces* had served, in Lefschetz's words, to "dispel as rapidly as possible" the "widespread attitude of doubt" about some of the results of particularly the Italian school of algebraic geometry, it had driven home to Zariski just how compromised by a lack of rigor the entire theory actually was.[70] As he put it, although his exploration of the work of the Italian masters had in no way diminished his "admiration for the imaginative geometric spirit that permeated the[ir] proofs," he had "bec[o]me convinced that the whole structure must be done over again by purely *algebraic* methods."[71] To him, those were the more intrinsic methods that seemed most desirable.

At the Institute for Advanced Study during the 1934–1935 academic year, Zariski had the chance not only to focus on putting the finishing touches on *Algebraic Surfaces* but also to immerse himself in modern algebra. That same

68. Zariski, *Algebraic Surfaces*, pp. 18–19. The quotation that follows is on p. 19. A Cremona transformation is a birational transformation of projective *n*-space (over some field) into itself.

69. Ibid., p. 23, note 1. Walker's paper was "Reduction of the Singularities of an Algebraic Surface," *AM* 36 (1935), 336–365.

70. Solomon Lefschetz, "Zariski on Algebraic Surfaces," *BAMS* 42 (1936), 13–14 on p. 14.

71. Oscar Zariski, Response in "1981 Steele Prizes," *NAMS* 28 (1981), 504–507 on p. 505 (my emphasis).

year, Brauer served as Weyl's assistant and Emmy Noether came over from Bryn Mawr to give a series of lectures on ideal theory in algebraic number theory.[72] He was thus exposed to some of algebra's best. By 1938, Zariski had begun to work fruitfully on what would be the next phase of his research—the arithmetization of algebraic geometry through the use of the theory of ideals—that would last until the early 1950s (see chapter ten). He would begin training graduate students in this research vein at Hopkins in the 1940s and continue following his move to Harvard in 1947.

Whereas algebraic geometry may have been overshadowed in the 1920s by its differential counterpart, the tables had turned by the 1930s. The program in differential geometry applied to the theory of relativity that Veblen and Eisenhart had animated at Princeton was still generating results, but it was "winding down."[73] Eisenhart's energies had been diverted first by his post as Dean of the Faculty from 1925 to 1933 and then by his appointment as Dean of the Graduate School in 1933, while Veblen poured himself into setting up and guiding the Institute's School of Mathematics as well as placing mathematicians displaced by events in Europe. Still, both men published texts in 1933 reflective of their shared agenda.

For his part, Veblen worked up into book form a series of lectures he had given at Göttingen in the summer of 1932. His *Projective Relativitätstheorie*, appearing in Springer's Ergebnisse series a year before Zariski's text, laid out a projective unified theory of gravitation and electromagnetism not dissimilar to ideas then issuing from the parallel Dutch school of Jan Schouten.[74] Eisenhart took a different tack, publishing an advanced textbook on the theory of continuous groups of transformations as it had developed at the hands of Sophus Lie, Wilhelm Killing, and others. In it, he specifically highlighted how Lie groups relate to differential geometry as well as how they can be applied in dynamics.[75] The work of both men was duly noted at the time; both

72. Parikh, pp. 74–76.

73. Ritter, p. 173.

74. Oswald Veblen, *Projective Relativitätstheorie*, Ergebnisse der Mathematik und ihrer Grenzgebiete, vol. 2, no. 1 (Berlin: Springer-Verlag, 1933). See also Ritter, pp. 173–175. Recall from chapter three, that Veblen and his student, Henry Whitehead, had published *The Foundations of Differential Geometry* a year earlier in 1932.

75. Luther Eisenhart, *Continuous Groups of Transformations* (Princeton: Princeton University Press and Oxford: Oxford University Press, 1933).

men continued to push the ideas that they presented in their books through the 1930s.[76]

Tracy Thomas and Bob Robertson were also still at Princeton training a trickle of students in their brand of differential geometry. Thomas produced what was considered a significant book on differential invariants in 1934, while Robertson gave a solution of the two-body problem in general relativity in 1938.[77] Despite this activity, the more ambitious project of the unification of mathematics and physics—characteristic of Eisenhart and Veblen's vision and in which Thomas, Robertson, and others had shared—ultimately went unrealized.

Be that as it may, the *mathematical* development of differential geometry had in no way been impeded by the shortfall. For example, on the East Coast, Dirk Struik at MIT continued work with Schouten in the Netherlands to produce the two-volume *Einführung in die neuren Methoden der Differentialgeometrie* in 1935 and 1938.[78] A substantial reworking, or indeed "reincarnation" in the words of one reviewer, of work the two authors had done in the 1920s, the new volumes abandoned earlier methods in favor of an exclusive reliance on tensor analysis and succeeded, among other things, in extending "Riemannian geometry to the general case in which the metric is not necessarily definite."

76. See, for example, Abraham Taub, Oswald Veblen, and John von Neumann, "The Dirac Equation in Projective Relativity," *PNAS* 20 (1934), 383–388 and Luther Eisenhart and Morris Knebelman, "Invariant Theory of Homogeneous Contact Transformations," *AM* 37 (1936), 747–765. Taub and Knebelman had both been students of the "Princeton School." Taub earned his Ph.D. in 1935 under the direction of mathematical physicist Bob Robertson, before proceeding first to a year at the Institute and then to an instructorship at the University of Washington, while Knebelman earned his in 1928 under Veblen and Eisenhart before joining the Princeton faculty in 1929 and serving there until his move to what is now Washington State University in 1939.

77. Tracy Thomas, *The Differential Invariants of Generalized Space* (Cambridge: Cambridge University Press, 1934) and Howard Robertson, "Note on the Preceding Paper: The Two Body Problem in General Relativity," *AM* 39 (1938), 101–104. For the assessment of Thomas's book, see Dirk Struik, Review of "*The Differential Invariants of Generalized Spaces.* By T. Y. Thomas," *BAMS* 41 (1935), 477–478.

78. Jan Schouten and Dirk Struik, *Einführung in die neuren Methoden der Differentialgeometrie*, 2 vols. (Groningen: Noordhoff, 1935 and 1938). See also the review by William Graustein, "Schouten and Struik on Differential Geometry," *BAMS* 45 (1939), 649–650. The quotations that follow in the next sentence are on pp. 650 and 649, respectively.

Other differential geometric techniques developed on the West Coast. Aristotle Michal had written his 1924 Rice dissertation on "Invariant Functionals" under Evans's watchful eye and clearly under the sway of the then-recent work in functional analysis that his advisor had presented in Colloquium Lectures before the AMS in 1916.[79] He followed an NRC fellowship at Princeton in 1927 with an assistant professorship at Ohio State before moving on to Caltech in 1929. By that point, he had begun to delve deeply into work emanating from France and Poland by Maurice Fréchet and Stefan Banach, respectively. The result was the development of his earlier research in n-dimensional differential geometry into a more general and abstract theory of differential geometry in function space. Michal highlighted what he termed "general differential geometry" in an invited address before the AMS at its April meeting in Berkeley in 1938.[80]

In his view, the area had grown from his own extension—to geometries over Banach spaces—of work he had done between 1927 and 1931 on Riemannian and non-Riemannian geometries over the space of real continuous functions. He and his students—among them, Angus Taylor later at UCLA and Donald Hyers later at the University of Southern California—had subsequently explored the ramifications of this more abstract approach, developing, through their studies especially of polynomial and analytic functions, the differential calculus in Banach space that Fréchet had initiated in 1925.[81] Michal augured big things for his line of work, immodestly stating that it "is destined to become one of the great branches of mathematics, comparable to the present status of general (abstract) algebra and general analysis" and counseling mathematicians-in-training that "[t]here is still time for a whole army of young mathematicians to earn their first laurels in general differential geometry while the subject is still in its infancy."[82]

79. Griffith Evans, *Functionals and Their Applications: Selected Topics, Including Integral Equations*, AMS Colloquium Lectures, vol. 5 (New York: American Mathematical Society, 1918).

80. Aristotle Michal, "General Differential Geometries and Related Topics," *BAMS* 45 (1939), 529–563. Donald Hyers provides an overview of Michal's life and work in "Aristotle D. Michal 1899–1953," *Mathematics Magazine* 27 (1954), 237–244.

81. Maurice Fréchet, "La différentielle dans l'analyse générale," *Annales scientifiques de l'École Normale Supérieure* 42 (1925), 293–323. For a sense of the "Pasadena school's" results, see Michal.

82. Michal, "General Differential Geometries," p. 529.

While Michal's Pasadena school pursued differential *geometry* on the West Coast, Morse (at Harvard until 1935 and then at the Institute) and Whitney (at Harvard in the 1930s and 1940s) were pioneering highly original, but not fundamentally unrelated, research in what would come to be called differential *topology* on the East Coast. Both areas deal with the behavior of differentiable manifolds, that is, topological spaces that not only resemble Euclidean space locally (or near each point) but on which additional structure can be defined so that differential calculus can be performed on the spaces. Differential topologists, however, have a more inherently global view, focusing on structures on manifolds that are only interesting "in the large" as opposed to being interesting locally or "in the small."

One manifestation of this point of view was the line of research Morse launched in the late 1920s, and on which he had given AMS Colloquium Lectures in 1931 and spoken at the Zürich ICM in 1932. By 1934, that work had reached a first culmination with *The Calculus of Variations in the Large*. As Morse explained, his research had been "oriented by a conception of what might be termed macro-analysis."[83] That meant making "radical additions to the methods of what is now strictly regarded as pure analysis," since "[a]ny problem which is non-linear in character, which involves more than one coordinate system or more than one variable, or whose structure is initially defined in the large, is likely to require considerations of topology and group theory in order to arrive at its meaning and its solution." Morse's book highlighted those "considerations of topology." In so doing, it blurred yet another mathematical boundary—this one between analysis and topology—a fact quickly appreciated by mathematicians in Europe. By 1938, the German topologists Herbert Seifert and Wilhelm Threlfall had published *Variationsrechnung im Grossen (Theorie von Marston Morse)*, described as a text "excellent both for the calculus of variations student who wishes to learn about the modern developments in his field, and for the topologist who wishes to become acquainted with an important and fascinating application of his field to analysis without becoming involved in too many analytic details."[84] Morse continued to pursue this fertile line of research throughout the 1930s and, indeed, for the rest

83. Morse, *Calculus of Variations in the Large*, p. iii. The quotations that follow in the next sentence are also on this page.

84. Herbert Seifert and Wilhelm Threlfall, *Variationsrechnung im Grossen (Theorie von Marston Morse)* (Leipzig and Berlin: B. Teubner Verlag, 1938). For the quotation, see Sumner Myers, "*Variationsrechnung im Grossen (Theorie von Marston Morse)*. By H. Seifert and W. Threlfall. Leipzig and Berlin, Teubner, 1938. 115 pp.," *BAMS* 46 (1940), 390.

of his career.[85] Moreover, before his move to the Institute, he trained a number of students—like Gustav Hedlund first at Bryn Mawr, then at Virginia beginning in 1939, and finally at Yale after 1948 and Sumner Myers at Michigan beginning in 1936—in his differential topological brand of the calculus of variations.

Like Morse, Whitney had come to differential topology from analysis albeit following a fruitful encounter with graph theory in his 1932 Ph.D., also like Morse, under Birkhoff at Harvard.[86] Casting around for a postdoctoral research project that would return him to the mathematical roots in analysis of the physics major he had completed at Yale, he decided to explore the properties of differentiable functions in Euclidean n-space $\mathbb{E}^n$. In particular, he asked two important questions: first, "if, in some sense, a function is n-times differentiable on any closed set of $\mathbb{E}^n$, can it be extended to an n-times differentiable function on all of $\mathbb{E}^n$," and second, "[i]f the function is analytic on the closed set, is there an analytic extension to $\mathbb{E}^n$?"[87] By 1934, he had answered both of these questions in the affirmative.

This success led him two years later to "a general study of the topology of differentiable manifolds, and maps of them into other manifolds" using "tools of a purely analytic character."[88] As he noted, there were two definitions of a differentiable manifold then commonly in use. One was abstract and due to Veblen and Whitehead in their 1932 treatise on *The Foundations of Differential Geometry*, namely, a differentiable manifold is "a point set with neighborhoods homeomorphic with Euclidean space $\mathbb{E}^n$, coordinates in overlapping neighborhoods being related by a differentiable transformation."[89] The other was more concrete: a differentiable manifold is simply "a subset of $\mathbb{E}^n$, defined near each point by expressing some of the coordinates in terms of others by differentiable functions." Whitney showed that, in fact, "the first definition is no more general than the second; any differentiable manifold may be imbedded

85. Bott gives a modern overview of Morse's work in this vein on pp. 920–935.

86. Whitney sketches his Ph.D. work in "The Coloring of Graphs," *PNAS* 17 (1931), 122–125.

87. Keith Kendig, "Hassler Whitney: 1907–1989," *Celebratio Mathematica* (2013), 1–19 on p. 11 at https://celebratio.org/Whitney_H/245/. Compare Hassler Whitney, "Analytic Extensions of Differentiable Functions Defined in Closed Sets," *TAMS* 36 (1934), 63–89 on p. 63. Kendig also discusses this work in *Never a Dull Moment: Hassler Whitney, Mathematics Pioneer*, AMS/MAA Spectrum, vol. 93 (Providence: American Mathematical Society, 2018), pp. 143–159.

88. Hassler Whitney, "Differentiable Manifolds," *AM* 37 (1936), 645–680 on p. 645. For more on this work, see Kendig, *Never a Dull Moment*, pp. 161–173.

89. Ibid. The quotations in the next two sentences are also on this page.

in Euclidean space," and, in particular, any n-dimensional compact connected differentiable manifold can be embedded in $\mathbb{E}^{2n+1}$. In so doing, "he broke entirely new ground," marking the "beginnings of differential topology" as a well-defined subdiscipline.[90]

Although different in thrust, the "geometrical" work of Morse and Whitney, as well as that of Lefschetz and Zariski, reveals a thread common to the 1930s. With the rise of the structural approach to algebra and the evolution of topology, a cross-fertilization of these approaches occurred with much of geometry, both algebraic and differential, as well as with other areas of what may be considered "classical" mathematics. Morse brought topology more fully into analysis; Whitney demonstrated how analytic tools could prove useful in the topology of differentiable manifolds; Lefschetz, but more critically Zariski, brought the structures first of topology and then of modern algebra to bear on algebraic geometry. In the 1930s, subdisciplinary boundaries became less sharp. Or, put another way, boundaries between mathematical subdisciplines became more permeable, often resulting in new and deeper mathematical insights.

Strides in Algebraic Research

The structural approach to algebra, key to many of these changes, had been evolving at least since the closing decades of the nineteenth century. Reflected in the United States of the 1920s in the group-theoretic work of Dickson and Miller as well as in the researches of Dickson and Wedderburn in the theory of linear associative algebras, it received its first, broad codification as the 1930s opened. In 1930–1931, Dutch mathematician Bartel van der Waerden published his two-volume textbook, *Moderne Algebra*, inspired in large part by the algebraic researches and approach of his mentors, Emmy Noether in Göttingen and Emil Artin in Hamburg.[91] This work went on to shape, in large measure, the algebra of the 1930s. The structures—groups, rings, ideals, fields, etc.—that he codified continued to be developed in and of themselves at the same time that they were identified, explored, and exploited within other, ostensibly non-algebraic mathematical areas.[92] Albert's research in the 1930s exemplifies both kinds of developments.

90. Dieudonné, *A History of Algebraic and Differential Topology*, pp. 60–62 on p. 60.

91. Bartel van der Waerden, *Moderne Algebra*, 2 vols. (Berlin: Julius Springer, 1930–1931).

92. Leo Corry provides a broader historical contextualization in *Modern Algebra and the Rise of Mathematical Structures*, Science Networks–Historical Studies, vol. 17 (Basel: Birkhäuser Verlag, 1996).

As noted in chapter one (and recall the various definitions and references given there), the classification of semi-simple linear associative algebras, thanks to the work of Wedderburn, boiled down to the classification of division algebras. This fact had stimulated work on both sides of the Atlantic on understanding the structure and properties of division algebras and, in particular, Albert's 1928 dissertation research on central division algebras of dimension 16 over their base fields. As Albert continued to tweak and expand his result into the 1930s, he became involved specifically in the determination and classification of finite-dimensional central division algebras over algebraic number fields, that is, over fields $\mathbb{Q}(\theta)$, where $\mathbb{Q}$ denotes the rational numbers and θ is a root of an equation irreducible in $\mathbb{Q}$ with rational coefficients.[93] Since this had come to define "[t]he main goal of the structure theory of algebras of the period 1929–1932" and since mathematicians on both sides of the Atlantic were actively involved in its pursuit, it comes perhaps as no surprise that an American, Albert, quickly found himself in direct competition with a German, Helmut Hasse.[94] Well grounded, unlike Albert, in algebraic number theory and, in particular, in the class field theory that he and others were developing, Hasse brought arithmetic techniques to bear on the problem, while Albert's approach, initially at least, was more purely algebraic. Both mathematicians made progress.

Albert and Hasse had been in correspondence about their mutual mathematical interests at least as early as February 1931.[95] By September 1931, Albert had shown that central division algebras of dimension 16 over $\mathbb{Q}(\theta)$ were cyclic;[96] in a paper that he completed in February 1930 but that only appeared in print in 1931, Hasse had explored central division algebras over p-adic number fields. By May 1931, the German had also established and submitted for publication to the AMS's *Transactions* a number of general results on cyclic central division algebras. That work, however, was only presented

93. Adrian Albert, "Normal Division Algebras of Degree Four over an Algebraic Field," *TAMS* (1932), 363–372 on p. 363.

94. Nathan Jacobson, "Abraham Adrian Albert 1905–1972," *BAMS* 80 (1974), 1075–1100 on p. 1079.

95. On Albert and Hasse's competition, see Della Fenster and Joachim Schwermer, "A Delicate Collaboration: Adrian Albert and Helmut Hasse and the Principal Theorem in Division Algebras in the Early 1930s," *Archive for History of Exact Sciences* 59 (2005), 349–379.

96. Adrian Albert, "New Results in the Theory of Normal Division Algebras," *TAMS* 32 (1930), 171–195 and "Normal Division Algebras of Degree Four," respectively.

FIGURE 6.3. A. Adrian Albert (1905–1972) and his daughter, Nancy, at the Institute for Numerical Analysis (see chapter 10) in 1952. (Photo courtesy of Marion Walter Photograph Collection, 1952–1980s, 1952, e_math_01120, Archives of American Mathematics, Dolph Briscoe Center for American History, University of Texas at Austin.)

to the Society in September and published several months later in 1932.[97] In it, he acknowledged that "[t]he theory of linear algebras has been greatly extended through the work of American mathematicians." Yet, "[o]f late, German mathematicians have become active in this theory. In particular, they have succeeded in obtaining some apparently remarkable results by using the theory of algebraic numbers, ideals, and abstract algebra, highly developed in Germany in recent decades." Unfortunately, he sensed, those results were not "as well known in America as they should be on account of their importance," owing "perhaps, to the language difference or to the unavailability of the widely scattered sources." For these reasons, he had decided to "present [his] paper for publication to an American journal and in English."[98] It seemed clear that a rivalry had been sparked. In retrospect, it is equally clear that by October

97. Helmut Hasse, "Über p-adische Schiefkörper," *Mathematische Annalen* 104 (1931), 495–534 and "Theory of Cyclic Algebras over an Algebraic Number Field," *TAMS* 34 (1932), 171–214, respectively.

98. Hasse, "Theory of Cyclic Algebras," p. 171.

1931 both Albert and Hasse were closing in on the principal theorem, that is, on proving that every central division algebra over an algebraic number field of finite degree is cyclic.

Albert wrote to Hasse in November 1931 to share the results he had obtained and communicated to the AMS in September, but which had not yet been published; Hasse had obtained the results that had also gotten him to within a hair's breadth of the prize, but they had not yet appeared.[99] In November, their respective new work still unknown to the other, Hasse, together with his German colleagues Brauer and Noether, published a joint paper that actually proved the principal theorem.[100] No mention of Albert's independent work was made; word of it had still not reached Germany. Albert, according to one of his biographers, had been "nosed out in a photo finish" by his German competitors.[101] In full recognition of that fact, Albert, "at the suggestion of H. Hasse and with his cooperation," published a paper, also in the AMS's *Transactions*, that set the record straight.[102] It had been a spectacular race in which the American had almost bested the Germans. It was yet another signal that American mathematics was becoming—or, indeed, had already become?—an international force to be reckoned with, but it equally underscored the barriers still inherent in mathematical communication in the 1930s.

Whereas Albert's work on central division algebras represents an example of how algebraic structures were studied and developed per se on American shores, his research on Riemann matrices exemplifies how algebraic structures were surfacing in other mathematical areas. Albert had followed his doctoral work under Dickson with an NRC postdoctoral fellowship for the 1928–1929 academic year quite naturally at Princeton, where Wedderburn, like Dickson a guru of the theory of linear associative algebras, was on the faculty. Interestingly, though, the fresh Chicago Ph.D. also caught Lefschetz's attention.[103]

99. Fenster and Schwermer lay out the exchange of letters in their paper.

100. Richard Brauer, Helmut Hasse, and Emmy Noether, "Beweis eines Hauptsatzes in der Theorie der Algebren," *Journal für die reine und angewandte Mathematik* 167 (1931), 399–404.

101. Irving Kaplansky, "Abraham Adrian Albert: November 9, 1905–June 6, 1972," *Biographical Memoirs*, vol. 51 (Washington, D.C.: National Academy of Sciences, 1980), pp. 3–22 on p. 5.

102. Adrian Albert and Helmut Hasse, "A Determination of All Normal Division Algebras over an Algebraic Number Field," *TAMS* 34 (1932), 722–726 on p. 722, note †. Although most sources today refer to the Brauer-Hasse-Noether Theorem, some rightly label it the Albert-Brauer-Hasse-Noether Theorem in recognition of Albert's contributions.

103. Daniel Zelinsky, "A. A. Albert," *AMM* 80 (1973), 661–665 on p. 664 and Jacobson, "Abraham Adrian Albert," pp. 1076–1077.

When Albert arrived in Princeton, the algebraic geometer had just finished writing his part of the 1928 report on topics in algebraic geometry for the National Research Council. There, he had not only surveyed results on but also detailed the utility of so-called Riemann matrices in uncovering and understanding the properties and interrelations of algebraic curves and their associated Riemann surfaces and abelian integrals. Given an algebraic curve, there is a Riemann surface K associated with it of genus p, that is, with p holes, as well as an abelian integral, the path of integration of which is over K. A $p \times 2p$ matrix ω with complex entries is called the Riemann matrix of the algebraic curve if there exists a $2p \times 2p$ matrix $C = -C^T$ with rational entries such that $\omega C \omega^T = 0$ and $\sqrt{-1}\ \omega C \bar{\omega}^T$ is a positive definite Hermitian matrix of rank p.[104] (Here, C^T and ω^T denote the transposes of C and ω, respectively, while $\bar{\omega}$ denotes the matrix the entries of which are the complex conjugates of the corresponding entries in ω.)

Now, for a given Riemann matrix ω, consider a $p \times p$ matrix α with complex entries and a $2p \times 2p$ matrix A with rational entries such that $\alpha\omega = \omega A$. This is called a multiplication of ω. The set D of all multiplications of ω is a linear associative algebra over the rationals and is termed the multiplication algebra of ω. A Riemann matrix ω is then called pure if its multiplication algebra is a division algebra.[105] Poincaré and Picard had already noted the importance of these concepts, and the Italian algebraic geometers Gaetano Scorza and Carlo Rosati had further explored especially the role of the multiplications in determining the singular correspondences on the underlying algebraic curve.[106] In Albert, Lefschetz recognized an algebraist perfectly suited to investigating "for their own sake" these matters "of considerable interest" in algebraic geometry.[107]

By November 1929, from a two-year instructorship at Columbia, Albert had already proven "that there exist no pure Riemann matrices whose multiplication algebras are normal division algebras of order sixteen over the field

104. Snyder, Coble, Emch, Lefschetz, Sharpe, and Sisam, pp. 310–315. Compare Adrian Albert, "The Non-Existence of Pure Riemann Matrices with Normal Multiplication Algebras of Order Sixteen," *AM* 31 (1930), 375–380 on p. 375.

105. Albert defines a multiplication algebra in "On the Construction of Riemann Matrices I," *AM* 35 (1934), 1–28 on p. 1 and a pure Riemann matrix in "The Non-Existence of Pure Riemann Matrices with Normal Multiplication Algebras of Order Sixteen," p. 375.

106. The particulars are given in the chapter "Correspondences between Algebraic Curves" in Snyder, Coble, Emch, Lefschetz, Sharpe, and Sisam, pp. 331–348.

107. Ibid., p. 314.

of all rational numbers."[108] Fifteen months later, he was presenting work to a meeting of the AMS toward a classification of all pure Riemann matrices of genus $p \leq 8$, and by March 1933, from the University of Chicago where he would remain for the rest of his career, he was formulating an even more unified theory. As he noted, although the problems of classification and construction "have been reduced to the case of pure Riemann matrices," "the methods of treatment . . . have not been quite as effective as may be desired."[109]

Interest was most definitely piqued by Albert's work in this direction. Spending the 1933–1934 academic year at the Institute, Albert interacted in Fine Hall not only with Wedderburn, who was in the final stages of completing his *Lectures on Matrices* based on courses he had given on the topic since 1920,[110] but also with Weyl as an auditor of his lectures on Lie algebras in the spring of 1934.[111] In a paper "On Generalized Riemann Matrices," Weyl built on the research that Albert, in his view, had "recently treated with such conclusive success" by suggesting a generalized definition that, while it affected none of Albert's results, struck Weyl as somewhat more intrinsic to "the nature of the subject."[112] Weyl explored his generalization in two more papers in 1936 and 1937;[113] Albert developed his version further in a series of papers in 1934 and 1935.[114] In 1939, the American mathematical research community recognized the quality of Albert's research, in general, and of his results on Riemann matrices, in particular. It selected him as one of the two AMS Colloquium Lecturers in September and awarded him its Cole Prize in Algebra in December. Albert codified not only his work on Riemann matrices

108. Albert, "The Non-Existence of Pure Riemann Matrices with Normal Multiplication Algebras of Order Sixteen," p. 375 and "The Structure of Pure Riemann Matrices with Non-Commutative Multiplication Algebras," *Rendiconti del Circolo matematico di Palermo* 55 (1931), 1–59 (the latter paper was dated 8 November, 1929).

109. Albert, "On the Construction of Riemann Matrices I," p. 1.

110. Wedderburn acknowledges Albert's contributions on p. iii. Ohio State's Cyrus Mac-Duffee had preceded Wedderburn by one year with an American book-length treatment of matrices: *The Theory of Matrices*, Ergebnisse der Mathematik und ihrer Grenzgebiete, vol. 2, no. 5 (Berlin: Julius Springer, 1933).

111. Kaplansky, "Abraham Adrian Albert," p. 7.

112. Hermann Weyl, "On Generalized Riemann Matrices," *AM* 35 (1934), 714–729.

113. Hermann Weyl, "Generalized Riemann Matrices and Factor Sets" and "Note on Matric Algebras," *AM* 37 (1936), 709–745 and 38 (1937), 477–483, respectively.

114. Adrian Albert, "On the Construction of Riemann Matrices I"; "On the Construction of Riemann Matrices II," *AM* 36 (1935), 376–394; and "A Solution of the Principal Problem in the Theory of Riemann Matrices," *AM* 35 (1934), 500–515. Jacobson gives a nice overview of this work in "Abraham Adrian Albert," pp. 1082–1085.

but also the rapid development in the 1930s of the theory of linear associative algebras in the Colloquium publication, *Structure of Algebras*, based on his lectures.[115]

Another algebraist in Princeton during the 1933–1934 academic year, graduate student Nathan Jacobson, was five years Albert's junior and presented results on the so-called Wedderburn decomposition of a cyclic algebra to the AMS in June 1933 that constituted his 1934 doctoral dissertation under Wedderburn's direction.[116] Having largely satisfied his doctoral obligations, Jacobson, like Albert, was free to attend Weyl's spring 1934 lectures on Lie algebras and was actually charged with writing up the course's official lecture notes.

Jacobson had been particularly struck by Weyl's suggestion that the properties of Lie algebras could and should be derived rationally, "that is, without recourse to extension of the base field to its algebraic closure—as had been done by Wedderburn in 190[7] for associative algebras."[117] Almost immediately, Jacobson, in conversation with Albert, succeeded in developing a method "of studying by elementary and direct means the relation between a Lie algebra and an enveloping associative algebra generated by it."[118] This marked Jacobson's first foray into a theory that would dominate his research in the 1930s and into the 1940s and come to define a major focus of his life's work.

Following a year at Bryn Mawr during which he stepped in to cover the courses that Noether would have taught had death not intervened, Jacobson spent the 1936–1937 academic year as an NRC fellow at Chicago working with Albert. A string of papers ensued, in which the younger algebraist began a systematic attack on the structure of Lie algebras, analyzing, in particular, simple Lie algebras of type A and normal simple Lie algebras of characteristic zero.[119] He continued to push this line of research after accepting, in 1937,

115. Adrian Albert, *Structure of Algebras*, AMS Colloquium Publications, vol. 24 (New York: American Mathematical Society, 1939).

116. Nathan Jacobson, "Non-Commutative Polynomials and Cyclic Algebras," *AM* 35 (1934), 197–208.

117. Nathan Jacobson, "A Personal History and Commentary 1910–1943," in *Nathan Jacobson: Collected Mathematical Papers*, ed. Nathan Jacobson, 3 vols. (Boston: Birkhäuser, 1989), 1: 1–11 on p. 3. For more on Wedderburn's work in this vein, see Parshall, "Wedderburn and the Structure Theory of Algebras," pp. 325–331.

118. Nathan Jacobson, "Rational Methods in the Theory of Lie Algebras," *AM* 36 (1935), 875–881 on p. 875.

119. See, for example, Nathan Jacobson, "Simple Lie Algebras of Type A," *AM* 39 (1938), 181–189 and "Simple Lie Algebras over a Field of Characteristic Zero," *DMJ* 4 (1938), 534–551.

the position at the University of North Carolina that he would hold until his 1943 move to Johns Hopkins. Interestingly, UNC also hired émigré algebraist Reinhold Baer that year in an effort "to upgrade" the department.[120] Had Baer not left for Illinois after just one year, UNC, with two young algebraists of such demonstrated talent, would likely have put the South on the mathematical map as early as the 1930s as an American center for algebraic research.

In addition to their work in the theory of associative and non-associative algebras, American algebraists were also instrumental in developing—at the intersection of algebra and logic—the related areas of lattice theory and Boolean algebra. Lattice theory, with its roots in Dedekind's turn-of-the-twentieth-century work, came into its own in the 1930s United States at the hands particularly of Harvard's Garrett Birkhoff and Yale's Oystein Ore.[121] Boolean algebra, on the other hand, had developed from George Boole's nineteenth-century efforts to formulate an algebra of logic and found an American champion in functional analyst Marshall Stone, at Yale from 1931 to 1933 and then at Harvard from 1933 until his move to Chicago in 1946.[122] The impetus behind both developments was the structural approach to abstract algebra as laid out in van der Waerden's *Moderne Algebra,* and both complemented the concomitant growth of symbolic logic.

The younger Birkhoff had come to lattices through the interest in group theory that he had developed under Philip Hall at Cambridge University in 1932–1933, the academic year following his Harvard A.B. As Birkhoff realized, the fact that lattices are defined in terms of a binary operation not only suggested that they be regarded "as *abstract algebras*" but also reflected their relationship to groups, rings, and linear associative algebras, among other constructs.[123] In particular, Birkhoff showed that lattices have applications to logic via Boolean algebras, to group theory given that the normal subgroups of a group constitute a lattice, and to geometry since the family of linear

120. Jacobson, "A Personal History and Commentary 1910–1943," p. 4.

121. Herbert Mehrtens details the history of lattice theory in *Die Entstehung der Verbandstheorie* (Hildesheim: Gerstenberg Verlag, 1979).

122. Corry provides a nicely self-contained overview of this work and the connections between lattice theory and Boolean algebras on pp. 263–292. On Stone at Chicago, see chapter nine.

123. Garrett Birkhoff, "Lattices and Their Applications," *BAMS* 44 (1938), 793–800 on p. 795 (his emphasis).

manifolds in n-space that pass through the origin also form a lattice.[124] His work in these directions resulted in the monograph, *Lattice Theory*, published in 1940 in the AMS's Colloquium series and described as "the first book" on that "far-reaching subject."[125]

Ore approached the theory from a different philosophical point of view. He opened his 1935 paper "On the Foundation of Abstract Algebra" by calling attention to the fact that, through "the study of the structure of the principal domains of algebra like group theory, ideal theory, hypercomplex systems, rings, moduli, etc. one has arrived at a great number of results showing close relationship and similarity. It is natural to expect," then, ". . . that these results may all be derived from a common source."[126] For him, that "common source" was the theory of lattices (he called them structures, in retrospect an unfortunate choice of terminology) which allowed for a shift of analytical focus from the *elements* in a particular domain to *substructures* within it. In particular, Ore developed lattice theory as a way to show how various of the decomposition theorems that had been proved—such as the Jordan-Hölder theorem in group theory—could be interpreted as arising naturally from a lattice-theoretic approach.[127] In order to do this, his lattices had to satisfy one more condition than Birkhoff's, the so-called Dedekind axiom.[128] Ore chose no less a venue than his invited address at the Oslo ICM in 1936 to highlight and bring to the attention of a wider mathematical audience all of these "very recent," "very fertile" lines of investigation.[129] For him, lattice theory was a

124. Garrett Birkhoff, "On the Combination of Subalgebras," *Proceedings of the Cambridge Philosophical Society* 29 (1933), 441–464. In a subsequent paper, he gave additional realizations of lattices in group theory and in the theory of probability (Garrett Birkhoff, "Applications of Lattice Theory," op. cit., 30 (1934), 115–122).

125. Garrett Birkhoff, *Lattice Theory*, AMS Colloquium Publications, vol. 25 (New York: American Mathematical Society, 1940). For the quotation, see the review by Lee Wilcox in *BAMS* 47 (1941), 194–196 on p. 194.

126. Oystein Ore, "On the Foundation of Abstract Algebra," *AM* 36 (1935), 406–437 on p. 406.

127. Recall that the Jordan-Hölder theorem states that given a group G and a composition series $1 = H_0 \lhd H_1 \lhd \cdots \lhd H_n = G$ of normal subgroups of G, any two composition series of G are equivalent in the sense that they have the same composition length $n + 1$ and the same composition factors H_i, up to permutation and isomorphism.

128. Ore gives the precise definition in "On the Foundation of Abstract Algebra," p. 412.

129. Oystein Ore, "On the Decomposition Theorems of Algebra," *Comptes rendus du Congrès international des mathématiciens Oslo 1936*, 1: 297–307 on p. 301.

sort of metatheory that could reveal structural similarities between seemingly disparate mathematical constructs.

Stone was one contemporary who needed no convincing on this point. Having published his magisterial treatment of linear transformations in Hilbert space in 1932, he had almost immediately embarked on a new line of research suggested by his analytic work on "the spectral theory of symmetric transformations in Hilbert space."[130] As he explained, in the theory of operator rings in Hilbert space, as well as in other rings and linear algebras, "the 'spectral' representation as a 'direct sum' of irreducible subrings reposes in essence upon the construction of an abstract Boolean algebra."[131] This fact prompted him to explore Boolean algebras per se, and by February 1933, his paper on the topic had been read by title at a meeting of the AMS in New York City.[132] When, by 1935, that research had been completed and accepted for publication in the AMS's *Transactions*, it had been critically informed by the work on lattices of both Birkhoff and Ore.

Stone had realized several things: first, that the Boolean algebras he had encountered in his work in analysis had long been studied by logicians from a purely postulate-theoretic point of view, but, second, and more importantly, that, "if one reflects upon various algebraic phenomena occurring in group theory, in ideal theory, and even in analysis, one is easily convinced that a systematic investigation of Boolean algebras, together with still more general systems, is probably essential to further progress in these theories."[133] What Stone proposed and proceeded to carry out, then, was "a study of Boolean algebras by the methods of *modern algebra*." He thus recognized that "Boolean algebras are identical with those rings with unit in which every element is idempotent," but his work also showed that a Boolean algebra is actually a special type of lattice.[134]

130. Marshall Stone, "The Theory of Representations for Boolean Algebras," *TAMS* 40 (1936), 37–111 on p. 37.

131. Marshall Stone, "The Representation of Boolean Algebras," *BAMS* 44 (1938), 807–816 on p. 808.

132. John Kline, "The February Meeting in New York," *BAMS* 39 (1933), 313–315 on p. 315.

133. Stone, "The Theory of Representations for Boolean Algebras," p. 37. The quotation that follows is also on this page (with my emphasis). Compare Corry, p. 288.

134. Stone, "The Theory of Representations for Boolean Algebras," p. 38. On the lattice-theoretic interpretation, compare Ore, "On the Decomposition Theorems," p. 305 and Birkhoff, *Lattice Theory*, pp. 56, 74, and 88.

Indeed, lattice theory seemed to be popping up in many new and exciting places in the 1930s. Indicative of the perception that it was then a "hot" area in, especially, American mathematics, the AMS, as it had done with topology in Durham in 1936, targeted lattice theory for a symposium held in conjunction with its spring meeting in Charlottesville, Virginia, in 1938.[135] Perhaps not surprisingly, the three main speakers were Garrett Birkhoff, Ore, and Stone. Their talks, however, were augmented by discussions led by the recent émigrés Baer, Menger, and von Neumann, as well as by Stone's first doctoral student, Holbrook MacNeille. All of these men had engaged or were engaging in research with lattice-theoretic consequences. Baer focused specifically on its applications to group theory; Menger worked on providing an "algebraic treatment" in the lattice-theoretic sense for projective, affine, and non-Euclidean geometry; MacNeille drew on his doctoral work on partially ordered sets to show how "the construction of the number system and the theories of measure and integration can be subsumed under a single abstract development in which partially ordered sets and Boolean rings play important roles"; von Neumann reinforced Boolean algebra's ring-theoretic side.[136] It was a thought-provoking event that underscored the reach of lattice theory into fields as seemingly disparate as algebra, set theory, probability, functional analysis, topology, and geometry. Perhaps not surprisingly, lattice theory—together with algebraic geometry and the theories of associative and non-associative algebras, among others—was cited as one of the exciting, then-recent advances in algebra by the algebraists and algebraic geometers who met at the University of Chicago in the summer of 1938 to discuss their work and to identify fruitful areas for future research.[137]

135. Temple Hollcroft, "The April Meeting in Charlottesville," *BAMS* 44 (1938), 463–470 on pp. 463–464.

136. Reinhold Baer, "The Applicability of Lattice Theory to Group Theory," *BAMS* 44 (1938), 817–820 and "The Significance of the System of Subgroups for the Structure of the Group," *AJM* 61 (1939), 1–44; Karl Menger, "New Foundations of Projective and Affine Geometry," *AM* 37 (1936), 456–482 and "Non-Euclidean Geometry of Joining and Intersecting," *BAMS* 44 (1938), 821–824 (the quotation is on p. 821); Holbrook MacNeille, "The Application of Lattice Theory to Integration," *BAMS* 44 (1938), 825–827 (the quotation is on p. 825) (see also his "Extension of a Distributive Lattice to a Boolean Ring," *BAMS* 45 (1939), 452–455); John von Neumann and Marshall Stone, "The Determination of Representative Elements in the Residual Classes of a Boolean Algebra," *Fundamenta Mathematicae* 25 (1935), 353–378 and John von Neumann, "On Regular Rings," *PNAS* 22 (1936), 707–713.

137. Adrian Albert, "The Chicago Conference and Seminar on Algebra," *BAMS* 44 (1938), 756–757 and Saunders Mac Lane, "Some Recent Advances in Algebra," *AMM* 46 (1939), 3–19.

As with American work in topology and geometry, American algebraic research in the 1930s reflected both the impact of the structural approach and the increasing awareness of the presence of algebraic structures through-out mathematics. Linear associative algebras in Albert's hands, Lie algebras in Jacobson's work, and lattices through the efforts of Garrett Birkhoff, Ore, Stone, and others were studied in and of themselves and fuller theories of them generated. From the position he took at Toronto in 1935, Brauer, too, did important work in this structural vein with his development of the mod-ular representation theory of finite groups.[138] Yet, algebraic structures were also isolated and exploited in perhaps less expected realms, such as Albert's development of Riemann matrices in algebraic geometry, Ore's illumina-tion of the commonalities among decomposition theorems in a wide range of settings via lattice theory, and Stone's appropriation of Boolean algebras from logic for algebraic purposes. If group theory per se had been one of the main thrusts of American algebraic research in the 1920s, what Gar-rett Birkhoff termed its "vigorous and promising younger brother," lattice theory, was seen as having overtaken it in the 1930s as algebraic structures in general swept across the American mathematical landscape.[139] And, if Dickson, Wedderburn, and Miller had led American algebra in the 1920s, a new generation comprised, among others, of Albert, Jacobson, Ore, Gar-rett Birkhoff, and Stone as well as recently arrived émigrés like Baer, Menger, and Brauer was taking the reins in the 1930s and guiding the area in new directions.

138. See, for example, his joint work with his first Toronto student, Cecil Nesbitt, "On the Modular Representation of Groups of Finite Order I" (*University of Toronto Studies* 4 (1937)). Charles Curtis gives a nice account of this work in historical context in *Pioneers of Representation Theory: Frobenius, Burnside, Schur, and Brauer*, HMATH, vol. 15 (Providence: American Math-ematical Society and London: London Mathematical Society, 1999), pp. 235–271. Brauer also continued his work in the theory of algebras with papers such as "On the Regular Represen-tations of Algebras" (also with Nesbitt) (*PNAS* 23 (1937), 236–240) and "On Modular and p-adic Representations of Algebras" (*PNAS* 25 (1939), 252–258). See also Paul Fong and War-ren Wong, ed., *Richard Brauer: Collected Papers*, 3 vols. (Cambridge: The MIT Press, 1980), 1: 336–354, 190–194, and 199–205, resp.

139. Birkhoff, "Lattices and Their Applications," p. 793. Group theory by no means disap-peared from the American mathematical landscape in the 1930s. In particular, Hermann Weyl's *Classical Groups: Their Invariants and Representations* (Princeton: Princeton University Press, 1939) appeared in 1939 as the inaugural volume in the Princeton Mathematical Series and also reflected the structural turn in algebraic research.

Analysis in the 1930s

Of all the areas of American mathematical expertise in the 1930s, however, it was analysis that continued to be the most variegated, attracting relatively large numbers of practitioners and embracing numerous subfields. At least two informal metrics gauge its "share" of the American mathematical scene: nine of the fourteen AMS Colloquium volumes published in the 1930s focused on analysis, and, of the nine addresses presented on the occasion of the semicentennial of the AMS in 1938, topology, geometry, and algebra were represented by one address each, while five treated what American mathematicians at the close of the 1930s deemed "representative subjects" in analysis.[140] In particular, Joseph Ritt's semicentennial address on "Algebraic Aspects of the Theory of Differential Equations" underscored the fact that while "[s]imilarities between linear differential equations and algebraic equations [had been] observed by the early analysts"—Laplace, Lagrange, Poisson—that work could be significantly extended and completed using the modern algebra of the twentieth century—results like those of Hilbert, Noether, and van der Waerden.[141] In other words, modern algebra could, in Ritt's view, inform analysis as meaningfully as it could topology and geometry.

Ritt had earned his Ph.D. at Columbia under Edward Kasner in 1917 for a dissertation on differential equations of infinite order and remained associated with the faculty at his alma mater for his entire career. By the 1930s, his research had come to center on "systems of differential equations, ordinary or partial, which are algebraic in the unknowns and their derivatives," and he had expounded on his ideas in a 1932 AMS Colloquium volume.[142] In particular, he sought to apply the well-developed theory of systems of *algebraic* equations—for example, Hilbert's finite basis theorem on the existence of finite bases for infinite systems of polynomials—to systems of *differential*

140. For the quotation, see The Committee on Publications, "Preface," *Semicentennial Addresses*. The ninth address, moreover, George Birkhoff's "Fifty Years of American Mathematics," devoted some fifteen pages to analysis, but seven to axiomatics and symbolic logic, five-and-a-half to algebra and number theory, six to geometry and topology, and three to applied mathematics. It is fair to say, however, that Birkhoff's overview was also strongly colored by his personal mathematical interests.

141. Joseph Ritt, "Algebraic Aspects of the Theory of Differential Equations," *Semicentennial Addresses*, ed. The Committee on Publications, pp. 35–55 on p. 36 (note †).

142. Joseph Ritt, *Differential Equations from the Algebraic Standpoint*, AMS Colloquium Publications, vol. 14 (New York: American Mathematical Society, 1932), p. iii.

equations.[143] Not only would this bring "some of the completeness enjoyed by the" former theory into the analogous differential equation setting, but it also promised to uncover new types of problems and questions.[144] Ritt strove toward these ends over the course of the last twenty years of his life (he died in 1951), bringing a number of graduate students into his mathematical fold in the 1930s—among them, Eli Gourin, Henry Raudenbush, Fritz Herzog, and Walter Stroudt.

In particular, Ritt considered "differential polynomials or forms [denote them by A] in the unknown functions $y_1, y_2, \ldots, y_n$ and their derivatives with coefficients which are functions meromorphic in some domain. Given a system Σ of such forms, he show[ed] that there exists a finite subsystem of Σ whose manifold of solutions is identical with that of Σ."[145] From there, he constructed a basis for Σ, defined the notion of reducibility, and began the development of an ideal theory in the context of which he was able to prove a theorem analogous to the unique factorization theorem of ideals in a Dedekind domain. This structure then allowed him to formulate a precise characterization of the general solution of $A = 0$.[146]

Although not a member of Ritt's intellectual circle at Columbia, differential geometer Joseph Thomas, on the faculty at Duke beginning in 1930, also embraced an algebraic approach to the theory of systems of partial differential equations. He, however, came to the topic from the perspective of his postdoctoral year in Paris in 1926–1927 and his exposure there to the work of both Élie Cartan and Charles Riquier (recall chapter three). In his own AMS Colloquium volume, *Differential Systems*, published five years after Ritt's book, Thomas, unlike Ritt, quite self-consciously employed "[t]he ideas and nomenclature of modern algebra, as developed, for instance, in van der Waerden's admirable treatise."[147] In particular, he adopted a postulational approach in order to build the theory of the non-commutative ring/algebra he called a Grassmann ring/algebra and to develop "ideas introduced by Grassmann and brought to such a high degree of perfection by Cartan."[148] To his mind,

---

143. Ritt, "Algebraic Aspects," p. 35.

144. Ritt, *Differential Equations*, p. iv.

145. Edgar Lorch, "Joseph Fels Ritt," *BAMS* 57 (1951), 307–318 on p. 313.

146. For more details and the full notational scheme, see Ritt, *Differential Equations*, pp. 1–46.

147. Joseph Thomas, *Differential Systems*, AMS Colloquium Publications, vol. 21 (New York: American Mathematical Society, 1937), p. v.

148. He gives his set-up for Grassmann rings/algebras in ibid., pp. 10–33. The quotations here and in the remainder of the paragraph are on p. v.

moreover, his "combination of Cartan's notation, the tensor calculus, and modern algebraic concepts seem[ed] very effective" and "radically different" from Ritt's approach in two key ways: his method was "algebra, rather than analysis," and while Ritt relied on reducibility to effect his results, he was concerned with proving theorems about when solutions for particular systems actually exist. For that, in Thomas's view, reducibility "is of little importance," "the important thing" being "for us to eliminate multiple roots." That, after all, had been Riquier's focus and an idea that had been guiding Thomas's research for a decade.

Like Ritt an analyst, Gilbert Bliss also exploited algebra in the 1930s and also in the context of an AMS Colloquium volume. He, however, streamlined largely late-nineteenth-century techniques of mathematicians like Dedekind and Weber as well as Kurt Hensel and Georg Landsberg to study algebraic functions and their integrals. If $f(x, y)$ is a polynomial in y of the form $f_0(x)y^n + f_1(x)y^{n-1} + \cdots + f_n(x)$, where the $f_i(x)$ are polynomials in x with complex coefficients, then an algebraic function $y(x)$ is one "defined for values x in the complex x-plane by an equation of the form $f(x, y) = 0$."[149] Bliss published his 1933 Algebraic Functions in order to mediate between what he saw as the three different ways of treating such functions—algebro-geometric, transcendental, and arithmetic—that had evolved over the course of the nineteenth and early twentieth centuries.

In Bliss's view, the algebro-geometric method hinged on the theory of plane algebraic curves; in the transcendental method, abelian integrals played the major role; and the arithmetic method turned on "the construction and analysis of the rational functions which are the integrands of Abelian integrals."[150] If this three-pronged attack sounds familiar, it is because the cast of characters behind each included many of the same players—for example, Brill, Noether, and Severi relative to algebro-geometric methods—involved in the closely related development of algebraic surfaces that Zariski had treated in his algebraic-geometric work. Yet, whereas Zariski also placed the work of Kronecker, Dedekind, and Weber in that category, Bliss saw it, along with the work of Hensel and Landsberg, as part of what he termed an arithmetic approach. Moreover, while Bliss's transcendental techniques reached all the way back to work of Abel and Riemann, the roots of Zariski's were more recent and found in papers by Picard.

149. Gilbert Bliss, Algebraic Functions, AMS Colloquium Publications, vol. 16 (New York: American Mathematical Society, 1933), p. 24.
150. Ibid., p. iii.

Algebraic Functions aimed to showcase the virtues especially of the arithmetic viewpoint by excising—from what Bliss viewed as the large and complicated body of *algebraic* literature—just what an *analyst* like Bliss might need. That he succeeded in this goal is reflected in the positive review Ritt published of the book in the *Bulletin*. Singling out its third chapter as its most "distinguishing feature," Ritt lauded the fact that it set up "the integrands of the three elementary types of abelian integrals . . . by the arithmetic methods of Dedekind and Weber" and acknowledged that Bliss had demonstrated that "[t]he arithmetic treatment is definitely simpler and more elegant than the potential-theoretic method of Riemann, or the geometric method of Brill and Noether."[151] Ritt also praised Bliss for his treatment of the theory of divisors, a topic "hitherto available chiefly in the ponderous classic treatise of Hensel and Landsberg" but which Bliss presented "in hardly more than thirty pages."

The work of Ritt, Thomas, and Bliss further underscores the cross-fertilization taking place between mathematical "areas" in the 1930s. At the same time, it highlights the increasing difficulty of delineating between mathematical fields. Solutions of systems of partial differential equations could be seen, à la Ritt, as squarely situated within analysis or, as Thomas interpreted them, as part of differential geometry. The theory of algebraic functions, approached by Bliss from a fundamentally analytic point of view, could nevertheless be improved by greater attention to some of its algebraic, or in his nomenclature "arithmetic," underpinnings. All three men thus realized that algebraic constructs could deepen and sharpen their mathematical understanding of the various problems that shaped their research interests.

As modern developments in algebra had an impact on analysis so, too, of course, did modern developments in analysis itself. In particular, the elaboration by Henri Lebesgue and others of a theory of measure and integration in the opening years of the twentieth century produced, in Wiener's view, "an explosive expansion in the theory of harmonic analysis," given the existence of a "perfect equivalence between the class of Lebesgue measurable functions with Lebesgue integrable square moduli, and the class of functions

---

151. Joseph Ritt, "*Algebraic Functions*. By Gilbert Ames Bliss. American Mathematical Society Colloquium Publications, Volume 16, New York, 1933. vi+218 pp.," *BAMS* 41 (1935), 9–10 on p. 9. The quotations that follow in this paragraph are also on this page.

which can be represented by Fourier series for which the sum of the squares of the moduli of the Fourier coefficients converges."[152] This opened up a widely ranging set of questions about Fourier series and integrals as well as about related topics in real and complex analysis that engaged—among others, in the 1930s—Hille, Wiener, Tamarkin, David Widder, and Aurel Wintner in the United States; G. H. Hardy and John Littlewood as well as Littlewood's student, Raymond Paley, in England; and 1933 émigrés to the United States Salomon Bochner and Otto Szász.[153] This list bears testament to Wiener's characterization of research in harmonic analysis as "highly international" by the 1930s with "no inconsiderable part" of the work having been done in the United States. Some of that research—Wiener's on so-called Tauberian theorems, that is, "theorems which assert that the convergence of a series implies its summability by a certain method to the same sum"—garnered the fourth Bôcher Memorial Prize of the AMS in 1933 jointly with Morse's explorations of the calculus of variations in the large.[154] Other research, especially the collaborative work that Wiener did with Paley when the latter was on a Rockefeller fellowship at MIT, "led successively to the study of quasi-analytic functions, of entire functions of order one-half, and of many related questions."[155] Wiener sketched these complex analytic results in the AMS Colloquium Lectures he gave in 1934, following Paley's untimely death in a skiing accident a year earlier, and further detailed them in their jointly authored AMS Colloquium volume, *Fourier Transforms in the Complex*

152. For the quotation, see Norbert Wiener, "The Historical Background of Harmonic Analysis," *Semicentennial Addresses*, ed. The Committee on Publications, pp. 56–68 on p. 64. The equivalence was established independently by Frigyes Riesz and Ernst Fischer in 1907. Thomas Hawkins puts Lebesgue's work in historical context in *Lebesgue's Theory of Integration: Its Origin and Development* (Madison: University of Wisconsin Press, 1970; reprint of 2d. ed, New York: AMS Chelsea Books, 2001).

153. Wiener, for example, published both "Generalized Harmonic Analysis," *Acta Mathematica* 55 (1930), 117–258 and *The Fourier Integral and Certain of Its Applications* (Cambridge: Cambridge University Press, 1933). For overviews of his work in this area, see Jean-Pierre Kahane, "Norbert Wiener et l'analyse de Fourier," *BAMS* 72 (1966), 42–47 and Pesi Masani, "Wiener's Contributions to Generalized Harmonic Analysis, Prediction Theory and Filter Theory," *BAMS* 72 (1966), 73–125.

154. Norbert Wiener, "Tauberian Theorems," *AM* 33 (1932), 1–100 (the quotation is on p. 3) and Morse, "The Foundations of a Theory of the Calculus of Variations in the Large."

155. Norbert Wiener, "R. E. C. Paley–In Memoriam Jan. 7, 1907-April 7, 1933," *BAMS* 39 (1933), 476.

Domain.[156] Other significant complex analytic and classical work was done in the 1930s in the context of AMS Colloquium volumes by Harvard's Joseph Walsh and émigré Gábor Szegő, respectively.[157]

The theory of harmonic functions also continued to attract serious research attention in the United States, although with Oliver Kellogg's death in 1932, the American potential-theoretic mantle had passed, at least symbolically, to Griffith Evans. Indeed, whereas the AMS had tapped Kellogg in 1926 to survey the field, it called on Evans to update the situation when it hosted its summer meeting in Cambridge on the occasion of Harvard's tercentenary in 1936 (see chapter seven for more on the latter event).[158] Evans next highlighted potential-theoretic advances on "Dirichlet Problems" in particular in the address he gave at the AMS's semicentennial celebrations in New York City two years later (recall the description of the problem given in chapter one).[159]

In Evans's view, of especial interest by the 1930s was what he termed the generalized Dirichlet problem, "in which the harmonic function is uniquely determined by continuous boundary data but cannot itself remain continuous at the boundary."[160] However, "problems in which the boundary data themselves are not continuous but where again the solution is unique within a certain class" also proved captivating. Besides Evans himself, these issues engaged his students, Alfred Maria, George Garrett, and John Green, as well as Robert Martin, one of Michal's Ph.D.s at Caltech.[161]

The related, long-standing Plateau problem, which had been cracked independently by Douglas and Radó in 1930 (recall chapter three), also continued to generate interest. For example, Courant, deeply impressed by Douglas's work and spurred by the talk he had invited the young American to give at

156. Raymond Paley and Norbert Wiener, *Fourier Transforms in the Complex Domain*, AMS Colloquium Publications, vol. 19 (New York: American Mathematical Society, 1934).

157. See Joseph Walsh, *Interpolation and Approximation by Rational Functions in the Complex Domain*, AMS Colloquium Publications, vol. 20 (New York: American Mathematical Society, 1935) and Gábor Szegő, *Orthogonal Polynomials*.

158. Griffith Evans, "Modern Methods of Analysis in Potential Theory," *BAMS* 43 (1937), 481–502.

159. Griffith Evans, "Dirichlet Problems," *Semicentennial Addresses*, ed. The Committee on Publications, pp. 185–226.

160. Ibid. p. 186. The quotation that follows may also be found here.

161. For a technical overview of this work, see ibid., pp. 206–215. Evans gives specific references to the pertinent papers on pp. 224–226.

NYU in the spring of 1935, nevertheless sought a more intrinsic, less "round-about" proof of the result.[162] In his seminar that fall, Courant "glimpsed a way in which the minimum problem related to that of Dirichlet could be employed in a straightforward manner for the solution of Plateau's problem" and actively continued to develop the connections between it, harmonic functions, and the Dirichlet problem through the 1930s and 1940s.[163] Courant's approach to Plateau's problem, in particular, and to the calculus of variations, in general, was transplanted to the West Coast when Lewy gave a lecture course on it following his 1935 move from Brown to Berkeley.[164] It was, however, Douglas's continuing attack on the problem and its ramifications at the end of the 1930s that won for him the 1943 Bôcher Prize.[165]

Although connected to potential theory as well as to the theory of harmonic functions, Plateau's problem dwells, most fundamentally, in the calculus of variations, an area of American analytic research as strong in the 1930s as it had been in the 1920s and in which Chicago continued to dominate. In fact, the program there had expanded—under Bliss as department chair—to include not only Bliss and his student-turned-colleague Lawrence Graves, but also another Bliss Ph.D., Magnus Hestenes, and William Reid, who had earned his doctoral degree at the University of Texas in Austin under the direction of George Birkhoff's student Hyman Ettlinger. As Saunders Mac Lane put it in a retrospective on mathematics at Chicago, "the calculus of variations was evidently a major issue there" in the 1930s.[166] So much was that

162. Reid, *Courant*, p. 181. The quotation that follows is also on this page.

163. See his 1936 talk at the Oslo ICM in 1936 (mentioned in the previous chapter) as well as "Plateau's Problem and Dirichlet's Principle," *AM* 38 (1937), 679–724 and *Dirichlet's Principle, Conformal Mapping, and Minimal Surfaces* (New York: Interscience Publishers, 1950).

164. Lewy's lectures were written up by John Green, a 1938 Ph.D. under Griffith Evans's direction, and published. The short book was favorably reviewed in the *Bulletin* by William Reid in "*Aspects of the Calculus of Variations*. Notes by J. W. Green after lectures by Hans Lewy. Berkeley, University of California Press, 1939. 6+96 pp.," *BAMS* 46 (1940), 595–596.

165. Jesse Douglas, "Green's Functions and the Problem of Plateau," *AJM* 61 (1939), 545–589; "The Most General Form of the Problem of Plateau," *AJM* 61 (1939), 590–608; and "Solution of the Inverse Problem of the Calculus of Variations," *PNAS* 25 (1939), 631–637.

166. Mac Lane, "Mathematics at the University of Chicago," pp. 139–140. Mac Lane, who, as noted, was a graduate student at Chicago for one year from 1930 to 1931, is quite disparaging in this article about what he called "the Bliss Department" as well as about much of the work that came out of it in the calculus of variations in the 1920s and, perhaps especially, in the 1930s. While it is true, as Mac Lane claims, that Bliss largely continued to follow in the research

the case that the university gathered together—in lithographed book form—collections of dissertations in the area throughout the 1930s and up to the outbreak of World War II.[167] According to Arnold Dresden, the reviewer of one of those volumes, "[a]nyone who looks through this book must be impressed by the active developments in the calculus of variations which are going forward at the University of Chicago." Indeed, he continued, "[t]he volume under review shows that the modern theories are being cultivated as well as the problems and methods of the classical period."[168]

One of those "modern theories"—presented by the Italian analyst Leonida Tonelli in a two-volume treatise published in 1921 and 1923—fundamentally shifted the calculus of variations from the dependence on the theory of differential equations characteristic of the "classical," Weierstrassian approach perpetuated by Bolza and others to "an entirely different procedure" that analyzes "the line integral whose extreme values are sought . . . by means of its direct functional dependence upon the curve along which it is taken."[169] "By putting into the foreground the dependence of the integral upon the entire curve," Tonelli successfully "focuse[d] attention upon the 'integral' aspects of the theory, rather than upon its 'differential' characters" and so effected "a calculus of variations which is concerned primarily with neighborhoods of curves, rather than with those of points." Bliss was apparently so struck by these ideas that he not only arranged for Tonelli to sketch them for the *Bulletin* in an article that Bliss himself translated from Italian into English but also encouraged his multilingual doctoral student, McShane, to read Tonelli's treatise.[170]

---

footsteps of his advisor Bolza and that he trained a large number of students accordingly, he also conducted, by Mac Lane's own admission, an advanced seminar on Morse's very new work on the calculus of variations in the large (p. 142) and encouraged students like McShane to look at other, new developments in the field. See later in this section.

167. *Contributions to the Calculus of Variations 1930, 1931–1932, 1933–1937, 1938–1941* (Chicago: University of Chicago Press, 1931, 1933, 1937, 1942; reprint ed., New York: Johnson Reprint Corp., 1965).

168. Arnold Dresden, Review of "*Contributions to the Calculus of Variations, 1933–1937.* (Theses submitted to the Department of Mathematics at the University of Chicago.)," *BAMS* 44 (1938), 604–609 on p. 609.

169. Arnold Dresden, Review of "*Fondamenti di Calcolo delle Variazioni.* By Leonida Tonelli," *BAMS* 32 (1926), 381–386 on p. 381. The quotations that follow are also on this page.

170. Leonida Tonelli, "The Calculus of Variations," *BAMS* 31 (1925), 163–172 and Leonard Berkovitz and Wendell Fleming, "Edward James McShane 1904–1989," *Biographical*

McShane earned his Ph.D. in 1930 and, in the earliest years of the Depression, was forced to take a string of short-term, albeit significant, appointments—a two-year NRC fellowship, a year at Göttingen, and two years at Princeton—before finally landing the professorship at Virginia in 1935 that he would hold until his retirement in 1974. Indicative of the stature McShane quickly attained, the AMS semicentennial subcommittee on invited speakers tapped him to give the overview of the calculus of variations at its 1938 gala. In his characterization, "the most inclusive variations problem involving single integrals" was the so-called problem of Bolza. "In (x, y, r)-space (the y and r being n-tuples) a region R and functions $f(x, y, r)$, $\theta_\beta(x, y, r)$, $(\beta = 1, \ldots, m < n)$ are given and in $(2n + 2)$-dimensional space a function $g(x_1, y_1, x_2, y_2)$. We seek to minimize the sum

$$J[y] = g(x_1, y(x_1), x_2, y(x_2)) + \int_{x_1}^{x_2} f(x, y(x), y'(x)) dx$$

in the class of functions $y^i = y^i(x)$, $x_1 \leq x \leq x_2$, which satisfy the differential equations

$$\theta_\beta(x, y(x), y'(x)) = 0,$$

and whose endpoints determine a point $(x_1, y(x_1), x_2, y(x_2))$ on a point set S in $(2n + 2)$-space."[171] With this set-up, the Lagrange problem was then a special case of the Bolza problem in which $g \equiv 0$, while in what was termed the Mayer problem, $f \equiv 0$. All three of these more classical problems, under various sets of assumptions, continued to generate much research activity in the 1930s. Among the Americans who focused on them were Bliss, Graves, Reid, and Hestenes, all at Chicago, as well as Morse at the Institute.[172]

Memoirs, vol. 80 (Washington, D.C.: National Academy of Sciences, 2001), pp. 227–239 on p. 229.

171. Edward McShane, "Recent Developments in the Calculus of Variations," *Semicentennial Addresses*, ed. The Committee on Publications, pp. 69–97 on pp. 70–71. For the quotations that follow in the next paragraph (with McShane's emphasis), see p. 78.

172. See, for example, Gilbert Bliss, "The Problem of Lagrange in the Calculus of Variations," *AJM* 52 (1930), 674–744; Lawrence Graves, "On the Weierstrass Condition for the Problem of Bolza in the Calculus of Variations," *AM* 33 (1932), 747–752 and "The Existence of an Extremum in Problems of Mayer," *TAMS* 39 (1936), 456–471; William Reid, "A Direct Sufficiency Proof for the Problem of Bolza in the Calculus of Variations," *TAMS* 42 (1937), 141–154; Magnus Hestenes, "Sufficient Conditions for the Problem of Bolza in the Calculus of Variations," *TAMS* (1934), 793–818; Marston Morse and Sumner Myers, "The Problems of Lagrange and Mayer with Variable End Points," *PAAAS* 66 (1931), 235–253; and Marston

FIGURE 6.4. Edward J. "Jimmy" McShane (1904–1989) ca. 1953.
(Photo courtesy of Mathematical Association of America Records,
1916-present, e_math_00993, Archives of American Mathematics,
Dolph Briscoe Center for American History, University of Texas at Austin.)

As McShane explained, though, these problems all "required the finding
of that one of a given class of functions or curves for which a certain integral

---

Morse, "Sufficient Conditions in the Problem of Lagrange without Assumptions of Normalcy,"
TAMS 37 (1935), 147–160. See also these more expository articles: Gilbert Bliss, "The Evo-
lution of Problems of the Calculus of Variations," *AMM* 43 (1936), 598–609 and Magnus
Hestenes, "The Problem of Bolza in the Calculus of Variations," *BAMS* 48 (1942), 57–75.

assumes its least value" and consideration was given to "functions or curves which give a *relative* minimum to the integral. It is quite another thing to ask if there actually exists a function or curve in the class for which the integral takes on its least value." Tonelli's work both addressed the technical aspects of this more modern problem and attracted the research attention in the 1930s of McShane and others.[173] McShane rounded out his survey by noting that Birkhoff and Hestenes both continued to explore the ramifications of the minimax principle that had grown initially from Birkhoff's work in dynamical systems, while Morse further developed his ideas on the calculus of variations in the large.[174]

As should now be clear, Americans were doing much analytical work in the 1930s, but perhaps their flashiest analytic result was George Birkhoff's 1931 proof of the so-called ergodic theorem, another line of research that had grown directly from his work on dynamical systems. Let T be a one-to-one, measure-preserving (in the sense of Lebesgue) transformation of the line segment $[0, 1]$ into itself. The ergodic theorem states that for any such T "and for each individual point P (except possibly an exceptional set of measure 0), there is a definite probability that its iterates under T, from P on, namely $P, T(P), T^2(P), \ldots$ and $P, T^{-1}(P), T^{-2}(P), \ldots$ fall in a given measurable set M."[175] In other words, "for a discrete measure-preserving transformation or measure-preserving flow of a finite volume, probabilities and weighted means tend toward limits when we start from a definite state P (not belonging to a possible exceptional set of measure 0), and, furthermore, the limiting value is the same in both directions."

While the forty-seven-year-old Birkhoff was working toward this result, the twenty-eight-year-old von Neumann, then recently arrived in the United States, was pursuing similar lines of thought from the point of view of operators on Hilbert space as opposed to that of dynamical systems. Indeed, von

173. See, for example, Edward McShane, "Semi-continuity of Integrals in the Calculus of Variations," *DMJ* 2 (1936), 597–616 and his series of five papers on "Some Existence Theorems in the Calculus of Variations," *TAMS* 44 (1938), 429–438, 439–453, and 45 (1939), 151–171, 173–196, 197–216.

174. George Birkhoff and Magnus Hestenes, "Generalized Minimax Principle in the Calculus of Variations," *DMJ* 1 (1935), 413–432 and Marston Morse, "Functional Topology and Abstract Variational Theory," *AM* 38 (1937), 386–44.

175. George Birkhoff, "What Is the Ergodic Theorem," *AMM* 49 (1942), 222–226 on p. 223. The quotation that follows is on p. 226. Birkhoff gave the proof in "Proof of the Ergodic Theorem," *PNAS* 17 (1931), 656–660.

Neumann had proved what was termed the mean ergodic theorem by September 1931, while Birkhoff had gotten his result, the ergodic theorem, only by the end of November. Birkhoff's paper, however, appeared in the December 1931 number of the *Proceedings of the National Academy of Sciences*, while von Neumann's did not come out until its January 1932 number.[176] A priority dispute could well have erupted had the settings and motivations not been substantially different and had Birkhoff and his 1926 Ph.D. student, Bernard Koopman, not been immediately involved in setting the record straight in print.[177] Indeed, it had been a paper of Koopman's that had spurred von Neumann on to his result.[178] Adding to this flurry of results in the early 1930s was Eberhard Hopf, first from Harvard where he had taken a Rockefeller fellowship in 1930 and then from his home base, until his return to Germany in 1936, at MIT.[179]

All of this activity quickly attracted others to ergodic theory per se as well as to related questions in the theory of probability. One, Hedlund, exploited links between dynamical systems, differential topology/geometry, and ergodic theory to understand the dynamics of geodesic flows.[180] Another, his slightly younger contemporary Joseph Doob, reoriented his research from the doctoral work in analysis he had done under Walsh at Harvard in 1932 to probability. In his first paper in his new field, Doob used the correspondence between the law of large numbers and Birkhoff's ergodic theorem to prove various probabilistic results.[181] He followed this with important work

176. Compare John von Neumann, "Proof of the Quasi-Ergodic Hypothesis," *PNAS* 18 (1932), 70–82. See also John von Neumann to Marshall Stone, 22 January, 1932, in *John von Neumann: Selected Letters*, ed. Miklós Rédei, HMATH, vol. 27 (Providence: American Mathematical Society and London: London Mathematical Society, 2005), p. 225 for von Neumann's read on the sequence of events.

177. George Birkhoff and Bernard Koopman, "Recent Contributions to the Ergodic Theory," *PNAS* 18 (1932), 279–282. Joseph Zund gives the fuller story of the intertwined chronology of the work of Birkhoff and von Neumann in "George David Birkhoff and John von Neumann: A Question of Priority and the Ergodic Theorems, 1931–1932," *Historia Mathematica* 29 (2002), 138–156.

178. See Bernard Koopman, "Hamiltonian Systems and Transformations in Hilbert Space," *PNAS* 17 (1931), 315–318 as well as Bernard Koopman and John von Neumann, "Dynamical Systems with Continuous Spectra," *PNAS* 18 (1932), 255–263.

179. See, for example, Eberhard Hopf, "On the Time Average Theorem in Dynamics," *PNAS* 18 (1932), 93–100.

180. See, for example, Gustav Hedlund, "Dynamics of Geodesic Flows," *BAMS* 45 (1939), 241–260.

181. Joseph Doob, "Probability and Statistics," *TAMS* 36 (1934), 759–775.

on the measure-theoretic foundations of probability after taking up the position at Illinois in 1935 that he would hold until his retirement in 1978. Doob's students in the 1930s, Paul Halmos and Warren Ambrose, both went on to distinguished careers, with Halmos joining the Department of Mathematics at Chicago in 1946 (see chapters nine and ten) and Ambrose taking an assistant professorship at MIT in 1947.

Analysis continued both to range widely and to attract new mathematical talent in the American mathematical research community of the 1930s. At the same time that its practitioners pushed established research agendas in, for example, harmonic analysis, potential theory, and the calculus of variations, they explored new areas like ergodic theory and probability and actively exploited new developments in both algebra and measure theory.[182] They also spread their analytic expertise from the traditional centers of strength—Harvard and Chicago—to institutions that had begun consciously to strengthen their mathematical research profiles in the 1930s—Berkeley, Duke, MIT, NYU, Virginia, and others. Not just the analysts but the American mathematical research community as a whole had managed to maintain into the 1930s—and against a backdrop of economic depression and geopolitical turmoil—the course set a decade earlier.

———

All of this American work took center stage in New York City in September 1938 as over 400 AMS members and some 700 people in all fêted the AMS on its golden jubilee. An excursion by steamer up the Hudson River to West Point and trips to Long Island's Jones Beach as well as to the grounds of the then soon-to-be-opened New York World's Fair provided extra-mathematical entertainment for mathematicians and their guests alike. Teas, toasts, receptions, and a "gala dinner" in the "beautiful banquet room of the Hotel Astor" also helped set a suitably festive tone.[183] At a more formal level, speeches were given, individuals honored, and congratulatory letters read, while exhibits of rare mathematical books, manuscripts, instruments, and models further celebrated the mathematical endeavor.

182. Some, like Caltech's Harry Bateman, also continued to exploit analytic results in mathematical physics as exemplified by his *Partial Differential Equations of Mathematical Physics* (Cambridge: Cambridge University Press and New York: The Macmillan Co., 1932).

183. Richardson et al., "The Semicentennial Celebration September 6–9, 1938," p. 7. This article provides a detailed account of the various events.

These events were ancillary, however, to the celebration of the actual research that had issued from the American mathematical community in the fifty years since 1888. The ten plenary addresses that were given over the course of the meeting's final three days took stock of that work. One speaker, Brown's Raymond Archibald, chronicled the history of an AMS that had supported research output through its publications as well as through its general advocacy of mathematicians and their work; another, American mathematical leader George Birkhoff, provided a broad overview "of *American* activity in mathematical research during this period"; the eight others presented "a conspectus of the contributions made on the *American continent* to [the] various special fields of" algebra, the algebraic aspects of the theory of differential equations, harmonic analysis, the calculus of variations, geometry, topology, Dirichlet problems, and hydrodynamical stability.[184] If these talks aimed secondarily "to acquaint mathematicians with current problems and research," their primary function was to trumpet American achievements.[185] The semicentennial was a boosteristic affair, and the message it served to hone—that American mathematical output was on a par with the best internationally—would be one that the American hosts of the International Congress of Mathematicians were already planning to convey to an international audience two years later in 1940.

184. Raymond Archibald, "History of the American Mathematical Society 1888–1938," *BAMS* 45 (1939), 31–46. This was a brief overview of his book-length *A Semicentennial History of the American Mathematical Society.* The other nine addresses were published in hard covers under the auspices of the Committee on Publications as the *Semicentennial Addresses of the American Mathematical Society.* This book, like Archibald's *Semicentennial History,* was in print in time for the 1938 festivities. For the quotations, see Arnold Dresden, Review of "*Semicentennial Addresses.* (American Mathematical Society Semicentennial Publications, vol. 2)," *BAMS* 45 (1939), 50–51 (with my emphases).

185. The Committee on Publications, "Preface," *Semicentennial Addresses,* p. i.

7

Looking beyond the United States

If the members of the American mathematical research community closed the 1930s in a spirit of inward-looking self-congratulation, that decade also found them focused both mathematically and politically on points beyond their national borders, and particularly on Europe. The mathematical successes they chronicled in 1938 at the AMS semicentennial suggested, to them at least, that they were finally poised to assume international mathematical leadership. As noted, their moment would come, they hoped, in two years' time. Their offer to host the 1940 International Congress of Mathematicians in Cambridge, Massachusetts, had been accepted at the ICM in Oslo in 1936.

As the Americans began planning for this cooperative, international event, mathematicians continued to flee Europe for the United States and elsewhere. As noted in chapter five, not only did active aggression against Jews continue with the Kristallnacht pogrom throughout Nazi Germany in November 1938, but Hitler's annexation of Austria in March 1938 had also been followed in September by Czechoslovakia's surrender of the Sudetenland in the wake of the Munich Agreement between Hitler, Britain's Neville Chamberlain, France's Édouard Daladier, and Italy's Benito Mussolini. Then, despite his promise to the contrary, Hitler had invaded and occupied Czechoslovakia in March 1939, and on 1 September, his army had invaded Poland as well. The latter act of aggression finally resulted in declarations of war on Germany by both Great Britain and France on 3 September, marking the beginning of World War II.

Further emboldened, Hitler had ordered the invasions of Norway and Denmark and the Blitzkrieg of Belgium and the Netherlands by the end of the spring of 1940. In June, Italy had entered the war on the German side, and

FIGURE 7.1. Forty-second Summer Meeting of the Mathematical Organizations of America in connection with the Harvard Tercentenary Celebration, Cambridge, MA, 31 August–5 September, 1936. (Photo courtesy of Harvard University Archives.)

France had been occupied. Almost nowhere in Europe seemed safe by 1940. This had prompted a second wave of mathematical refugees to come ashore in a United States that, at least from George Birkhoff's vantage point, had finally "reached [the] point of saturation" in September 1938 that Veblen had been warning about since the close of 1933.[1] American mathematicians thus looked beyond the United States in at least two senses as the 1930s passed into the 1940s.

Hosting the International Congress of Mathematicians?

Just six weeks after the Oslo ICM adjourned, the AMS and the MAA held their joint summer meeting in Cambridge in conjunction with the Harvard tercentenary. As would be the AMS semicentennial two years thence, it was a time of great celebration. Harvard invited some seventy-five speakers—forty-seven from Europe, twenty-one from the United States, nine from elsewhere, and, explicitly, none from Harvard itself—to "address themselves chiefly to

1. Birkhoff, "Fifty Years of American Mathematics," p. 277.

FIGURE 7.1. Continued.

the fundamental problems of science and society."[2] The mathematical sciences were represented, from England, by statistician Ronald Fisher, mathematician G. H. Hardy, and astronomer Arthur Eddington, by logician Rudolf Carnap from Prague, and by mathematicians Élie Cartan, Leonard Dickson, and Tullio Levi-Civita from Paris, Chicago, and Rome, respectively. All of these men, eminent in their fields, were instructed simply "to contribute a paper on a topic of their own choosing."[3] Each spoke in a conventional lecture setting before audiences in joint sessions of the American Mathematical Society and, depending on the topic, the Mathematical Association of America, the Institute of Mathematical Statistics, the Association for Symbolic Logic, and the American Astronomical Society. All of those groups had opted to link regular meetings in 1936 to the Harvard event. Dinners, a garden party, and numerous excursions in and around Cambridge rounded out a week-long mathematical gala that was immortalized in a panoramic photograph of the participants and their guests. Altogether, "more than one thousand persons

2. "The Harvard Tercentenary Conference of Arts and Sciences," *Science* 83 (1936), 311–313 on p. 311.

3. Clark Elliott, "The Tercentenary of Harvard University in 1936: The Scientific Dimension," in *Commemorative Practices in Science: Historical Perspectives on the Politics of Collective Memory*, ed. Pnina G. Abir-Am and Clark A. Elliott, *Osiris* 14 (1999), 153–175 on p. 164.

attend[ed] the meetings arranged by the" AMS, and, of those, 443 were AMS members.[4] This "constitute[d] a record attendance, an excess of fifty per cent [*sic*] more than at the largest meeting hitherto."

Amid all of this mathematical activity, the Society also managed to conduct its business. In particular, it formed a "Special Committee on the International Mathematical Congress of Mathematicians" with Eisenhart as chair. Charged with "map[ping] out a general plan for the Congress," by March 1937, Eisenhart's group had made an extensive report to the AMS Council, detailing how an undertaking of that magnitude might be arranged.[5] American mathematicians—under the aegis of their *research* organization, the AMS, and most definitely not under that of their *teaching* organization, the MAA— wanted to put their own unique stamp on the event. The 1940 ICM would be primarily an AMS affair.[6]

As they acknowledged, earlier ICMs had achieved their scientific aims by combining a series of "long summary addresses by leaders in various branches of mathematics" with short communications on individual research.[7] The major addresses had successfully highlighted the latest movers and shakers in the field, for example, Hilbert at the 1900 ICM in Paris. The shorter communications, on the other hand, had served the "essential" purpose of providing "a forum free to any person interested in mathematics," despite being "unorganized and uncoördinated." In short, Eisenhart and his fellow planners accepted those features that had come to characterize the ICMs, but they wanted to do more.

The committee recognized that then-recent international conferences like the one in topology held in Moscow in 1935 and like the AMS's experiment in Durham in 1936 had generated interest and excitement precisely because

4. Richardson, "The Summer Meeting in Cambridge," p. 762. The quotation that follows is also on this page.

5. Roland Richardson, "The Summer Meeting in Cambridge," *BAMS* 42 (1936), 761–776 on p. 769. The other members of this pre-planning committee were: Gilbert Bliss, Edward Chittenden, Arthur Coble, Griffith Evans, Earle Hedrick, Theophil Hildebrandt, Einar Hille, John Kline, Marston Morse, Paul Smith, Marshall Stone, Jacob Tamarkin, Joseph Walsh, and Norbert Wiener.

6. Temple Hollcroft, "The March Meeting in New York," *BAMS* 43 (1937), 307–314 on pp. 310–311.

7. The quotations in this and the next three paragraphs are from the "Report of the Committee on the International Congress of Mathematicians," March 1937, pp. 2–4, Box 5, Folder 20: Lectures and Meetings–1940: International Congress of Mathematicians, Whyburn Papers.

they targeted a specific area of mathematics. It thus proposed to incorporate this innovation into the American ICM as a way to increase the congress's "scientific effectiveness." Still, it would be important to choose the topics for such focused conferences carefully. They should treat areas of mathematics "in which a vigorous advance has just been made or is currently in progress." Moreover, the individual programs should "be firmly managed in order to secure the desired articulation" of emerging lines of research. The committee members felt that it would be "reasonable to hope for three to six" such conferences, but they also thought that "it would be unwise to set any definite quota" and that "it would seem better to err in organizing too few rather than too many." The idea would be for the congress, through this innovation, actively to stimulate and guide research internationally rather than to serve merely as a vehicle for the communication of results already obtained.

Much power, then, would be in the hands of those few selected as organizers. They would be "persons recognized as authorities on the conference topic," and they would be "given full responsibility" for all of the arrangements. In the view of Eisenhart's committee, "an effective conference program can be obtained only by restricting formal communication to the invited speakers and discussion to the members of the conference." The invited speakers would also be among the principal speakers of the congress as a whole, a number, in the committee's view, that should be "around twenty." The "members" of the conferences would be expected to engage actively in the proceedings, asking questions, probing ideas more deeply, and suggesting additional lines of inquiry related to the conference topic. These would thus *not* be the usual sorts of encounters in which the audience listens passively. They would be carefully engineered brain-storming sessions designed to accelerate mathematical research in fruitful new directions, and they would be guided by members of the North American mathematical research community.

On submitting its report, Eisenhart's committee requested "that it be dissolved with the understanding that material and plans in the hands of the sub-committees . . . is at the disposal of the appropriate new committee to be organized." That new committee was in place and its membership announced six months later at the AMS's summer meeting in 1937.[8] Under

8. Temple Hollcroft, "The Summer Meeting in State College," *BAMS* 43 (1937), 745–757 on p. 749. See also Everett Pitcher, *A History of the Second Fifty Years: American Mathematical Society 1939–1988* (Providence: American Mathematical Society, 1988), p. 147.

Harvard's William Graustein as chair and with members chosen both from Eisenhart's committee and anew, the "Organizing Committee" built solidly on the groundwork already laid.[9]

By September 1938, four conference fields had been determined and their organizing committees named.[10] In "algebra," Adrian Albert served as chair of a committee consisting of Garrett Birkhoff and Saunders Mac Lane, both then at Harvard, Richard Brauer, the recent German émigré by then at the University of Toronto, and Yale's Oystein Ore. A conference on the "theory of measure and integration, probabilities, and allied topics" was planned under Wiener's leadership together with Evans, Tamarkin, and von Neumann. In "topology," Lefschetz joined forces with Kline, Wilder, Columbia's Paul Smith, and Zariski. And, finally, in "mathematical logic," Haskell Curry headed a committee comprised of Church, Stephen Kleene of Wisconsin, and Mac Lane, and his Harvard colleague, Willard Quine. These were deemed the four areas—and these among the movers and shakers on North American shores within them—that were at the right stages in their developments for the stimulation that a successful ICM conference could provide.[11] The inclusion among these organizers of Brauer and von Neumann, moreover, reflected the acculturation then under way of 1930s émigrés into the North American mathematical community. Zariski's presence, moreover, suggested that that process need not take long.

Six sections outside the four conferences were also planned in algebra; analysis; geometry and topology; probability, statistics, actuarial science, and economics;[12] mathematical physics and applied mathematics; and logic, philosophy, history, and didactics. The ICM would thus clearly be geared toward the research aims of the AMS, but its organizers nevertheless recognized the

9. Bliss, Evans, Kline, Stone, Tamarkin, and Wiener provided continuity from the earlier committee. The new members were George Birkhoff, Solomon Lefschetz, Oystein Ore, Roland Richardson, John Synge, and Joseph Thomas.

10. John Kline, Minutes of the Meetings of the Organizing Committee, 5–6 September, 1938, and Minutes of the Meeting of the Organizing Committee, 29 October, 1938, both in Box 93-372/3, Folder: International Congress of Mathematicians 1940, Price Papers. The quotations that follow in this paragraph are also from the minutes of 5–6 September.

11. For the specific rationales for conferences in these four areas, see Appendix G of Hollings and Siegmund-Schultze (pp. 295–296) which gives a transcription of part of the report of the subcommittee, chaired by Marshall Stone, on conferences, dated 15 April, 1938.

12. There was some discussion of actually mounting an ICM conference on mathematical statistics and mathematical economics in addition to the regular section (Kline, Minutes, 5–6 September, 1938), but no such conference materialized in the planning.

"whole problem" "of the extent to which the" MAA would or should "assist in the organization of Section VI."[13] It was ultimately decided that it should play a role limited to "organizing, *part* of Section VI (say, History, Didactics)."[14] The duly appointed committee fairly quickly had two symposia on didactics in the works. One, by van der Waerden, would be on the "Modern Work in Algebra" that had stemmed from the structural approach his *Moderne Algebra* had helped to introduce into the subject. The other, by the New Zealand–born Greenwich astronomer Leslie Comrie, would detail his pioneering work in both mechanical computation and the teaching of numerical analysis.[15]

These were ambitious plans that strongly reflected the purist orientation of American mathematics, but they would come to naught without adequate financial support. Fortunately, efforts in that direction had continued literally to pay off. In addition to the $15,000 that had been pledged as early as the summer of 1936 by the Rockefeller Foundation and the Carnegie Corporation, the financial committee had also secured $2,500 from the Institute for Advanced Study, a pledge of $1,000 for the year 1938 from the National Research Council with a strong indication of two more $1,000 allocations in 1939 and 1940, and $3,000 from private donors, for a total of at least $21,500. In reporting for the Finance Committee, Morse calculated that "[a]dding to this the estimates for sums to be received from fees and the sale of publications, there now seems to be $29,500 available for the purposes of the Congress," that is some $6,500—and likely $8,500—more than the initial cost estimate of $23,000.[16]

Eight months later, in May 1939, the full list of invited speakers for the four plenary sessions as well as for the six conference sessions had finally

13. John Kline, Minutes of the Meeting of the Organizing Committee International Congress of Mathematicians, Charlottesville, Virginia, 15 April, 1938, Box 93-372/3, Folder: International Congress of Mathematicians 1940, Price Papers. The quotations in this paragraph are also from these minutes.

14. Roland Richardson to the members of the Organizing Committee of the International Congress and its four subsidiary committees, 12 January, 1938, Box 93-372/3, Folder: International Congress of Mathematicians 1940, Price Papers (my emphasis).

15. Arnold Dresden to the members of the Committee on Didactic Section of the International Congress, 12 December, 1938, Box 93-372/3, Folder: Didactic Section Committee 1940 Int. Congress, Price Papers.

16. Kline, "Minutes of the Meetings of the Organizing Committee," 5–6 September, 1938. In 2021 dollars, $29,500 was equivalent in "real wealth" to some $559,100 (https://www.measuringworth.com). Michael Barany also discusses the funding for this

been drawn up (see fig. 7.2).[17] It was an international line-up that neverthe-less highlighted the best American talent. For example, Albert was shoulder to shoulder with then-recent German émigré Artin as an algebra plenary speaker, while Lefschetz stood beside Moscow's Alexandroff in topology. In the conference sessions, Jacobson then at the University of North Carolina was, along with Paris's Claude Chevalley, one of the up-and-comers in algebra, while Alexander would share the limelight in topology with Lev Pontrja-gin of Moscow's Steklov Institute. In the theory of measure and integration, probabilities, and allied topics, both Hedlund, then at Bryn Mawr but within months of his move to the University of Virginia, and Moscow State Univer-sity's Kolmogoroff would be featured speakers, while in mathematical logic, Church would share that honor with Göttingen's Gerhard Gentzen.

With these decisions made, the last major hurdle was actually getting all of the speakers, and especially those from Germany, to the United States. Given the political problems surrounding German participation in the ICMs since 1920, the last thing the Americans wanted was the appearance of discrimina-tion against their German counterparts. Still, in November 1938, Richardson, for one, was not sanguine about the chances for German attendance, given the deteriorating geopolitical situation and the fact that Harvard was to be one—along with MIT—of the congress's hosts.

His reason for skepticism relative to Harvard was quite targeted. When Erich Hecke, Professor of Mathematics at the University of Hamburg, visited the United States in the winter of the 1937–1938 academic year, he had been under "the strictest orders not to accept any invitation to lecture at Harvard," since Harvard had refused "to accept from a pro-Nazi a sum of money which would assist an American student to carry on study at German universities as administered by the present regime."[18] There was no doubt that late-1930s

---

congress in "Remunerative Combinatorics: Mathematicians and Their Sponsors in the Mid-twentieth Century" in *Mathematical Cultures: The London Meetings 2012–2014*, ed. Brendan Larmor (Cham: Springer Basel, 2016), pp. 329–346 on pp. 331–333.

17. "International Congress of Mathematicians," May 1939, Box 5, Folder 20: Lectures and Meetings–1940: International Congress of Mathematicians, Whyburn Papers. Fig. 7.2 is a retyped version of the original that preserves its layout, spacing, etc.

18. Richardson to the members of the Committee on Didactic Section of the Interna-tional Congress, 23 November, 1938, Box 93-372/3, Folder: Didactic Section Committee 1940 Int. Congress, Price Papers. Hollings and Siegmund-Schultze also quote from this let-ter (p. 133) in a copy they found in a different archive. Their volume appeared just as I was putting the finishing touches on a complete draft of the manuscript of the present book. Not surprisingly, they and I came upon some of the same sources in our respective research.

INTERNATIONAL CONGRESS OF MATHEMATICIANS
Cambridge, Massachusetts, September 4-12, 1940

Speakers Invited for Plenary Sessions

Albert, A. A.
Alexandroff, Paul
Artin, Emil
Cramér, Harald
Denjoy, Arnaud
Dirac, Paul Adrien Maurice
Gödel, Kurt
Hodge, William Vallance Douglas
Hopf, Heinz
Khintchine, A.
Lefschetz, Solomon
Moore, R. L.
Morse, Marston

von Neumann, John
Pearson, Egon Sharpe
Rademacher, H. A.
Riesz, Marcel
Stone, M. H.
Synge, J. L.
Tarski, Alfred
Taylor, Geoffrey Ingram
Tinbergen, Jan
Tonelli, Leonida
Vinôgradoff, Ivan M.
van der Waerden, B. L.
Weyl, Herman

Additional Speakers Invited for Sessions of Conferences

Conference on Algebra

Aitken, Alexander Craig
Birkhoff, Garrett
Brauer, Richard
Chevalley, Claude
Deuring, M.
Hall, Philip
Jacobson, Nathan
Krull, Wolfgang
MacDuffee, C. C.
MacLane, Saunders
Ore, Oystein
Schilling, O.F.G.
Zariski, Oscar

Conference on Theory of Measure and Integration, Probabilities, And Allied Topics

Banach, Stefan
Cantelli, Francesco Paolo
Hedlund, G. A.
Hostinsky, Bohuslav
Kolmogoroff, A.
Lévy, Paul
von Mises, Richard
Ridder, J.
Riesz, Friedrich
Steinhaus, Hugo
Ulam, S. M.
Young, Laurence Chisholm
Zygmund, Antoni

Conference on Mathematical Logic
Bernays, Paul
Carnap, Rudolf
Church, Alonzo
Chwistek, Leon
Gentzen, Gerhard
Heyting, Arend
Kalmár, László
Quine, W. V.
Scholz, Heinrich
Skolem, Thoralf Albert

Conference on Topology
Alexander, J. W.
Borsuk, Karl
Čech, Eduard
Hurewicz, Witold
Kline, J. R.
Kuratowski, Kazimierz
Pontrjagin, Lev Semiovich
Stoïlow, Siméon
Whitney Hassler

May 1939

FIGURE 7.2. Draft list (dated May 1939) of invited speakers for the ICM, Cambridge, MA, 1940. (Typed Facsimile of the document in Whyburn Papers, University of Virginia.)

geopolitics would affect plans for a successful congress, but the congress organizers had recognized this fact from the start by forming what they called the
"Cooperation Committee" to deal with such sticky matters.[19]

Under Stone as chair, the Cooperation Committee was particularly dependent on the many connections of his fellow committee member Veblen. Of
primary concern were securing travel subventions for foreign speakers and
assuring that those foreigners obtained the necessary visas for entry into the
country. In the spring of 1938, for example, Veblen and George Birkhoff had
tried to convince the National Academy of Sciences to partner with the AMS
in sponsoring the ICM. They argued that formal Academy sponsorship would
carry more weight both with foreign governments and with the State Department than would sponsorship by a mere scientific society. While that may well
have been true, the Academy flatly refused "on the grounds that no precedent
existed [for Academy co-sponsorship], that such action might tend to embroil
the Academy, that many congresses in other fields had been held without
such sponsorship, and finally that nothing could be done without the formal
approval of the State Department."[20]

With this rebuff, Veblen and Birkhoff took a different tack. Citing a then-
recent precedent set by the International Union of Geodesy and Geophysics,
they lobbied and drafted the language for an actual Congressional bill that
would "authorize and request the President of the United States to invite
such international scientific congresses as have been approved by the National
Academy of Sciences to hold their meetings in the United States at such times
and places as have been designated, and to invite foreign governments to
participate in such meetings."[21] Moreover, it called for "an appropriation of
$5,000 to assist in meeting the expenses necessary for participation by the
United States in the meetings." Perhaps not surprisingly, this initiative also

19. Hollings and Siegmund-Schultze note that the official German position about allowing
German participation in the congress shifted over the months between June 1939—Germany
would send a delegation—and December 1939—perhaps it would not. By December, moreover, there was discussion about finding it "reasonable to bring Italy, Russia and Japan to the
same opinion." See pp. 132–133 (the quotation is on p. 133).

20. John Walsh, "Minutes of the Mathematics Section of the National Academy of Sciences,"
28 April, 1938, Box 9, Folder: National Academy of Sciences 1933–49, Veblen Papers.

21. Letter dated 10 May, 1938, from Birkhoff to Paul Brockett recorded in full in the minutes
of the Executive Committee of the National Academy of Sciences, 7 June, 1938, Box 9, Folder:
National Academy of Sciences 1933–49, Veblen Papers. The next quotation is also from this
letter.

failed. It is nevertheless telling that mathematicians in the 1930s found taking their cause directly to Congress a viable option.

The Cooperation Committee next focused its attentions on the State Department, relative to what would inevitably be visa issues, as well as on the nation's colleges and universities, relative to travel subsidies for foreign speakers. The latter initiative had, by March 1939, netted pledges of $100 each from Brown, Columbia, Duke, and Yale, and Veblen was personally soliciting aid from other schools.[22] These efforts, however, soon became moot.

When France and Great Britain declared war on Germany in September 1939, Griffith Evans had been in the AMS presidency for nine months. Just days after the declaration, the Berkeley analyst and some 350 of his mathematical colleagues came together at the University of Wisconsin in Madison for the AMS's forty-fifth summer meeting, fully cognizant of what the minutes officially described as "the distressful world situation."[23] In Madison, in fact, it was resolved that the Cambridge ICM would have to be postponed indefinitely. Despite its best-laid plans, the American mathematical community would not test its mettle before the international mathematical community at an ICM on its home turf in 1940.

Confronting a Second Wave of Mathematical Émigrés

Indeed, the "world situation" *had* become increasingly "distressful." The Nazi aggressions of 1938 had prompted a second wave of mathematical émigrés beginning at the end of that year and continuing beyond the United States' entry into the war in December 1941. As the members of the Emergency Committee in Aid of Displaced German Scholars watched these events unfold, they recognized the need to expand their mission "to include refugee professors from all countries of Western Europe overrun by the Nazi armies."[24] They made their newly enlarged mission manifest on 9 November, 1938, by changing their name to the Emergency Committee in Aid of Displaced *Foreign* Scholars.

Veblen continued to be a key player in the EC, but his efforts relative to mathematics per se were significantly reinforced by others outside the

22. Veblen to Gordon Whyburn, 2 March, 1939, Box 5, Folder 20: Lectures and Meetings– 1940: International Congress of Mathematicians, Whyburn Papers.

23. William Ayers, "The September Meeting in Madison," *BAMS* 45 (1939), 801–811 on p. 804.

24. Duggan and Drury, p. 7 (note 2).

organization. In particular, his two émigré colleagues at the Institute, Weyl and von Neumann, also embraced the cause, while in England, Hardy remained active as a liaison between Great Britain and the United States as did the Academic Assistance Council, which had changed its name to the Society for the Protection of Science and Learning in 1936. After 1938, the EC also collaborated even more robustly with Alvin Johnson of the New School for Social Research and its associated graduate school, the University in Exile, both located in New York City, as well as with Hertha Kraus and the American Friends Service Committee (AFSC) based in Philadelphia.

The University in Exile, founded in 1933 and renamed the Graduate Faculty of Political and Social Science in 1934, provided a safe haven primarily for social scientists fleeing Nazi Germany and later Fascist Italy and occupied France,[25] while the AFSC was a Quaker organization founded in 1917 to assist the civilian victims of World War I. In the years leading up to and during World War II, the AFSC focused specifically on helping non-religious Jews and Jews married to non-Jews escape from Nazi Germany.[26] As Veblen put it to George Rainich, it had become increasingly important by 1939 not "to cross wires" with these various groups trying to help, given the dramatic increase in the volume of academics seeking refuge at the close of the 1930s.[27]

Of the 613 men and women in all fields whose cases came before the EC, more than half actually made it to the United States. Their arrival dates mark well the storm surge of academic emigration: "30 arrived in 1933; 32 in 1934; 15 in 1935; 20 in 1936; 15 in 1937; 43 in 1938; 97 in 1939; 59 in 1940; 50 in 1941; and 10 during 1942 and 1943."[28] The case of Beniamino Segre amply illustrates the extent of international cooperation in trying to find places for late-1930s mathematical émigrés.

25. After the German occupation of France, DeGaulle's Free French government chartered the Université Libre des Hautes Études, which was also located at the New School, and where exiled French scholars taught in French. Claus-Dieter Krohn discusses the New School and its University in Exile in *Intellectuals in Exile: Refugee Scholars and the New School for Social Research*, trans. Rita and Robert Kimber (Amherst: University of Massachusetts Press, 1993), pp. 59–91.

26. Allan Austin treats the history of the American Friends Service Committee in *Quaker Brotherhood: Interrracial Activism and the American Friends Service Committee, 1917–1950* (Urbana: University of Illinois Press, 2012).

27. Veblen to Rainich, 6 February, 1939, Box 33, Folder: Refugees, Segre, Beniamino 1938–41, Veblen Papers.

28. Duggan and Drury, p. 25.

On 5 September, 1938, Mussolini's Fascist government passed laws that resulted in a purge of Italian Jewish academics that paralleled the Nazi purge five years earlier.[29] The thirty-four-year-old Segre, since 1931 Professor of Analytical and Projective Geometry at the University of Bologna, found himself "expelled from all Italian mathematical organizations, . . . dismissed from his duties as co-editor [with his co-religionist Levi-Civita] of the *Annali di mathematica,*" and denied "the right to compensation or to pension."[30] Almost immediately, Segre had written to fellow algebraic geometer Zariski at Johns Hopkins to see whether there might be hope of a position in the United States, and Zariski had sought Veblen's counsel. "As a friend of Segre," Zariski confided, "I should like to help him. . . . I have a great respect for his mathematical abilities and I feel that it would be a pity if he was lost to mathematics." Still, Zariski "realize[d] that [Segre']s poor knowledge of English would handicap him here in the beginning," so he wondered whether "the Institute would be willing to make some arrangements by which he could spend a semester or two in Princeton." That would give him time to improve his English, to acculturate to American academic society, and to participate mathematically in the main center of algebraic geometric research in the United States. It seemed like a good idea, but in order for it to work, it would need Lefschetz's blessing. He, after all, was the head of Princeton's group in algebraic geometry.

Lefschetz proved less than enthusiastic. As a Jew himself, he held that "the situation in this country is becoming so difficult for the Jews that we ought not to aggravate it by bringing any more."[31] Moreover, in his view, "there are a number of mathematicians who either are, or are soon likely to be, as much in

29. For more on this in a mathematical context, see Angelo Guerraggio and Pietro Nastasi, *Italian Mathematics Between the Two World Wars* (Basel: Birkhäuser Verlag, 2005), pp. 251–268.

30. Zariski to Veblen, 3 November, 1938, Box 33, Folder: Refugees, Segre, Beniamino 1938–41, Veblen Papers. The quotations that follow in this paragraph are also from this letter. For more on Segre's life, see Enzo Martinelli, "Beniamino Segre: His Life, His Work," *Rendiconti dell'Accademia nazionale delle scienze detta dei XL* 4 (1979–1980), 1–12 and Patrick Du Val, "Beniamino Segre," *Bulletin of the London Mathematical Society* 11 (1979), 215–235.

31. Veblen to Zariski, 8 November, 1938, Box 33, Folder: Refugees, Segre, Beniamino 1938–41, Veblen Papers. The following notation appears in pencil at the bottom of this letter: "This page *not* mailed at Professor Lefschetz's request." The quotations that follow are also from this unsent letter. As André Weil put it, "Lefschetz was reputed to be among those Jews who, to avoid any accusation of favoring their co-religionists, go to the other extreme and display patently anti-Semitic behavior. A conversation I had with him on this subject . . . left me dumbfounded." See André Weil, *The Apprenticeship of a Mathematician* (Boston: Birkhäuser Verlag, 1992), p. 183.

difficulty as Segre, and who are more valuable scientifically." Veblen disagreed with Lefschetz on both these scores, but he nevertheless "doubt[ed] whether it is advisable to bring an algebraic geometer to Princeton as long as Lefschetz takes this point of view."

Lefschetz's attitude may have made Princeton an unrealistic fit, but that did not deter Veblen. He suggested that Zariski write to Tracy Thomas, Veblen's former doctoral student and colleague who had just moved to UCLA, certain that "Thomas would be interested in getting other men who are active scientifically into his environment" but less sure that Thomas, then still an associate professor, "would be sufficiently influential to accomplish anything."[32] At least as Veblen (and others) saw it, the ongoing refugee situation should be used strategically to boost the profiles of schools, like UCLA in the 1930s, that had not yet moved fully into the research ranks mathematically. This strategy, after all, seemed to be at work already at, for example, NYU, Penn, Notre Dame, Kentucky and elsewhere.

By February 1939, Veblen had sounded out possibilities for Segre at both Michigan and Cornell. The most promising plan seemed to come from the latter quarter, with Cornell's Virgil Snyder working with Illinois's Arthur Coble to arrange a semester for Segre at each of their respective institutions. In the meantime, Zariski had heard from British geometer William Hodge at Cambridge that although Segre had applied unsuccessfully for assistance from the then severely strapped Society for the Protection of Science and Learning, "about ten of us geometers raised a private subscription for him, hoping to be able to provide for him for two years."[33] When that subscription came up short, the Society agreed to make up the difference, and Segre and his family made their way to England. In order to safeguard Segre's status in Great Britain, however, Hodge offered the opinion that "Coble could be of most assistance . . . if he would postpone his plan . . . for a year. If Segre has not found a permanent post by 1941 he may find himself in serious difficulty, and Coble's plan might then save the situation."

Veblen continued actively to seek a position for Segre in Australia at the University of Melbourne and in the United States at the University in Exile

---

32. Veblen to Zariski, 8 November, 1938, Box 33, Folder: Refugees, Segre, Beniamino 1938–41, Veblen Papers. Typed at the bottom of this expurgated (and presumably delivered) version of the 8 November letter is: "COPY HANDED TO PROF. LEFSCHETZ."

33. Hodge to Zariski, 5 August, 1939, Box 33, Folder: Refugees, Segre, Beniamino 1938–41, Veblen Papers. The quotation that follows is also from this letter.

and UCLA. Writing in June 1941 directly to Thomas, by then Professor of Mathematics at UCLA and a member of the National Academy of Sciences, Veblen confided that "I think that [Segre] would have found a place in this country before now if it were not for the opposition of Lefschetz. I cannot help thinking," he continued, "that this opposition is due to the fact that Segre does not go wholeheartedly in the Lefschetz direction for algebraic geometry. . . . [I]t seems to me that it would be a healthy thing to have a competing point of view in algebraic geometry maintained in this country."[34] To that end, Veblen urged Thomas to have UCLA request funding from the EC to bring Segre to California. In his view, "[i]t would . . . be a good deed to relieve the English mathematicians of his support." In 1942, the efforts of Veblen and others were finally mooted when Segre took the position at the University of Manchester that he would hold until his return to the University of Bologna in 1946. Segre's case thus brings into stark relief the network that linked—even more tightly than in the mid-1930s—individual mathematicians and relief agencies on both sides of the Atlantic during the second wave of mathematical emigration.

His case also underscores the sustained role of strong and energetic advocates, although many cases that came before the EC and other agencies did not have such support. Given that the United States, in particular, had already absorbed some thirty mathematicians prior to January 1938,[35] the new surge in potential émigrés beginning in 1938 put increasing pressure on those, like Veblen, who were trying to place them. Was there a way actually to rank-order them? Was it really possible to be more strategic—rather than reactive—in their placement? These questions persisted as the young generation of American mathematical aspirants continued to watch foreign mathematicians getting jobs while they struggled. It is perhaps not surprising that resentment surfaced.

Consider the case of Dick Wick Hall. A young topologist who had earned a 1938 Ph.D. at Virginia under Whyburn, Hall had taken an NRC fellowship to Penn for the 1938–1939 academic year to work with Kline before accepting a two-year instructorship at Brown. By 1941, however, he had still not found a permanent position. "It used to be that we had to compete with

---

34. Veblen to Thomas, 25 June, 1941, Box 33, Folder: Refugees, Segre, Beniamino 1938–41, Veblen Papers. The quotation that follows is also from this letter.

35. This count is based on the data given in Siegmund-Schultze, *Mathematicians Fleeing from Nazi Germany*, Appendix 1.1, pp. 343–357.

smart Europeans like Sammy [Eilenberg]," he lamented to Whyburn, "but now their dumb assistants get the jobs."[36] Eilenberg, a Polish Jew, had emigrated to the United States in 1939 and, with the help of Lefschetz and Veblen, had landed an instructorship at Michigan in 1940.[37] He went on to become, with Saunders Mac Lane, the founder of category theory (see chapter ten), and while Hall certainly recognized Eilenberg's mathematical abilities and did not begrudge him a permanent job, he was less generous when it came to those he deemed lesser European talents.[38] If some system could finally be devised to place in positions only those émigrés, like Eilenberg, who were unquestionably among the very best, then jobs for home-grown mathematicians would be less scarce, and the kind of resentment that Hall, by no means an isolated case, expressed might largely be avoided. By 1939, mathematical activists like Veblen and Weyl were trying to do precisely that.

An unsigned, two-page memorandum—dated 2 May, 1939, and entitled "Mathematicians and Physicists whose records have been sent to [the] American Friends Service Committee" (fig. 7.3)—provided a roster of mathematicians seeking to emigrate with actual grades of A, A-, B+, B, B-, C+, and C assigned for "scholarship," "personality," and "teaching ability/adaptability," where, as the author (or authors) of the list explained, " 'B' indicates successful research work of considerable originality."[39] While it could hardly have been

36. Hall to Whyburn, 10 March, 1941, Box 1, Folder 70: Hall, Dick Wick, Whyburn Papers. The assistant in question was Olaf Schmidt. See below.

37. Hyman Bass, Henri Cartan, Peter Freyd, Alex Heller, and Saunders Mac Lane, "Samuel Eilenberg (1913–1998)," NAMS 45 (1998), 1344–1352 on p. 1344 and compare cattell, American Men of Science, 9th. ed.

38. Hall was able to stay on at Brown until 1943. He secured a permanent job at the University of Maryland that year and had risen to the rank of full professor by 1947. In 1956, he moved from Maryland to what is now the State University of New York at Binghamton and remained there for the rest of his career.

39. The grading, guided by their own personal senses of and research into each individual case, was likely the joint effort of Veblen and Weyl. See Box 30, Folder: Refugees, General 1935–41, Veblen Papers. Indicative of the difficulty in securing information on the refugees, not all of the dates recorded are accurate: Hilda Geiringer was born in 1893 not in 1895; Friedrich Kottler was born in 1886 not in 1887. Fig. 7.3 is a retyped version of the original that preserves layout, spacing, etc. as much as possible.

There is evidence that such a grading scheme may actually have been in use as early as 1933. The undated (but from 1933) fact card in Richard Courant's Emergency Committee file was, at some point, attached to a sheet of paper on which the notation "A" appears. See Series I: Grantees, Box 5, Folder: 13 Courant, Richard, Emergency Committee Papers.

an objective rating system, nevertheless, among the forty-eight scholars listed, the research of seven—Guido Fubini, Hans Hamburger, Felix Hausdorff, Karl (later Charles) Loewner,[40] Hermann Muentz, Issai Schur, and Beniamino Segre—earned an "A," while that of nine—among them, Peter Scherk—were rated "C." In terms of both its grade and its particulars, Scherk's case provides a counterpoint to that of Segre.

Scherk was born into a Jewish family in Berlin in 1910, received mathematical training first in Berlin and then in Göttingen, and took courses at the latter with Weyl and others. He earned his Göttingen doctoral degree in February 1935 under the direction of Werner Fenchel for a dissertation "On Real Closed Space Curves of the Fourth Order" and spent his first postdoctoral summer back in Berlin.[41] There, he served as private assistant to Edmund Landau, the Jewish number theorist who, after being ousted from his Göttingen chair in the 1933 purge, had returned to his hometown to live. Scherk worked with Landau again in the spring of 1936, following a sojourn during the winter semester of 1935–1936 at the German university in Prague, where Loewner was a senior member of the faculty and Scherk's host. The young mathematician's brief stay in Prague, however, aroused the suspicion of the Gestapo, and he was forced to leave Germany in October 1936. The next two years found him back in Prague and elsewhere in Czechoslovakia, barely eking out a living.

As early as 28 October, 1936, Scherk had written to his former teacher, Weyl, in Princeton, asking him whether it might be possible for him to come to the United States. Weyl, who "naturally want[ed] to help," could not be optimistic.[42] As he explained, "all of my American friends, even the most liberal, have the impression that, with regard to the absorption of German émigrés, the saturation point in mathematics is practically attained." Saturation point.

---

40. Following his emigration to the United States, Loewner, whose name in his native Czech was Karel and in German Karl, changed it to Charles.

41. Peter Scherk, "Über reelle geschlossene Raumkurven vierter Ordnung," *Mathematische Annalen* 112 (1936), 743–766. Scherk thanks Fenchel for his guidance on p. 745. The Mathematics Genealogy Project also lists Gustav Herglotz as an advisor. See http ://www.genealogy.ams.org. For Scherk's handwritten curriculum vitae, see "Vita," Box 33, Folder: Refugees, Scherk, Peter 1939–41, Veblen Papers. The brief biography that follows is taken from this source.

42. Weyl to Scherk, 17 November, 1936, Box 33, Folder: Refugees, Scherk, Peter 1939–41, Veblen Papers (my translation). The quotation that follows is also from this letter with my translation.

MATHEMATICIANS AND PHYSICISTS

whose records have been sent to American Friends Service Committee

-in the column "Acad. Position" means: "held no academic position proper", Privatdocentur ie, Assistantship is not counted as such. For exact positions see individual records.

The significance of scientific output is the basis for rating under column "scholarship". In Judging scholarship age has to be taken into account. "B" indicates successful research work of considerable originality.

	Schol-arship	Person-altiy	Teaching Ability Adapta-bility	Acad. Posi-tion	Birth Date	Remarks
Bernard Baule	C	A	A		1891	Catholic, in jail and concentration camp for more than a year
Gustav Bergmann	C+	–	–	–	1906	interesting for combining math., psychology, Philosophy and law practice
Paul Bernays	B+	B	B		1888	
Ludwig Dorwald	B+	A	B		1883	record will follow
Otto Blumenthal	B	B	A		1874	
Alfred T.Brauer	B	B	A		1896	
Max Dehn	B	A	A		1878	
Felix Ehrenhaft	B	C	C		1879	
Robert Frucht	B-	–	–	–	1906	good actuarial work
Guido Fubini	A	A	B		1879	
Paul Funk	B-	B	B		1886	
Hilda Geiringer	B+	B	A		1895	
Wilhelm Gross	–	–	–		1883	Prof. of mining engineering
Hans L.Hamburger	A	A	B		1889	
Felix Hausdorff	A	A	–		1868	
Ernst Hellinger	C+	B	A		1883	
Edward Helly	C	B	B		1884	
Paul Hertz	B	C	C-		1881	Impossible as a teacher for undergraduates
Ludwig Hopf	B	B	A		1884	
Ernst K. Jacob-stall	B	B	A		1882	
George Jaffe	B	A	A		1890	
Stanislaus Jolles					1857	too old to be placed
Friedrich Kottler	B	B	B		1887	
Heinrich Loewig	C	–	–		1904	

FIGURE 7.3. The first page of a two-page memo on "Mathematicians and Physicists whose records have been sent to American Friends Service Committee," 2 May, 1939. (Typed Facsimile of the document in Veblen Papers, Library of Congress.)

It had become a refrain over the course of the 1930s. Weyl counseled his former student to be in touch with the Academic Assistance Council in London. Still, if Scherk could get to the United States, Weyl assured him that he would try to help, even though, at that moment, the small German Mathematicians' Relief Fund that he oversaw—made up of private donations—had run completely dry.[43]

Scherk finally arrived in the United States, but not until March 1939 and "practically penniless."[44] Despite the fact that he was "not a particularly strong or independent mathematician," that is, despite their grading scale and Scherk's placement on it, Weyl, Veblen, and their network worked hard to find some sort of work for him, even if they did not actively propose him for college or university posts. They were trying to uphold their principle of placing only the very best in the few academic posts available. From the by then somewhat replenished German Mathematicians' Relief Fund, Weyl did send Scherk $150 to help him get settled, and Veblen counseled him not to look only for work in mathematics but "at least for the next months [to] be ready to make the best of whatever work you can find."[45] "At present," Weyl insisted, "your need of making a living should prevail over all other considerations."

Scherk took a job tutoring one of Kurt Friedrichs's students at NYU; he graded papers for a professor at Hunter College; and he did kitchen work for the family of a Columbia University professor.[46] By December 1939, though, he had landed a job at a high school in Connecticut teaching mathematics as well as French and Latin, and four months later, he had begun to find his mathematical footing, submitting a number-theoretic paper for consideration by

43. Weyl and Emmy Noether had begun soliciting funds early in 1935 from German scholars already in the United States, proposing that they donate between 1% and 4% of their salary "to support refugees still without situations." See Norbert Wiener to Edward R. Murrow, 25 February, 1935, Series I; Grantees, Box 2, Folder 13: Bernstein, Felix, Emergency Committee Papers. Weyl continued this effort following Noether's untimely death in April 1935, just months after the two had begun their initiative.

44. Weyl to Charlotte Salmon of the American Friends Service Committee, Box 33, Folder: Refugees, Scherk, Peter 1939–41, Veblen Papers. The quotation that follows in the next sentence is also from this letter.

45. Weyl to Scherk, 14 April, 1939, Box 33, Folder: Refugees, Scherk, Peter 1939–41, Veblen Papers. The quotation that follows in the next sentence is also from this letter.

46. See the various letters in Box 33, Folder: Refugees, Scherk, Peter 1939–41, Veblen Papers.

the *Annals*.[47] Weyl had also recommended him to Hertha Kraus of the American Friends Service Committee for a place in one of the Committee's summer "American Seminars" designed to help émigrés hone their English and navigate American society. After the seminar's end, Kraus reported back to Weyl that Scherk's health had seemed so precarious that the Committee, fearing a brain tumor, had had him thoroughly examined by a Viennese neurologist in New York City. Thankfully, the doctor concluded "that Scherk's troubles [were] largely nervous disorders, and likely to disappear if and when he has satisfactory work in the only line he is interested in, mathematics."[48]

To that end, Kraus had "arranged with Yale University to have [Scherk] accepted as an honorary fellow in mathematics and as a member of our small Yale Friends University Center, which provides social contacts and guidance for honorary fellows." While at Yale, he would live and have almost all of his meals in a dormitory, but, as Kraus explained, he would still need $45.00 a month "to get along," and she hoped that she could count on Weyl's fund to help "make the Yale venture possible." "He is really quite a pathetic case," Kraus confided, ". . . and, after several years of fruitless struggle in exile, very much at the end of his rope."

Although it is not clear exactly how Scherk's year-long stay in New Haven was paid for—Weyl had written to Kraus that his fund was again exhausted— Scherk went to Yale and, thanks to Artin and his colleagues, next took the temporary teaching assistantship at Indiana University that he held from 1941 to 1943. Scherk finally found his footing when the University of Saskatchewan in Saskatoon hired him as an instructor in 1943. By 1955, he had been promoted to a full professorship there.

The stories of Segre, the A-rated mathematician who never came to the United States, and Scherk, the C-rated mathematician who did, underscore the difficulties inherent in trying strategically to place only the best émigré mathematicians in American academe in the late 1930s. That may have been the naïve ideal, an effort at disinterestedness in the face of grim realities, but personal connections and basic human compassion fundamentally

---

47. Peter Scherk, "Two Estimates Connected with the (α, β) Hypothesis," *AM* 42 (1941), 538–546. The paper had been received by the journal on 22 April, 1940, and presented two months earlier at a meeting of the AMS.

48. Kraus to Weyl, 24 September, 1940, Box 33, Folder: Refugees, Scherk, Peter 1939–41, Veblen Papers. The quotations that follow in this paragraph are also from this letter. As Scherk had written to Weyl on 17 April, 1939, "without mathematics . . . existence is meaningless." See Box 33, Folder: Refugees, Scherk, Peter 1939–41, Veblen Papers (my translation).

affected the process. It was all complicated further by the facts that money and positions were in too short supply and the volume of asylum seekers too great. Still, that at least some mathematicians in the United States had the confidence in the 1930s to rate the Europeans as a way of addressing these problems was another indication—like the ultimately aborted 1940 ICM— that the American mathematical community viewed itself on an equal footing with that of its European counterparts.

Accommodating a Second Wave of Mathematical Émigrés in the Northeast

As during the five years from 1933 to 1938, European mathematicians who emigrated after 1938 found academic refuge throughout the country. Some institutions—like Bryn Mawr, Brown, and the Institute—that had already provided a foothold for displaced foreign scholars earlier in the 1930s continued to be supportive, while others that either had not been able or had not chosen to accommodate those in the first wave made room for some in the second. Conspicuous in the latter category had been Harvard, despite the efforts of some on its faculty, like astronomer Harlow Shapley, to work with the EC and others to place academic émigrés.[49]

As noted in chapter five, the Harvard administration under Lawrence Lowell until 1933 and James Conant from 1933 until 1953, had been less than open to calls for helping refugee scholars. Lowell, in particular, had viewed such initiatives as a way of co-opting Harvard's reputation for a cause in which Lowell, for one, did not believe.[50] Lowell's foremost mathematician, the elder Birkhoff, did not disagree. By 1939, however, the Harvard Engineering Department had invited the independently wealthy Richard von Mises—applied mathematician, Vienna Circle member, and former director (from 1919 to 1933) of the University of Berlin's Institute of Applied Mathematics—to join its faculty without pay.

Von Mises, a Catholic, had left Berlin following the purge in 1933; owing to his family's Jewish heritage, he was non-Aryan by the Nazi definition regardless of his actual religious beliefs and saw the writing on the wall. He quickly accepted a position in Istanbul that involved not only studying Turkish but

49. Bessie Jones, "To the Rescue of the Learned: The Asylum Fellowship Plan at Harvard, 1938–1940," *Harvard Library Bulletin* 32 (1984), 205–238.

50. Norwood, p. 33.

also adapting to Turkish culture. In particular, he set up a mathematics institute from scratch as part of President Mustafa Kemal Atatürk's efforts, beginning in 1933, to transform Istanbul University into a modern institution.[51] Atatürk's death in 1938, however, brought fears both of Nazi aggression in the region and of instability for foreign workers. When the university failed to meet von Mises's demand that the contract of his long-time assistant and future wife, applied mathematician Hilda Geiringer, be renewed, von Mises made inquiries about emigrating to the United States that ultimately resulted in his acceptance of the unpaid Harvard position.[52]

Harvard also served briefly as a temporary home for two Polish Jewish mathematicians, Stanislaw Ulam and Alfred Tarski. Ulam had earned his Ph.D. in 1933 from the Lwów Polytechnic Institute (now Lviv Polytechnic National University) under Kuratowski's direction and had made a key American connection late in 1934 when he wrote to von Neumann about some of his mathematical work. Impressed by what he read, von Neumann invited Ulam to visit the Institute.[53] The young mathematician arrived in Princeton in December 1935 and spent that spring term actively participating in its mathematical environment. He made a particular point to acquaint himself with its mathematical scientists: Lefschetz and Bochner at the university and Veblen, Weyl, Morse, Alexander, Einstein, and, of course, von Neumann at the Institute. As luck had it, he also met Birkhoff *père* at a tea hosted by the von Neumanns. The Harvard scion, who had already heard good things about Ulam from both his son, Garrett, and Kuratowski, sounded Ulam out on the possibility of a three-year position in Cambridge. The Society of Fellows had been founded at Harvard in 1933 thanks to a gift from Lowell on the occasion of his retirement as president, and Birkhoff proposed that Ulam apply for one of its junior fellowships. Given Lowell's and Birkhoff's feelings about the refugee situation, it is ironic that one of the early junior fellows would be

51. Fritz Neumark discusses these early initiatives in *Zuflucht am Bosporus: Deutsche Gelehrte, Politiker und Künstler in der Emigration, 1933–1953* (Frankfurt am Main: Verlag Josef Knecht, 1980), pp. 11–27. Neumark, a German economist and himself one of the émigrés, provides a first-person account in this book of the experiences of German intellectuals in Turkey.

52. On these aspects of von Mises's story, see Siegmund-Schultze, *Mathematicians Fleeing from Nazi Germany*, pp. 141–146 and 383–387. Von Mises was not listed on the émigré "grade sheet" of 2 May, 1939 (fig. 7.2), but Geiringer was. She was rated a B+.

53. Stanislaw Ulam, *Adventures of a Mathematician* (New York: Charles Scribner's Sons, 1976), pp. 65–66. He wrote of his time in Princeton in 1936 on pp. 65–83.

a Polish Jew, but, at least in 1936 when Ulam was appointed, he was not a refugee.

After each year of his fellowship, Ulam, like von Mises independently wealthy, returned to Poland for the three summer months to visit his family, although he had begun to prize what he termed "the free and hopeful 'open-ended' conditions of life in America."[54] The trip home for the summer of 1939 would have meant the end of his American sojourn had Birkhoff not offered him a lectureship in the Mathematics Department for the 1939–1940 academic year. Ulam jumped at the opportunity, even though he realized that his chances for "more permanent prospects" for employment in the United States "were not promising." It was on 21 August, 1939, during his return voyage to the United States, that he heard the breaking news of what seemed to be an imminent non-aggression pact between Germany and Russia.[55] Two days later such a pact was signed, and nine days after that, Germany invaded Poland. Ulam had a one-year post, but, at that moment, he essentially became a refugee.

On Birkhoff's advice and thanks to his intervention, Ulam followed his extra year at Harvard with an instructorship at Wisconsin in Madison. A year later in 1942, he was promoted to an assistant professorship, a step that, in his view, "gave [him] hope and some confidence in the material aspects of the future."[56] He left Madison in 1943, however, to join the team of scientists working to develop the atomic bomb at Los Alamos (see the next chapter).

As Ulam discovered on his first night at sea, onboard with him was logician Alfred Tarski, also en route to Harvard but to participate in the Fifth International Congress for the Unity of Science. An outgrowth of the philosophical ideas of the Vienna Circle in the late 1920s, the Unity of Science Movement sought to establish all of the sciences on a unified philosophical and methodological foundation and, in the process, to create one universal language of science.[57] Tarski had been introduced to the movement's way of thinking—and it to his—as early as the winter of 1930; Vienna Circle member, Karl Menger, had invited Tarski to speak in his Mathematical Colloquium on the

54. Ibid., p. 106. The quotations that follow in this paragraph are on p. 113.

55. Anita Burdman Feferman and Solomon Feferman, *Alfred Tarski: Life and Logic* (Cambridge: Cambridge University Press, 2004), p. 126.

56. Ulam, *Adventures of a Mathematician*, p. 133.

57. On this movement, see, for example, Harmke Kamminga and Geert Somsen, ed., *Pursuing the Unity of Science: Ideology and Scientific Practice from the Great War to the Cold War* (Abingdon and New York: Routledge/Taylor and Francis Group, 2016).

logical work that he and his Polish colleagues had been pursuing. There, Tarski became personally acquainted with fellow logicians, Rudolf Carnap and Kurt Gödel, and began what eventually became the association with the Unity of Science Movement that would bring him to the United States.[58]

When his ship docked in New York City after the transatlantic voyage, Tarski found himself stranded abroad with little money and only a suitcase full of summer clothing, while his wife, two young children, and other relatives were in a Poland under German attack.[59] Despite his precarious position, Tarski did about all that he could do: work on a way to get his family out of Poland and proceed to Cambridge to participate in the congress. For a week, he immersed himself in the meeting, at the same time that he renewed his acquaintance with Harvard's Willard Quine and met many of America's other top logicians, among them, the then President of the Association for Symbolic Logic, Haskell Curry, as well as Church and Church's two students, Kleene and Rosser. This network of logicians, together with the indefatigable Harlow Shapley, immediately went to work to find at least a short-term solution to Tarski's predicament.[60]

At Harvard, Quine, Shapley, and Stone wondered about some sort of temporary lectureship for Tarski, but the fact that Ulam already had such a position for the year was a complicating factor. Stone also contacted Veblen to see if perhaps the Institute could accommodate Tarski on short notice, especially given that the outbreak of the war might have forced some to forgo their prearranged visiting positions.[61] As it became clear, moreover, that Tarski would need to obtain a permanent visa for the United States and that the most expeditious way for him to do that would be to leave and reenter the country via Cuba, Curry asked Veblen if he would address a letter to the American Consul in Havana "testifying to Tarski's ability."[62] He also wondered whether Veblen might enlist Lefschetz's support. In the end, not only did the letter-writing campaign work—Tarski got the visa and entered the country legally on 1 January, 1940—but he was also offered a semester-long, visiting

58. Feferman and Feferman, pp. 81–98.

59. Ibid., p. 127.

60. See, in particular, Shapley to Betty Drury of the EC, 26 October, 1939, Series I: "Grantees," Box 33, Folder 3: Tarski, Alfred.

61. Stone to Veblen, 11 and 17 September, 1939, Box 34, Folder: Refugees, Tarski, Alfred 1939–40, Veblen Papers.

62. Curry to Veblen, 15 November, 1939, Box 34, Folder: Refugees, Tarski, Alfred 1939–40, Veblen Papers. See also Feferman and Feferman, p. 130.

professorship of philosophy at the City College of New York that was reported in the *New York Times*.[63]

He followed this teaching stint, first, with a research appointment at Harvard and a concurrent series of paid lectures at the Young Men's Hebrew Association in New York City and, then, with a Guggenheim fellowship from April 1941 to October 1942. An offer of a lectureship at Berkeley materialized for the fall of 1942, partially funded by the Emergency Committee.[64] Tarski spent the rest of his career in Berkeley successfully building a program there in mathematical logic; in 1946, after an almost seven-year separation, he finally succeeded in getting his wife and children out of Europe and once again by his side.

If the Harvard administration as well as certain members of its faculty had initially been cool to the plight of the refugees and had only acted in response to the second wave, others in the Northeast had persisted in their efforts throughout the 1930s and into the 1940s. In Princeton, Veblen and his colleagues kept the Institute open to as many short-term visitors as they could accommodate. For example, number theorists Alfred Brauer and Carl Siegel found places there thanks to EC funding. As his younger brother Richard had done in 1934–1935, Brauer served as Weyl's assistant, in his case from 1939 to 1942, while Siegel participated as one of the IAS's "members" from 1940 until he was named an actual professor there in 1945. They were joined from 1939 to 1942 by the sixty-year-old analyst Guido Fubini, who, with his family, had fled Fascist Italy, and in 1940 by the EC-supported Kurt Gödel. In 1953, two years after Siegel's return to Germany, the IAS named Gödel to a professorship.[65]

Brown University also continued to do what it could. Thanks to Richardson's active interventions, it had already made room, as noted, for Hans Lewy and Otto Szász when the first wave of mathematical refugees had washed over the United States. With the second, it also accommodated William Feller, Otto Neugebauer, Courant's former assistant and the then-current editor of Springer Verlag's *Zentralblatt für Mathematik und ihre Grenzgebiete*,

---

63. Feferman and Feferman, p. 131 and "Mead Not a Candidate to Head City College," *New York Times*, 21 November, 1939, p. 16. Its title aside, this article also announced Tarski's appointment and was actually illustrated with his photograph. Feferman and Feferman reproduce the photo and the relevant part of the article on p. 133.

64. Feferman and Feferman, pp. 139 and 147.

65. See Mitchell, ed., *A Community of Scholars*.

and Neugebauer's assistant, Olaf Schmidt.[66] In fact, it had been Schmidt's appointment as an instructor—at a time when his own job prospects were so tenuous—that had prompted Dick Wick Hall's bitter complaint about posts going to "dumb assistants."[67] Consonant with Richardson's interest in fostering a program in applied mathematics, moreover, Brown engaged other refugee mathematicians during World War II: Stefan Bergman, Lipman Bers, Charles (formerly Karl) Loewner, and William "Willy" Prager, among others.[68] (For more on this, see the next chapter.)

Bryn Mawr, too, despite its limited resources, tried to leave open its doors to female refugee mathematicians. In the spring of 1939, Veblen received letters from Istanbul from Hilda Geiringer and from a Richard von Mises soon to be en route to Harvard as well as from Hadamard in Paris, urging him to find Geiringer a position.[69] Quite naturally, he wrote to Anna Pell Wheeler, the chair of Bryn Mawr's Department of Mathematics, asking for her help.[70] As had been the case with Emmy Noether in 1933, Bryn Mawr, with its graduate program in mathematics, seemed the best option in the United States for an active female researcher like Geiringer, but although Wheeler wanted to accommodate the request, money was just too tight. "All American colleges which depend upon income from endowments are suffering from the effects of the falling rate of interest," Veblen explained to Geiringer, and Bryn Mawr, in particular, had "to curtail their expenses in all directions" making it "very hard for them to undertake any new commitments."[71]

Given this less than optimistic response, Geiringer pursued other avenues, securing a grant from the English Society for the Protection of Science and Learning and making her way to Portugal from Istanbul for the trip to

66. Brown accommodated others in the second wave with temporary appointments. For example, it hosted George Pólya for the two years from 1940 to 1942 when efforts between the Institute and Stanford to cobble together the two-year appointment needed for non-quota visa status fell through. On Pólya's case, see Box 33, Folder: Refugees, Pólya, George 1940–41, Veblen Papers.

67. Hall to Whyburn, 10 March, 1941.

68. Bers, "The Migration of European Mathematicians to America," p. 240.

69. Joan Richards gives a brief account of Geiringer's life and work in "Hilda Geiringer von Mises (1893–1973)," in *Women of Mathematics: A Biobibliographical Sourcebook*, ed. Louise Grinstein and Paul Campbell, pp. 41–46.

70. Veblen to Wheeler, 17 April, 1939, Box 32, Folder: Refugees, Geiringer, Hilda 1939–44, Veblen Papers.

71. Veblen to Geiringer, 4 May, 1939, Box 32, Folder: Refugees, Geiringer, Hilda 1939–44, Veblen Papers.

FIGURE 7.4. Hilda Geiringer (1893–1973) teaching at Wheaton College in 1953. (Photo from "Overdue Recognition," *Wheaton News*, November 5, 2019.)

England. Unfortunately, given the outbreak of the war in the interim, the English had forbidden entry to holders of German passports. That development effectively left Geiringer and her teenaged daughter stranded in Lisbon.[72] Von Mises, already in the United States and beside himself with worry, tried everything to get his former assistant a position on his side of the Atlantic.[73] Finally, through the intervention of his acquaintance, the Princeton physicist Rudolf Ladenburg, the National Refugee Service (NRS) had come through with funding that it would make available to an educational institution that would agree to pay Geiringer's salary.[74] Veblen wrote

72. Veblen to Wheeler, 25 September, 1939, Box 32, Folder: Refugees, Geiringer, Hilda 1939–44, Veblen Papers.

73. Siegmund Schultze gives a poignant excerpt (translated from the German) from von Mises's diary in *Mathematicians Fleeing from Nazi Germany*, pp. 383–387.

74. The NRS was an American aid agency that had been founded in New York in June 1939. A reorganization of the National Coordinating Committee, a group funded by various Jewish organizations that had been formed in 1934, the NRS aimed to help immigrants "in meeting

to Wheeler at once with word of this new development. Bryn Mawr apparently agreed to the NRS's stipulation, since Geiringer took up a temporary post there.

Over the course of the next four years, the Bryn Mawr administration, Veblen, Weyl, von Mises, and others worked to secure the money, from the Emergency Committee and elsewhere, that allowed Geiringer to stay on at Bryn Mawr.[75] In 1943, however, she married von Mises and wanted to be closer to him in Cambridge. That opportunity came in the winter of 1944, when she was offered a position at Wheaton College in Norton, Massachusetts, as Professor of Mathematics and head of the department. She remained on the faculty there until her retirement in 1959, despite efforts—largely thwarted owing to her sex—to move to a more research-oriented school.

Another women's college, Mt. Holyoke—and also with the help of Emergency Committee funding—provided a job for the Polish analyst Antoni Zygmund, when, in 1940, he fled Wilno (present-day Vilnius) and the professorship he held there. Zygmund made it to the United States thanks to the visa and offer of a visiting professorship at MIT arranged for him by Tamarkin, Wiener, and Jerzy Neyman and served on the Mt. Holyoke faculty until his move to the University of Pennsylvania in 1945. Two years later, he left for Chicago where he remained until his retirement in 1980 and not only continued his research particularly in harmonic analysis but also saw some forty students through to the Ph.D. (see chapter ten).[76]

Elsewhere in the Northeast, two Frenchmen, among a number of other mathematicians, also found temporary situations. The seventy-six-year-old Hadamard, who had associations with the United States going back to 1901

the requirements of the immigration laws," to provide "legal protection of immigrants in jeopardy of deportation," and to aid in both their "resettlement throughout the country" and their "adjustment to American life." Lyman White discusses this group in *300,000 New Americans: The Epic of a Modern Immigrant-Aid Service* (New York: Harper & Brothers, 1957), pp. 51–76. According to Bryn Mawr President Marion Park, the $2,500 for Geiringer's salary had actually been provided by Geiringer's England-based brothers, presumably via the NRS (Park to Laurens H. Seelye of the EC, 4 March, 1941, Series I: Grantees, Box 11, folder 8; Geiringer, Hilda, Emergency Committee Papers).

75. See the letters contained in Series I: Grantees, Box 11, folders 8–9; Geiringer, Hilda, Emergency Committee Papers.

76. Series I: Grantees, Box 36, Folder 10: Zygmund, Antoni, Emergency Committee Papers contains primary sources pertinent to his emigration. See also, among other secondary sources, Ronald Coifman and Robert Strichartz, "The School of Antoni Zygmund," in *A Century of Mathematics in America*, ed. Peter Duren et al., 3: 343–368 on p. 346.

and particularly to the 1920s, was a special case of an EC-supported mathematician who was well past retirement age. Revered as one of the greatest mathematicians then living, Hadamard, a Jew, had been forced to flee his home outside Paris as the Germans advanced on the capital city in the spring of 1940. By June, he and his family had arrived in Toulouse, and thanks to the intervention of biochemist and social activist Louis Rapkine, made their way, not without incident, to the United States in 1941. Hadamard took up a year-long position at Columbia at the invitation of its president, Nicholas Murray Butler,[77] but that year was followed by frustrations and feelings of increasing isolation as Hadamard tried unsuccessfully to secure a permanent job while making ends meet through paid, invited lectures.[78] With Rapkine's help, he moved to London in 1944, finally returning to a war-ravaged Paris in the spring of 1945.[79]

The thirty-five-year-old André Weil represented a very different case. A member of the French mathematical collective known as Nicolas Bourbaki, Weil was officially a reserve officer in the French Army.[80] He had decided, however, that should war break out, he would seek asylum in some neutral country rather than serve. In Finland when war was declared in September 1939, he was suspected as a spy, thrown in jail, and ultimately shipped off by railroad and released at the Swedish border as the Russians advanced on Helsinki. When it was no longer possible for him to remain in Sweden, he surrendered to the French Legation in Stockholm and eventually made his way via the British Isles to a military prison in France. He stood trial in May 1940 for "failing to report, rather than [for] desertion," was found guilty, and was given the maximum sentence of five years in jail.[81] Fortunately, there was an option to exercise: to "petition for a suspended sentence, in exchange for serving in a combat zone." A series of wartime experiences ended with his

77. Recall from chapter five that Butler had been one of the original founders of the organization out of which the EC had grown.

78. The various letters in Box 32, Folder: Refugees, Hadamard, Jacques 1936–42, Veblen Papers as well as Series I: Grantees, Box 13, Folders 9–11: Hadamard, Jacques document his travails.

79. Vladimir Maz'ya and Tatyana Shaposhmikova, *Jacques Hadamard: A Universal Mathematician*, HMATH, vol. 14 (Providence: American Mathematical Society and London: London Mathematical Society, 1998), pp. 229–249.

80. Liliane Beaulieu provides an account of the origins of Bourbaki in "A Parisian Café and Ten Proto-Bourbaki Meetings (1934–1935)," *MI* 15 (1993), 27–35.

81. Weil, *Apprenticeship*, p. 151. He recounts the dramatic story of his wartime experiences on pp. 123–174. The next quotation is on p. 152.

discharge late in 1940 and his departure for New York City early in 1941 with his wife and her son. His first two years in the United States—spent at the Institute and at nearby Haverford College in Pennsylvania—were funded by the Rockefeller Foundation thanks, as in Hadamard's case, to Rapkine's powers of persuasion.

In Princeton, Weil was reunited with his friends, analytic number theorist Carl Siegel at the Institute and algebraic geometer and group theorist Claude Chevalley at Princeton,[82] while at Haverford, he gained his initial exposure to teaching American undergraduates. An instructorship—Weil managed to parlay it into an assistant professorship—at Lehigh University in eastern Pennsylvania followed. After two miserable years at what he described as "a second rate engineering school attached to Bethlehem Steel," Weil accepted a visiting professorship at the University of São Paulo in Brazil in 1944, arriving in January 1945.[83] Two years later, he was back in the United States, this time in a professorship at the University of Chicago (see chapter nine).

Accommodating a Second Wave of Mathematical Émigrés outside the Northeast

If, by December 1941 and the United States' entry into the war, many schools in the Northeast *had* effectively reached what Veblen and his colleagues had long been terming "a saturation point" in mathematics, other regions of the country still had the capacity for absorption. The University of Chicago, for example, had not been aggressive in hiring during the first wave, but, unlike Harvard, that had not owed to a pro-American, anti-refugee stance. Like its main East Coast rivals, the other two of the "top three" Harvard and Princeton, Chicago had been hit hard financially by the Depression, but relative to the émigré situation, its administration had tried but largely failed in its efforts to enlist the philanthropic support of Chicago's Jewish community.[84] At the

---

82. Chevalley had been a "member" at the Institute during the 1938–1939 academic year and was still in the United States, with the blessing of the French Embassy, when war was declared in Europe. He accepted an assistant professorship at Princeton in 1939 thanks to Lefschetz's intervention and remained on the faculty there until his move to Columbia in 1949. Jean Dieudonné and Jacques Tits, "Claude Chevalley (1909–1984)," *BAMS* 17 (1987), 1–7. They give the year of Chevalley's move to Columbia as 1948, but his entry in Jaques Cattell, ed., *American Men of Science*, 9th ed. vol.1 gives the year as 1949.

83. Weil, *Apprenticeship*, p. 180.

84. Norwood, pp. 29–30 and 32.

time of the second wave, however, Chicago's Jewish algebraist, Adrian Albert, was able successfully to make the case, but only on the occasion of Dickson's retirement in 1939, to hire fellow algebraist, the twenty-seven-year-old Otto Schilling, as an instructor in mathematics.[85] Schilling, Noether's last student at Göttingen, subsequently rose through the ranks at Chicago, becoming a professor there in 1958, although he trained few students.

Nearby in Evanston, the fifty-six-year-old Ernst Hellinger managed to eke out a series of short-term appointments at Northwestern, owing partly to EC support and partly to the financial interventions of his sister who had already emigrated to the United States.[86] A 1907 student of Hilbert, Hellinger had established a reputation in the theory of integral equations while a member of the faculty at the University of Frankfurt. There, he, together with Max Dehn, animated the vibrant mathematical community that, in the 1920s, also included Szász, Siegel, and number theorist Paul Epstein. By the end of 1935, however, Hellinger had been forced to retire from his position; as a Jew who had fought for Germany in World War I, he had initially been exempted from the 1933 purge. However, following Kristallnacht in 1938, he was sent to the concentration camp in Dachau where he spent six unbearable weeks before being released on the condition that he emigrate immediately. Dutch mathematician Hans Freudenthal arranged for the temporary permit that allowed Hellinger to transit through the Netherlands on his way to the United States.[87]

Hellinger, rated a C+ in the May 1939 assessment, spent the decade from 1939 to 1949 at a Northwestern that was only just beginning to develop a graduate program at the close of the 1930s. As Veblen saw it, Hellinger "must be regarded primarily as a scholar rather than a research man, although he has some first-rate research to his credit." Still, he continued, "[t]o establish him at Northwestern would . . . be a really significant step in the process of

85. Nancy Albert, A^3 & His Algebra: How a Boy from Chicago's West Side Became a Force in American Mathematics (New York: iUniverse, 2005), p. 122. Compare also William McNeill, Hutchins' University: A Memoir of the University of Chicago, 1929–1950 (Chicago: University of Chicago Press, 1991), pp. 41–42 on the impact of the Depression on faculty hiring at Chicago.

86. Series I: Grantees, Box 14, Folder 6: Hellinger, Ernst, Emergency Committee Papers documents the many and sustained efforts to secure funding for Hellinger.

87. Box 32, Folder: Refugees, Hellinger, Ernst, 1938–41, Veblen Papers. The most thorough account of Hellinger's life and work to date is James Rovnyak, "Ernst David Hellinger 1883–1950: Göttingen, Frankfurt Idyll, and the New World," in Topics in Operator Theory: Ernst D. Hellinger Memorial Volume, Operator Theory, Advances and Applications, vol. 48 (Basel: Birkhäuser Verlag, 1990), pp. 1–44.

building our scientific situation by judicious transplanting."[88] Indeed, in his ten years on the Northwestern faculty, Hellinger directed the research of five of the roughly fifteen Ph.D. students who successfully defended theses there, but this productive period was cut short, when on reaching the age of sixty-five, he was forced to retire.[89] Unfortunately, having almost no pension on which to live, he died, "troubled by financial insecurity," of cancer in 1950. His was exactly the sort of case that the EC had foreseen when it had initially made the abstract decision to privilege established, midcareer scholars over either younger, unproven or older scholars when making academic placements.

Hellinger's friend, Dehn, also managed to emigrate to the United States but to the Far West and after, in some sense, an even more harrowing escape.[90] Dehn, five years older than Hellinger and rated a B, had made a splash in the mathematical world in 1900 when he solved the third of the problems that his *Doctorvater* Hilbert had articulated in his famous address at the Paris ICM earlier that same year, namely, is it possible to prove "the equality of the volumes of two tetrahedra of equal bases and equal altitudes" without invoking "the axiom of continuity (or . . . the axiom of Archimedes)"?[91] Dehn showed, as Hilbert had suspected, that the answer was "no." His "brilliantly simple" argument hinged on the definition of what has since been called the Dehn invariant and involved an exploration of it using little more than linear algebra and elementary number theory.[92] While at Frankfurt in the 1920s and

88. Veblen to Warren Weaver, 13 November, 1939, Series I: Grantees, Box 14, Folder 6: Hellinger, Ernst, Emergency Committee Papers.

89. Rovnyak lists Hellinger's doctoral students on p. 35. The quotation that follows is on p. 27.

90. Another émigré who could be mentioned in connection with the Far West is the geometer Arthur Rosenthal. A professor at the University of Heidelberg until 1935, Rosenthal emigrated to the United States in 1940 and held a lectureship at the University of Michigan for the 1940–1941 academic year. In 1942, he moved to the University of New Mexico in Albuquerque with support from the EC, rising from a lecturer (1942–1943) to assistant professor (1943–1946) to associate professor (1946–1947). In 1947, he moved to a professorship at Indiana's Purdue University where he remained for the rest of his active career. See Betty Drury to Arnold Dresden, 26 November, 1941, Series V: General Correspondence, Box 177, Folder 10: Dresden, Arnold (1934, 1940–1944), Emergency Committee Papers and Siegmund-Schultze, *Mathematicians Fleeing from Nazi Germany*, p. 468.

91. David Hilbert, "Mathematical Problems," trans. Mary Winston Newson, *BAMS* 8 (1902), 437–479 on p. 449.

92. John Stillwell provides more on Dehn's work on this and other aspects of geometry and topology in the biographical context of his contribution, "Max Dehn," in *History of Topology*, ed. Ioan James, pp. 965–978. The quotation is on p. 967.

1930s, Dehn continued his mathematical researches in topology and combinatorial group theory. He also turned his attention seriously to the history of mathematics, however, animating the seminar in which Hellinger, Siegel, and Epstein actively participated and that attracted visitors such as Weil.[93]

When the Nazi's purged the universities in 1933, Dehn, like Hellinger, was initially able to keep his position on account of his service in World War I, but, also like Hellinger, he was fired in 1935. He, too, was also arrested following Kristallnacht, but, in his case, the jails were too full to accommodate him so he was set free. The Dehns fled Germany, first to Copenhagen and then to Trondheim in 1939; they had sent their three children to live abroad before the pogrom.

In the United States, Hellinger—whose own position at Northwestern was so tenuous—nevertheless worked actively on Dehn's behalf to raise money for his support and to secure the promise of a position that would qualify him for a non-quota visa.[94] That promise and that position, an assistant professorship of mathematics and philosophy, ultimately came from the Southern Branch of the University of Idaho in Pocatello, but the dean there, John Nichols, wanted Dehn to understand what he would face. As he explained to Hellinger, "we are teaching only freshman and sophomore mathematics which would be about the equivalent of the first year in the German Gymnasium."[95] He feared that Dehn "would find the students woefully unprepared, generally disinterested in mathematics, and possibly, from his point of view, exceedingly lazy and slipshod in all their work and thinking. On the other hand," he allowed that Dehn "might arise to the challenge and see what he could do with the type of students who come to an American junior college." Hellinger immediately assured him that Dehn "would enjoy thoroughly working with that type of student . . . and that he would get along extremely well with them."[96] The position became a reality in June 1940 thanks both to $1,200 raised by Hellinger and to support from the EC.[97] Dehn and his wife finally arrived on 1 January, 1941, after an exodus that began when the

93. Rovnyak, p. 13. See also Carl Ludwig Siegel, "On the History of the Frankfurt Mathematics Seminar," *MI* 1 (1979), 223–230.

94. Hellinger to Dean John Nichols of the University of Idaho, Southern Branch, 28 May, 1940, Dehn Papers.

95. Nichols to Hellinger, 31 May, 1940, Dehn Papers.

96. Hellinger to Nichols, 4 June, 1940, Dehn Papers.

97. Nichols to Betty Drury, 6 June, 1940, Series I: Grantees, Box 6, Folder 5: Dehn, Max, Emergency Committee Papers.

Germans invaded Norway in the spring of 1940 and that ultimately took them on a harrowing trip through Stockholm, Moscow, Siberia, and Japan, and across the Pacific Ocean to San Francisco.[98]

Although the Dehns received a warm welcome in Pocatello, their arrival in the state of Idaho was not celebrated by all. One John Kreupper of Rigby, a small town some sixty-five miles northeast of Pocatello, wrote disparagingly to the State Superintendent of Public Instruction about the hiring of a non-American. "Why is it that some one had to go across the ocen to find a man cabel to teach in our school," he asked (with his orthography), "have not we men heare at home that could have fullfild this place; kindly advise."[99] Kreupper also inquired "who were the men that had the athoriety to go over and secure this man as one of our teachers; a man of forn birth a man not a citizen of the contry." He had apparently also been informed that Dehn was being "paid by some philanthropic organazation" and he wanted to know "who are the men of that organization; are they men of American burth or where does thoos men of that organizacion live heare in our united State"? And, Kreupper was just one of the "super-patriots" who protested the Dehns' arrival.[100] Rumors had run rampant that Dehn "was or might be (1) a Nazi spy or (2) a communist, since he travelled across Russia and got here via the Pacific!," while "[o]ne member of [Idaho's] Legislative Committee on Finance even objected to [the University's] budget on the basis of hiring such a person."[101] These aspects of Dehn's reception make manifest the xenophobia that the EC and others well knew was present in the United States of the 1930s and early 1940s as they tried to place and distribute refugees across the country.

Dehn remained in Pocatello through the end of the 1941–1942 academic year, tendering his resignation in the summer of 1942 to take a visiting lectureship at the Illinois Institute of Technology in Chicago where he "hope[d] to contribute rather directly to the war effort."[102] After just a year, he moved to

98. Dehn gave an account of this exodus in a public lecture in Pocatello in the spring of 1941. See the Dehn Papers for a typescript. See also John Dawson, Jr., "Max Dehn, Kurt Gödel, and the Trans-Siberian Escape Route," NAMS 49 (2002), 1068–1075.

99. Kreupper to C. E. Roberts, 7 February, 1941, Dehn Papers. The quotations that follow, also with Kreupper's orthography, are from this letter.

100. Nichols to Laurens Seelye, 6 March, 1941, Series I: Grantees, Box 6, Folder 6: Dehn Max, Emergency Committee Papers.

101. Nichols to Stephen Duggan, 3 March, 1942, Series I: Grantees, Box 6, Folder 6: Dehn Max, Emergency Committee Papers.

102. Dehn to Nichols, 20 August, 1942, Dehn Papers.

St. John's College in Annapolis, Maryland, but philosophical differences with the administration prompted yet another move in 1945 to Black Mountain College in North Carolina. His friends—and especially Weyl—had tried in vain to find a more suitable post, but as Weyl lamented to the Executive Secretary of the EC, Dehn's "age seems an almost insurmountable obstacle, and I feel that we individuals are powerless to grapple with this problem of providing for refugee scientists who have reached retirement age."[103] Dehn spent the rest of his life at Black Mountain College, dying there in 1952 at the age of seventy-three.

Other schools in the South also made room for refugee mathematicians in the second wave, but, as during the first, southern placements were not numerous. German-Jewish number theorist Alfred Brauer, like Polish-Jewish topologist Witold Hurewicz, followed a three-year stint at the Institute with an assistant professorship at UNC in Chapel Hill, although, in Brauer's case, the EC had tried to secure a position for him at a number of schools in the Midwest and West before UNC, in the South, came through.[104] It was precisely schools in these regions that had not yet reached the "saturation point."

Hurewicz held his UNC position until his move to MIT in 1945, while Brauer remained on the UNC faculty until his retirement as Kenan Professor of Mathematics in 1966 and served a second southern school, Wake Forest University in Winston-Salem, North Carolina, as a visiting professor until his second retirement in 1975. As noted, Brauer began his research career in number theory, but he moved into matrix theory in 1946 as a result of teaching a course on the subject at UNC. Together with his colleague, analyst William (brother of Gordon) Whyburn, Brauer was instrumental in building a graduate program at UNC in the 1950s and in establishing it as a new program to be reckoned with in the South alongside those formed in the 1930s at Duke and Virginia.[105]

103. Weyl to Frances Fenton Park, 14 November, 1944, Box 31, Folder: Refugees, Dehn, Max 1938–45, Veblen Papers.

104. See Series I: Grantees, Box 4, Folder 5: Brauer, Alfred, Emergency Committee Papers.

105. For more on Hurewicz's life, see Solomon Lefschetz, "Witold Hurewicz, in Memoriam," *BAMS* 63 (1957), 77–82. On Brauer's life and accomplishments, see Richard Hudson and Thomas Markham, "Alfred T. Brauer As a Mathematician and Teacher," *Linear Algebras and Its Applications* 59 (1984), 1–17 and Richard Carmichael, "Alfred T. Brauer: Teacher, Mathematician, and Developer of Libraries," *The Journal of the Elisha Mitchell Scientific Society* 103 (1986), 88–106.

Brauer's experiences in the South sharply contrasted with those of another mathematical refugee, Karl Loewner. A complex function theorist who had been one of the handful of A-rated mathematical refugees, Loewner left his professorship in Prague in 1939 just after war was declared in Europe and had a difficult transit to the United States. He arrived "just about penniless" in November to take up a lectureship funded by the EC at the University of Louisville in Kentucky.[106] By 1942, he had been promoted to Assistant Professor, but the level of mathematics instruction he endured was low and the incentives for more advanced teaching nonexistent.[107] After war work took him to Brown in 1944–1945, he moved to Syracuse University in upstate New York in 1946. There, he joined his former student, Lipman Bers, on the faculty and once again had the possibility of doing more advanced teaching and of actually training graduate students. While at Syracuse, Loewner supervised the work of at least four Ph.D. students, while Bers guided the research of at least two more. Both men moved on in 1951: Loewner crossed the country to take up a position at Stanford where he joined his friend Szegő as well as fellow second-wave émigré, the Hungarian mathematician and former professor at the ETH in Zürich, George Pólya; Bers augmented Courant's team at NYU. (On their work in the 1940s, see chapter ten.)

These examples of second-wave mathematical refugees paint a complex picture. Some like von Mises, Ulam, Tarski, Siegel, Gödel, Neugebauer, and Alfred Brauer fairly quickly found stable and congenial homes in American academe that allowed them to continue their own research and to contribute to the strengthening of American institutions already well under way. Others like Hadamard, Hellinger, and Dehn were near or past retirement age when they arrived. They succeeded neither in regaining their mathematical footing

---

Analyst William Whyburn, elder brother of University of Virginia topologist Gordon Whyburn, had moved to UNC in 1948 as chair of the department. Whyburn had risen through the ranks at the University of California in Los Angeles, starting out in 1928 as an assistant professor and ending by serving as department chair from 1937 to 1944. A native Texan, he accepted the presidency of the Texas Technological College (now Texas Tech University) in Lubbock, before answering UNC's call.

106. Betty Drury to Stephen Duggan, 8 November, 1939, Series I: Grantees, Box 22, Folder 2: Loewner, Karl, Emergency Committee Papers. The letters in this folder detail Loewner's odyssey as well as the efforts of Veblen, von Neumann, and others to secure a job for him and to get him safely to American shores.

107. Lipman Bers, "Editor's Introduction," in *Charles Loewner: Collected Papers*, ed. Lipman Bers (Boston: Birkhäuser Boston, Inc., 1988), pp. vii-x on p. viii.

on this side of the Atlantic nor in making an impact on the American mathematical scene. Still others, like Weil, Loewner, Zygmund, and Geiringer, after her move from Bryn Mawr to Wheaton, found employment but at a tier of schools that provided basic education for American undergraduates and to which the research ethos of the Harvards, Princetons, and Chicagos had not penetrated. By March 1939, Veblen had come around to the view that while "we are not far from the saturation point in the more prominent universities ... there are still a great many less well known academic institutions in which refugees could be placed with substantial advantage both to the individual and to the institution."[108] Many of these placements distinctly reflect those advantages. Still, regardless of the institutions at which they found themselves, these and other mathematical émigrés were accepted into a multi-tiered American mathematical community that assessed their accomplishments against a backdrop of its own. It was a mature community increasingly recognizing its changing place in the international mathematical scene.

Geopolitics and Mathematical Reviewing in the Late 1930s

That recognition was starkly in evidence when the AMS Council convened its winter meeting in Williamsburg, Virginia, at the end of December 1938. It had a lot on its plate. The various committees concerned with what would have been the 1940 International Congress of Mathematicians met for more than ten hours to hash out, among other things, the proposed list of plenary speakers (recall fig. 7.2); the Council convened a marathon session that only adjourned at 1:30 in the morning; "meetings of committees galore" generally left AMS Secretary Richardson "a wreck."[109] Dealing with the final planning of the Congress would certainly have been enough to exhaust the seemingly indefatigable Richardson, but in December 1938, he was also doing his best "to direct the discussion into profitable channels" of whether the AMS should actually assume responsibility for the publication of a new international journal.

At issue was the *Zentralblatt für Mathematik und ihre Grenzgebiete*, the abstracting journal edited by Neugebauer that had been founded by Germany's

108. Veblen to Stephen Duggan, 13 March, 1939, Series V: General Correspondence, Box 185, Folder 6: Veblen, Oswald (1934–1944), Emergency Committee Papers.

109. Richardson to Veblen, 31 December, 1938, Box 11, Folder: Richardson, R.G.D. 1938, Veblen Papers. The quotation that follows in this paragraph is also from this letter.

Springer-Verlag in 1931.[110] An international undertaking from the start, the *Zentralblatt* had sought to engage not only German mathematicians but also mathematicians from other countries, and particularly from the United States, in surveying current mathematical literature. Ferdinand Springer, the company's owner, wanted his new publication venture—unlike the venerable German abstracting journal, the *Jahrbuch über die Fortschritte der Mathematik* begun in 1868—to bring out its abstracts in a timely fashion and to cover the literature in the field more fully.[111]

Neugebauer had been particularly anxious to enlist American support. As he saw it, "mathematical production [had] increased considerably in America in the recent past," so it was only natural actively to engage the members of that thriving mathematical community.[112] As early as March 1931, he had approached Veblen about "the idea of an American branch of the *Zentralblatt*," that is, a well-placed cadre of Americans on the editorial board to coordinate coverage of what were acknowledged as the rapid developments on the western side of the Atlantic.[113] This was yet another sign of America's growing international presence.

Veblen was enthusiastic and recommended that Neugebauer also contact Bliss as a possible board member. Although "much interested in the *Zentralblatt*" and with no doubt "that cooperation with the Germans in the publication of the abstract journal is our best policy," Bliss declined Neugebauer's invitation on the grounds that it would be likely "to eat into one's time and energy more than" anticipated.[114] Quite naturally, Neugebauer next approached his consultant—and perhaps the American he would most have

110. Siegmund-Schultze chronicles the early history of this journal in " 'Scientific Control' in Mathematical Reviewing and German-U.S.-American Relations between the Two World Wars," *Historia Mathematica* 21 (1994), 306–329, especially pp. 317–323. On Neugebauer's life in general, and on his association with the *Zentralblatt* in particular, see Noel Swerdlow, "Otto E. Neugebauer (26 May 1899–19 February, 1990)," *Proceedings of the American Philosophical Society* 137 (1993), 138–165, especially pp. 148–151.

111. Siegmund-Schultze gives much more on the history of this journal in *Mathematische Berichterstattung in Hitlerdeutschland: Der Niedergang des 'Jahrbuchs über die Fortschritte der Mathematik'* (Göttingen: Vandenhoeck & Ruprecht, 1993).

112. Neugebauer to Wilhelm Blaschke, 19 March, 1938, Box 9, Folder: Neugebauer, Otto 1932–38, Veblen Papers (my translation). See also Siegmund-Schultze, " 'Scientific Control' in Mathematical Reviewing," p. 322.

113. Neugebauer to Courant, 3 March, 1931, as translated and quoted in Siegmund-Schultze, " 'Scientific Control' in Mathematical Reviewing," p. 318.

114. Bliss to Veblen, 28 November, 1932, Box 3, Folder: G. A. Bliss 1923–40, Veblen Papers.

liked to have had as an active collaborator, anyway—Veblen. Thinking aloud in a letter to a Courant then still in Göttingen, Veblen confessed that he was "very strongly tempted to accept the offered position" since, in his view, "Neugebauer is doing a very good editorial job and so I should feel proud to be associated with his journal."[115] Still, he initially hesitated for the same reason Bliss had. Neugebauer finally secured Oliver Kellogg to serve on the *Zentralblatt's* editorial board in 1931, but following his untimely death in 1932 at the young age of fifty-four, Kellogg was replaced by Jacob Tamarkin in 1933. Veblen ultimately did come on board, but only in 1937.

Tamarkin initially weathered the geopolitical storms that began to buffet the *Zentralblatt* beginning in 1933 with Hitler's rise to power. In that year, Neugebauer was asked—and refused—to sign a loyalty oath to the new regime. When this cost him his position at Göttingen, he relocated to the University of Copenhagen with the help of his friend, Harald Bohr. From there, he continued to perform his editorial duties largely unencumbered, although, interestingly, he and Veblen had been in discussion off and on about the journal's possible move to the United States since at least the summer of 1936. As Veblen explained, they had agreed at that time on "the desirability of doing nothing to disturb *Zentralblatt* unless we are forced to."[116] That eventuality unfortunately became a reality in October 1938, when Neugebauer realized that, given its German roots, it would no longer be possible to conduct the business of the journal in the necessary spirit of international cooperation.

Mathematicians back in Germany like Wilhelm Blaschke had been lobbying hard to force the *Zentralblatt* to conform to Nazi policies, and in October 1938, they finally succeeded in a one-two punch. They convinced Springer-Verlag to strike the name of the Italian Jewish mathematician Tullio Levi-Civita from the list of editorial board members, without consulting Neugebauer in advance. They also demanded that Neugebauer "give a binding promise that, in future, articles written by Germans would not be assigned to emigrés for review."[117] This action at a distance was the last straw, given

115. Veblen to Courant, 6 December, 1932, Box 4, Folder: Courant, Richard 1923–38, Veblen Papers.

116. See Swerdlow, p. 148 as well as Veblen to Neugebauer, 15 April, 1937 (for the quotation), Neugebauer to Veblen, 2 April, 1937, and Veblen to Neugebauer, 19 April, 1938, the latter all in Box 9, Folder: Neugebauer, Otto 1932–38, Veblen Papers.

117. C. Raymond Adams to the Members of Council of the American Mathematical Society, undated (but sometime shortly after 6 May, 1939), Box 6, Folder 12: AMS–Committee on Abstract Journal (1937–1938), Whyburn Papers. See also Swerdlow, pp. 148–150.

that "émigré" was a code word for "Jew."[118] Neugebauer tendered his resignation from the Zentralblatt's editorship effective 1 December, 1938, and his example was followed by all of the American-based members of the editorial board in addition to Hardy in England and Harald Bohr in Denmark. It also almost immediately prompted Veblen to engage vigorously in discussions in the United States about "starting a new abstracting journal in this country and importing Neugebauer to edit it."[119]

Those discussions were one of the reasons for the AMS's Council's lengthy deliberations in Williamsburg. As Veblen reported to Courant, by then a fellow colleague in the American mathematical community, the idea "seems to be well received by all those to whom I have spoken about it."[120] Still, he had to acknowledge that it was by no means universally embraced. Arthur Coble, AMS President in 1933 and 1934, voiced the major concern: money. He remembered well the various initiatives in the 1920s and earlier 1930s to get "the financial affairs of the Society on a sound foundation," and he, for one, could not see his way clear to take on yet another publication unless "the Society had the assurance that some responsible agency would take care of possible deficits."[121]

Other, less savory undercurrents of dissent also flowed. Richardson had predicted that "political, religious, and racial questions ... were bound to come to the surface," and surface they did.[122] In describing the Council's deliberations to Veblen, Richardson noted that "[o]ur beloved friend Wiener exploded a few times, but when he was sat back on his haunches by R. L. Moore and [Warren] Weaver, he took it good naturedly. I am sure you would have been amused by the discussion, if you could feel that your own projects [like the EC's émigré placement efforts] were not too much involved."[123]

118. Compare Siegmund-Schultze, "'Scientific Control' in Mathematical Reviewing," p. 322.

119. Veblen to Courant, 22 November, 1938, Box 4, Folder: Courant, Richard 1923–38, Veblen Papers. Courant had continued to serve on the Zentralblatt editorial board following his move to the United States in 1934.

120. Veblen to Courant, 22 November, 1938, Box 4, Folder: Courant, Richard 1923–38, Veblen Papers.

121. Coble to C. Raymond Adams, 12 January, 1939, Box 6, Folder 13: AMS–Committee on Abstract Journal (1939), Whyburn Papers.

122. Richardson to Veblen, 31 December, 1938, Box 11, Folder: Richardson, R.G.D. 1938, Veblen Papers. The quotation that follows is also from this letter.

123. Although the exact cause of the fireworks is not clear, two things are well-known: 1) in the immediately prewar years Wiener was, by his own admission, "subject to a great number of

All of this debate finally resulted in the naming of an ad hoc committee—chaired by C. Raymond Adams and comprised of George Birkhoff, Coble, Thornton Fry, Morse, and Gordon Whyburn—to study the issue. These men were by no means chosen at random. Richardson had given serious thought to the politics of the matter and particularly to the need for the "right" constituencies to be aware of the committee's deliberations and to be involved behind the scenes. In articulating his reasoning to Veblen, Richardson provided a telling snapshot of the American mathematical community as the 1930s drew to a close:

> What would you think of the following names: Morse (who could keep in touch with you all at Princeton), Birkhoff (to represent Harvard and to put the brakes on at appropriate and inappropriate places), C. R. Adams (as vice-president and as representing our local situation at Brown), Coble (as a mid-westerner whose judgment would be welcome), G. T. Whyburn (as a younger man of promise and in touch with the Polish School), Fry and Weaver (to help delimit the field in the direction of applied mathematics and to keep us in touch with important related groups)?[124]

As this list makes clear, Richardson deemed it essential for the groups in Princeton and at Harvard to be represented. They, after all, formed two of the three leading centers in the country. Richardson also wanted an inside track, so having someone on the committee from his own department at Brown was critical. Then, there were the regional concerns. At Illinois, Coble could represent the Midwest, although Veblen, in the margin of Richardson's letter, suggested Albert, Bliss, and Lawrence Graves, all at Chicago, the other top-three center. Whyburn, who, as noted in chapter four, was in the process of building a research department at the University of Virginia, could represent a growing AMS constituency in the South. Whyburn would also serve the purpose of providing insight into and contact with a key foreign constituency that could ultimately prove useful to the endeavor: the Polish topologists. As for the West, Evans at Berkeley would officially assume the AMS presidency on 1 January, 1939, so Richardson perhaps figured that Evans's very position would provide sufficient representation from that coast, since, in practical

---

separate emotional strains," among them, the angst he felt over the plight of would-be Jewish émigrés as he strongly advocated for them and 2) R. L. Moore, for one, did not see eye-to-eye with Wiener on the matter. See Wiener, *I Am a Mathematician*, p. 212.

124. Richardson to Veblen, 18 December, 1938, Box 11, Folder: Richardson, R.G.D. 1938, Veblen Papers. The quotation that follows is also from this letter.

terms, those on the West Coast were just "too far away to be useful in a hurry-up job." Finally, the inclusion of Fry, an applied mathematician employed by Bell Laboratories then located in New York City, and Weaver, another applied mathematician who had assumed the directorship of the Rockefeller Foundation's Division of the Natural Sciences in 1932, demonstrates the AMS's continuing concern about its relationship to both applied mathematics and the foundations on which it had come to count for support.[125]

This carefully chosen committee confronted a three-pronged charge: "to ascertain whether the time is favorable for starting an abstract journal in America under the auspices of the Society, whether international cooperation for such a venture can be obtained, and whether money is available from sources outside the Society for an initial period of five years."[126] For its part, the Council gave the committee the power "to set up such a journal for an experimental five-year period" provided "studies show that conditions in these respects are favorable." In Richardson's view, "this is as important a matter as has come to the attention of American mathematicians in decades."[127] After all, abstracting journals had long provided mathematicians with a means of keeping abreast of mathematical developments elsewhere, and that was particularly critical to an American mathematical community that viewed itself as poised for international leadership.

While the committee got to work immediately, it was building on efforts already under way by Veblen and others both to establish an abstracting journal in the United States and to find a permanent position for Neugebauer there.[128] Veblen had already secured fellowship support for him from Weaver at the Rockefeller Foundation, provided a university home could also be found, and Veblen had put out feelers to Princeton, Columbia, and Brown to that end. By 20 December, just days before the committee was constituted, Veblen was able to write to George Mullins, Professor of Mathematics at

---

125. Weaver, although ultimately not an official member of the committee, was mentioned along with Veblen as having "cooperated with the committee" (Adams to the Members of Council of the American Mathematical Society, undated (but sometime shortly after 6 May, 1939)).

126. This and the next two quotations are on p. 203 of Temple Hollcroft, "The Annual Meeting of the Society," 45 BAMS (1939), 197–208 on p. 203. See also Pitcher, A History of the Second Fifty Years, pp. 69–70.

127. Richardson to the Committee on Abstract Journal, 9 January, 1939, Box 6, Folder 13: AMS–Committee on Abstract Journal (1939), Whyburn Papers.

128. Box 9, Folder: Neugebauer, Otto 1938–48, Veblen Papers contains the extensive correspondence.

Columbia's Barnard College, that "the way seems clear [for Neugebauer] at Brown."[129] By January 1939, Neugebauer had accepted Brown's offer and had made arrangements to come to the United States for a preliminary ten-week stay beginning in mid February.[130]

Veblen had also been in direct contact with Ferdinand Springer, sounding him out about, among other things, the possibility of the AMS actually purchasing the *Zentralblatt*. Recognizing the import of such a suggestion and trying his best to maintain good relations with the American mathematical community, Springer proposed early in January to send Friedrich Karl Schmidt to the United States as his representative to talk over the various issues face-to-face just as soon as such a trip could be arranged.[131] Schmidt, who had replaced Courant as editor of Springer's famous "Yellow Series," was a confidant of Springer's, a friend of the émigré Courant, and a political moderate who was less than sympathetic to the hard-line Nazi agenda. He was thus well-suited to serve as a negotiator.[132]

Meanwhile, the AMS-appointed committee was doing its best to determine whether the *Zentralblatt* would—or could—move to the United States with Neugebauer, or whether the AMS would begin its own, new abstracting journal. As early as 9 January, 1939, Adams and Birkhoff had drafted a letter to Springer on the committee's behalf in which they proposed to ask if he "found [him]self in a position to assure us that the policy of the *Zentralblatt* will for a substantial period of time remain as it has been except for the requirement that abstracts of articles under German authorship are not to be reviewed by German emigrés."[133] Richardson had been smart to assure that he and Veblen were as directly plugged into the committee as non-committee members could be. Both were concerned, on reading the draft, that it would

129. Veblen to Mullins, 20 December, 1938, Box 9, Folder: Neugebauer, Otto 1938–48, Veblen Papers.

130. Swerdlow, p. 150.

131. Reingold, p. 193 (in the reprinted edition).

132. Wilhelm Süss, President of the Deutsche Mathematiker-Vereinigung and a committed Nazi, led a failed campaign to prevent Schmidt from journeying to the United States in this capacity. Volkert Remmert details Schmidt's attempts to steer mathematical publishing during the Nazi period in "Mathematical Publishing in the Third Reich: Springer-Verlag and the Deutsche Mathematiker-Vereinigung," *MI* 22 (2000), 22–30, especially pp. 24–26.

133. Quoted in a letter from Richardson to Veblen, 18 January, 1939, Box 11, Folder: Richardson, R.G.D. 1939, Veblen Papers. The quotations that follow in this paragraph are also from this letter. Additional drafts of a possible letter to Ferdinand Springer are contained in Box 18, Folder: American Mathematical Society, Committee to consider establishing an abstracting

be interpreted as tacitly condoning Springer's action relative to Levi-Civita as well as the publisher's new Nazi-inspired policy. Richardson, in fact, had begun to rue some of his suggestions for members of the committee. "Doubtless you will say," he wrote wistfully to Veblen, "that with ultra-conservatives like Birkhoff and Coble on the committee I have got myself (or rather this whole question) into a mess." Still, he had "faith that arguments will prevail."

That faith proved justified, for by 18 January another draft, written this time by Fry, had been circulated. In that version, the committee diplomatically and gratefully acknowledged the service to the international community of mathematicians provided by Springer-Verlag through its publication of the *Zentralblatt*. It nevertheless informed Springer of the AMS's intentions to launch its own abstracting journal in light "of the recent change of editorial policy of the *Zentralblatt*," a policy that, "in all probability," would prevent the journal from "provid[ing] an unbiased critical review of the world's mathematical literature."[134]

The committee continued to argue about suitable wording for a letter to Springer, and, indeed, about the suitability of writing to Springer at all.[135] Finally, after all of the back-and-forth in January, a letter very different from the one earlier contemplated was sent to Springer on 13 February, 1939. It simply informed the publisher that the AMS had formed a committee "to study the abstract journal situation" and that "the Society ha[d] empowered this committee to proceed with the inauguration of an abstract journal under the sponsorship of the Society for a preliminary period, in case certain conditions are satisfied." It closed with the committee's assurance of "the friendliest feelings for their colleagues in Germany and for the firm of Julius Springer" and of its eagerness to discuss "problems of mutual interest" with Springer or his emissary.[136]

---

journal in America 1938–39, Veblen Papers. The latter folder also documents the opinions— almost all in support of establishing an abstracting journal in the United States—of many mathematicians in both the United States and Europe.

134. "Draft" (of the Committee on Abstract Journal) to Ferdinand Springer, 18 January, 1939, Box 6, Folder 13: AMS–Committee on Abstract Journal (1939), Whyburn Papers.

135. Despite trying his hand at a draft, Thornton Fry consistently argued against the wisdom of writing to Springer. See, for example, Fry to Adams, 10 January, 1939, Box 18, Folder: American Mathematical Society, Committee to consider establishing an abstracting journal in America 1938–39, Veblen Papers.

136. Adams to Ferdinand Springer, 13 February, 1939, Box 18, Folder: American Mathematical Society, Committee to consider establishing an abstracting journal in America 1938– 39, Veblen Papers.

With the matter of the letter settled, the committee prepared for a meeting scheduled for 24 February in New York at which Neugebauer would be present to answer questions. In its subsequent report to the AMS Council, the committee not only announced that the Carnegie Corporation would pledge $60,000 "for the support of 'an international mathematical journal'" should one be launched but also recommended that the AMS contribute $1,000 a year for five years toward such a journal's expenses.[137] Obviously, this was far from suggesting that the AMS underwrite a new publication venture, but, in approving the recommendation, the Council took a tentative first step toward greater AMS involvement in the business of international abstracting.

By April, the Council had met in Durham, North Carolina, and had directed the committee to provide it with a concise statement of the issues, as it saw them, on or before 1 May. A decision, one way or the other, needed to be made. In the meantime, arrangements had finally solidified for Schmidt's visit to the United States as Springer's representative, and an appointment was scheduled with him and the committee—to be augmented by Veblen, Richardson, and Weaver (who, finally, was not an official committee member)—in New York City on 6 May. Clearly, the committee would need to take that meeting before issuing its statement to the Council. To make sure that everyone was on the same page, moreover, Adams provided a detailed account of the exchange with Schmidt in a letter to Coble, who had been unable to make the trip east from Champaign-Urbana.

Adams opened by describing Schmidt as a "thorough gentleman" and by acknowledging that all present appreciated "the difficulty and embarrassment" of his position.[138] He then summarized the main points that Schmidt had communicated. The German had first explained that it would be impossible for Springer to sell the *Zentralblatt*, as had been suggested, since "both the German mathematicians and those in governmental positions would regard it as a disgraceful procedure for him to withdraw his hand from the plow in

137. For the quotation, see Pitcher, *A History of the Second Fifty Years*, p. 70. Veblen had approached the President of the Carnegie Corporation with a proposal as early as 26 November, 1938 (Veblen to Frederick Keppel, 26 November, 1938, Box 6, Folder 12: AMS–Committee on Abstract Journal (1937–1938), Whyburn Papers). In 2021 dollars, $60,000 is equivalent in "real wealth" to some $1,153,000 (https://www.measuringworth.com).

138. Adams to Coble, 8 May, 1939, Box 6, Folder 16: AMS–Committee on Colloquium Publications (1941–1943), Whyburn Papers. The quotations that follow in this paragraph are also from this letter.

which it has been set." Relative to the new restriction on reviewers, moreover, Schmidt stressed that it "is solely that papers under German authorship must not be reviewed by German emigrants," but he was quick (too quick?) to emphasize that that "distinction . . . is entirely between German emigrants and others, not at all between Jew and Gentile." Schmidt contextualized his explanation this way: "the German idea is that mathematics, like everything else, exists in a real world in which political considerations play a part; and that, like everything else, mathematics must expect to be affected in some measure by political considerations."

Following this discussion and a collegial dinner together with their German colleague, the committee went into a closed session that made it clear that nothing had been heard that had changed minds already made up. Some were still in favor of a new journal; some were opposed. A draft report was circulated and discussed, but dissension persisted. In the end, Adams and Weaver were charged with writing what would be the final report, one that would reflect the various points made and positions taken.

Sometime after this 6 May meeting, that report went to the Council. It provided evidence—gathered from a massive letter-writing campaign—of support, both foreign and domestic, for an American abstracting journal. It also gave a detailed analysis of the sobering financial implications of such a venture, estimating an annual deficit of some $10,000 per year when income from sales and subscriptions was subtracted from the total production costs. Indeed, at an annual estimated cost of over $16,000, it would be "a project whose annual budget will be about 50% of the Society's present yearly expenditures."[139]

Nor did the report fail to lay out the various objections—in addition to these purely financial ones—that had been raised during the course of the committee's deliberations and fact-finding. First and foremost among them was the issue of launching a journal that would be in direct competition with the *Zentralblatt* and that would thus be viewed as a move antithetical to a spirit of international cooperation. Arnold Dresden had captured this sentiment clearly when he had written to Richardson early in January that "[t]he worst thing that could happen . . . would be the establishment of two competing abstract journals, each with an incomplete international board, reflecting in mathematics the division of the 'civilized world' into two hostile

---

139. Adams to the Members of Council of the American Mathematical Society, undated (but sometime shortly after 6 May, 1939).

camps."[140] Moreover, given that the Americans were in the process of organizing what they hoped would be a highly successful ICM in 1940, this was
not a criticism to be taken lightly. They wanted strong participation from
abroad. Their goal of showcasing the strength of the American mathematical
endeavor could not be achieved otherwise.

On the basis of both the committee's findings and its careful consideration
of them, however, the report concluded that

> if an abstract journal is to be started in this country, now is a time at which
> it would be possible to secure a good deal of cooperation by capitalizing
> on the general feeling of irritation toward Germany which is so prevalent
> outside of the sphere of German influence. Statements . . . seem to indi-
> cate that in some minds there is a definite desire for haste in order that an
> effective and well-understood gesture of protest against recent happenings
> in Germany may be made by mathematicians in America.[141]

The committee closed by asking each member of Council to weigh in with an
opinion—essentially "yes" or "no"—on the question of whether to go forward
with the establishment of a new abstracting journal. On 25 May, Richardson
informed the senior Birkhoff, who had remained steadfast in his opposition
to an American competitor to the *Zentralblatt*, that the Council had voted
twenty-two for, five against, and four undecided.[142] As Richardson proudly
announced to the AMS membership, "the Society establishes a new periodical," and Veblen was charged to chair the committee (also consisting of Fry
and Weaver) that would set it up.[143]

The *Mathematical Reviews* began operations immediately, in the summer of
1939, with Neugebauer and his colleague, Tamarkin, at Brown, as executive
editors. In addition to the $60,000 subvention from the Carnegie Corpo-
ration, the $1,000 per year for five years from the AMS, and $1,000 for
the first year and $500 per year for four years thereafter from the MAA,

140. Dresden to Richardson, 4 January, 1939, Box 18, Folder: American Mathematical Society, Committee to consider establishing an abstracting journal in America 1938–39, Veblen
Papers.

141. Adams to the Members of Council of the American Mathematical Society, undated
(but sometime shortly after 6 May, 1939).

142. Richardson to Birkhoff, 25 May, 1939, as recounted in Reingold, p. 195 (in the
reprinted edition).

143. Roland Richardson, "The Society Establishes a New Periodical," BAMS 45 (1939),
641–643 on p. 641.

the venture started off with $12,000 from the Rockefeller Foundation and $3,000 from the American Philosophical Society. Its first volume appeared in January 1940, contained 2120 reviews by 350 reviewers, and went to 700 subscribers, 200 more than the *Zentralblatt*. At 400 pages, it was an impressive achievement, especially given the six-month time frame in which it had been produced.[144]

The founding of the *Mathematical Reviews* represented a turning point for the American mathematical community.[145] Not only had it categorically assumed a leadership role internationally, it had plunged itself even farther into the geopolitics of the late 1930s. In a letter to Courant in November 1938, just as the *Zentralblatt* issue was really heating up, George Birkhoff had confided that "I dislike intensely mixing up political and scientific matters."[146] This issue of abstracting was just the start of such an admixture. Less than a year later, Birkhoff's vision for the first American ICM had also fallen victim to geopolitics. And, in the summer of 1940, the American mathematical community had actually begun preparing for war.

———

The last half of the 1930s and the 1940s found an American mathematical community focused on assuming a leadership role on the international stage. It worked, in particular, to bring the International Congress of Mathematicians to the United States in 1940 and, in so doing, to make manifest, especially to the Europeans, the quality of its mathematical output. That ICM promised, once and for all, to establish the United States as the mathematical equal of, as Birkhoff saw it, "the whole of Europe."[147]

144. For the numbers, see Swerdlow, p. 151; Pitcher, *A History of the Second Fifty Years*, p. 73; Elton Moulton, "Mathematical Reviews," *AMM* 46 (1939), 523; and Boyer, p. 46.

145. Interestingly, in his otherwise excellent account of Springer-Verlag during what he termed the "Years of Danger (1933–1945)," Heinz Sarkowski did not mention the formation of the *Mathematical Reviews* as one of the many consequences of Nazi policies on the publishing house (Heinz Sarkowski, *Springer-Verlag: History of a Scientific Publishing House*, part 1, trans. Gerald Graham (Berlin: Springer-Verlag, 1996), pp. 325–385). The point *is* made in Michael Knoche, "Scientific Journals under National Socialism," *Libraries & Culture* 26 (1991), 415–426 on pp. 422–423.

146. Birkhoff to Courant, 28 November, 1938, Box 4, Folder: Courant, Richard 1923–38, Veblen Papers.

147. Birkhoff to Veblen, 17 February, 1927, Box 2, Folder: Birkhoff, George D. 1912–47, Veblen Papers.

The outbreak of World War II in Europe in the fall of 1939 left that promise unfulfilled, but two events—totally unforeseen and totally out of the control of American mathematicians—gave the American mathematical community the opportunity to assert its leadership in other ways: the absorption of refugee mathematicians into a community that had matured over the course of the first three decades of the twentieth century and the assumption of the responsibility for overseeing and coordinating the abstracting of the fruits of mathematical research internationally. Political events in Germany in the 1930s had caused both the mathematical exodus and the undermining of faith in the Germans' ability objectively and comprehensively to survey the field. In both cases, the American mathematical community was able to step into the breach. Its networks were in place from coast to coast thanks to the establishment of professional organizations, both research- and teaching-oriented. It had attained a critical mass of active and talented researchers in a range of subfields. It had developed the resources—positions at colleges and universities, journals, philanthropic and other financial support—not only to provide for its material needs but also to seize new opportunities as they arose.

Into this environment came more than 120 émigrés. Many of them—algebraists, geometers, analysts, topologists—found their areas of mathematical expertise well-developed in their new home, providing a ready-made audience for their work. Some, however, did not. Probabilists, statisticians, more applied mathematicians found only scattered pockets of expertise. They had the potential, at least, meaningfully to contribute to the diversification of the American mathematical community, especially in time of war. Yet, just how—or if—that community would be able to contribute to a war effort remained to be seen.

1941–1950: The "center of gravity of mathematics has moved more definitely toward America."

—Roland Richardson, 25 April, 1939

8

Waging War

The generational shift that was already under way as the first European mathematical émigrés reached the United States in the early 1930s was largely
complete as the American mathematical community first prepared for and
then engaged in World War II. Vassar's Henry White, Wisconsin's Edward Van
Vleck, and Harvard's William Osgood had all been part of that first generation
of American research mathematicians who had sought advanced training in
Germany, and particularly in Klein's lecture hall in the closing decades of the
nineteenth century. They had each brought what they had learned back home
as they pursued their research and worked to create the infrastructure of the
American mathematical research community. Each died in 1943.

When the United States entered World War II, the members of the second generation who had been so instrumental in their community's growth in
the 1920s and 1930s—among them, the senior Birkhoff, Veblen, Richardson,
R. L. Moore, Bliss, Lefschetz, Dickson, Hedrick—had already or were about
to turn sixty. Dickson had retired from Chicago in 1939; Bliss followed him
two years later in 1941; Hedrick and Birkhoff died unexpectedly at ages sixty-
six and sixty in 1943 and 1944, respectively. Although other members of the
second generation—Veblen, Lefschetz, and Richardson, to name just a few—
continued to play vital roles in American mathematics, it was largely members
of the third generation—Griffith Evans, John Kline, Marston Morse, émigré
John von Neumann, Marshall Stone, and others—who saw the community
through the war years. These, and perhaps especially Morse and Stone, were
the new mathematicians as politicians who had assumed that mantle from
Birkhoff and Veblen in the 1940s.

The United States' entry into World War I in April 1917—almost three
years into the fighting in Europe—had caught America's scientists, in general, and its mathematicians, in particular, off guard. Mechanisms that would

have allowed them to contribute quickly and effectively to a war effort had not been in place. That wake-up call, as noted, had resulted in the founding of the National Research Council within the National Academy of Sciences as a permanent organization for the encouragement and support of research. America's mathematicians had worked successfully within that context as well as within the AMS and the MAA to foster research in addition to teaching at their respective colleges and universities throughout the 1920s and 1930s. They had also demonstrated the strength of their resolve and of their lines of communication in dealing with the influx of mathematical refugees. Unlike at the outbreak of the First World War, they were part of a mature—as opposed to to a newly formed—mathematical community. Moreover, as the Second World War loomed, they were focused beyond their national borders on the international scene and their role in it. This time they would not be caught unawares. Starting in 1939 and through to the end of the war in September 1945, the American mathematical community vigorously engaged in service to a nation in wartime.

Mobilizing American Mathematics for War

Evans assumed the AMS presidency in January 1939 at a moment when the situation in Europe looked increasingly dire. Already active in the effort to compile, through the NRC, a *National Roster of Scientific and Specialized Personnel,* he immediately pushed for the concurrent formation of a War Preparedness Committee (WPC) as a joint initiative of the AMS and the MAA.[1] As he and MAA President Walter Carver understood, in time of war, technical expertise might well be called upon to solve particular mathematical problems, but basic mathematical training would most definitely be needed—as it had been some twenty-five years earlier—for those like gunnery recruits who would have to perform routine calculations. When Morse, chair of the duly created WPC and soon-to-be President-elect of the AMS, delivered his committee's organizational report to the AMS Council on 9 September, 1940, it reflected a keen awareness of the challenges organized mathematics would likely face in wartime.

1. I have also discussed the American mathematical war mobilization efforts in Parshall, " 'A New Era,' " pp. 295–298. On the early days of the roster, see Leonard Carmichael, "The National Roster of Scientific and Specialized Personnel: A Progress Report," *Science* 93 (1941), 217–219.

Morse was an excellent choice for chair. A man of "vast energy," he had served with distinction in the U.S. Army's Ambulance Corps in World War I, winning the Croix de Guerre with Silver Star, and had returned from Europe to pick up the career in mathematics he had begun as a doctoral student under George Birkhoff.[2] Another star, his in American mathematics, had consistently risen throughout the 1920s and 1930s as a result of his innovative work on the calculus of variations in the large. He was literally "starred" in the fourth edition of Cattell's *American Men of Science* in 1927; he had been invited to give an hour address at the ICM in Zürich in 1932; and he had won one of the two Bôcher Prizes awarded by the AMS a year later. His stature as one of the American mathematical research community's leaders was further cemented in 1935 with his appointment to the permanent faculty of the Institute. Like Veblen and Birkhoff, moreover, Morse was committed to the common good of his community, serving, for example, as a representative for mathematics on the National Research Council from 1934 to 1937. He had learned his way around Washington's scientific circles.

Morse's WPC isolated three well-defined goals: 1) "[t]he solution of mathematical problems essential for military or naval science, or rearmament, 2) [t]he preparations of mathematicians for research essential for the preceding objective, and 3) [t]he strengthening of undergraduate mathematical education to a point where it affords adequate preparation in mathematics for military and naval service of any nature."[3] Three subcommittees—on Research, on Preparedness for Research, and on Education for Service—were promptly formed to determine how best to realize these objectives.

Dunham Jackson chaired the first—Subcommittee on Research—that also included McShane, Stone, Wiener, Wilks, and Stone's Harvard colleague, the physicist John (son of Edward) Van Vleck.[4] Conceived as a kind of mathematical clearinghouse, this subcommittee would receive problems from the various military branches and would then oversee the assignment of mathematical consultants—either from the committee itself or from the

2. The characterization is from Bott, p. 908.

3. "Minutes of Council, 9 September, 1940," Ms.75.2, Box 15, Folder 29: War Preparedness Committee, AMS Papers. Morse also reported on his committee's work more publicly in "Report of the War Preparedness Committee of the American Mathematical Society and Mathematical Association of America at the Hanover Meeting," *BAMS* 46 (1940), 711–714 on p. 711.

4. William Ayers lists these and the subcommittee members below in "The Annual Meeting of the Society," *BAMS* 47 (1941), 175–187 on p. 182.

FIGURE 8.1. Marston Morse (1892–1977) in 1965. (Photo courtesy of Archives
of the Mathematisches Forschungs institut Oberwolfach.)

broader community of experts—to work on the necessary solutions. This
kind of arrangement, however, seemed to overlook matters like security
clearance, which would have to be addressed in advance for it to work.
The second—Subcommittee on Preparedness for Research—was formed in
acknowledgment of the fact that the predominantly purist American mathematical research community had, by and large, not explicitly fostered applied
mathematics, despite recurrent calls from those of influence like Veblen
and despite the presence in the community of isolated applied mathematicians like Stephen Timoshenko at Stanford, Ivan Sokolnikoff at Wisconsin,
Theodore von Kármán at Caltech, and others. Under Stone as chair, this
committee also included Columbia's Bernard Koopman, Sokolnikoff's colleague Rudolph Langer, émigré and Berkeley professor Hans Lewy, Frank
Murnaghan of Hopkins, and Princeton's Bob Robertson. Its charge was to

assess the availability and depth of applied mathematical training and then to coordinate instruction in key courses. Finally, the third—Subcommittee on Education for Service—canvassed, under the direction of its chair, Minnesota's William Hart, the nation's service academies to survey the mathematical materials then in use there and to recommend how best to structure mathematical preparation "for workers in government and industry, and for officers and enlisted men in the Army and Navy."[5]

A select group of "chief consultants"—designated for each of six areas deemed critical—supplemented the subcommittees. Caltech's Harry Bateman would field questions specifically concerning aeronautics; von Neumann would be in charge of ballistics; Yale's Howard Engstrom would handle cryptanalysis; Thornton Fry of Bell Labs would use his connections to coordinate with industry; Wiener would deal with mechanical and electrical aids to computation; and Wilks would treat matters involving probability and statistics. At least as best they could foresee, applications of mathematics to the war effort would arise in these areas.

Three months after the formation of this infrastructure, in December 1940, the WPC delivered its second report. It had been making progress. As early as June, the committee had begun approaching key members of the military and had offered its assistance in any mathematical problem-solving that might arise. For example, in a letter dated 18 June, 1940, General George Marshall, then the Chief of Staff at the War Department, thanked the WPC for its willingness to help and predicted that "the Air Corps, Ordnance Department, Signal Corps, Engineering Corps and Coast Guard may find your services of definite value in the field of ballistics, fire control methods, research and development."[6]

If Marshall's response was somewhat perfunctory, that of Colonel W. S. Bowen of the Coast Artillery Corps reflected an astute read of the American mathematical community. "Your letter indicates," he wrote, "that you are primarily concerned with the problem of peacetime education of your men,

5. William Hart, "On Education for the Service," *AMM* 48 (1941), 352–362 on p. 354. The other members of Hart's subcommittee were Richard Burington of the Case School of Applied Science (now Case Western Reserve University), Harvard's Julian Coolidge, Haskell Curry of Penn State, Berkeley's Elmer Goldsworthy, Frank Griffin of Reed College, Wisconsin's Mark Ingraham, and Elton Moulton of Northwestern.

6. Marston Morse, "Report of the War Preparedness Committee [24 December, 1940]," 26 pp., Ms.75.4, Box 27, Folder 80, AMS Papers. The quotations that follow in this paragraph are also from this report with my emphasis.

but that you are also concerned with research problems." Interestingly, he foresaw that it would be in "the latter field that we [would] have the greatest need for advanced mathematical analysis," and he anticipated, "that we might be able to indicate problems, the study of which will be of considerable interest and, at the same time, productive of valuable results." Bowen appreciated, in other words, that the mathematicians' actual research skills could not only positively affect a war effort but also generate new, intellectually engaging results. (For some examples of this, see chapter ten.)

The WPC's individual subcommittees had also been hard at work. Hart's group had been gathering and "studying military and naval manuals and related texts," while Stone's "ha[d] been busy with organization work and with the collection of preliminary data and opinions."[7] To Morse's way of thinking, Stone's committee had particular potential; it might finally effect the attitudinal change needed in order actively to foster applied mathematics in the United States.[8]

Morse elaborated on this point in a talk on "Mathematics in the Defense Program" that he gave in February 1941, just weeks after he assumed the AMS presidency. Interestingly, he chose to address his remarks not to his fellow research mathematicians in the AMS but rather to the precollegiate mathematics teachers represented by the National Council of Teachers of Mathematics.[9] They were in a position to orient students at a more formative, younger age. As Morse saw it, "North America leads the world in pure mathematics. We are also strong in the simpler applications appearing in ordinary engineering or industrial practice; but we have preferred experiment to theory and have tended to use the laboratory to obtain results which might

7. Ibid.

8. On the development of applied mathematics in the United States in the twentieth century, see, for example, Amy Dahan-Dalmedico, "L'essor des mathématiques appliquées aux États-Unis: L'impact de la Seconde Guerre Mondiale," *Revue d'histoire des mathématiques* 2 (1996), 149–213 as well as Reinhard Siegmund-Schultze, "The Late Arrival of Academic Applied Mathematics in the United States: A Paradox, Theses, and Literature," *NTM International Journal of the History and Ethics of Natural Sciences, Technology, and Medicine* 11 (2003), 116–127 and "The Ideology of Applied Mathematics within Mathematics in Germany and the U.S. until the End of World War II," *LLULL* 27 (2004), 791–811.

9. The NCTM was founded in Cleveland, Ohio, in 1920 in an effort to give secondary school mathematics educators their own voice in matters such as the curriculum and curricular reform. Eileen Donoghue discusses this in "The Emergence of a Profession: Mathematics Education in the United States, 1890–1920," in *A History of School Mathematics*, ed. George Stanic and Jeremy Kilpatrick, 1: 159–193 on pp. 186–188.

have been predicted."[10] Although, in his view, "[w]e are beginning to correct this situation," it arose because of what he termed "our national suspicion of theory." "We are perilously lowbrow," he continued. "One result [of this] has been a lack of coöperation between the theoretically-minded scientist and the practically-minded scientist. The pure scientists have intensified their study of science for science's sake, and the applied scientists have adhered to 'common sense' and the laboratory. It is one of the problems of education to show that the more mature and socially-minded way is to respect both theory and practice, and particularly their combination."

Morse was not alone in trying to drive home this point to the broader mathematical audience. At Richardson's urging, Fry published the report he had made on "Industrial Mathematics"—part of a broader study on "Research–A National Resource" to the 77th Congress—in a supplement to the *American Mathematical Monthly*. In so doing, he took the case to another key constituency: college mathematics teachers.[11] "Though the United States holds a position of outstanding leadership in pure mathematics," he stated categorically, "there is no school which provides adequate mathematical training for the student who wishes to use the subject in the field of industrial applications rather than to cultivate it as an end in itself."[12] Making a direct plea, he went on to argue that "[b]oth science generally, and its industrial applications in particular, would be advanced if a group of suitable teachers were brought together in an institution where there was also a strong interest in the basic sciences and in engineering." The problem that both Morse and Fry highlighted was, after all, what had necessitated the creation of Stone's subcommittee in the first place, and it was precisely what both Courant's strategy at NYU and Richardson's own strategy at Brown (see the next section) aimed to address.

The WPC had recognized and made provisions to confront this issue of applied mathematics from its inception. What it had not foreseen was something much more fundamental to the American mathematical community's ability to aid in a war effort: the availability of mathematicians actually to do mathematical work as opposed to serving in combat roles. Congress had

10. Marston Morse and William Hart, "Mathematics in the Defense Program," *AMM* 48 (1941), 293–302 on p. 294. The quotations that follow in this paragraph are also on this page.

11. See Richardson to William Cairns, 25 June, 1940, Box 86-14/68, Folder 2, MAA Records and Thornton Fry, "Industrial Mathematics," *AMM* 48 (June-July 1941), 1–38. Although Morse's talk was given to the NCTM, it, like Fry's paper, was published in the MAA's *Monthly*.

12. Fry, p. 2. The quotation that follows may also be found here.

passed the Selective Training and Service Act on 16 September, 1940. In providing for the first peacetime draft in the country's history, it initially required all men between the ages of twenty-one and thirty-five to register with their local draft boards. Later in the fall of 1940, however, the American Council on Education, with the endorsement of General Lewis Hershey, Director of the Selective Service System, issued a bulletin advising local Selective Service Boards "to give special consideration to men in chemistry, physics, engineering, biology, et cetera, with reference to the desirability of granting 'individual occupational deferment.' "[13] Conspicuously absent from this list? Mathematicians. By June 1941, the WPC had been augmented by a new subcommittee on Supply and Demand for Mathematicians headed by UCLA's Tracy Thomas and charged with gathering the data that would make the case for the explicit occupational deferment of mathematicians. By 1943, that objective had been achieved thanks both to the work of Thomas's subcommittee and to Stone's lobbying efforts in Washington.

Stone, a son of the Chief Justice of the United States, was no stranger to the corridors of power in Washington. On leave from Harvard and home-based in the capital for much of the war, he aimed to use both his own and his father's connections in the service of mathematics as AMS President. His colleagues recognized the potential for mathematics of his insider's access to the Washington establishment. As the elder Birkhoff saw things at the close of 1942, "[i]t may well turn out that Marshall could 'do a job' for the A.M.S. and for American mathematics in quite an effective way in two or three months." Richardson concurred, reporting to Wisconsin's Mark Ingraham early in January 1943 that "Stone is taking hold in Washington to organize mathematics for the war effort and I think he will do a splendid job."[14]

That mathematicians had not been singled out as among those initially eligible for occupational deferment was a bitter pill.[15] The American mathematical community had, after all, been consistently laboring since the

13. Marston Morse, "Report of the War Preparedness Committee of the American Mathematical Society and Mathematical Association of America at the Chicago Meeting," *BAMS* 47 (1941), 829–831 on p. 830.

14. Birkhoff to Richardson, 29 December, 1942, Box 15: Correspondence, 1943–1945 'A-H,' Folder: Birkhoff, George D., Richardson Papers and Richardson to Ingraham, 2 January, 1943, Box 16: Correspondence, 1943–1945 'I-R,' Folder: Ingraham, Mark, Richardson Papers, respectively.

15. Interestingly, both Veblen and Morse opposed the move for occupational deferment for mathematicians. According to George Birkhoff, their attitude was one of " 'Let

FIGURE 8.2. Marshall Stone (1903–1989) in his office at the University of Chicago (see chapter 9) in 1952. (Photo courtesy of Hanna Holborn Gray Special Collections Research Center, University of Chicago Library.)

1920s to elevate its position within the American professional landscape. It had successfully become part of the NRC's fellowship program; it had attracted Rockefeller and Carnegie funding; it had put on its best face for the broader public at the "Century of Progress" World's Fair in Chicago in 1933.

---

the mathematicians take their chances along with the others.' " Birkhoff to Richardson, 29 December, 1942. They, however, were in the minority.

Why then had it been overlooked in 1940 and at such a critical historical juncture? Stone had an answer. In his view, it owed to "the failure of the various *government agencies* which are handling the scientific aspects of the present emergency to make effective use of the *mathematical skills* which are available in the country."[16]

On 27 June, 1940, and almost at the same time that the mathematicians were organizing their WPC, Franklin Roosevelt issued a presidential decree forming the National Defense Research Committee (NDRC) "to conduct research for the creation and improvement of instrumentalities, methods and materials of warfare" and so "to aid and supplement the experimental research activities of the War and Navy Departments."[17] Vannevar Bush, the then President of the Carnegie Institution of Washington and formerly professor of electrical engineering, Vice President, and Dean of the Engineering School all at MIT, had used his influence directly to lobby the President and to see himself installed at the head of the NDRC.

With the goal of effectively connecting the military with the civilian scientific community, the NDRC was comprised of Bush and seven strategically chosen men. Three came from outside academe: the Commissioner of Patents and one representative each from the Army and the Navy. The other four were among Bush's close friends: Karl Compton, a physicist and the President of MIT; James Conant, a chemist and the President of Harvard; Frank Jewett, a fellow electrical engineer and the President of both the National Academy of Sciences and Bell Labs; and Richard Tolman, a physicist and professor at Caltech. With engineers, physicists, and chemists in control, it is perhaps little wonder that mathematics per se did not really figure in the NDRC's agenda. Having engineers, physicists, and chemists in control, however, was exactly what Bush had wanted. "There were those," he acknowledged, "who protested that the action of setting up NDRC was an end run, a grab by which a small company of scientists and engineers, acting outside established channels, got hold of the authority and money for the program of developing new weapons. That, in fact, is exactly what it was."[18]

16. Stone to Gilbert Bliss and Griffith Evans, 15 September, 1941, Box 37, Folder 3: Scientific Mobilization, Stone Papers (my emphases).

17. Irvin Stewart, *Organizing Scientific Research for War: The Administrative History of the Office of Scientific Research and Development* (Boston: Little, Brown and Company, 1948), p. 8.

18. Vannevar Bush, *Pieces of the Action* (New York: William Morrow and Company, Inc., 1970), pp. 31–32.

A year after its establishment, the NDRC, which, as its name implied focused on war *research*, was subsumed under the broader Office of Scientific Research and Development (OSRD), also under Bush and also with the continued cooperation of his handpicked friends. Bush had quickly realized that while research was one thing, it was vitally important to be able to transform results in the abstract quickly and effectively into actual devices or procedures for use in the field. Being in control of both aspects of the process and having not only a direct pipeline to the President but also what was essentially an open Federal bank account on which to draw, the OSRD under Bush wielded great power and influence.[19]

Stone, for one, recognized how critical it was for mathematics to be part of this new Federal presence. The WPC had made, as early as 1940, what he later called a "first tentative offer of [its] services to the then newly created National Defense Research Committee," but "[a]part from receiving acknowledgments of its offer, the committee made little progress towards a more specific understanding of its potential usefulness in the national emergency."[20] The creation of the OSRD in June 1941 only heightened the sense of urgency and prompted Stone to approach Morse, in his capacity as AMS President, about the matter of establishing lines of communication between the WPC and what seemed to be a nascent yet quickly growing scientific bureaucracy. As Stone put it in a letter dated 15 September, 1941, he and Morse "reached agreement on the conclusion that it would be appropriate at this time for a small informal group of mathematicians to take counsel with one another and to make representations of a kind and manner most likely to produce desirable changes in the present unsatisfactory situation."[21] What he was proposing was the creation of a mathematical clique—of which he would, of course, be a member—that would devise a strategy for convincing the clique that was the OSRD that mathematics was critical to its objectives. This mathematical politicking—dealing directly with the Federal government—was thus of a kind different from what Veblen and Birkhoff had engaged in in the 1920s and 1930s.

Gilbert Bliss objected to Stone's proposed strategy. "The move to have them [mathematicians] given a hearing and whatever responsibility they can

19. Stewart, pp. 35–40.
20. Marshall Stone, "American Mathematics in the Present War," *Science* 100 (1944), 529–535 on p. 530.
21. Stone to Bliss and Evans, 15 September, 1941.

effectively take, should come through our mathematical societies, preferably through the Council of the A.M.S.," Bliss argued.[22] "It seems to me that the purpose of your informal committee should be first to stir up the Council," since "[a] committee appointed by Council to wait upon Bush, Conant, or Jewett should have much more influence than an informally self-selected one." Perhaps, but Stone, like Bush, saw the end run as quicker and potentially more effective in this case.

In 1941, it was just not yet clear how best to get the American mathematical community's message across in Washington, but figuring that out became all the more pressing following the United States' declarations of war on Japan on 8 December, 1941, and on Germany and Italy three days later. Looking beyond the United States as it had been at the close of the 1930s, that community now sensed a deep patriotic need to contribute critically to the defense of the nation. It had been forced to engage in geopolitics in the 1930s. In December 1941, it confronted the American political beast that was Washington, D.C., at the same time that it deployed for war.

A New Educational Initiative: Brown's Program of Advanced Instruction and Research in Mechanics

The AMS was by no means alone in prewar planning for mathematics. At Brown University, Roland Richardson, long an advocate of building up applied mathematics in the United States,[23] Secretary of the AMS from 1921 to 1940, and dean in charge of graduate studies at Brown from 1926 until his retirement in 1948, realized that the time had finally come to act on what American mathematicians had long rued, namely, the dearth of instruction and research in applied mathematics. Armed with arguments like Fry's and with the support of Brown President Henry Wriston, Richardson threw himself into the task of setting up an "educational experiment of high order" in the spring of 1941.[24] His efforts resulted in the creation first of a twelve-week

22. Bliss to Stone, 16 December, 1941, Box 37, Folder, 3: Scientific Mobilization, Stone Papers. The following quotation is also from this letter.

23. At least as early as 1924, Richardson had advocated "the building up of a school of applied mathematics" in the United States, calling that "our most pressing need at present." See Richardson to Veblen, 19 July, 1924, HUA 4213.2, Box 4, Folder: Correspondence 1924 N–Q, Birkhoff Papers.

24. Richardson to Luther Eisenhart, 18 August, 1941, Box 1, Folder I.23: E, Division of Applied Mathematics Papers. Claire Kim wrote her Brown senior thesis, "Math Derived, Math

summer program of Advanced Instruction and Research in Mechanics and then of two more sessions in each of the semesters of the 1941–1942 academic year, all "designed for the purpose of increasing the effectiveness of research in essential American industries."[25] As Richardson explained to von Kármán, in the short term, "Brown wishes to do its share in building up a strong program for defense," but, he freely acknowledged, "our hopes run beyond 1942."[26] "If we can develop along these lines from the small beginnings now planned, Brown University may be able to make a notable contribution to the building-up of applied science" in the United States. Under Richardson's leadership, Brown was taking immediate action on the issue that Stone's WPC subcommittee had been formed to analyze.

Brown's "small beginnings" were made financially possible by support for student fellowships from the Rockefeller Foundation and for general operating expenses, such as summer salaries, from both the Engineering, Science, and Management War Training Program of the U.S. Department of Education and the Carnegie Corporation of New York.[27] The entering summer session class of over fifty students, roughly a quarter of whom had completed a Ph.D., came primarily from graduate programs in mathematics, physics, and engineering but also from industry. They sought to learn more about the mathematics underlying mechanics—that is, to Richardson's way of thinking, the mathematics of "fluid dynamics, elasticity, plasticity, aerodynamics, theory of vibrations, theory of structures, and so forth."[28]

To instruct them, Richardson assembled a group of recognized experts. Among others, Harvard's Richard von Mises and NYU's Kurt Friedrichs

Applied: The Establishment of Brown University's Division of Applied Mathematics, 1940–1946," on the evolution of this program. Surprisingly, her study seems to have left largely untapped the Division of Applied Mathematics Papers, a rich source from which the present account draws extensively. I thank Brown archivist, Raymond Butti, for informing me about this thesis. For its URL, see the References.

25. "Notes," *BAMS* 47 (1941), 548–552 on p. 548.

26. Richardson to von Kármán, 31 March, 1941, Box 2, Folder I.37: Von Kármán, Theodore, Division of Applied Mathematics Papers. The quotation that follows is also from this letter.

27. A year into the experiment, the program had secured some $90,000 from these three sources. See Richardson to Frank Jewett, 19 March, 1942, Box 2, Folder I.31: H, Division of Applied Mathematics Papers. The "real wealth" of $90,000 in 1941 dollars is equivalent to $1.63M in 2021 dollars (https://www.measuringworth.com).

28. Richardson described the program in his article, "Applied Mathematics and the Present Crisis," *AMM* 50 (1943), 415–423. The quotation is on p. 422.

collaborated on lectures on fluid dynamics; Wisconsin's Ivan Sokolnikoff gave a course on elasticity; and Brown's own Jacob Tamarkin as well as its new émigré hire, Stefan Bergman, covered aspects of partial differential equations.[29] NYU's cooperation with Brown's initiative was remarkable, since it had announced the establishment of its own Institute for Applied Mathematics, despite the fact that it was effectively comprised of just Courant, Friedrichs, and Stoker. Courant's simultaneous, but abortive, agitation for the creation of a free-standing National Institute for Advanced Instruction in Basic and Applied Sciences further suggested NYU's attempt to be the American leader in applied mathematics. Ultimately, NYU would make its wartime mark, but through its participation in the work of the Applied Mathematics Panel (see the next section).[30]

The 1941 summer instruction in which Friedrichs participated was supplemented by two additional features. One was a series of stand-alone special lectures from those inside as well as outside academia, among them, Wiener on "Operational Calculus Methods in Problems of Mechanics," General Electric's Hillel Poritsky on numerical and graphical methods for solving partial differential equations, and, from the National Advisory Committee on Aeronautics, Theodore Theodorsen on wing flutter in aircraft.[31] The other was a three-day conference on "Non-Linear Vibrations" that featured invited addresses by five recognized experts: from MIT, mathematician Norman Levinson as well as former MIT M.S. in electrical engineering and soon-to-be (with von Neumann) computer architect Julian Bigelow; Harvard electrical

29. On the offerings, see Box 1, Folder I.2: Miscellaneous Notices sent out 1941–42, Division of Applied Mathematics Papers. Richardson had tried to entice von Kármán to come to the East Coast for the summer, but press of commitments had forced him to remain in Pasadena. See von Kármán to Richardson, 12 April, 1941, Box 2, Folder I.37: Von Kármán, Theodore, Division of Applied Mathematics Papers. He did serve on the committee selected to evaluate the summer program.

30. Reid, *Courant*, pp. 234–235. See also Richard Courant, "National Institute for Advanced Instruction in Basic and Applied Sciences" (draft), undated but spring 1941, Box 1, Folder: Courant, Richard, Division of Applied Mathematics Papers. After the war, Courant continued to lobby for variants of his mathematical center idea. Although an Institute for Mathematics and Mechanics was nominally created in 1946, NYU's Institute of Mathematical Sciences—the present-day Courant Institute of Mathematical Sciences—finally came into existence "with all the multifarious activities to be housed under one roof" in 1953. Reid follows the convoluted story in *Courant*, pp. 251–286. The quotation is on p. 286.

31. Richardson to Courant, 29 June, 1941, Box 1, Folder I.19, Courant, Richard, Division of Applied Mathematics Papers

engineer Philippe LeCorbeiller; Friedrichs and his NYU colleague James Stoker; and control theory pioneer Nicholas Minorsky, then of the David Taylor Model Basin in Maryland.[32] Also in attendance were the summer program students and "about a dozen engineers and mathematicians interested in these questions," who had been specifically invited to participate. The students thus "had the opportunity to meet these guests and to discuss and come in personal contact with an important field of application."

That most of the experts brought in for the summer program were either refugees or foreign born could, however, hardly have gone unnoticed. The Emergency Committee's Stephen Duggan, for one, found "surprising" the large number of refugee scientists on the roster.[33] This drove home the fact, Morse lamented, "[t]hat there is a serious shortage of good men in applied mathematics" in the United States.[34] That shortage was further highlighted by Richardson's appointment of European refugee Willy Prager to a permanent position in applied mathematics.

Prager had earned his Ph.D. in engineering at the Technical University of Darmstadt in 1926 and had served as the acting director of Göttingen's Institute for Applied Mathematics before taking a professorship in applied mechanics at the University of Karlsruhe. An opponent of the Nazi regime, he had left Germany for a position in Istanbul in 1934. Like von Mises, he, too, recognized that the situation in Turkey had become unstable with Atatürk's death and so responded favorably to Richardson's overtures. Although the plan had been for Prager to join the Brown program in time for the beginning of the 1941 summer session, his transit from Istanbul to Providence proved both politically and logistically complex. He finally arrived only in time to teach in the spring of 1942 after overcoming convoluted visa problems and enduring what was described as "a hazardous journey two-thirds around the war-torn globe."[35]

32. Rough draft of a letter from Richardson to von Kármán and others, undated but spring 1942, Box 2, Folder I.37: Von Kármán, Theodore, Division of Applied Mathematics Papers. The quotations that follow in this paragraph are also from this draft. The David Taylor Model Basin is an important test facility for ship design.

33. Duggan to Richardson, 15 October, 1941, Box 13: Correspondence, 1940–1942 'A-M,' Folder: D, Richardson Papers.

34. Morse to Frank Jewett, 7 May, 1941, Box 13, Folder: J, Richardson Papers.

35. David Dietz, "'Scientist in Exile' Now Aids U. S. Defense," *New York World Telegram*, 3 December, 1941, in Box 14: Correspondence, 1940–1942 'N-Y,' Folder: Willy Prager, Richardson Papers.

As Prager slowly made his way to the United States, a five-person commit-tee was at work evaluating, at Brown's request, its summer "experiment."[36] Chaired by Morse, its other members were von Kármán; Weaver; Mervin Kelly, physicist, engineer, and from 1936 to 1944, director of research at Bell Labs; and George Pegram, physicist and dean of the Faculty of Applied Sci-ences at Columbia. Richardson had secured a distinguished group, and it provided an honest assessment of the Brown experiment at the same time that it placed it in the broader context of "the general problem of applied mathematics in the United States."

In the committee's view, Brown had done an admirable job with the resources at its disposal, among them, a talented and energetic administra-tor in Roland Richardson and a strong department of mathematics. "On the other side of the ledger," however, the committee judged that "Brown clearly lacks certain elements of strength and certain advantages of location."[37] In particular, Providence was not "in the neighborhood" of "important engi-neering industries" that would potentially populate its applied mathematics courses with "employed engineers seeking part-time intensive supplementa-tion of their previous mathematical training." It was also the case that "the graduate and research aspects of engineering training ha[d] not been devel-oped at Brown" and to do so adequately would likely "take something of [sic] the order of ten years of time and millions of dollars." That said, the commit-tee deemed that Brown had performed "an excellent experiment," so excellent, in fact, that they thought the next step should be the formation of an actual Graduate School of Applied Mathematics under a director and with an advi-sory board chosen "on a national basis, from experienced engineers, scientists, and mathematicians." A frank but ultimately very encouraging document, the committee's report meaningfully informed Brown's subsequent plan of action.

The two sessions in the 1941–1942 academic year built on the summer's success. Year-long courses were given by Tamarkin on partial differential equa-tions, Brown's recent émigré hire, William Feller, on numerical and graphical methods in applied mathematics, and Bergman on fluid dynamics. In the fall, these were supplemented by classes on the theory of air flight by von Mises

36. "A Report on Advanced Training in Applied Mathematics, with Special Reference to the School of Mechanics at Brown University," undated but marked "rec'd Nov. 25 [1941]," 22 pp. on p. 1, Box 3, Folder I.49: Marston Morse, Division of Applied Mathematics. The quotation that follows is also on this page.

37. Ibid., p. 16. For the quotations that follow, see pp. 16–17 and p. 20.

and on elasticity as well as on advanced dynamics by John Synge, whom Richardson had attracted as a short-term visitor from his regular position at the University of Toronto. Prager's lectures on the theory of plasticity and on engineering mechanics as well as Bergman's course on special topics in elasticity followed in the spring. Synge, Prager, and Bergman, moreover, led advanced research seminars on topics in elasticity, fluid dynamics, and the theory of structures that aimed to get students who were sufficiently advanced up to the research level.[38] The second summer and year-long programs in 1942–1943 followed the same general pattern as the first, although with a greater number and variety of courses.[39]

Results over the first seven terms were tangible. Richardson reported in the summer of 1943 that "[m]ore than 25 research papers have been completed and others are under way; some deal with immediate practical problems which have arisen in the prosecution of the war, while others are of a more fundamental character. As a significant feature of the program," he continued, "a limited number of research problems are being investigated for government agencies and war industries."[40] By that point, some 200 students— among them mathematicians Saunders Mac Lane and Garrett Birkhoff, both of whom opted to "brush up on their skills in applications" for the war effort—had participated in the Brown program.[41] Forty of those students had "gone into research in government agencies concerned with aeronautics, ship construction, gun construction, radar, etc.," twenty more were "engaged in research in industries connected with the war," ten had "special assignments in connection with the armed forces," and twenty-five were "continuing as engineers in industry." The rest were either "instructors or graduate students in universities."

38. "Material Made Available to Advisory Committee," Box 1, Folder I.5: Material Made Available to Advisory Committee, Division of Applied Mathematics Papers.

39. See "Notes," *BAMS* 48 (1942), 206–209 on p. 206.

40. Richardson, "Applied Mathematics and the Present Crisis," p. 423. The quotations that follow in this paragraph are also on this page.

41. For the quotation, see Saunders Mac Lane, *Saunders Mac Lane: A Mathematical Autobiography* (Wellesley: A K Peters, 2005), p. 113. Clifford Truesdell, who, in 1942, had recently taken his Caltech M.A. under Harry Bateman and who would go on to a career in applied mathematics and the history of science, attended the second Brown summer program and left a vivid—and rather acerbic (as could be his way)—account of his perceptions of its faculty and course work. See David Rowe, "Mathematics in Wartime: Private Reflections of Clifford Truesdell," *MI* 34 (2012), 29–39.

FIGURE 8.3. Advanced students in Applied Mathematics at Brown University with their professor, William Feller (center) (ca. 1945). (Photo courtesy of Rockefeller Foundation Records, Photographs, Series 100–1000 (FA003), Series 244D: Rhode Island–Natural Sciences and Agriculture, Box 59, Folder 1343, "Brown University Applied Mathematics.")

As these data suggest, and as Richardson modestly but proudly offered, "[t]he experiment at Brown University is a lively one; it has already evoked much interest; and it permits a hope that it will prove a contribution to the advancement of American science." Morse's evaluating committee had hit the nail on the head when it concluded, as early as November 1941, that "Brown University has actually *done* something about the problem" of applied mathematics in the United States.[42] "Brown has *acted.* And this action has, of itself, given them a very real and ponderable advantage." They used that leverage in 1943.

Like the possibility of greater emphasis on teaching and research in applied mathematics in American institutions of higher education, the possibility of a journal to support such research had also long been under discussion but

42. "A Report on Advanced Training in Applied Mathematics," p. 15 (my emphasis). The quotation that follows is also on this page (also with my emphasis).

had never materialized.[43] The apparent success of the Brown experiment suggested, at least to Richardson and his colleagues, that the time was finally ripe for establishing such a periodical. Brown acted once again. Its *Quarterly of Applied Mathematics* aimed "to fill" the perceived "gap between purely engineering journals and purely mathematical ones."[44]

Perhaps informed by the example of the *Mathematical Reviews* under the expert guidance of his recently hired Brown colleague, Otto Neugebauer, Richardson first sought a trusted advisor to "handle [the] business end of [the] inauguration of [the] journal" and to "assist [in the] administration [of the] Mechanics School" prior to Prager's arrival.[45] By October 1942, he had secured the services of his old friend, Earle Hedrick, for a year-long visiting professorship that would involve teaching, politicking, and administrative organization. Given the many leadership and editorial roles he had played in both the MAA and the AMS and given the fact that he had then just retired as Vice President and Provost at UCLA, Hedrick, as a specialist in the theory of partial differential equations and their applications, most definitely knew his way around applied mathematics, the American mathematical community, journals, and academic administration.[46] He threw himself into the work.

Wartime censorship confronted him from the moment he arrived in the East. One key question was whether a journal like the one proposed would actually be allowed—given security concerns—to publish all of the research papers it deemed worthy. Although ultimately not a problem, it required some legwork to get the facts straight.

Hedrick also talked up the new journal to members of the AMS in addition to making numerous contacts with and advertising the new journal to professional societies with applied mathematics potential like the Optical Society of America.[47] His efforts, however, came tragically to an end in February 1943 when he succumbed to a lung infection just prior to the appearance of the journal's first number in April. Still, that number attested to the efforts of all

43. Raymond Archibald sketches the AMS's efforts in this direction, unfortunately just at the time of the stock market crash in 1929, in *Semicentennial History*, pp. 17–18.

44. Earle Hedrick, "A Proposed Journal of Applied Mathematics," *BAMS* 48 (1942), 791.

45. "Day letter" from James Adams, Vice President of Brown, to Hedrick, 5 October, 1942, Box 2, Folder I.31: H, Division of Applied Mathematics Papers.

46. Walter Ford, "Earle Raymond Hedrick," *AMM* 50 (1943), 409–411.

47. See Hedrick to Thornton Fry, 2 November, 1942, Box 1, Folder I.27: Thornton C. Fry, and Hedrick to Robert H. Kent, consultant and ordnance engineer at the Aberdeen Proving Ground, 23 October, 1942, Box 2, Folder I.36: K, Division of Applied Mathematics Papers.

involved in getting the venture off the ground. With Prager as managing editor, the editorial team included applied mathematicians Fry, von Kármán, Synge, and Sokolnikoff together with National Bureau of Standards physicist Hugh Dryden, and MIT mechanical engineer John Lessells. Bateman, Friedrichs, Murnaghan, and Timoshenko were among the eleven editorial "collaborators." As Hedrick had put it in announcing the new journal to the members of the AMS, these "names will carry a message of confidence of high standards to all who know the field."[48]

One name was conspicuously missing from this list: Richard von Mises. As the founder in 1921 of the *Zeitschrift für angewandte Mathematik und Mechanik*, von Mises would have been a natural choice but for the perception that he would have had such decided opinions as to adversely affect the new journal's development. As Richardson confided to Fry, "'[t]his business of keeping von Mises off the Board of Editors is naturally having its repercussions."[49] "We will have to stand by our guns," he pronounced, "but on the other hand I feel we should mitigate the results as far as possible." To that end, he wrote diplomatically to von Mises inviting him both to serve as one of the collaborating editors and to be one of the inaugural authors.[50] The tactic did not work. Von Mises never published in or was officially associated with the *Quarterly*, but his absence seemed in no way to affect the journal's success.

At over 350 printed pages, the inaugural volume carried papers by Bers, Friedrichs, Stoker, Synge, and von Kármán as well as, interestingly, by four Chinese applied mathematicians who had been trained in North America: Hsue-Shen Tsien (Qian Xuesen), a 1938 Ph.D. student of von Kármán; his academic "brothers," Chia-Chiao Lin (Lin Jiaqiao) and Yung-Huai Kuo (Guo Yonghuai), both of whom earned their doctorates in 1944; and Wei-Zang Chien (Qian Weichang), a 1942 Toronto Ph.D. under Sygne. Indeed, over the course of its first decade, the *Quarterly* published papers by others from China in addition to mathematicians from Japan, India, and elsewhere.

Tsien and Lin had both come to the United States as part of the Boxer Indemnity Scholarship Program, an interesting result of the Boxer Uprising

48. Hedrick, "A Proposed Journal," p. 791.

49. Richardson to Fry, 16 November, 1942, Box 1, Folder: Thornton C. Fry, Division of Applied Mathematics Papers. The quotation that follows is also from this letter.

50. Richardson to von Mises, 12 November, 1942, Box 3, Folder: Richard von Mises, Division of Applied Mathematics Papers.

against foreigners that had taken place in China around 1900.[51] Tsien had become part of von Kármán's team at Caltech after earning his Ph.D. and worked on the Manhattan Project. When, in 1950, he was accused of pro-Communist sympathies, he was stripped of his security clearance and detained for five years. Following his return to China in 1955, he assumed the directorship of the Fifth Academy of the Ministry of National Defense, the group that ultimately developed China's atomic bomb. Lin's trajectory was smoother. He taught first at Caltech and then at Brown before moving to MIT in 1947. A full professor there by 1953, he had found a congenial home in the United States in which to pursue his ultimately award-winning research in fluid dynamics.

Like Tsien, Kuo and Chien also both ultimately returned to China. Kuo started out on the Cornell faculty but left in 1956 to take up the deputy directorship of the Institute of Mechanics of the Chinese Academy of Sciences. Chien went back a decade earlier in 1946, ultimately becoming a member of the Chinese Academy of Sciences and the President of the Shanghai University of Technology. The distinguished careers of all four of these Chinese applied mathematicians foreshadowed Richardson's conviction that, after the war, it would be incumbent upon the United States to be "a great center of scientific culture" that drew in students from Asia, Latin America, and Europe.[52] The United States had an international role to play.

The founding at Brown in 1946 of its Graduate Division of Applied Mathematics under Prager as chair represented a step in that direction. In five years and during wartime, Brown had carefully laid the groundwork for and created

51. Harvard had been a magnet for aspiring Chinese mathematicians on Boxer Indemnity scholarships as early as 1909 when Wang Renfu went there to pursue the bachelor's degree he earned in 1913. After its founding in 1933, the Institute for Advanced Study was as well (see this chapter's final section). Yibao Xu detailed the implications of this program for mathematics in China and the United States in "Chinese–U. S. Mathematical Relations, 1859–1949," in *Mathematics Unbound: The Evolution of an International Mathematical Research Community, 1800–1945*, ed. Karen Hunger Parshall and Adrian Rice, HMATH, vol 23 (Providence: American Mathematical Society and London: London Mathematical Society, 2002), pp. 287–309.

Probablilist Kai-lai Chung also came to the United States, in his case to Princeton, on a Boxer Indemnity scholarship. He earned his Ph.D. in 1947 under John Tukey and Harald Cramér and remained in the United States for the remainder of a long and highly productive career spent, after 1961, at Stanford.

52. Richardson to Stephan Duggan, 22 October, 1941, Box 13: Correspondence, 1940–1942 'A-M,' Folder: D, Richardson Papers.

a free-standing graduate department of applied mathematics as well as a new research journal for the field that attracted those at home and from abroad. In so doing, it realized one of the American mathematical research community's longstanding desiderata.

War Work

As the activities at Brown and NYU as well as of the joint AMS-MAA War Preparedness Committee exemplify, that community had begun gearing up even before the United States' formal declaration of war. Some young men had jumped at the chance to enlist. William Claytor, the African-American topologist who had made such a splash with his work in the mid-1930s and who had gone to Michigan, on his own dime, for a research year with Wilder in 1936–1937, was one of them. Claytor had been able to stay on at Michigan for two additional research years thanks to fellowships from the Julius Rosenwald Fund, but he had only managed to hang on in Ann Arbor from 1939 to 1941 through piece work associated with the university's correspondence division.[53] With American engagement seemingly imminent in the spring of 1941, Claytor enlisted in the Army and, by the summer of 1942, had been accepted on Wilder's strong recommendation to Officer Candidate School. He spent the rest of the war teaching in the Anti-Aircraft Artillery Schools at Fort Eustis in Virginia and Fort Stewart in Georgia and never regained his research footing when once again a civilian.[54]

Unlike Claytor, many young mathematicians were drafted before the occupational deferment issue had been settled. Still others left school or their academic posts for war-related jobs. The result was dramatically decreased enrollments and a deficit of mathematics instructors in the nation's colleges and universities. One year into the war, it had become painfully evident to both the Army and Navy that this had resulted in a severe deficit of adequately trained men.

53. For more on this phase of Claytor's life, see Parshall, "Mathematics and the Politics of Race," pp. 229–233.

54. Claytor left the Army at war's end. He spent the remainder of his career teaching mathematics at historically black universities: first at Southern University in Baton Rouge, Louisiana, then at Virginia's Hampton Institute, and, finally, at Howard University in Washington, D.C., beginning in 1947. He died in Washington at the young age of fifty-nine having published no mathematical research after his Michigan years.

For the Army, as well as for the Air Forces that were then a part of it,[55] that meant, first and foremost, men with specific technical training, while the Navy, Marine Corps, and Coast Guard faced a dearth of officer candidate material. In particular, the Army had immediate need for some 90,000 men; the Air Corps 70,000; the Navy another 40,000.[56] Analogous to the Student Army Training Corps established during World War I, the Army Specialized Training Program (ASTP) and the Navy V-12 Program were created at the end of 1942 to address these critical personnel issues through targeted, accelerated training.[57] Two smaller-scale training initiatives were also conceived in 1942. The wheels were set in motion for the formation in 1943 of the Army Air Forces Pre-Meteorology Program to address the shortage of trained weather forecasters, while the Air Corps Flying Training Command issued guidelines, as opposed to a free-standing program, for academic pre-training for aviation cadets.[58]

The implications of these new initiatives for higher education in general and for mathematics in particular were immediate. Colleges and universities around the country—227 for the ASTP, 131 for the Navy V-12, and twenty-seven for the Pre-Meteorology Program—were contracted to take in students

55. In 1941 when the U.S. Army Air Forces (USAAF) were created, the U.S. Army Air Corps, which had grown out of World War I, became the training and logistical arm of the USAAF. The U.S. Air Force became a separate branch of the military in 1947.

56. John Kline, Minutes of the War Policy Committee, 21 February, 1943, Ms.75.5, Box 28, Folder 110, AMS Papers. The joint AMS-MAA War Preparedness Committee had become the War Policy Committee in January 1943.

57. The Navy V-12 Program superseded the earlier V-1 and V-7 Programs in which colleges and universities had been participating. The former had been comprised of college freshmen and sophomores "procured" from some "1,000 institutions, both junior colleges and four-year colleges," while students in the latter had been "procured largely from upper classes in a smaller number of colleges." The Navy had other V classifications, such as V-5 for naval aviation cadets and V-7 for officer trainees. "Outline of Information for the Board of Counselors Regarding the Navy V-12 Unit at Brown and the Navy College Training Program," 30 September, 1943, 14 pp. on p. 2, Box 17: Correspondence, 1943–1945 'S-Y,' Folder: D & P War Training at Brown, Richardson Papers.

58. See "Training in Meteorology," AMM 49 (1942), 697–698 and R. E. Rowland, "The Premeteorology Program of the Army Air Force, 1942–1945" (see the References for the URL) as well as Barton K. Yount, Maj. Gen. U.S. Army, Memorandum on "Academic pre-training for aviation cadets," undated but around March 1942, Box 17: Correspondence, 1943–1945 'S-Y,' Folder: D & P War Training at Brown, Richardson Papers (the memorandum was published as "Pre-Training of Aviation Cadets," AMM 49 (1942), 274–276).

and to staff a sequence of prescribed courses for them.[59] The ASTP generally, although not exclusively, tapped larger, public or land-grant institutions; the Navy V-12 favored smaller, private schools; and the Pre-Meteorology Program opted for a limited number of public and private schools. Pre-training for aviation cadets could, for most candidates, be met through existing, basic high school work and introductory college courses in mathematics and physics at their undergraduate institutions.

For students in the ASTP, mathematics was required in each of the three, twelve-week terms of the basic program as well as in the first term of the advanced engineering programs that could follow: one, a three-term program in civil engineering and the others, four-term programs in chemical, mechanical, and sanitary engineering. Outlines for the courses—intermediate algebra and basic trigonometry (six hours weekly), analytic geometry (five), and differential and integral calculus (five) for the basic program and, for the engineers, a more advanced calculus course that included first- and second-order differential equations (five)—had been "drawn up by an advisory committee of mathematicians" consisting of Penn's John Kline, James (Henry) Taylor of George Washington University, and West Point's Harris Jones.[60]

The Navy V-12 program was similar in spirit. It encompassed programs of lengths varying from two, sixteen-week terms for aviation candidates to eight, sixteen-week terms for engineer specialist, Civil Engineer Corps, and Construction Corps candidates.[61] Depending on placement, there were four levels of a course that combined college algebra, trigonometry (including some spherical trigonometry), and analytic geometry (five hours weekly) aimed at overcoming high school mathematical deficiencies, followed by four-hour courses in each of differential calculus, integral calculus, and differential equations, and two, three-hour courses in navigation and nautical astronomy.

59. V. Ray Cardozier, *Colleges and Universities in World War II* (Westport: Praeger, 1993), pp. 25, 51, and 148–149.

60. Carroll Newsom, ed. "War Information," *AMM* 50 (1943), 464–470. For the ASTP, see pp. 466–470. The quotation is on p. 468; the four course outlines are on pp. 469–470. The members of the advisory committee are listed in William Hart, et al., Subcommittee on War Training of the War Policy Committee, "Report on Mathematics in the Army Specialized Training Program," Box 8, Folder 3: AMS–Correspondence regarding the War Policy Committee (1943–1944), Whyburn Papers, 8 pp. on p. 3.

61. *The Navy College Training Program: V-12: Curricula, Schedules, Course Descriptions* (n.p.: Training Division, Bureau of Naval Personnel, U.S. Navy, 1943), pp. 1–2. For the courses that follow, see pp. 33–35.

Finally, the Pre-Meteorology Program was divided into three categories. Those in the "A" schools—Caltech, MIT, NYU, Chicago, and UCLA, the homes of the nation's five, pre-existing programs in meteorology—were college graduates and pursued a nine-month cadet meteorological program. Students at the "B" schools—among them, Brown, MIT, NYU, Michigan, UNC, and Wisconsin—had had one or more years of college training and required an additional six months of more targeted pre-meteorological course work to qualify them for the "A" schools, while students at the "C" schools—among them, Reed College, Kenyon College, Berkeley, Chicago, and Washington University of St. Louis—had limited or no college background and required twelve months of preparation before qualification. For students at the latter schools, mathematics was covered in two courses that ran simultaneously, one that moved from college algebra to analytic geometry and finally to differential and integral calculus and one that began with trigonometry before taking up vector analysis, vector calculus, and mechanics.[62]

All of this new coursework and all of these new students taxed mathematics teaching resources around the country. Already in the spring of 1942, University of Virginia-trained topologist John Kelley, who was then at Notre Dame, told his doctoral advisor, Gordon Whyburn, that even though enrollments were low overall, "the math department is extra busy. Math isn't a required course," he explained, "but since the war everyone wants it. . . . As a consequence we're all overloaded. We'll work straight through the summer on a three semester plan."[63] Things had only gotten worse in this regard after the ASTP and Navy V-12 Programs got into full swing early in 1943. "The Army program calls for mathematics six days a week," Kline told Whyburn in January before the first students had arrived, "and I am certain that it will make demands on our staffs so that not only will every available mathematician need to be used, but we must also call on anyone else with mathematical training in other departments."[64] Indeed, that had quickly come to be the case.

62. This overview has been compiled from "Training in Meteorology" as well as from these three reminiscences: Rowland; Henry Bernat, "Overview of the Army Meteorological Program and the Program at Reed College"; and Herman Lunden Miller, "First-Hand: My Experience in the Army Air Force, 1943 to 1946 (for the URLs, see the References).

63. Kelley to Whyburn, undated but the spring semester of 1942, Box 1, Folder 96: Kelley, John L., Whyburn Papers. Baley Price gives an early indication, based on a survey conducted between November and December 1942 of sixty-seven colleges and universities, of the toll that these extra students was taking on mathematics programs. See his "Adjustments in Mathematics to the Impact of War," *AMM* 50 (1943), 31–34.

64. Kline to Whyburn, 28 January, 1943, Box 1, Folder 99: Kline, J. R., Whyburn Papers.

FIGURE 8.4. A class of Pre-Meteorology Program students at the University of Chicago under Herman Meyer (1915–1993) (back row in civilian clothes) in 1943. (Photo courtesy of the private collection of Prof. Elizabeth Meyer.)

From Raymond Wilder's perspective in March, the University of Michigan was "rapidly turning" its "campus into army and navy camps" with the effect that "the major part" of his teaching was directed to soldiers, and he feared that soon he would be "doing nothing but teaching elementary mathematics day and night."[65] By the summer, that was pretty much the case at the University of Illinois, where "some three thousand service men [were] taking mathematics."[66]

And, Illinois was by no means alone. In all, some 200,000 men successfully went through the ASTP; 50,000 naval officers commissioned during the war had at least part of their training through the Navy V-12 Program; and over 5,500 had received meteorological training.[67] Taking into account those who started but did not finish one of these programs, the numbers of students

65. Wilder to Whyburn, 25 March, 1943, Box 2, Folder 96: Wilder, Raymond L., Whyburn Papers.

66. Coble to Whyburn, 16 July, 1943, Box 2, Folder 47: Ritt, Joseph F., Whyburn Papers.

67. For the numbers, see Cardozier, pp. 25 and 71 and Rowland, respectively.

in military training in the nation's classrooms was markedly higher than even these numbers would suggest.[68]

From a mathematical point of view, this teaching, while critical to the war effort, could in no way be construed as glamorous. It did not have the potential—that deeper mathematical expertise was assumed to have—dramatically to impact the course of the war. Although necessarily in fewer numbers than their teaching colleagues, American mathematicians were also called on to put their problem-formulating and -solving skills to work in military contexts.

The Aberdeen Proving Ground in Maryland—which had been established in October 1917 and which had served as a key locus of mathematical activity during the First World War—had been quick to expand after England and France declared war on Germany in 1939.[69] More land was acquired for ordnance and other testing, and new facilities were built for the Ballistic Research Laboratory (BRL) as well as for the Ordnance Training School. As during World War I when Oswald Veblen headed up work at Aberdeen on ballistics, a group of scientific experts, once again including Veblen as section head, was mobilized to supplement the BRL's regular personnel.[70] Not surprisingly, when it was "realized that the scientific staff of the laboratory was not of sufficient caliber to carry out their assigned duties," the BRL turned to Veblen for advice in its effort to "get in scientists of greater stature to do the work."[71] For external ballistics, that is, for studying the behavior of ammunition once it has left the gun barrel but before it has hit its target, Veblen tapped astronomer Edwin Hubble of the Mount Wilson Observatory in California to head up the experimental section and Virginia's Jimmy McShane to lead its mathematical section.[72]

68. As Herman Miller put it in his memoir on his experiences in the Pre-Meteorology Program at Washington University, "[s]ome washed out because they could not stand the stress. One washout said he just wanted to wash airplanes."

69. On mathematical work at Aberdeen during World War I, see Archibald, Dumbaugh, and Kent, pp. 241–244.

70. Grier, "Dr. Veblen" and Saunders Mac Lane, "Oswald Veblen: June 24, 1880–August 10, 1960," *Biographical Memoirs*, vol. 37 (Washington, D.C.: National Academy of Sciences, 1964), pp. 324–371 on p. 333.

71. Edward J. "Jimmy" McShane, unpublished autobiographical notes, chapter one, pp. 10–11. I thank McShane's daughter and son-in-law, Jennifer and Thann Ward, for making the first two chapters of these notes available to me.

72. Another mathematician "of greater stature" who was called to the BRL was Berkeley's Charles Morrey, although he was not in McShane's group.

McShane immediately called in John Kelley, whom he, like his colleague Whyburn, had taught in the graduate program at Virginia.[73] By 1943, they had been joined by John Green, then at the University of Rochester (but at UCLA immediately after the war), Berkeley's Anthony Morse, and 1941 NYU Ph.D. (under Courant) Arthur Peters. Two mathematics graduate students in uniform joined these civilians: Morse's 1944 Ph.D. student, Herbert Federer, and Kirk Fort, a student of McShane's who went on to earn his Ph.D. in 1948. Together, these mathematicians worked to provide the best possible numerical data for field use.

For example, bombardiers were trained at airfields at differing elevations above sea level, yet the bombing tables they used to calculate when and where to drop their payloads in order to hit their desired targets were prepared only for targets at sea level. Since "a systematic error is introduced by the use of such tables at the fields having high altitudes," the BRL was asked to "prepare tables permitting correction for height of target above sea level."[74] McShane tackled the problem.

He began his analysis by looking at the "pure-drag solution" by which the sea-level values were generated. This involved setting up and solving a particular pair of differential equations. He next incorporated height above sea level into those same equations and analyzed the mathematical implications of that additional datum. Finally, he considered what happened if lift were taken account of separately from drag. He concluded that "there is no point in adopting any procedure of greater theoretical refinement than the pure-drag solution."

If a target's height above sea level ultimately did not introduce mathematical complications into bombing accuracy, at least two other things did. One was the actual sighting device used. During World War II that was the Norden bombsight, named after its developer, Dutch-born American engineer Carl Norden. McShane and John Synge, who had been called from his new position at the Ohio State University to serve as a ballistics mathematician in England with the U.S. Army, realized that "by far the greatest error" in bombing "was introduced by the fact that the Norden bombsight had built into

---

73. For the list of personnel, see ibid., p. 11.

74. E. J. McShane, "Corrections of Elements of Bombing Tables for Effects of Height of Target," Report No. 374, Aberdeen Proving Ground, 2 July, 1943," 13 pp. on p. 1 (see the References for the URL). The quotations that follow in the next paragraph are also on this page.

it a correction for wind at altitude, but nothing to take account of the fact that wind velocities changed, sometimes drastically, between [an] airplane's altitude and the ground."[75] What was needed, then, were "bombing tables that would provide corrections for winds that varied with altitude." In the late spring of 1945, McShane and Synge were "directed to" and did "produce an acceptable version of such a modification of the tables."[76]

Another was the flying altitude of the aircraft. Until 1940, bombing tables had been calculated based on test drops from altitudes up to 18,000 feet under the assumption that aircraft would be flying no higher than 20,000 feet. The latter assumption was revised upward at least twice: first, to 35,000 feet in 1940 and then to between 50,000 and 60,000 feet in 1942. These changes necessitated the recalculation of bombing tables and eventually brought about the standardization of altitudes for regular range bombing at 2,000, 10,000, 25,000, and 35,000 feet.[77]

In addition to refining bombing tables, the BRL was charged with the production of accurate artillery firing tables. Given a particular "weapon firing a given projectile at a given velocity under arbitrarily chosen 'standard' conditions," they needed to calculate the range-elevation relationships that would allow the gunner to angle his weapon so as to hit his target a given distance away, while accounting for "corrections . . . due to nonstandard conditions of weather and materiel."[78] As with the calculation of bombing tables, the underlying mathematics involved little more than algebra, trigonometry, calculus, and some differential equations. Taking account, however, of the fact that both gunner and target may be in motion, the number of calculations needed to produce one artillery firing table greatly increased. Indeed, since it took some 750 calculations to compute a single trajectory, and since each table had around 3,000 trajectories, the total number of computations involved

75. McShane, unpublished autobiographical notes, chapter one, p. 12. The quotations that follow in this paragraph are also on this page.

76. McShane and two of his BRL colleagues, John Kelley and Frank Reno, later published a textbook, *Exterior Ballistics* (Denver: University of Denver Press, 1953), which was "regarded as the definitive work on the subject." For the quotation, see Berkovitz and Fleming, p. 4.

77. Gordon Barber et al., *Ballisticians in War and Peace*, vol. 1, *A History of the United States Army Ballistic Research Laboratories, 1914–1958* (Aberdeen, MD: Aberdeen Proving Ground, [1956]), p. 54. For the URL, see the References.

78. Elizabeth R. Dickinson, "The Production of Firing Tables for Cannon Artillery," Report No. 1371, U.S. Army Materiel Command, Ballistic Research Laboratories, Aberdeen Proving Ground, November 1967, p. 15 (for the URL, see the References).

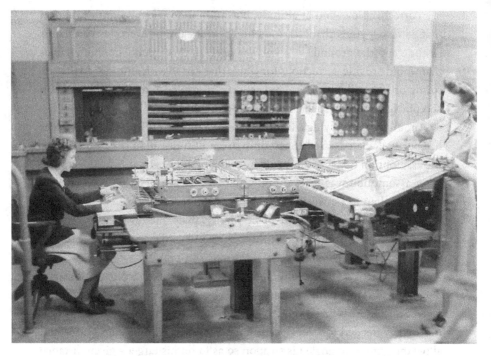

FIGURE 8.5. Kay McNulty (left) and others operating Bush's differential analyzer at the University of Pennsylvania's Moore School (ca. 1945). (Photo from Wikimedia Commons.)

is well over two million. In human terms, a calculator, many of whom were women, required at least seven hours to compute one trajectory using the best calculating device then available, namely, Vannevar Bush's electromechanical differential analyzer.[79] This time-intensive task spurred the staff of the BRL's computing laboratory, which was located off-site at the University of Pennsylvania's Moore School of Electrical Engineering, to seek a speedier way to get the job done.

Enter Herman Goldstine, an instructor at Michigan, who enlisted in the Army at the outbreak of the war (ultimately rising to the rank of Lt. Colonel). A 1936 Chicago mathematics Ph.D. (under Lawrence Graves and William Reid), Goldstine was well versed in ballistics computations as a result of

79. Herman Goldstine, "Remembrance of Things Past," in *A History of Scientific Computing*, ed. Stephen Nash (Reading: Addison-Wesley Publishing Co., 1990), pp. 5–16 on p. 9. The differential analyzer was augmented by punch-card machines as well as by relay calculators made by both IBM and Bell Laboratories over the course of World War II. See Barber et al., pp. 38–39 (for the URL, see the References).

having taught Bliss's course on the subject at Chicago in the late 1930s. He was thus a natural for a BRL assignment at Penn in the summer of 1942.[80]

In discussions with one of his new colleagues, Penn engineering graduate student J. Presper Eckert, Goldstine became convinced that the physics-trained, Penn electrical engineering instructor, John Mauchly, had figured out how to make an electronic digital computer that would both significantly speed up and markedly improve the accuracy of the BRL's calculations. Aberdeen Proving Ground consultant and BRL scientific board member John von Neumann was also quickly sold on the idea as critical for both the BRL and the Manhattan Project for which he also consulted (see later in this section). Indeed, von Neumann, a sort of universal consultant during World War II, was also involved in work for the Navy and its Bureau of Ordnance on mine warfare in general and shock waves in particular, for the NDRC on shaped charges as a way to control the physical effects of detonations, for the Applied Mathematics Panel on various questions, for Los Alamos, and for other agencies.[81]

In the spring of 1943, Goldstine successfully pitched the idea to the Army and development got under way. The end result was the construction at Penn—owing to the combined efforts of Eckert, Mauchly, Goldstine, von Neumann, and others—of the Electronic Numerical Integrator and Computer (ENIAC), America's "first electronic, general purpose, large scale, digital computer."[82] Although the ENIAC was completed too late in 1945 to have an impact on the war effort, its creation not only helped usher in the digital age but also spurred research in numerical analysis after the war (see chapter ten). Adele Goldstine, Michigan M.A. in mathematics, instructor for the Moore School's women computers, and wife of Herman Goldstine, not only wrote ENIAC's operators' manual but was also a computer programming pioneer.[83]

80. Ibid., p. 8. Bliss had worked with Veblen at Aberdeen during World War I. He later authored *Mathematics for Exterior Ballistics* (New York: J. Wiley and Sons, Inc. and London: Chapman and Hall, Ltd., 1944) based on his researches.

81. Norman Macrae, *John von Neumann* (New York: Pantheon Books, 1992), pp. 200–245.

82. Thomas Haigh, Mark Priestley, and Crispin Rope, *ENIAC in Action: Making and Remaking the Modern Computer* (Cambridge: The MIT Press, 2016), p. 7. See also Herman Goldstein, *The Computer from Pascal to von Neumann* (Princeton: Princeton University Press, 1972; reprinted with a New Preface, 1993).

83. Adele Goldstine, "Report on the ENIAC (Electronic Numerical Integrator and Computer)," Technical Report 1, Moore School of Electrical Engineering, Philadelphia, Pennsylvania, 1 June, 1946 (for the URL, see the References).

While the precedent for the participation of mathematicians at the Aberdeen Proving Ground during wartime had been set during World War I, other avenues for their active engagement were slow to be identified during World War II. As noted, Bush and his associates wanted central control of the mobilization of science, that is, under their OSRD and its NDRC, while the joint AMS/MAA War Preparedness Committee, and Stone in particular, held that mathematicians could mobilize themselves more effectively than could some centralized group of non-mathematicians. A bureaucratic impasse. Still, in a speech given before the Institute of Radio Engineers in January 1942, Bush crony Jewett had publicly acknowledged that "[w]ithout insinuating anything as to guilt, the chemists declare this is a physicist's war. With about equal justice one might say that it is a mathematician's war."[84] Did this augur well for the mathematicians' cause?

By March of 1942, Stone and Morse finally succeeded in getting the face-to-face meeting with Bush, Conant, and Jewett that had been under discussion at the close of 1941. It would be their chance to drive home how important "the employment of mathematicians in the professional service of our country" would be in confronting "the realities of total war and its aftermath."[85] They made their case not only in person but in the form of a tightly argued memorandum written by Stone with the input of Morse and others.

At that case's heart was the acknowledgment that "[t]he specialist tendencies of scientific development in recent decades have created a considerable gulf between mathematics on the one hand and the exact sciences on the other."[86] "[T]his means," they continued, "that mathematicians, even though they may have had basic training and a continued general interest in the exact sciences, have been absorbed especially in the internal development of their own field as pure mathematics, while physicists and engineers by and large have lost touch . . . with modern mathematics." The import of this state of affairs was that "[i]n terms of problems . . . in all probability many substantial advances in applied mathematics (especially in aerodynamics, hydrodynamics, and elasticity) have been retarded because the circumstances did not favor the application of the newer mathematical techniques to problems which

---

84. Frank Jewett, "The Mobilization of Science in National Defense," *Science* 95 (1942), 235–241 on p. 240.

85. Unattributed but by Marshall Stone, Memorandum: "Mathematics in War," undated but spring 1942, Box 37, Folder 1: Correspondence with Warren Weaver, Stone Papers.

86. Ibid. The quotations that follow in this paragraph are also from the memorandum.

were known to defy the older methods tried upon them." While this blunt assessment may not have made for the most politic argument to present to two electrical engineers and a chemist, the meeting resulted in Jewett's formation of a Joint Committee on Mathematics (JCM) under the aegis of both the National Academy of Sciences and its National Research Council ostensibly to coordinate the mathematical community in the war effort.

Chaired by Morse and with an executive committee comprised of Morse, Stone, and Evans, the JCM united these three men in common cause with Chicago astronomer and applied mathematician Walter Bartky, as well as Bateman, the senior Birkhoff, Jackson, Robertson, Veblen, and Weaver.[87] In retrospect, the problems with this arrangement are apparent. The group was geographically dispersed. Administratively, it was an arm of not one but two bodies, the National Academy and its subsidiary National Research Council. And, although the provision was made for Morse to serve as liaison to the NDRC then under Conant's direction, there was no *effective* connection between that branch of Bush's OSRD and the committee, despite the fact that Robertson and Weaver were part of Conant's NDRC team. Little wonder, then, that Morse was soon frustrated. "The situation is not yet clear with regard to whether or not they [the NDRC] will use our committee," he wrote in exasperation to Stone in August 1942. "My conclusion is either that Conant is *too busy to use us*, or that his subordinates such as Robertson of Princeton *don't want to use us*, or that the Army and Navy *don't wish to go further* with the deeper scientific investigations." "In the N.D.R.C," he reported disgustedly, "there are deadly and stultifying rivalries beyond belief."[88]

Morse's committee had been mooted by November when the NDRC moved to establish its own free-standing mathematics unit. Without consulting the JCM, an Applied Mathematics Panel (AMP) was organized under Weaver's direction. At least in Stone's view, Weaver did not meet all three of the qualifications that had been explicitly laid out in the spring memorandum to Bush and company, namely, "a broad knowledge of modern *pure* mathematics," "recognized high professional standing," and "a wide acquaintance

87. Larry Owens discusses this committee in "Mathematicians and the War: Warren Weaver and the Applied Mathematics Panel, 1942–1945," in *The History of Modern Mathematics*, ed. David Rowe and John McCleary, 2 vols. (Boston: Academic Press, Inc., 1989), 2: 287–305 on pp. 294–295.

88. Morse to Stone, 14 August, [1942], Box 37, Folder 1: Correspondence with Warren Weaver, Stone Papers (Morse's emphases).

among professional mathematicians in the United States."[89] For Stone, only a pure mathematician could champion and represent the interests of pure mathematics during wartime: a political defeat in that one of the nation's best *pure* mathematicians was not put in charge; a victory in that mathematics was officially recognized within the government's wartime science organization.

Stone, who became President of the AMS for a two-calendar-year term beginning in January 1943 and on whose watch the War Preparedness Committee was converted into the War Policy Committee, pressed the issue of the AMP's leadership throughout his presidency and openly criticized the quality of some of the AMP's work. He continued to argue that the best pure mathematicians made the best mathematicians in wartime.[90] Regardless, it had been with great efficiency that proven science administrator Weaver had gotten the AMP up and running in New York City early in 1943.

The Panel functioned, in consultation with an advisory committee, through contracts to groups at eleven hand-picked universities. With Weaver as Chief, Fry as his Deputy, and Mina Rees, 1931 Ph.D. under Dickson at the University of Chicago, as his executive assistant and the AMP's chief technical aide, the AMP had as advisors former AMS Presidents Veblen, Evans, and Morse, NYU's Courant, Chicago mathematician Lawrence Graves, and Princeton statistician Wilks.[91] Among the universities with which they entered into contracts were, not surprisingly given the committee's make-up, NYU, Princeton, and Berkeley, but also Columbia, where—as Weaver

89. Memorandum: "Mathematics in War" (emphasis in the original). Stone's view of Weaver was far from universally shared. W. Allen Wallis, who headed up the AMP's Statistical Research Group at Columbia (see later in this section), exuberantly described Weaver as "one of the most remarkable, admirable, brilliant, sagacious, and civilized human beings on the American scene in the past half-century" in "The Statistical Research Group, 1942–1945," *JASA* 75 (1980), 320–330 on p. 322.

90. Stone began his campaign against Weaver and his leadership of the AMP and, by association, of mathematics in wartime as early as the late fall of 1943, keeping it up in earnest through the fall of 1944. He clearly believed that the War Policy Committee, of which he was a key part, should have served in the coordinating role. See the various letters in Box 13, Folder: Stone, Marshall 1943–44, Veblen Papers as well as Stone's speech (cited above), "American Mathematics in the Present War," delivered before the AMS at its summer meeting in Wellesley, MA, in August 1944 and published in no less prominent a place than *Science*. It also seems clear that AMS President Stone viewed AMP Chief Weaver as a leadership rival.

91. Mina Rees, "The Mathematical Sciences and World War II," *AMM* 87 (1980), 607–621; reprinted in *A Century of Mathematics in America*, ed. Peter Duren et al., 1: 275–289 on p. 277.

knew—physicists were already hard at work on the Manhattan Project, and Brown with its new program in applied mathematics.

In particular, the AMP sponsored so-called Applied Mathematics Groups (AMGs) at Columbia first under E. J. Moulton of Northwestern and then under Saunders Mac Lane, at the Institute under von Neumann, at NYU under Courant, at Northwestern under Moulton after his move back to the Midwest from New York, at Brown under Richardson and Prager, and at Harvard under George Birkhoff; Statistical Research Groups (SRGs) at Columbia under economist and statistician W. Allen Wallis, and at Princeton under Wilks; and Ballistics Research Groups at Columbia under astronomer Jan Schilt and at Berkeley under statistician Jerzy Neyman.[92] As chief technical aide, Rees oversaw the work of all of these groups, making site visits and generally keeping Weaver apprised of the progress being made.[93] When needed, moreover, the Panel's groups—as well as the NDRC and other units—tapped the services of the Mathematical Tables Project for help with their purely calculational work. Located in New York, the latter project, under the direction of Gertrude Blanch, a 1935 Cornell Ph.D., had been created in 1938 by Roosevelt's Works Progress Administration in its effort to provide work for Americans still affected by the Depression.[94]

The AMGs, SRGs, and BRGs worked on projects that had been "formulated by the Panel in response to requests for help from the military services or their contractors."[95] For example, Courant and his NYU team, that included his colleagues Friedrichs and Stoker, worked on phenomena associated with

92. Owens lists these and the other major AMP contractors—the University of New Mexico, the Carnegie Institution of Washington, and the National Bureau of Standards—together with the number of personnel funded and the funding level of each on p. 288.

93. Kathleen Broome Williams, *Improbable Warriors: Women Scientists and the U. S. Navy in World War II* (Annapolis: Naval Institute Press, 2001), p. 168–169 and Amy Shell-Gellasch, *In Service to Mathematics: The Life and Work of Mina Rees* (Boston: Docent Press, 2000), pp. 57–63.

94. David Alan Grier, *When Computers Were Human* (Princeton: Princeton University Press, 2005), pp. 212–219 and 247–250. Blanch had earned her Ph.D. under Virgil Snyder in his brand of algebraic geometry but had been unable to get a academic job in the tight, Depression-Era job market. For more on her life, see, for example, Green and LaDuke, pp. 144–145.

95. Saunders Mac Lane, "The Applied Mathematics Group at Columbia in World War II," in *A Century of Mathematics in America*, ed. Peter Duren et al., 3: 495–515 on p. 496.

The actual work done by the AMP was surveyed in 1946 in three volumes entitled *Summary Technical Report of the Applied Mathematics Panel* (Washington, D.C.: NDRC, 1946). The first

aerial and underwater explosions and did important work on the theory of shock waves.[96] Prager's group at Brown focused, quite naturally given the orientation towards mechanics of Brown's new applied mathematics program, on problems that could be cast in terms of classical dynamics as well as on the mechanics of deformable media, while at Harvard, Birkhoff, his son, Garrett, and others worked on underwater ballistics and especially on the problem of water entry.[97] Those working on statistical issues—like Churchill Eisenhart, Harold Hotelling, Milton Friedman, Abraham Wald, and Jacob Wolfowitz at Columbia—developed statistical methods for use in inspection, research, and development work, while the teams under Wilks at Princeton and Neyman at Berkeley focused primarily on probabilistic and statistical aspects of bombing problems.[98] In particular, in 1943, Wald developed the time-saving sequential probability ratio test that uses data as they are being gathered in order to determine when to stop an inspection, experiment, or test as opposed to continuing through to the end of some, potentially large, predetermined sample.[99] This proved to be one of the major, theoretical outcomes of mathematicians' wartime engagement.

By far the largest of these teams was the AMG at Columbia (AMG-C) with, over the course of its almost three-year existence from 1943 to 1945, some thirty mathematicians serving on the research staff, a group of twenty female computers, and a secretarial and administrative support staff numbering almost two dozen more women.[100] At first, though, it was unclear

---

volume, edited by Ivan Sokolnikoff, treats *Mathematical Studies Relating to Military Physical Sciences*; the second, edited by Mina Rees, deals with *Analytical Studies in Aerial Warfare*; and the third, edited by Samuel Wilks, covers *Probability and Statistical Studies in Warfare Analysis*. The URLs for these volumes are given in the References.

96. Reid discusses the AMP work of the NYU group as well as the impact that the contract had on Courant's efforts to build his program there in *Courant*, pp. 236–245.

97. Rees, "The Mathematical Sciences in World War II," p. 280.

98. Ibid., pp. 281–284.

99. While still engaged in its war work, the Statistical Research Group began writing up the applications side of the new theory in *Sequential Analysis of Statistical Data: Applications* (New York: Columbia University Press, 1945). Patti Hunter discusses Wald's discovery in "Connections, Context, Community: Abraham Wald and the Sequential Probability Ratio Test," *MI* 26 (2004), 25–33. For more details on the work of and personnel associated with the SRG-C, see Wallis, "The Statistical Research Group."

100. Mac Lane, "The Applied Mathematics Group at Columbia," pp. 511–514. Among the female computers, Anna Merjos was an undergraduate at Hunter College who earned her B.A. in 1944 and went on to a successful career as a financial analyst. Evelyn Garbe had done

precisely what sort of questions the AMG-C would tackle. When Mac Lane accepted Weaver's invitation to come down from Harvard to be one of the group's first members, for example, Weaver wrote him "a long letter to get [him] thinking about a problem in which various gases and liquids circulate in an elaborate arrangement of pipes and tubes."[101] Yet, once Mac Lane arrived in New York, he "never again heard of this particular problem" and only later realized that it "had to do with the gaseous diffusion process for separating isotopes of uranium" that was going on clandestinely at Columbia as part of the Manhattan Project.[102] In retrospect, he realized that "the secrecy about the problem was such that nothing about it would have been delegated to a bunch of mathematicians in a project housed in a converted apartment building in Morningside Heights next to Columbia University."[103]

The AMG-C's work focused instead on "all the varieties of airborne fire control," that is, on "how best to make use of the various lead computing sights" that equipped U.S. aircraft. In simplified terms, "[t]he gunner on a bomber tracks an approaching fighter; a gyroscope on his gun measures the rate of change of angle and multiplies this by 'time of flight' (of the bullet) to determine the angle by which the gun direction should 'lead' the fighter in order to score a hit." In the air, this is complicated by the difference between the lead computed by the bombing sight and the "true lead," that is, the ballistic lead determined from ballistic tables plus the lead arising from the relative motion of the two aircraft, the so-called kinematic lead.

To cite just one problem assigned to the AMG-C, University of Chicago mathematician Magnus Hestenes "did real mathematics" on the problem of calculating true lead by "appealing to ballistic theory (the differential equations) rather than to the derivative ballistic tables."[104] After the war, he garnered an individual citation for this work in the Naval Ordnance Development Award that the AMG-C collectively received. Other of his AMG-C colleagues so honored were Irving Kaplansky (an algebraist and

---

graduate work at Chicago, earning her M.S. in mathematics and completing all of the coursework for the doctorate. After the war, she had a long career at the National Security Agency. See their obituaries, the URLs for which are in the References.

101. Ibid., p. 496.

102. Mac Lane, *A Mathematical Autobiography*, p. 119.

103. Mac Lane, "The Applied Mathematics Group at Columbia," p. 497. For the quotations that follow in the next two sentences, see pp. 497–498.

104. Ibid., p. 502. For the list of awardees that follows, see pp. 502–503 as well as "Notes," *BAMS* 52 (1946), 979–1002 on p. 979.

1941 Harvard Ph.D.), Walter Leighton (of the Rice Institute), Donald Ling (a 1944 Columbia Ph.D.), the AMG-C's director Mac Lane, and Hassler Whitney (an Associate Professor at Harvard). Although, as at the Aberdeen Proving Ground, the mathematics needed even in this explicitly cited work was little more than plane and spherical trigonometry, differential calculus, and differential equations, the results were critical to aerial gunners' success and hence to the war effort as a whole.

Little wonder, then, that the AMG-C had close connections with the Army Air Forces (AAF). Gustav Hedlund, who had been called to the AMG-C from Virginia in the summer of 1944, described one aspect of the relationship this way. "The A.A.F. demands an answer, even if you can't get an honest one, but it is understandable that decisions must be made on the basis of whatever information is available, even though it is incomplete."[105] Indeed, the AAF relied on the AMG-C not only for down-and-dirty problem-solving but also for training, in particular, of the operations analysts who joined its ranks during the second half of the war.

By the end of 1942, the AAF had, following closely behind the Navy, established its first research group in what the Americans called operations analysis and the British—as well as those Americans working in Britain—termed operations research. Pioneered by the British Royal Air Force (RAF) not long after Great Britain's declaration of war, operations research involved using "scientists and other skilled civilians in the application of scientific methods to military operations, such as early warning systems, bombing, gunnery, and antisubmarine warfare."[106] As the war progressed, the AAF incorporated operations analysis groups into various of its air commands both at home and abroad, but the first and ultimately the largest was with the AAF's Eighth headquartered just outside London. Its mission was to analyze "bombing operations with a view to finding weak points in our method of attack in bombing and also any weaknesses in the employment of the enemy defense system" as well as "[t]o assess and evaluate the effectiveness of bombing attacks and also to investigate various communication problems relating to this command." In the two-and-a-half years it functioned, the Eighth's Operational Research

---

105. Hedlund to Gordon Whyburn, Box 1, Folder 75: Hedlund, Gustav A., 30 July, 1944, Whyburn Papers.

106. Charles McArthur, *Operations Analysis in the U. S. Army Eighth Air Force in World War II*, HMATH, vol. 4 (Providence: American Mathematical Society and London: London Mathematical Society, 1990), p. 1. The quotations that follow in this paragraph are also from this source on p. 19.

Section (ORS) hosted some four dozen people with scientific and/or technical training, among them eighteen mathematicians, fifteen of whom were analysts and thirteen of whom remained with the section for six months or more.[107]

In the beginning, these civilian mathematicians were recruited for the AAF directly by the military, but it was quickly realized that that was perhaps not the best approach. Before the end of 1943, the AMP, with its knowledge of the American mathematical landscape, had taken charge of recruitment with the AMG-C and SRG-C collaborating to give crash courses in the mathematics of gunnery. These supplemented the training in gunnery per se that the men received at various AAF gunnery schools stateside before taking up their posts in England.[108] Although the experiences in the Eighth's ORS of Purdue's William Ayres, the University of Kansas's Baley Price, and UCLA's Angus Taylor are more typical of the mathematician's role in the ORS—Ayres was a gunnery analyst while Price and Taylor worked, among other things, on analyzing bombing data and making specific recommendations on how to improve accuracy—those of Edwin Hewitt, a 1942 Harvard Ph.D. under Stone, illustrate just how diverse such experiences could be.[109]

Hewitt was one of the longest serving mathematicians with the Eighth's ORS, joining the group in April 1943 and remaining with it until August 1944. An instructor at Harvard when he was approached by ORS Chief Col. John Harlan, Hewitt "knew nothing about gunnery when he arrived in England" but was nevertheless assigned the so-called "gunnery problem."[110] At issue were the disturbing losses of Eighth AAF aircraft in their sorties against the German Luftwaffe. On the Eighth Bomber Command's first four missions to targets in Germany in 1943, for example, some 11% of its bombers were lost, most as a result of attacks from German fighters.[111] Something needed to be done.

107. Ibid., pp. xviii and xxi. The eighteen were: James Alexander, William Ayres, Blair Bennett, Harry Carver, James Clarkson, Robert Dilworth, Ray Gilman, Edwin Hewitt, Forrest Immer, Ralph James, George Mackey, John Odle, Baley Price, Bob Robertson, Frank Stewart, Angus Taylor, John Youden, and J.W.T. "Ted" Youngs.

108. Ibid., pp. 124–125.

109. McArthur discusses the work of all four of these men in some detail throughout his book. Other mathematicians who receive extensive coverage, although less coverage than these four, are: Clarkson, Immer, Odle, Stewart, Youden, and Youngs.

110. Ibid., p. 53.

111. Ibid., p. 50.

FIGURE 8.6. Members of the Bombing Accuracy Subsection of the Eighth Air Force
Operations Analysis Section: Left to right: James A. Clarkson, Lt. Col. Philip C. Scott,
G. Baley Price, Forrest R. Immer, William John 'Jack' Youden, and Charles R. Darwin in 1944.
(Photo courtesy of Air Force Operations Analysis Section Collection, 1942–1948, 1985, and
undated, e_math_00678, Archives of American Mathematics, Dolph Briscoe Center for
American History, University of Texas at Austin.)

Hewitt immersed himself in the problem, reading all of the material on
aerial bombing and exterior ballistics that he could get his hands on. "As he
began to understand the mathematics of the gunnery problem," he realized
that he "needed tables giving the time of flight of the bullet as a function
of range and muzzle velocity" but that such tables were unavailable in England.[112] After an inquiry at the American Embassy in London proved unpro-
ductive, he sought the counsel of gunnery experts at the British Ministry of
Aircraft Production, who produced "some elegant ballistic charts, from which
it was possible to compute the time of flight and bullet drop for .50 caliber

112. Ibid., p. 53. For the quotations that follow in this and the next two paragraphs, see
pp. 53–55 (unless otherwise indicated). The latter are taken from the account Hewitt gave in
the 1945 unpublished memoir, "A Sketch of Gunnery Activities in the Operational Research
Section Eighth Army Air Force, from June 1943 to August 1944," on his wartime service.

ammunition." Using these charts, Hewitt computed a full set of time of flight tables and ultimately succeeded in computing "the deflection necessary to hit a fighter attacking the bomber, under the assumption that the fighter's path is rectilinear during the time of flight of the bullet."[113] "Having computed the bomber's deflections for ordinary B-17 and B-24 altitudes and airspeeds, for several fighter speeds, and representative ranges and angles off," he realized that the angle off, that is, the angle of the target relative to the aircraft, was "the principle variable to consider." From this, he concluded that the British zone system of firing—a system whereby a gunner's field of fire was divided into zones as opposed to aiming at one particular point—was the best operating practice. The problem then became one of getting this information out to American gunners in a user-friendly way.

To help effect that, Hewitt actually enrolled in an RAF gunnery school in order to understand the problem from the gunner's—as opposed to the mathematician's—point of view. While he reported learning much about the mechanics and maintenance of the .50 caliber machine gun there, he got "absolutely no information on deflection shooting." In frustration, he concluded that "[g]unners, bombardiers, and navigators were going into combat without any knowledge whatever of how to aim their guns so as to hit attacking fighters." Indeed, something *did* need to be done . . . and quickly.

Like a good mathematician, Hewitt broke the problem down into its component parts. He needed "to work the zone system into a form so that all gunners, even the least intelligent, could understand it and apply it readily." To that end, he divided "each gunner's field of fire into zones, which were distinguished by reference marks on the bomber itself, in each of which zones the gunners were to keep a constant deflection." He next wrote down step-by-step rules for each type of gunner—waist gunners, nose gunners, radiomen gunners, tail gunners—"in terms of the particular sight he was using and for each of the zones encountered in shooting from his position." Together with explanatory diagrams, these rules were then reproduced on "poop sheets" that "were distributed literally to thousands of gunners." For even greater assurance that the gunners got the message, Hewitt personally went to various of the bomber groups to present and discuss the material. So well was his program received that Harlan arranged for Hewitt to detail his findings to the commanding general of the Eighth Bomber Command. In this way, the

113. The deflection of a shell is a calculable function of the wind velocity, the time of flight of the shell, the length of the range traversed by the shell, and the muzzle velocity of the gun.

"[t]wenty-three-year-old Hewitt, new mathematics Ph.D. from Harvard and neophyte gunnery expert . . . was in the position to tell the commanding general" in the fall of 1943 "how the gunners *should have been* aiming their guns for the past year"![114]

Hewitt's work earned him the confidence and respect of men at all levels in the chain of command from the gunners, to their instructors, to the top brass. Indicative of this, he was sent back to the United States in February 1944 "to assist in writing a gunners' manual and to reach an agreement on a single system of deflection shooting to be taught in the United States and in the ETO [European Theater of Operations]."[115] Following his return to England in April, he became the de facto spokesperson for the gunners at the Eighth's headquarters. He also got back to work on operational questions.

During his absence over the course of the spring, some gunners had noticed a disturbing phenomenon that required not only explanation but countermeasures. Their planes had been attacked in ways that could not have been the result of a pursuit curve, that is, the course a plane takes while following a target. Because of this, their firing rules had been rendered ineffectual. Two possible explanations for this had been posited, but which, if either, was the cause? As Hewitt later explained, "[w]e made large numbers of theoretical calculations concerning both types of attack and ran a number of air tests," which seemed to answer the question.[116] But, was their theoretical conclusion a conclusion in practice?

Hewitt argued that he needed to witness these non-pursuit attacks firsthand in order to be sure they had the right answer for what was actually happening in the air. Only after taking his case all the way to the very top of the chain of command, Eisenhower's office, did he gain approval for his plan. In all, he flew seven missions in May and June 1944, one of which was over Cherbourg on D-Day. "Officially," Hewitt related, "I rode as an observer, but I found it useful (and by no means repugnant to the bombardier) to ride in the bombardier's seat, and consequently, on four missions, acted as bombardier-gunner." Although Hewitt never did observe one of the non-pursuit curve attacks, he found that his "[p]articipation in these combat missions was an extremely valuable experience." Apparently, the AAF did, too. Hewitt was the only operations analyst to earn its Air Medal.

114. McArthur, p. 55 (my emphases).

115. Ibid., p. 123 (quoting Hewitt's unpublished memoir).

116. Ibid., p. 161 (quoting Hewitt's memoir). For the quotations that follow in the next paragraph, also from Hewitt's memoir, see pp. 162–163.

By August 1944, the stresses of his work had begun to take their toll. Hewitt, suffering from exhaustion, was sent back home to rest. Although he did not return to England after his forced hiatus, he did take up a new AAF post as a "trouble-shooting gunnery analyst" for the Twentieth headquartered at the Pentagon. He remained there until the end of the war.[117]

Hewitt, an early civilian operations analysis recruit, did not have the advantage of going through the joint AMG-SRG training courses at Columbia before going to England, but he did pass through that training ground subsequently. Later recruits, like fellow Eighth AAF operations analysts, Ayres and Penn State topologist John Odle, did. Tulane analyst William Duren did as well, although he served as an operations analyst not in England but with the Second AAF headquartered stateside in Colorado Springs.[118] Still others, like Columbia ergodic theorist Bernard Koopman, and the then Berkeley graduate student George Dantzig, got their exposure to operations analysis in other sorts of war work, Koopman at the Antisubmarine Warfare Operations Research Group in Washington and Dantzig at the Office of Statistical Control at AAF headquarters also in the capital city. Dantzig's war work, in particular, involved the development of "statistical factors, such as sortie rates," needed for combat planning by the AAF and primed him for the articulation in 1947 of the simplex method, "a model based on linear inequalities . . . used for planning activities of large-scale enterprises," that became so essential in the evolution of linear programming after the war.[119] The efforts of all of these mathematicians, as well as of other men and women, served to create a brand-new field of active inquiry, operations research, in the immediately postwar period.

Just as mathematicians cum operations analysts plied their developing trade in a variety of wartime contexts so, too, did other mathematicians at all levels of their professional careers. Donald May, a 1941 Princeton Ph.D. under

117. Ibid., p. 251.

118. Mac Lane, "The Applied Mathematics Group at Columbia," p. 504. See also William Duren's handwritten memoir, "Operations Analyst, US Army Air Force in WWII," 75 pp., Box 2010-192/9, Duren Papers.

119. For the quotations, see George Dantzig, "Impact of Linear Programming on Computer Development," Technical Report SOL 87-5, Systems Optimization Laboratory, Department of Operations Research, Stanford University, June 1985, p. 1 and George Dantzig, "Origins of the Simplex Method," Technical Report SOL 87-5, Systems Optimization Laboratory, Department of Operations Research, Stanford University, May 1987, p. 2, respectively. For the URLs, see the References.

Salomon Bochner, served at the Navy's Bureau of Ordnance.[120] Notre Dame instructor and 1940 Rice Ph.D. John Nash (not to be confused with game theorist John Forbes Nash later of Princeton), as well as MIT professor Wiener, worked with the OSRD-sponsored Radiation Laboratory at MIT. The so-called Rad Lab was the site of major breakthroughs during the war in microwave and radar technologies such as Wiener's development—aided on the engineering side by Julian Bigelow—of the Wiener filter, a device in signal communication that reduced the amount of noise present in a signal based on a statistical treatment of anti-aircraft fire control.[121] Institute professor Morse served as a consultant to the Office of the Chief of Ordnance of the Army on matters of terminal ballistics, like bomb damage and how to maximize it, and was awarded both the Ordnance Department Meritorious Service Award in 1944 and the Army-Navy Certificate of Merit in 1948.[122] Princeton stalwart and editor of the *Annals of Mathematics* Lefschetz consulted at Maryland's David Taylor Model Basin. There, he met Nicholas Minorsky and became aware of the "importance of the applications of the geometric theory of ordinary differential equations to control theory and nonlinear mechanics," the topic that defined Lefschetz's research from 1943 to the end of his life.[123]

Harvard professor Stone had perhaps even more varied wartime experiences. He was, first, from 1942 to 1943, a contract employee working on mine warfare at the Navy's Bureau of Ordnance and attached to the Office of the Chief of Naval Operations in Washington and, then, from 1944 to 1945, a civil service employee in the Office of the Chief of Staff of the War Department. In the latter post, he was part of the Military Intelligence Service and sent on an extended clandestine mission to British radio intelligence units in India and Ceylon (modern-day Sri Lanka) in order to provide "an extensive analytic

120. Jaques Cattell, ed., *American Men of Science*, 9th ed., vol. 1.

121. On Nash, see John Kelley to Gordon Whyburn, undated but 1942, Box 1, Folder 96: Kelley, John L., Whyburn Papers. After the war, Nash served as an applied mathematician at the University of Illinois's Digital Computing Laboratory, where two computers, first ORDVAC and then ILLIAC, were built using the architecture developed by von Neumann, Goldstine, Bigelow, and others at the Institute after the war. Nash also did research in numerical analysis under contract with the Office of Naval Research (ONR). On the ONR, see the next chapter. On Wiener, see his *I Am a Mathematician*, pp. 249–255.

122. Pitcher, "Marston Morse, March 24, 1890–June 22, 1977," pp. 226–227.

123. Phillip Griffiths, Donald Spencer, and George Whitehead, "Solomon Lefschetz: September 3, 1884–October 5, 1972," *Biographical Memoirs*, vol. 61 (Washington, D.C.: National Academy of Sciences, 1992), pp. 271–313 on p. 275.

treatment of the problems of the U. S. Army radio-intelligence service."[124] The activities of the listening posts that Stone visited were intimately related to cryptographic work being done back home by, for example, Chicago's Adrian Albert.[125] These—and other mathematicians who served with them and elsewhere during the war—demonstrated how mathematical training of all sorts could fruitfully be brought to bear on a war effort.

One last venue, although one generally associated with physicists, chemists, and engineers, also defined the war work of a few mathematicians: the Manhattan Project at Los Alamos. Perhaps the most visible of the mathematicians there was John von Neumann, who made repeated trips west beginning in early September 1943. Indeed, von Neumann estimated that thirty percent of his time from then until the end of the war was spent consulting at Los Alamos.[126] In particular, the expertise that he had gained in his work with the Navy and the NDRC on shock wave behavior informed his major contribution to the building of the plutonium bomb: the theoretical conclusion—reached in October 1943 in collaboration with fellow Budapest native and Hungarian émigré, physicist Edward Teller—"that implosion at more violent compressions than" had yet been "attempted should squeeze plutonium to such unearthly densities that a solid subcritical mass could serve as a bomb core."[127] This insight provided a (by no means trivial!) workaround to development problems that had seemed well-nigh intractable. The problem became one of translating the theory into a workable bomb, and that, in turn, required more theoretical work—the generation of equations and their solutions—as experiments turned up new complications. To aid in this, von Neumann enlisted Wisconsin's Stanislaw Ulam, another fellow Hungarian, for Los Alamos late in 1943. By the winter of 1944, von Neumann, Ulam,

124. Ronald Spector, *Listening to the Enemy: Key Documents on the Role of Communications Intelligence in the War with Japan* (Wilmington: Scholarly Resources Inc., 1988), p. 149. Transcriptions of Stone's reports back to headquarters are given on pp. 136–150. I thank Peter Donovan of the University of New South Wales for pointing this source out to me.

125. See Nancy Albert, pp. 126–127. Although they did not, for the most part, have formal mathematical training beyond whatever mathematics they might have had in high school and college, the so-called "code girls" working in Arlington Hall just across the Potomac from Washington, D.C., also played a key role in the United States' cryptographic efforts during World War II. See Liza Mundy, *Code Girls: The Untold Story of the American Women Code Breakers of World War II* (New York: Hachette Books, 2017).

126. Macrae, pp. 207–215.

127. Richard Rhodes, *The Making of the Atomic Bomb* (New York: Simon and Schuster, 1986), p. 480.

and John "Jack" Calkin, a 1938 Harvard Ph.D. under Stone, former assistant to von Neumann at the Institute, and member of the Los Alamos team, were discussing the hydrodynamical problems associated with implosion and making detailed calculations.[128] Less than a year later, their ideas had moved into the development phase.[129] The result was the bomb, called Fat Man, dropped on Nagasaki on 9 August, 1945.[130]

In addition to his discussions on implosion with von Neumann, Ulam worked in the group directed by Teller that was simultaneously developing the science behind a thermonuclear-fusion-based weapon.[131] In all of this work, the researchers at Los Alamos were fundamentally aided by yet another mathematician, the same Donald Flanders, who, as a fresh Penn Ph.D., was behind the push to build a modern mathematics department at NYU that netted Richard Courant. Flanders directed Los Alamos's computer department, where he and his team were instrumental in grinding out, among much else, solutions of hyperbolic partial differential equations associated with fluid flows.[132]

American mathematicians at all levels—from graduate students drafted into active military service, to mathematical enlistees, to fresh Ph.D.s "drafted" as civilians to do war work, to seasoned mathematicians engaged in both teaching mathematics to the military and solving the military's mathematical problems—participated in the United States' war effort. At the same time that they threw themselves into this work, they also recognized the importance of maintaining the forward momentum that their community had gained over the course of the 1920s and 1930s.

128. Ulam, *Adventures of a Mathematician*, pp. 153–154.

129. For the interested reader, Rhodes details von Neumann's development of the notion of an explosion lens, that is, an arrangement of explosives that initially produces a convergent instead of a divergent shock wave, on pp. 544–545 and 575–576.

130. Macrae, p. 241. Von Neumann was part of the team that calculated the optimal delivery specifications—height of release, etc.—for Fat Man.

131. Rhodes, pp. 545–546 and Ulam, *Adventures of a Mathematician*, pp. 152–153. Despite Saunders Mac Lane's recollection to the contrary, Ulam would only be joined at Los Alamos by C(ornelius) J. Everett in 1947. Compare Mac Lane, "The Applied Mathematics Panel," p. 496; Ulam, *Adventures of a Mathematician*, p. 131; and Jaques Cattell, ed., *American Men of Science*, 9th ed. vol. 1. Ulam and Everett, a 1940 Wisconsin Ph.D. under Cyrus MacDuffee, struck up a long-lasting friendship and mathematical collaboration on Everett's return to an instructorship in Madison in 1942.

132. Reid, *Courant*, pp. 241–242.

	1939	1940	1941	1942	1943	1944	1945
AJM	1008	912	888	772	736	648	616
AM	948	896	1239	831	810	800	718
BAMS	952	970	964	958	947	950	1008
AMM	674	734	726	708	664	624	610
TAMS	975	1074	1095	1191	1085	1095	964
DMJ	962	508	770	901	785	897	722
JSL	194	188	188	180	164	107	158
MR	—	399	419	375	339	327	334
QAM	—	—	—	—	355	353	384
MTOAC	—	—	—	—	132	203	143

FIGURE 8.7. Pages published per year per journal, 1939–1945.[134]

Trying to Maintain Professional Normalcy in Wartime

A little over a year into the war, John Kline put into words a concern then on the minds of many in the American mathematical community. "I am afraid," he wrote to Gordon Whyburn, "that all mathematicians are going to be so busy with either war research or with heavy teaching loads in their departments that it will be difficult to carry our normal publication program at anything like its usual pre-war level."[133] In retrospect, there was little need to worry, at least not relative to the journal publication of stand-alone research articles (see fig. 8.7). Although the numbers of pages published fluctuated somewhat from 1939 to 1945, all of the journals remained robust over the course of the war, and a new journal, the *Quarterly of Applied Mathematics*, firmly established itself following its 1943 launch, just as Richardson and his colleagues had felt sure it would.

Some of this research issued from schools like Michigan and Illinois with heavy wartime teaching commitments. Some was produced by mathematicians at schools less affected by the war. Consider the example of just two

133. Kline to Whyburn, 28 January, 1943, Box 1, Folder 99: Kline, J. R., Whyburn Papers.

134. The numbers for the *Mathematical Reviews* include the indices. Because of a somewhat late start in 1943, the *Quarterly of Applied Mathematics* published only three of its four issues in 1943, publishing the fourth in 1944. This was the pattern in its earliest years. The numbers given here are for the full, four-issue volumes. The journal *Mathematical Tables and Other Aids to Computation* was, like the *Quarterly of Applied Mathematics*, begun at Brown in 1943. The numbers are given here for completeness, but for more on the journal, see chapter ten.

journals—the more teaching-oriented *American Mathematical Monthly* and the research-oriented *Duke Mathematical Journal*—in one year, 1944.

On average, some four, full-length articles opened each of the *Monthly's* ten issues in 1944. These were followed by its sections on "Discussions and Notes," "Problems and Solutions," "News and Notices," "War Information," book reviews, and meeting reports to fill out its average of some sixty pages per issue. Angus Taylor, for one, submitted a study on the "Differentiation of Fourier Series and Integrals," while Thomas Wade and Richard Brunk teamed up to consider "Types of Symmetries."[135] It was not long after he had submitted his paper that Taylor was granted leave to assume his wartime post with the Eighth Army Air Force's ORS in England. Wade, a 1933 Ph.D. from the University of Virginia, and Brunk, who had worked with Richard Brauer at Toronto and earned his Ph.D. there in 1940, both taught stateside throughout the war, Wade at Florida State College for Women, a school which did not have the additional burden of targeted mathematical teaching for the military, Brunk at Wisconsin which did. Dunham Jackson, Raymond Wilder, and John Synge, each of whom participated in war work of one sort or another, shared the *Monthly's* pages with Taylor, Wade, Brunk, and others.[136]

At least two émigrés also published there in 1944: Max Dehn, by then at St. John's College in Maryland, had, as noted, emigrated late, only in 1941; Paul Halmos had come early, in 1929 at the age of thirteen, and had finally landed a more permanent position at Syracuse in 1943 after holding temporary posts at his alma mater, Illinois, and at the Institute.[137] Dehn was already in his mid-sixties in 1944 and at a small liberal arts college that was relatively insulated from the war. Halmos taught ASTP courses at Syracuse where he had an eighteen-hour teaching load "for several of the war semesters" and twenty-one hours when he taught the " 'super' ASTP course."[138] The second

135. Angus Taylor, "Differentiation of Fourier Series and Integrals," *AMM* 51 (1944), 19–25 and Thomas Wade and Richard Bruck, "Types of Symmetries," *AMM* 51 (1944), 123–129.

136. Dunham Jackson, "The Harmonic Boundary Value Problem for an Ellipse or an Ellipsoid," *AMM* 51 (1944), 555–563; Raymond Wilder, "The Nature of Mathematical Proof," *AMM* 51 (1944), 309–323; and John Synge, "Focal Properties of Optical and Electromagnetic Systems," *AMM* 51 (1944), 185–200.

137. Max Dehn, "Mathematics, 200 B.C.–600 A.D.," *AMM* 51 (1944), 149–157 and Paul Halmos, "The Foundations of Probability," *AMM* 51 (1944), 493–510.

138. Paul Halmos, *I Want to Be a Mathematician: An Automathography* (New York: Springer-Verlag, 1985), p. 112. The quotation that follows is also on this page.

year of his three-year affiliation with Syracuse, moreover, found him out of the classroom for eight months and at the Rad Lab in Cambridge. Despite his heavy teaching commitment and his other war work, Halmos managed to publish eight papers—including the one in the *Monthly*—between 1943 and 1946. As he later mused, "[b]ack then the days must have had 36 hours"!

In contradistinction to the *Monthly*, the *Duke Mathematical Journal* carried full-length research articles exclusively and averaged some 225 pages and nineteen articles per quarterly issue in 1944. Jackson also published there as did Morse, Hedlund, and émigré Bergman.[139] When Morse and Hedlund submitted their paper, the former was already in Washington working for Army Ordnance, but the latter had not yet left Virginia for the AMG-C; Bergman was fully engaged in Brown's year-round offerings in applied mathematics.[140]

Research of those not immediately participating in war-related work also appeared in Duke's journal in 1944. Morse's colleague at the Institute, émigré Carl Siegel, published there as did 1942 Chicago Ph.D.s Alice Schafer and J. Ernest Wilkins. Siegel, ensconced in the "paradise for scholars" and totally free of wartime commitments, contributed a clever observation about algebraic integers, while Schafer and Wilkins submitted results from their respective home bases of then all-women's Connecticut College and historically black Tuskegee Institute.[141] Schafer, a student of E. P. Lane, gave the journal a taste of her dissertation research on the singularities of space curves; Wilkins, a Hestenes student and the ninth African-American to earn a U.S. Ph.D. in mathematics, improved on the work of two of his former professors, Bliss and Reid, in his contribution on "Definitely Self-Conjugate

---

139. Dunham Jackson, "Boundedness of Orthonormal Polynomials on Loci of the Second Degree," *DMJ* 11 (1944), 351–365; Marston Morse and Gustav Hedlund, "Unending Chess, Symbolic Dynamics and a Problem in Semigroups," *DMJ* 11 (1944), 1–7; and Stefan Bergman, "Solutions of Linear Partial Differential Equations of the Fourth Order," *DMJ* 11 (1944), 617–649.

140. Bergman joined von Mises at Harvard in 1945 before accepting the professorship at Stanford that he would hold for the remainder of his career. See Box 31, Folder: "Refugees, Bergman, Stefan 1941–46," Veblen Papers as well as Menahem M. Schiffer, Robert Osserman, and Hans Samelson, "Memorial Resolution: Stefan Bergman (1895–1977)," Stanford University Faculty Senate Records, Stanford University Library (for the URL, see the References).

141. Carl Siegel, "Algebraic Integers Whose Conjugates Lie in the Unit Circle," *DMJ* 11 (1944), 597–602; Alice Schafer, "Two Singularities of Space Curves," *DMJ* 11 (1944), 655–670; and J. Ernest Wilkins, "Definitely Self-Conjugate Adjoint Integral Equations," *DMJ* 11 (1944), 155–166.

Adjoint Integral Equations." Not long after this paper appeared, Wilkins left Tuskegee for the position at Chicago's Metallurgical Laboratory, yet another part of the Manhattan Project, that he would hold through the end of the war. While there, he made fundamental contributions to the theory of neutron absorption.

Not surprisingly, some of this work, as well as some of that published during the war in the American mathematical community's other journals, had been presented as part of meetings of the AMS, the MAA, the Association of Symbolic Logic, and other societies. Indeed, the societies' professional activities continued largely uninterrupted throughout the war, although at least one meeting, of the AMS that had been scheduled for New York City in February 1945, was canceled due to travel restrictions.[142] In particular, wartime Presidents of the AMS—Morse (1941–1942), Stone (1943–1944), and Hildebrandt (1945–1946)—and of the MAA—Raymond Brink (1941–1942), William Cairns (1943–1944), and Cyrus MacDuffee (1945–1946)—guided their organizations at the same time that they either pursued their contract war work, in the cases of Morse and Stone, or their teaching in the cases of Hildebrandt, Brink, Cairns, and MacDuffee. As department chair at Michigan, Hildebrandt, for example, was responsible not only for teaching in but also for figuring out the staffing for the large number of ASTP and Pre-Meteorology Program courses that swamped his department, while Cairns, who had retired from his professor- and department headship at Oberlin in 1939, taught as a visiting professor in the University of New Mexico's Pre-Meteorology and Navy V-12 Programs in 1943 and 1944, often making it difficult for him to preside over MAA meetings back East.[143] Regardless, the mathematical "show" had to go on.

There were prizes to award. The AMS's Bôcher Memorial Prize went to Jesse Douglas in 1943 for his work in 1939 on Green's function and Plateau's Problem; its Cole Prize in Algebra was awarded to Oscar Zariski in 1944 for results on algebraic varieties published in 1939 and 1940 (see chapter ten);

142. John Kline, "Report of the Meeting of the Council of the Society in New York City on February 24, 1945," *BAMS* 51 (1945), 351–355 on p. 351.

143. On Hildebrandt's efforts, see Box 2, Folder: Mathematics Department Executive Committee Minutes, 1935–1946, University of Michigan, Department of Mathematics Records: 1913–1981 and Box 86-14/66, Folder 9, MAA Records. Cairns's friend, Carroll Newsom, a 1931 Michigan Ph.D. under Walter Ford, was professor and department head at New Mexico. Like others in his position, he cast a wide net in recruiting teachers for the crush of students in the military programs. Likely on Cairns's recommendation, Newsom took the professorship and department headship at Oberlin in the fall of 1944 that he then held until 1948.

the MAA's Chauvenet Prize honored MIT analyst Robert Cameron for the paper on "Some Introductory Exercises in the Manipulation of Fourier Transforms" that he published in the *National Mathematics Magazine* in 1941.[144]

Special lectures also continued to be sponsored. Although the AMS had no Gibbs Lecture in 1942, the series resumed in November 1943 when Harry Bateman spoke on "The Control of Elastic Fluids." He was followed twelve months later by von Neumann lecturing on "The Ergodic Theorem and Statistical Mechanics."[145] The war was over, however, by the time MIT physicist and Physics Department chair John Slater gave his 1945 talk on "Physics and the Wave Equation" inspired at least partly by the fundamental work on microwave transmission he had done during the war at both Bell Laboratories and MIT's Rad Lab.[146]

As for the AMS's series of Colloquium Lectures, it ultimately proceeded uninterrupted, even though the 1945 lectures were almost scuttled during the war's final months. Tibor Radó, who had accepted the invitation to speak, had been asked in June "by the Army Air Force Technical Service to go over to Germany for a period of about ninety days, in connection with a project that require[d] personnel thoroughly conversant with the German language and German conditions."[147] As he explained resignedly to John Kline, who had succeeded Richardson as AMS Secretary in 1941, "[t]here seemed to be no choice but to accept." When the AMS Colloquium Committee tried but failed to find a substitute on short notice, it moved that year's Colloquium Lectures from their customary summer meeting slot to a place on the program of the November meeting in Chicago.[148]

144. Robert Cameron, "Some Introductory Exercises in the Manipulation of Fourier Transforms," *NMM* 15 (1941), 331–356. The *National Mathematics Magazine* was begun in 1926 as the *Mathematics News Letter* of the Louisiana-Mississippi Section of the MAA. The name change took place in 1934, and the journal made explicit its dedication to the publication both of "papers on the cultur[al] aspects, humanism and history of mathematics" and "of expository mathematical articles" as well as to the promotion of "more scientific methods of teaching mathematics." See *National Mathematics Magazine* 9 (1934), 1. It became the *Mathematics Magazine* in 1947.

145. Harry Bateman, "The Control of Elastic Fluids," *BAMS* 51 (1945), 601–646. Von Neumann's lecture was not published, likely due to the time constraints his war work at Los Alamos and elsewhere imposed on him.

146. John Slater, "Physics and the Wave Equation," *BAMS* 52 (1946), 392–400.

147. Box 6, Folder 18: AMS–Committee on Colloquium Publications (1945), Whyburn Papers. The quotation that follows is also from this letter.

148. See various of the letters in Box 6, Folder 18: AMS—Committee on Colloquium Publications (1945), Whyburn Papers.

If the Colloquium Lectures themselves did not fall victim to the press of wartime activities, the intended publication in book form of several of them did. Whyburn's 1940 lectures on "Analytic Topology," Wilder's on the "Topology of Manifolds" in 1942, Hille's two years later on "Selected Topics in the Theory of Semi-Groups," and Radó's in 1945 on "Length and Area" all ultimately appeared. Whyburn's was "on time" in 1942, but Wilder's only came out in 1949 preceded by Hille's and Radó's in 1948 (see chapter ten). Books based on Ore's 1941 lectures on "Mathematical Relations and Structures" and on McShane's 1943 discussion of "Existence Theorems in the Calculus of Variations" never did materialize.[149] War work, although of very different sorts, occupied Ore and McShane, with Ore working tirelessly for organizations geared toward the relief of his native Norway and McShane, as noted, at the Aberdeen Proving Ground.[150] The war also prevented von Neumann from bringing to fruition the volume that would have been based on his 1937 presentation of "Continuous Geometry." As he ruefully admitted to the Colloquium Publications Committee in 1943, "[b]y procrastinating in the years 1937–1938 I delayed the project into the war years, during which it has been physically impossible for me to give the subject the attention which it deserved and to complete the book."[151] The same fate likely befell the book manuscript that should have followed from Stone's 1939 lectures on "Convex Bodies." Two volumes not associated with actual Colloquium Lectures did come out during the war years, however: one, Lefschetz's 1942 *Algebraic Topology*, completed just before the war and the other, Weil's 1946 *Foundations of Algebraic Geometry*, finished in 1944 just before his move from Lehigh to the University of São Paulo (see chapter ten).[152]

The war was even harder on the MAA's Carus Mathematical Monograph series, which had already been struggling. While three volumes had come out

149. All published by the AMS in New York City, the references are: Gordon Whyburn, *Analytic Topology*, vol. 28 (1942); Raymond Wilder, *Topology of Manifolds*, vol. 32 (1949); Einar Hille, *Functional Analysis and Semi-groups*, vol. 31 (1948); and Tibor Radó, *Length and Area*, vol. 30 (1948).

150. "Oystein Ore 1899–1968," *Journal of Combinatorial Theory* 8 (1970), i–iii on p. iii. Ore received the Knight Order of St. Olaf from the King of Norway in 1947 for his war work.

151. Von Neumann to Whyburn, 23 August, 1943, Box 6, Folder 16: AMS–Committee on Colloquium Publications (1941–1943), Whyburn Papers.

152. Both published by the AMS in New York City, the references are: Solomon Lefschetz, *Algebraic Topology*, vol. 27 (1942) and André Weil, *Foundations of Algebraic Geometry*, vol. 29 (1946).

in the 1920s following the series' inception in 1922, only two had appeared in the 1930s. Moreover, seven years elapsed between the publication in 1934 of David Smith and Jekuthiel Ginsburg's *A History of Mathematics in America Before 1900* and that of Dunham Jackson's *Fourier Series and Orthogonal Polynomials* in 1941.[153] The only monograph in the series actually to appear during the war was Cyrus MacDuffee's 1943 *Vectors and Matrices*. Almost ironically and at the same time, the AMS launched a new series of "Mathematical Surveys" with the publication in 1943 of *two* books: *The Problem of Moments* co-authored by James Shohat and Jacob Tamarkin and Nathan Jacobson's *Theory of Rings*.[154] This strong start was followed by a six-year hiatus. The next survey appeared only in 1949.

Of course, the AMS and the MAA were not the only publishers of mathematics books. Commercial publishers and university presses continued to support the field, bringing out both textbooks and research-oriented tracts. Just as the U.S. was about to enter the war, for example, Macmillan in New York City published Garrett Birkhoff and Saunders Mac Lane's *A Survey of Modern Algebra*, a textbook aimed specifically at the undergraduate and beginning graduate classrooms that was heralded as "an important contribution to the pedagogy of algebra."[155] A year later, Princeton University Press added Paul Halmos's *Finite Dimensional Vector Spaces* to its Annals of Mathematics Studies series. His text targeted graduate students exclusively and sought "to emphasize the simple geometric notions common to many parts of mathematics, and to do it in a language which gives away the trade secrets and tells the students what is in the back of the minds of people proving theorems about integral equations and Banach spaces."[156] Undergraduate and graduate teaching would also go on; a next generation of mathematicians would be trained. As for publications at the research level, they, too, continued to

153. See the bibliography for the full references.

154. Both published by the AMS in New York City, the references are: James Shohat and Jacob Tamarkin, *The Problem of Moments*, no. 1 (1943) and Nathan Jacobson, *The Theory of Rings*, no. 2 (1943).

155. Robert Thrall, Review of "*A Survey of Modern Algebra*. By Garrett Birkhoff and Saunders Mac Lane. New York, Macmillan 1941. 11+450 pp. $3.75.," *BAMS* 48 (1942), 342–345 on p. 342. See also Garrett Birkhoff and Saunders Mac Lane, "*A Survey of Modern Algebra*: The Fiftieth Anniversary of Its Publication," *MI* 14 (1992), 26–31.

156. Mark Kac, Review of "*Finite Dimensional Vector Spaces*. By Paul R. Halmos. (Annals of Mathematics Studies, no. 7.) Princeton University Press, 1942. 5 + 196 pp.," *BAMS* 49 (1943), 349–350 on p. 349 (quoting Halmos's preface).

be produced, among them, John von Neumann and Oskar Morgenstern's mammoth *Theory of Games and Economic Behavior* (see chapter ten).[157]

If research and publication were two of the hallmarks of professional normalcy for the American mathematical community, international cross-fertilization was another. In the 1890s, American mathematical aspirants had gone to Europe. Beginning in the 1920s, the Rockefeller Foundation had made it possible both for Americans to go abroad and for Europeans to come to the United States. The Visiting Lecturer program that the AMS had launched in the 1920s was sustained into the 1930s until the situation in Europe made it untenable. Despite the restrictions on travel that the war imposed, such international contact, albeit not with war-torn Europe but with other regions, continued to have interesting implications for mathematics in the United States. Indeed, the community's international focus of the 1930s was significantly broadened to include, in particular, China and Latin America.

As noted, the Boxer Indemnity Scholarship Program had already served to attract Chinese *students* with strong mathematical talent to the United States. For the most part, they had come, studied, and returned home to establish programs in mathematics around China. By the late 1930s, however, the Institute for Advanced Study had come to represent a *research-level* "bridge between the U. S. and China" as it had between the United States and mathematical émigrés fleeing Europe.[158] The outbreak of the Second Sino-Japanese War in 1937 had wreaked havoc on Chinese academe, resulting, for example, in the merger of the faculties of Peking, Tsinghua, and Nankai Universities and then of the relocation of the new conglomerate to Kunming in western China. Shiing-Shen Chern was one of the mathematicians affected by these events.

European-trained at the University of Hamburg where he had earned his doctorate in 1936 and at the Sorbonne where he had done postdoctoral work with Élie Cartan in 1936–1937, Chern returned from Europe to take up a university post at Tsinghua just following the relocation. Cut off from Western mathematical resources, he worked on the differential geometrical ideas to which Cartan had exposed him and began to write up his results.[159] He

157. John von Neumann and Oskar Morgenstern, *Theory of Games and Economic Behavior* (Princeton: Princeton University Press, 1944).

158. Xu, p. 296.

159. Nigel Hitchin, "Shiing-Shen Chern, 26 October 1911–3 December 2004," *Biographical Memoirs of Fellows of the Royal Society* 60 (2014), 75–85.

included some of this research—in particular, a paper that "he [had] offered for publication in one of the American mathematical journals"—in the 1941 letter to Veblen that he hoped would secure for him a position at the Institute for the 1942–1943 academic year.[160] Veblen was impressed. Chern's paper, Veblen told the Institute's then director Frank Aydelotte, "struck me as being extremely good, and the referee's report pronounces it first class."[161] In fact, he continued, Chern's "work altogether seems to establish" him as "the most promising Chinese mathematician who has thus far come to our attention." Yet, not even this endorsement was enough to secure the funding necessary to bring Chern to the Institute in 1942. It would take almost another year before he would make it to Princeton, but then he would have the opportunity to work there uninterrupted for two-and-a-half years. Eight more Chinese mathematicians followed him between 1945 to 1950.[162]

Like American mathematical relations with China, those with Latin America also stemmed from broader, extra-mathematical initiatives that went back at least to the turn of the twentieth century with the inclusion of the United States in the Pan-American Scientific Congress (recall chapter three). Things changed dramatically in 1933 when the development of a close relationship between the United States and the countries of Latin America actually became part of America's foreign policy. Then newly elected President Franklin Roosevelt pledged in his first inaugural address that "[i]n the field of world policy," he would "dedicate this nation to the policy of the good neighbor—the neighbor who resolutely respects himself and, because he does so, respects the rights of others—the neighbor who respects his obligations and respects the sanctity of his agreements in and with a world of neighbors."[163] A month later, Roosevelt explicitly singled out the countries of Latin America as among those to which the United States would be just such a "good neighbor."

160. Veblen to Aydelotte, 22 April, 1942, as quoted in Xu, p. 298. The quotation that follows is also from this letter and on this page.

161. See Shiing-Shen Chern, "On Integral Geometry in Klein Spaces," *AM* 43 (1942), 178–189. Interestingly, the paper was marked received by the *Annals* on 3 September, 1940, so one is left to wonder if Chern's letter to Veblen might have finally prompted its refereeing.

162. Xu, pp. 299–301.

163. Franklin Roosevelt, *Roosevelt's Foreign Policy 1933–1941: Franklin D. Roosevelt's Unedited Speeches and Messages* (New York: Wilfred Funk, Inc., 1942), p. 3. See also Karen Hunger Parshall, "A Mathematical 'Good Neighbor': Marshall Stone in Latin America (1943)," *Revista brasileira de história da matemática: Especial No 1—Festschrift Ubiratan D'Ambrosio* (Dec. 2007), 19–31 on p. 19.

Reciprocal trade agreements as well as policies both of non-intervention in Latin American political affairs and of the mutual defense of North, Central, and South American interests against external threats followed. By 1940, a governmental Office of Inter-American Affairs had been created, and the Rockefeller and Guggenheim Foundations had embraced the Good Neighbor Policy by providing financial and logistical support for inter-American intellectual exchange and cooperation. The first mathematicians to take part in this new initiative were George Birkhoff in the late spring and summer of 1942 and Marshall Stone a year later in the summer and early fall of 1943.[164]

Birkhoff's trip took him to Mexico, Peru, and Chile where he lectured in Spanish and where he was warmly welcomed by his hosts, among them, Godofredo García and his Polish-born colleague Alfred Rosenblatt, at the Universidad Nacional Mayor de San Marcos in Lima. As Birkhoff reported back to Richardson, García "and Rosenblatt make an effective pair. Of course what they need most is books and journals for the students, and then salaries from which they can live without seeking further jobs. The same holds in Mexico."[165] While Birkhoff could do nothing about the salary issue, he did arrange for subscriptions to the AMS's *Transactions* and the *American Journal* to be sent at his expense to the department in Lima beginning with the first numbers of 1942. Reflecting on the trip after his return, he found "[e]very minute . . . interesting" and "[t]he cordiality and hospitality . . . unbounded."[166] Less than two years later, he was back, lecturing during the winter semester of the 1943–1944 academic year at the University of Mexico.[167]

Stone's trip was somewhat more extensive and equally gratifying. It was also symbolic in that he was the then President of the American Mathematical Society, a fact that all of his Latin American hosts well recognized. Stone journeyed first to Lima, where he, too, lectured in Spanish at the Universidad Nacional, and then to Bolivia and Argentina with a scientific side trip to

164. Eduardo Ortiz provides detailed accounts of these trips in "La politica interamericana de Roosevelt: George D. Birkhoff y la inclusión de América Latina en las redes matemáticas internacionales: Primera Parte" and "Segunda Parte," *Saber y tiempo: Revista de historia de la ciencia* 15 (2003), 53–111 and 16 (2003), 21–70 and Parshall, "A Mathematical 'Good Neighbor.'"

165. Birkhoff to Richardson, 24 May, 1942, Box 13: Correspondence, 1940–1942 'A-M,' Folder: Birkhoff, G. D., Richardson Papers.

166. Birkhoff to Richardson, 28 August, 1942, Box 13: Correspondence, 1940–1942 'A-M,' Folder: Birkhoff, G. D., Richardson Papers.

167. "Notes," *BAMS* 50 (1944), 173–178 on p. 174.

Uruguay and a touristic excursion to Paraguay and Brazil to see the Iguazu Falls. At his home base—from July through early October—in Buenos Aires, he gave a two-month-long course on Boolean algebras and their connections to topology. He also participated in a meeting of the Unión Matemática Argentina at which he heard lectures by, among others, Spanish mathematician in exile Julio Rey Pastor, and the young Alberto Calderón, both then associated with the Universidad de Buenos Aires.[168]

Like Birkhoff, Stone came away from all of these experiences with clear impressions. In particular, he held that if Latin American mathematicians could come to the United States "with some knowledge of the English language," they would be able to "appreciate at first hand our very rich and active mathematical development" and to "see at first hand the structure of our scheme of instruction." In so doing, they would be able to "present to [their] fellow countrymen the advantages and disadvantages of our North American mathematical organization in a way which will command the closest attention."[169] Clearly, Stone believed that the American mathematical community was then in a position to serve as a model for developing mathematical communities elsewhere, just as Germany had done for the United States. Indeed, García was in the United States in the fall of 1943, visiting UCLA, Harvard, and Princeton,[170] and a number of talented Latin American students, among them, Calderón (see chapter ten), would come northward after the end of the war to pursue advanced mathematical training at American universities.[171] The Good Neighbor visits of Birkhoff and Stone helped pave the way for these interactions.

———

American mathematicians faced and rose to the challenges of war between 1939 and 1945. Quickly recognizing the many ways in which research mathematicians could prove essential in a war effort, they mobilized their talent only

168. Parshall, "A Mathematical 'Good Neighbor,'" p. 25 and "Cronica: Asamblea de la Unión Matemática Argentina," *Revista de la Unión Matématica Argentina* 9 (1943), 144.

169. Stone to Henry Moe of the Committee for Inter-Artistic and Cultural Relations, 13 April, 1944, Box 35, Folder 7, Stone Papers, quoted in Parshall, "A Mathematical 'Good Neighbor,'" p. 27.

170. William Cairns to Harry Carver, 23 October, 1943, Box 86-14/66, Folder 9, MAA Records.

171. See Ortiz, "Segunda Parte," pp. 43–61 for an account of some of these students.

to confront an emergent scientific bureaucracy in Washington that initially failed to appreciate what they could offer. How best to get that message across occupied America's new mathematical leaders, figures like Morse and Stone, throughout the war, and although they were ultimately successfully, the frustrations along the way had been palpable. "We can feel proud," Richardson told von Neumann in the summer of 1944, "that the mathematicians are playing such an important role in these preparations for war. No matter how unsatisfactory in some ways our relations to the government are, they are much better than we had a right to expect from previous experience and I believe that our services will be appreciated and that mathematics will have a higher standing than it did before the war."[172]

Of course, only time would tell how accurate Richardson's prediction would be, but mathematicians certainly did provide many "services" during wartime. They taught for the Army Specialized Training Program and other military programs. They developed new curricula like Brown's Program in Advanced Instruction and Research in Mechanics. They worked in ballistics at the Aberdeen Proving Ground and elsewhere. They contributed to the development of whole new fields: high-speed computing, sequential analysis, operations research, linear programming.

At the same time, they managed to sustain the vibrant community that they had built through their research, publications, and continued professional interactions. Despite their myriad wartime distractions, they seemed never to lose sight of their shared conviction that "[t]he center of gravity of mathematics ha[d] moved more definitely toward America" by the close of the 1930s,[173] and that, therefore, "with the return of peace," they would "be called upon to take the leadership in the adjustments necessary for the resumption of normal research activities *throughout the world*."[174] If they viewed the 1940s as their time, it remained to be seen whether they would succeed in rising to their perceived challenge.

172. Richardson to von Neumann, 25 July, 1944, Box 9, Folder II.111: von Neumann, John, Division of Applied Mathematics Papers.

173. Roland Richardson to "Dear Colleagues," 25 April, 1939, Box 11, Folder: Richardson, R.G.D. 1939, Veblen Papers.

174. John Kline to Einar Hille, Mark Ingraham, John Synge, and Gordon Whyburn, 22 September, 1943, Box 1, Folder 99: Kline, J. R., Whyburn Papers (my emphasis).

9

Picking Back Up and Moving On in the Postwar World

The war ended in Europe on 8 May, 1945. It went on longer in the Pacific but was over there on 2 September. For American mathematicians, as for the rest of the country, the fall of 1945 marked the beginning of the transition back to peacetime. But what exactly would that mean?

For those who had not left their colleges and universities? Carrying on with their usual teaching and administrative duties, especially for those who had shouldered the teaching overloads of special training courses for the military's recruits in arithmetic, trigonometry, and other basics. For others? Returning to academic positions left vacant while they participated in the war effort away from their campuses. Some would leave academia for different pursuits altogether, their tastes having been whetted for occupations more rewarding than the routine, low-level teaching that tended to define professional lives outside the more elite, research-oriented schools. Some would quickly pick their research back up where it had left off; others would shift into new mathematical areas; still others would move on, having been permanently diverted from the research ideal they had earlier embraced.

The fall of 1945 also marked the establishment of a new world order. The Allies had won the war; Germany had been crushed; the United States had emerged from the conflict as a major world power. The same seemed true of the American mathematical community. "The United States," asserted John Kline categorically from the dais at the MAA's annual meeting on 25 November, 1945, "has assumed world leadership in mathematics."[1] As he saw it, "[t]hat leadership brings responsibilities, both for imparting the results

---

1. Kline, "Rehabilitation of Graduate Work," p. 131. The next quotation is also on this page.

to coming generations and for ever widening the frontiers of mathematical truth." The Americans now had to set the bar for mathematical training as well as for original research. There was much to do.

Adjusting to New Political Realities

Jockeying for position relative to postwar scientific initiatives and funding began even before the war ended. In particular, in the summer of 1944, plans were being discussed for the establishment of a postwar Research Board for National Security (RBNS) that would replace Vannevar Bush's ad hoc Office of Scientific Research and Development but actively continue its function as a joint military and civilian research agency.[2] Scientists as well as military leaders hoped not to repeat the mistakes that had followed World War I relative to preparedness for war—or lack thereof—in peacetime. Yet, it was unclear just what an RBNS might look like and how it would be set up.

By 1944, there were at least two models. One would have it, like the National Research Council, within the National Academy of Sciences, reliant on Academy expertise and, in the new postwar context, dependent for its funding on the Army and the Navy. The other would have it, like the OSRD, an independent but government-funded agency controlled by civilian scientists. For various reasons, it had been decided by the spring of 1945 that an RBNS under physicist, MIT President, and OSRD member Karl Compton as chair would, until the OSRD was officially disbanded at the end of the war, start off within the Academy and only later become a free-standing, Congressionally approved agency.[3] Planning began in earnest, and, unlike at the creation of the NDRC, mathematicians were part of the process from the beginning.

On 20 March, 1945, the AMP's Warren Weaver wrote to eight leading mathematical scientists—SRG-C director W. Allen Wallis; BRG-C head Jan Schilt; mathematicians, all attached in one way or another to the AMP, Mac Lane, Courant, Richardson, and von Neumann; and statisticians, similarly connected, Wilks and Neyman—to inform them personally of plans under way to create a Research Board for National Security and to seek their advice. As he explained, Oswald Veblen was the (sole) mathematics representative

2. Daniel Kevles traces the twists and turns in the debates over the formation of the RBNS in "Scientists, the Military, and the Control of Postwar Defense Research: The Case of the Research Board for National Security, 1944–46," *Technology and Culture* 16 (1975), 20–47. See also Karl Compton, "Research Board for National Security," *Science* 101 (1945), 226–228.

3. Kevles, "Scientists, the Military, and Control of Postwar Defense Research," pp. 24–28.

on the forty-man board and had quite rightly recognized "that an opportunity exist[ed] to recommend just how mathematics [would] operate in
this new Board."[4] It would be a moment in which past oversights would be
corrected, for, Weaver admitted, "[a]ll of us, I think, regretted the fact that
mathematicians did not have much of a chance to advise concerning the set-
up of mathematics in the NDRC." He thus asked each man to share his views
directly with Veblen.

Strong, and sometimes differing, opinions emerged. "[N]early everyone
is in favor of having something which may be called a Mathematics Panel,"
Veblen related to Weaver on 4 April, but "[s]uch division of opinion as exists
is over the question of whether this Panel should administratively control any
considerable number of sub-sections. Von Neumann . . . would say that there
would be few, if any, sub-sections. Courant would tend to put most of the
activities in which mathematics figures prominently under the control of a
Mathematics Panel."[5] Unclear as to "where to strike the balance," Veblen also
sought the counsel of his fellow members of the Committee on Mathematics
of the National Academy and its NRC.

At a 22 April meeting, that nine-person group, chaired by Morse and
augmented by some eleven additional "mathematicians active in war work,"
aimed "to provide Veblen, informally, with the points of view of representative colleagues" as well as "to decide what action, if any, [should] be taken to
increase the representation of mathematics on the RBNS."[6] Relative to the
latter, perhaps not surprisingly, those in attendance overwhelmingly favored
a presence on the Board greater than one in forty for mathematics, citing
"matters of [future] financial expenditures on mathematics, [the] prestige of
mathematics, and [the] full use of mathematics" in defense-related research.
When push came to shove, though, and the question was posed directly, "[d]o
you think the RBNS would be improved if mathematics were represented by
two members?" the vote came back in the negative. One mathematician on
the RBNS was certainly better than none, and it would be impolitic in such

4. Weaver to Wallis, Schilt, Mac Lane, Courant, Richardson, von Neumann, Wilks, and Neyman, 20 March, 1945, Box 34, Folder: Research Board for National Security Apr.-July 1945,
Veblen Papers. The quotation that follows is also from this letter.

5. Veblen to Weaver, 4 April, 1945, Box 34, Folder: Research Board for National Security
Apr.-July 1945, Veblen Papers. The quotation that follows is also from this letter.

6. Report of a meeting of the Committee on Mathematics of the National Academy and
the NRC, 22 April, 1945, Box 34, Folder: Research Board for National Security Apr.-July 1945,
Veblen Papers. The quotations that follow in this paragraph are also from this letter.

early days to raise the issue of greater representation, especially since "the RBNS was organized on the basis of projects rather than academic subjects, and . . . in the projects the role of mathematics was largely subsidiary." Better to focus, then, on the bigger issue of how mathematics would be institutionalized within the new board. In the discussion of that point, von Neumann's less structured panel model prevailed over both Courant's more formally subdivided, AMP-like conception as well as a variant in which there was no overarching mathematics panel at all but in which mathematicians worked in ad hoc groups.

A month later, it had been decided that a five- to seven-person Mathematics Panel would be appointed within the RBNS and that Veblen should provide a list of some ten possible candidates from which to choose.[7] The question of what names to include was clearly a political one, and those from whom Veblen sought advice, once again, saw things somewhat differently.

When Veblen finally had a chance to weigh the suggestions he had received against his own knowledge of the various personalities as well as against the thoughts of Thornton Fry from his perspective as an industrial mathematician, he reported to Compton that he and Fry were "in good agreement about the following list of names: Warren Weaver, Marston Morse, Richard Courant, E. J. McShane, C. B. Morrey or G. C. Evans (both of [the] University of California), A. A. Albert or L. M. Graves (both of [the] University of Chicago), [and] E. Bode (Bell Laboratories)."[8] Interestingly, the two men had decided "to add to this Saunders Mac Lane or Marshall H. Stone (both of Harvard)," but while Veblen held that "it would be wise to include Stone," Fry was apparently "very dubious." Stone had clearly ruffled more than a few feathers while AMS President with his adamant advocacy of mathematics and his often strident opposition of the Applied Mathematics Panel, in general, and of Warren Weaver, in particular. Veblen counseled Compton to get Fry's "views and Weaver's as well as mine on this point" and then weigh them accordingly.[9]

7. Louis Jordan, Executive Office of the RBNS to Veblen, 22 May, 1945, Box 34, Folder: Research Board for National Security Apr.-July 1945, Veblen Papers.

8. Veblen to Compton, 13 June, 1945, Box 34, Folder: Research Board for National Security Apr.-July 1945, Veblen Papers. The quotations that follow in this paragraph are also from this letter. Although Veblen's typescript letter clearly reads "E. Bode (Bell Laboratories)," it must have been referring to Hendrik Bode, who did war work, among many other things, on automatic anti-aircraft control systems from his position at Bell Labs.

9. In a memorandum to Fry and Veblen dated 12 June, 1945, Mina Rees noted that, at a meeting of the AMP held the previous day, John Synge, mathematical physicist and then the

Veblen also noted that the omissions of the names of von Neumann and Wilks from the list owed to the fact that the former was being proposed as the head of what would be a Fluid Mechanics Panel and the latter as the head of a Statistics Panel. Finally, Veblen singled out "four younger men"—topologist John Kelley, Princeton-trained topologist (later turned statistician) John Tukey, Hassler Whitney, and Garrett Birkhoff—before closing with the recommendation of McShane as the first choice and Mac Lane the second for the Panel's head. In Veblen's view, both had "displayed all of the necessary qualities" for leadership—McShane at Aberdeen and Mac Lane within the AMP—and both had "acquired the right sort of background of experience." A Mathematics Panel so constituted would unquestionably have reflected some of the best technical and administrative talent then available within the American mathematical community for dealing with the applications of mathematics to matters of national security.[10] It would also have firmly established some of the members of the next generation of that community's leadership, among whom, notably, were 1930s émigrés.

Indeed, postwar leadership was key. As Wilder put it to Mac Lane in 1946 as the AMS sought to find Theophil Hildebrandt's presidential successor, "[u]sually the choice of president has been based on (a) scientific standing; he should be a man fit to assume leadership of what is now, I suppose, the greatest group of mathematicians in the world. With the advent of the war," however, "another factor entered, namely (b) capability of representing and guiding the Society during time of crisis."[11] The latter, according to Wilder, had actually been decisive in the nominating committee's choice for president in 1943 of the then thirty-nine-year-old Stone over James Alexander and Joseph Ritt, mathematicians, respectively, fifteen and ten years Stone's senior who were

---

chair of the Department of Mathematics at the Ohio State University, Stone, Wiener, and Lefschetz had all been "Blackballed" relative to candidacy for the Mathematics Panel. See Box 34, Folder: Research Board for National Security Apr.-July 1945, Veblen Papers.

10. In an undated but presumably later list of "Names Suggested for Consideration As Civilian Members for RBNS Panels," Box 34, Folder: Research Board for National Security 1945–1946, Veblen Papers, three additional names had been added: Howard Engstrom, a doctoral student of Oystein Ore who had taken a position alongside his advisor on the Yale faculty before resigning in 1941 to join the Navy; Princeton mathematical physicist Bob Robertson; and Cornell logician J. Barkley Rosser.

11. Wilder to Mac Lane, 16 February, 1946, Box 86-36/28, Folder 13: AMS Records, 1922–1948. The quotations that follow in this paragraph are also from this letter.

deemed less forceful and politically savvy.[12] For Wilder, (b) was still important in the postwar world, but he felt that the time was actually ripe for another
key change. In his view, it was "no longer necessary, as it was (for various reasons) during the war, to nominate an American by birth," and, in fact, "there
are good reasons for nominating non-native citizens." For his part, Wilder
thought "that nationalism is one of the surest ways to suicide, and perhaps
the present is a good time to show our foreign colleagues that we mathematicians recognize this (if we do)." Although the AMS did not elect an émigré
president until 1951, when von Neumann assumed the post, the representation of émigrés on the roster for possible RBNS membership demonstrated
that the American mathematical community had fully embraced its newest
members.[13]

As planning for and discussion of the RBNS continued into the fall of 1945,
funding proved to be a fatal sticking point. It eventually emerged that the
Army and Navy could not legally transfer funds to an agency that had not
been authorized by Congress, yet that authorization would not be forthcom-
ing until the OSRD was officially dissolved.[14] Catch twenty-two. With great
reluctance, the Secretaries of War and the Navy wrote in February 1946 to
Frank Jewett, in his role as President of the National Academy, requesting that
he "terminate the Research Board for National Security."[15] Jewett relayed the
bad news to Compton, the RBNS's putative chair, closing his letter ruefully
"RBNS – R.I.P.!!!"[16]

The OSRD persisted until December 1947. In the meantime, Congress
continued to debate the ever-evolving legislation that Harley Kilgore, Democratic Senator from West Virginia, had originally introduced in 1942 and
that would ultimately result in 1950 in the founding of the National Science
Foundation (see the coda). It acted, however, in the summer of 1946, when it

12. The twenty-seventh president of the AMS, Stone was then the youngest mathematician
to have been so chosen.

13. Einar Hille succeeded Hildebrandt as AMS President; next came Joseph Walsh in 1949–
1950; von Neumann followed him in 1951–1952. The next émigré to serve as AMS President,
Richard Brauer, held the office in 1957 and 1958.

14. Kevles, "Scientists, the Military, and Control of Postwar Defense Research," p. 43

15. Secretary of War Robert P. Patterson and Secretary of the Navy James Forrestal to Frank
Jewett, 28 February, 1946, Box 34, Folder: Research Board for National Security 1945–1946,
Veblen Papers.

16. Jewett to Compton, 25 March, 1946, Box 34, Folder: Research Board for National
Security 1945–1946, Veblen Papers.

authorized the creation of the Office of Naval Research (ONR), a body that, at least relative to the Navy, would serve a purpose similar to that of the ill-fated RBNS.[17]

As early as February 1945 in his annual report to the President, Secretary of the Navy James Forrestal had argued strenuously for "the establishment by law of an independent agency devoted to long-term, basic, military research, securing its own funds from Congress and responsive to, but not dominated by, the Army and Navy." "The Navy," he continued, "so firmly believes in the importance of this solution to the future welfare of the country that advocacy of it will become settled Navy policy."[18] When plans for an RBNS ultimately came to naught, the Navy moved quickly.

Under physicist Emanuel Piore, the ONR was "committed primarily to the support of fundamental research in the sciences," "a basic policy decision which recognize[d] that the United States must be strong scientifically if it is to be strong militarily."[19] Computer, Logistics, Mathematics, and Mechanics Branches defined its main foci, and Mina Rees, Weaver's former assistant at the Applied Mathematics Panel, was tapped to head Mathematics.

Rees assumed her new post in August 1946 and immediately began consulting with a wide range of members of the American mathematical community to get their ideas on how her new department could best aid them in their work, while, at the same time, fulfilling the needs of the Navy.[20] This dual purpose initially made it unclear whether the Navy would support research in what most American mathematicians actually did, namely, pure mathematics, since "the most obvious types of mathematical research which would seem to warrant Navy support would be research in applied directions."[21] Fortunately,

17. Harvey Sapolsky traces the history of the ONR, although with little reference to mathematics, in *Science and the Navy: The History of the Office of Naval Research* (Princeton: Princeton University Press, 1990).

18. Forrestal as quoted in Mina Rees, "Mathematics and the Government: The Post-War Years As Augury of the Future," in *The Bicentennial Tribute to American Mathematics, 1776–1976*, ed. Dalton Tarwater (n.p.: Mathematical Association of America, 1977), pp. 101–116 on p. 104.

19. Mina Rees, "The Mathematics Program of the Office of Naval Research," *BAMS* 54 (1948), 1–5 on p. 1.

20. Amy Shell-Gellasch, "Mina Rees and the Funding of the Mathematical Sciences," *AMM* 109 (2002), 873–889 on pp. 881–882. Shell-Gellasch also discusses this in the published version of her dissertation, *In Service to Mathematics*, p. 70.

21. Rees, "The Mathematics Program of the Office of Naval Research," p. 1. The quotation that follows in this paragraph is on p. 2 (with my emphasis).

FIGURE 9.1. Mina Rees (1902–1997). (Photo courtesy of the Graduate Center Library, City University of New York. Washington, D.C.)

that point was quickly clarified when "it was decided as *a matter of policy* that sound support of mathematical research in this country must include support of work in pure mathematics." The battles fought by Morse, Stone, and others during the war to win recognition for the critical role of mathematics writ large in the nation's service had apparently been won.

Rees then proceeded to craft and implement from scratch a program—based on the screening of proposals by a committee of mathematical peers—that funded basic research in both pure and applied mathematics, in addition to postdoctoral support, teaching release, and sabbatical leaves.[22] In another key innovation, it also provided "assistance to universities which [we]re taking steps to strengthen their research activity in either pure or applied mathematics, and particularly in parts of the country which lack[ed] strong research centers."[23] The ONR, through Navy contracts administered directly to institutions, thus became "virtually the only source of funds available for the support of basic research in the mathematical sciences" in the immediately postwar period.[24] More importantly, Rees had ushered the field into the brave new world of grant-funded research that has characterized it ever since. As one of her contemporaries put it, "[t]here could be no doubt . . . that mathematics had been recognized, once and for all, to be important—like physics, chemistry, and engineering mechanics—and was so treated."[25]

Reestablishing a Professional Rhythm

Just as those in Washington envisioned and created new administrative structures like the RBNS and the ONR in the months after the war, so, too, did those in the American mathematical community. The war's end had not only made superfluous the War Policy Committee (WPC) that the AMS and the MAA had jointly created in December 1942, but it had also raised questions about what roles the two organizations should play in the postwar world. From his vantage point as WPC chair, Stone, for one, held that "the postwar problems of mathematicians may be of such urgency and importance as to require very efficient joint handling by the Society and the Association."[26] Thus, two days after the end of the war in the Pacific when he formally requested of the then Presidents of the AMS and MAA, Hildebrandt and MacDuffee, respectively, that the WPC be dissolved, he also offered a suggestion. "In the opinion of your Committee," he wrote, it would be "highly

---

22. Among the pure mathematical fields supported were complex and functional analysis, group theory, the calculus of variations, and analytic number theory, while the applied fields included mathematical statistics, numerical analysis, and electronic computing.

23. Rees, "The Mathematics Program of the Office of Naval Research," p. 3.

24. Rees quoting Frederick Terman, Dean of Stanford's School of Engineering, in "Mathematics and the Government," p. 106.

25. F. Joachim Weyl, "Mina Rees, President-Elect 1970," *Science* 167 (1970), 1149–1151 on p. 1150.

26. Stone to "Colleagues," undated but from the late summer of 1945, Ms.75.5, Box 31, Folder 112, AMS Papers.

desirable that joint study of common policy problems be continued through a peace-time body representing your organizations and others such as the Institute of Mathematical Statistics, the Association for Symbolic Logic, and so forth."[27]

MacDuffee totally agreed. "It cannot be denied," he stated, "that mathematical organizations have suffered in comparison with the physical societies, for example, from an almost total lack of a united front. There is no organization which is able to speak for American mathematics *in all of its phases*, including pedagogical problems. In these days when propaganda and political action in a moderate degree are essential to the prosperity of an organization," he concluded, "we must give thought to these activities."[28]

Indeed, MacDuffee already had. He proposed a "United Front Committee" as an analog for mathematics of the American Institute of Physics (AIP), an organization formed in 1931 that had initially served as an umbrella for the American Physical Society, the Optical Society of America, the Acoustical Society of America, the Society of Rheology, and the American Association of Physics Teachers. The AIP had aimed, in the Depression Era, to advance and diffuse " 'knowledge of the science of physics and its application to human welfare' . . . by achieving economies in the publishing of journals and the maintenance of membership lists."[29] In spirit, at least relative to the "economies," it was similar to what the AMS and the MAA had considered in the 1920s, but the AIP actually had its own building in New York City as well as its own budget and a paid director.[30] MacDuffee was not suggesting that the American mathematical community go quite that far, but he was advocating that it appoint "a part-time director who may eventually become full-time" and possibly give the new entity oversight of the *Mathematical Reviews* although of none of the other society journals.[31] Such a "United Front Committee" would

27. Stone to Hildebrandt and MacDuffee, 10 September, 1945, Ms.75.2, Box 15, Folder 91, AMS Papers.

28. MacDuffee to Hildebrandt, John Kline, and Walter Carver, 1 September, 1945, Box 86-14/69, Folder 2, MAA Records (my emphasis). MacDuffee is explicitly replying to Stone's letter quoted in the previous paragraph, although his letter is dated ten days prior to Stone's. Perhaps he had an advance copy?

29. See https://www.aip.org/aip/history.

30. Henry Barton gives an accounting of its finances in 1943 in "American Institute of Physics: Report of Director for 1943," *Review of Scientific Instruments* 15 (1944), 130–133 on p. 133.

31. MacDuffee to Hildebrandt et al., 1 September, 1945. The quotation that follows is also from this letter.

be "responsible to all of the mathematical organizations but" it would have "enough authority and prestige so that it [could] handle the public relations problems connected with mathematics as they occur without having to refer minor problems back to the member organizations for approval."

This amount of organization seemed like overkill to some, like the MAA's then Secretary-Treasurer Walter Carver, especially given that the various mathematical groups had fairly well-defined spheres of influence by 1945.[32] Nor was it ultimately what Stone really had in mind. A Policy Committee for Mathematics comprised of six members was formed in April 1946: four represented the AMS, Evans, Morse, and Hildebrandt with Stone as chair (that is, the AMS Presidents who had led the Society just before, during, and after the war); one, Church, represented the Association for Symbolic Logic; and one, Feller, represented the Institute of Mathematical Statistics (for more on this committee, see the coda).[33] Noticeably absent was the MAA which had opted not to be officially associated with a Policy Committee that would have four members from the AMS but only a proposed two from the MAA. Carver, for one, felt "that in view of the way this Policy Committee was set up it would be better if the Association not recognize it in any way."[34] If Stone and the AMS wanted heavy-handedly to skew the committee, as some in the MAA perceived, toward the politics of mathematics solely at the research level, the MAA would push its own, educational agenda.[35] "We complain,"

32. Carver, also one of MacDuffee's predecessors as MAA President, was Secretary-Treasurer from 1943 to 1948. In a letter to MacDuffee dated 10 September, 1945 (Box 86-14/69, Folder 2, MAA Records), he cited what he viewed as the American tendency "to overorganize. We organize companies, then holding companies, and then higher holding companies to hold the holding companies." He felt that, in this instance, it would suffice if each society had a cooperation committee "whose business it would be to keep in touch with what the other organizations are doing."

33. Temple Hollcroft, "The April Meeting in New York," BAMS 52 (1946), 580–593 on p. 588. John Kline, AMS Secretary, was appointed as a non-voting, ex officio member to act as the committee's secretary. Evans was replaced by Wisconsin's Rudolph Langer beginning 1 January, 1947.

34. Carver to MacDuffee, 19 December, 1946, Box 86-14/69, Folder 2, MAA Records. The MAA ultimately decided to participate in the Policy Committee in the summer of 1948. Again, see the coda.

35. Three years later, this perception still persisted. As William Cairns wrote to Harry Gehman, "I feel there is a *strong* question about our accepting membership in the Policy Com., so long as the Soc. men (notably Marshall Stone, if I understand correctly) determine the ratio of 4:2:1:1." See 30 July, 1948, Box 86-14/66, Folder 9, MAA Records (his emphasis).

Carver told MacDuffee, "about what the education people [that is, those who theorize about the teaching of mathematics as opposed to those who actually teach it] are doing to secondary school education. . . . I believe that our inactivity with regard to these matters is responsible in some degree for the situation that exists," namely, the tension between what has been termed "academic mathematics" and more practical or vocational mathematics (recall chapter six).[36]

Carver was thus wholeheartedly in favor of MacDuffee's initiative to poll some 140 officers of the MAA and its sections for their opinions on whether the Association should focus on postwar educational problems. In a circular dated 1 February, 1945, MacDuffee framed the issue this way: "The newspapers and magazines are full of plans for the drastic revision of education in the post-war world. There are plans to increase the offerings of mathematics in the secondary schools, to make its study compulsory, to eliminate it entirely. . . . The situation is fraught with danger as well as with opportunity. It is a situation demanding the leadership of those persons and organizations who really know the answers and who can turn this fluid situation into victory for sound education."[37] When his colleagues overwhelmingly agreed, he acted accordingly, proposing the formation of various committees as well as the participation of the MAA in a number of curricular reform initiatives.[38] In particular, the MAA was well represented on the Cooperative Committee on Science and Mathematics Teaching that had been formed in 1941, but that made a substantial report on "The Present Effectiveness of Our Schools in the Training of Scientists" to the President's Scientific Research Board in 1947.[39] Indeed, throughout the 1940s and into the 1950s, the MAA continued to

36. Carver to MacDuffee, 24 January, 1945, Box 86-14/69, Folder 2, MAA Records. See Kliebard and Franklin for the distinctions between "academic" and "practical" or "vocational" mathematics.

37. MacDuffee to "the Officers of the Mathematical Association of America," 1 February, 1945, Box 86-14/69, Folder 2, MAA Records.

38. See, for example, Cyrus MacDuffee, "The Coordinating Committee," *AMM* 52 (1945), 294. This committee, chaired by Carroll Newsom, was charged "to keep in close touch with all educational movements in the United States and Canada, and to make the information thus acquired available to all committees of the Association and its Section who can profit" from it.

39. John Steelman, *Manpower for Research: Science and Public Policy: A Report to the President*, 5 vols., *The Crisis of Science in the United States*, vol. 4 (Washington, D.C.: Government Printing Office, 1947), Appendix II, pp. 47–149. Raleigh Schorling gives a synopsis of the report's recommendations in "A Program for Improving the Teaching of Science and Mathematics," *AMM* 55 (1948), 221–237. Schorling, the MAA's representative on the committee,

promote dialog on mathematics education and curricular reform even as the "wartime legitimation of school mathematics" waxed into mathematics' "relative curricular insignificance ... during the immediate postwar years."[40] This seemingly perennial issue would raise its head once again when the Cold War, ever-more-frigid by 1957, witnessed Russia's launch of Sputnik.

The societies as collectives also tried to do their parts to reestablish lines of mathematical communication severed by the war.[41] The AMS, for example, briefly reinstituted its Visiting Lecturer program with the 1948–1949 lecture tour of Polish topologist Casimir Kuratowski. It also hosted various foreign mathematicians, such as the University of Stockholm's Harald Cramér in December 1946, as attendees and invited speakers at its meetings and constituted a Committee on Aid to Devastated Libraries in May 1945 that, by 1949, had sent "packages of periodicals and of books to thirty-five institutions in fifteen countries in Europe and Asia."[42]

Still, immediately postwar mathematical relations were unclear on a number of levels. Relative to Germany and the other former Axis nations, Veblen offered this opinion, in response to a letter from the University of Marburg's Kurt Reidemeister in December 1945: "[w]hat would be desirable would be the resumption of the freest possible inter-communication as soon as the general political situation will allow. You know of course that you can count on me," he added, "for full cooperation in this direction, and I think that most of my colleagues will feel the same way."[43] While Veblen may well have been right about that sort of cooperation, nationalistic strains affected cooperation of other sorts.

Early in 1946, Harvard's Joseph Walsh, who had strong ties to France going back at least to 1920 and his foreign study tour in Paris, inquired

---

was professor of education at the University of Michigan and head of the Department of Mathematics in the University High School.

40. Alan Garrett and O. L. Davis, Jr. trace these developments in "A Time of Uncertainty and Change: School Mathematics from World War II until the New Math," in *A History of School Mathematics*, ed. George M. A. Stanic and Jeremy Kilpatrick, 1: 493–519. The quotations are on p. 515.

41. See, for example, "Notes," *BAMS* 51 (1945), 868–873 with news on mathematicians in the Netherlands and Poland and "Notes," *BAMS* 53 (1947), 268–271, reporting once again on mathematical news from the Netherlands.

42. Arnold Dresden, "Report of the Committee on Aid to Devastated Libraries," *BAMS* 55 (1949), 1–2 on p. 1.

43. Veblen to Reidemeister, 13 December, 1945, Box 10, Folder: Reidemeister, Kurt 1929–46, Veblen Papers.

of the AMS Colloquium Publications Committee about the possibility of publishing a work in French by analyst Arnaud Denjoy.[44] Writing to fellow committee members, Gordon Whyburn and Joseph Ritt, Cyrus MacDuffee rather callously "wonder[ed] if the United States was now going to be under the obligation of financing all mathematical publication of the European survivors."[45] Although he was quick to add that he did not object in principle to publishing Denjoy's book, he had "a strong suspicion that," if they did publish it, it "would be only the beginning, and the continuance of this policy would make it difficult for *American* mathematicians to publish worthy books." His fellow committee members apparently agreed. In conveying the committee's decision to Walsh, Whyburn echoed MacDuffee's nationalistic opinion that "[t]he end of the war has brought greatly increased activity on the part of past and future Colloquium Speakers and we anticipate all the manuscripts we can handle from these and other *American* mathematicians."[46]

Whyburn and MacDuffee might have been more charitable had they actually experienced postwar mathematical Europe firsthand. Saunders Mac Lane's account of his first postwar trip to Germany was not atypical in its description both of a physically damaged Europe and of struggling mathematical colleagues there. In the spring of 1948, Mac Lane and his wife, Dorothy, journeyed to Switzerland in hopes of traveling from there into Germany. Mac Lane had been invited to lecture at the University of Heidelberg, but he also wanted to take the opportunity to return to Göttingen, where his 1934 Ph.D. had been directed by Paul Bernays and Hermann Weyl (before the latter's assumption in January 1934 of a professorship at the Institute in Princeton), and to renew contact with mathematicians elsewhere. When permission was finally granted, the Mac Lanes set out, but the scene they encountered was grim. "Beyond Basel," Mac Lane told Raymond Wilder, "the train crosses an imaginary line—and immediately one is moved from prosperous, undamaged, well fed Switzerland to a shabby half starved Germany."[47]

44. At issue was presumably part of what was ultimately Denjoy's four-volume work, *L'énumération transfinie*, 4 vols. (Paris: Gauthier-Villars, 1946–1954).

45. MacDuffee to Whyburn and Ritt, 12 February, 1946, Box 6, Folder 19: AMS–Committee on Colloquium Publications (1946–1947), Whyburn Papers. The quotation that follows is also from this letter (with my emphasis).

46. Whyburn to Walsh, 19 February, 1946, Box 6, Folder 19: AMS–Committee on Colloquium Publications (1946–1947), Whyburn Papers (my emphasis).

47. Mac Lane to Wilder, 1 April, 1948, Box 86-36/8, Folder 3: General Correspondence Mac Lane, Wilder Papers. The quotations that follow in this paragraph are also from this letter.

He found, however, that Göttingen had been largely "undamaged in the war," although the mathematicians there were "practically starving for lack of scientific contacts with the outside world." Frankfurt, on the other hand, had been "completely bombed out," and in Heidelberg, he discovered topologists Herbert Siefert and William Threlfall living "in the Mathematics Institute in two miserable, small rooms, which formerly served to store baggage" and with "practically no new mathematical literature available and no time or energy left for mathematical work." There was no telling how long it would take for life, both mathematical and otherwise, to return to prewar levels of prosperity and productivity.

Moreover, although Whyburn's rather flip remark belied the fact, making sure that "future American Colloquium speakers" continued to be produced was actually of critical concern in the war's immediate aftermath. Indeed, John Kline focused his November 1945 address before the MAA on precisely the issue of the "Rehabilitation of Graduate Work" as a means of postwar "recovery and development of scientific talent."[48] The war, he documented, had resulted in a "very rapid decrease in the number of doctorates and master's degrees" in mathematics with Ph.D.s dropping from a high of 104 in 1939 and 1941 to only 39 in 1944 and Master's degrees falling from 303 in 1940 to 114 in 1944. This owed to several compounding factors: "the great numbers of graduate students and prospective graduate students, who were undergraduate majors in mathematics [but who had] entered the armed forces" thereby leaving mathematics, the fact that only after July 1942 did mathematicians qualify for deferment, and changes in April 1944 that adversely affected deferment across the board. In the "conservative opinion" of the government's Office of Scientific Personnel, moreover, "it will be 1955 before we will be producing the number of Ph.D.s that we would have expected to produce each year, if the growth of graduate work had continued at the same rate as in the decade 1930–1940."[49]

The country, so it seemed, faced dire consequences unless the rhythm of graduate work was quickly reestablished. A dearth of mathematicians could result both in industry and in the research and teaching staffs of colleges and

---

48. Kline, "Rehabilitation of Graduate Work," p. 121. The quotations and the data that follow in the next two sentences are on p. 122. Kline was echoing warnings that had started to issue from various quarters earlier in 1945. For an overview, see Carroll Newsom, ed., "General Information," *AMM* 52 (1945), 409–414.

49. Kline, "Rehabilitation of Graduate Work," p. 123.

universities. There could also be a concomitant lowering of the mathematical "competence of the nation" at a moment when mathematics had been shown to have "made tremendous contributions to our national war effort" and to have "been the means by which other sciences [had] been able to make much more significant contributions" than might otherwise have been possible.[50] These things could not be allowed to happen if the United States was going to retain what Kline and many of his colleagues saw as its newly gained position of "world leadership" in mathematics.[51] Much seemed to be riding on graduate education.

Building Programs

Graduate work, however, thrives in strong departments, and at least one traditionally trend-setting department had slipped over the course of the 1930s, despite continuing to train prodigious numbers of students: the University of Chicago. Instrumental in instilling the research ethos in the United States at the turn of the twentieth century and still widely deemed one of the "big three" prior to World War II, Chicago was producing research with a rather dated feel by the 1930s, especially in geometry and analysis, although not in algebra (recall chapter six). Its faculty and graduate students were publishing, but their research did not attract much contemporary attention outside extended Chicago circles. Mathematical inbreeding, that is, the hiring of students and then of students' students, had largely been the culprit.[52] That practice may have made sense as a strategy in the first decade of the century when Chicago was almost alone turning out quality Ph.D.s, but with the rise of other programs—at Harvard, at Princeton, at some of the land-grant institutions and elsewhere—over the course of the teens, twenties, and thirties, it had become less wise. Fresh perspectives and new ideas had tended to be closed out to the program's detriment. This began to change immediately after the war.

As is well-known, Chicago was the site both of the first sustained nuclear chain reaction (in December 1942) and, in its Metallurgical Laboratory, of other research—like Ernest Wilkins's on neutron absorption—critical to the Manhattan Project.[53] The concentration there during the war of

50. Ibid., p. 124.
51. Ibid, p. 131.
52. Mac Lane, "Mathematics at the University of Chicago," pp. 141–143.
53. Rhodes, pp. 394–442.

top-flight physical scientists like Enrico Fermi, James Franck, and Harold Urey prompted then University Chancellor Robert Hutchins to seize the opportunity of snagging some of that talent for his university's permanent faculty.[54] As at Berkeley in the 1930s, it was quickly realized that a world-class group in the physical sciences should be complemented by a world-class group in mathematics. Given that Dickson and Bliss had retired in 1939 and 1941, respectively, the timing was propitious for Hutchins to rebuild his mathematics department.

With the guidance, among others, of von Neumann whom he had gotten to know during the war, Hutchins had set his sights on Marshall Stone as early as May of 1945 and had entered into discussions with him that would last for a full year.[55] Stone, then the chair of the Mathematics Department at Harvard, drove a hard bargain. It would not be enough for him to come to Chicago as a chaired professor at the then princely salary of $15,000.[56] He would actually have to be named chair of the department. He would also need assurances from Hutchins that he would have the resources actually to build the department beyond its prewar average of eleven members and to move it into previously unrepresented areas, including applied mathematics, statistics, and electronic computing. These had been areas advocated by his WPC subcommittee on preparedness for research; they also aligned well with ONR priorities. Stone's back-of-the-envelope calculation for the cost of running the department with these and other improvements was some $150,000 per year, and he asked point-blank "*is the University—that is, the Chancellor—ready to approve expenditures for this purpose at a rate not to exceed that mentioned above,*

54. Compare Marshall Stone, "Reminiscences of Mathematics at Chicago," *The University of Chicago Magazine* 69 (Sept. 1976), 27–31; reprinted in *A Century of Mathematics in America*, ed. Peter Duren et al., 2: 183–190 on p. 183; Mac Lane, "Mathematics at the University of Chicago," p. 146; and McNeill, pp. 158–160.

55. See the letters in Series II, Appointments and Budgets, Box 990, Folder 14: Department of Mathematics, Appointments and Budgets, 1940–1947, Office of the President, Hutchins Administration, Records 1892–1951, University of Chicago. Initially, Hutchins approached Stone for the deanship of the Division of Physical Sciences, but their discussions quickly shifted to focus on the Department of Mathematics. Stone, "Reminiscences," p. 185.

56. In 2021 dollars, this converts to a "real wealth" salary of $222,500 (see https://measuringworth.com). This made Stone the highest paid member of the Department of Mathematics (by $4,000) and the second-highest paid member (after organic chemist Morris Kharasch who earned $18,000) in the Division of the Physical Sciences. See Series I, General Files, Box 190, Folder 1: Salary levels (1919–1947), Office of the President, Hutchins Administration, Records 1892–1951, University of Chicago.

namely, $150,000 per annum?"[57] After Hutchins personally assured Stone that he wanted Chicago's mathematics group to be "the best in the country" and that Stone would "have the complete support of the administration in [his] efforts to raise the work in mathematics here to the highest possible level," the deal was sealed.[58] Stone joined the Chicago faculty effective 1 July, 1946, and immediately got to work. "Big good things are happening at Chicago," reported then current member of the undergraduate teaching staff William Karush, "every bastard in the dep[artmen]t is quaking in his boots."[59]

In mathematics, that faculty comprised: at the rank of professor— geometer and former student of Wilczynski, Ernest Lane, who was Stone's immediate predecessor as chair, analyst and former student of Bliss, Lawrence Graves, and algebraist and former student of Dickson, Adrian Albert; at the associate professor level—analyst and another former student of Bliss, Magnus Hestenes, and émigré algebraist, Otto Schilling; at the assistant professor rank—topologist John Kelley; and at the rank of instructor—algebraists, Irving Kaplansky and former Albert student, Daniel Zelinsky.[60] Stone had immediately arranged for an offer of an assistant professorship to probabilist Paul Halmos, so he was also on the faculty at the start of Stone's first fall quarter in residence.

Mathematical inbreeding was in evidence from top to bottom, but Stone had big ideas that, perhaps not surprisingly, involved "some *immediate* moves designed to improve the quality of the Department."[61] In his view, it was

57. Stone to Walter Bartky, Dean of the Division of Physical Sciences, 11 April, 1946, Box 35, Folder 10: MHS–Chairmanship at Chicago, Stone Papers (Stone's emphasis). In 2021 dollars, $150,000 amounts to "real wealth" salaries totaling $2.05M (see https://measuringworth.com).

58. Hutchins to Stone, 13 May, 1946, Box 35, Folder 10: MHS–Chairmanship at Chicago, Stone Papers.

59. Karush to Herman Meyer, 23 May, 1946, Meyer Papers. At this point, Chicago had two departments of mathematics, one in the College, where Hutchins's undergraduate "Great Books" curriculum was taught, and one in the Division of Physical Sciences, where graduate instruction was given. Karush had been a student of Magnus Hestenes; Meyer was then working with Hestenes and would earn his Ph.D. in 1947. Both were associated with the college in the late 1940s; Meyer remained there for the rest of his career.

60. Budget and Appointment Recommendations, dated 16 January, 1947, Series II, Division of Social Sciences, Box 308, Folder 5, Office of the President, Hutchins Administration, Records 1892–1951, University of Chicago.

61. Stone to Ernest Colwell, President of the University of Chicago, 25 July, 1946, Series II, Division of Social Sciences, Box 308, Folder 2, Office of the President, Hutchins

"time to add two or three mathematicians of great power and distinction" and thereby both bootstrap the department into "a leading position" and make it "easier to carry out a long-range program of attracting younger men of unusual promise."

The short list he presented (in alphabetical order) for the university's consideration was stellar: Mac Lane, then an associate professor at Harvard; von Neumann, professor at the Institute; Weil, at that point a visiting lecturer at the University of São Paolo; Whitney, professor at Harvard; and Zariski, research professor at Illinois. Acknowledging that von Neumann, "without doubt the most remarkable mathematical figure of this generation," was "not immediately available," Stone made a pragmatic but nevertheless astute argument for approaching Whitney first and offering him a salary of $12,000.[62] "[T]he most profound and effective of the four mathematicians now under consideration" (that is, after removing von Neumann, at least temporarily, from the running), Whitney had just been promoted to professor at Harvard and was making $9,600. Harvard, Stone reasoned, would be highly unlikely to counteroffer at $12,000 because that would raise Whitney's salary to Harvard's maximum and so "over the heads of his seniors." It was thus a move with a reasonable chance of success. If Whitney accepted, moreover, Stone argued that the best one-two punch would be an immediate approach to Weil. The administration approved an offer to Whitney at the end of August. The waiting game began.

In the interim, Stone also wanted to augment the faculty at the junior level, and he received authorization for that as well. As noted, already in the lower ranks were Halmos, who began as an assistant professor in 1946, and Kaplansky, who had come to Chicago as an instructor in 1945 and whom Stone successfully put up for promotion to assistant professor in 1947. Kaplansky had taken his Ph.D. at Harvard in 1941, Mac Lane's first student, and had followed a Peirce Instructorship at his alma mater with a stint as a member, with his advisor, of Columbia's Applied Mathematics Group; Halmos had earned his Ph.D. under Doob at Illinois in 1938 for a thesis on measure-theoretic probability, had hung on there for an additional year during the "threadbare thirties," and had moved to the Institute for three years (two-and-a-half as

Administration, Records 1892–1951, University of Chicago (my emphasis). The quotations that follow in this and the next paragraph are also from this letter.

62. In 2021 dollars, this converts to a "real wealth" salary of $164,000 (see https://measuringworth.com).

von Neumann's assistant) before finally landing the assistant professorship at Syracuse University that he held for most of the war.[63] Fortunately, both men met with Stone's approval, but he felt that even more staffing at the junior level would allow the department to meet its teaching obligations better and to bring additional "promising young mathematicians" into it.[64] To that end, he proposed that an assistant professorship be offered to Irving Segal, a 1940 Ph.D. student of Hille's at Yale who was then serving as Veblen's assistant at the Institute and who was just beginning the mathematical work on quantum mechanics that would lead to deep applications of algebraic methods to analysis.[65] Although Segal accepted, a Guggenheim allowed him to stay one more year at the Institute, so he joined the department only in 1948, the same year as Edwin Spanier, algebraic topologist and 1947 Michigan Ph.D. under Norman Steenrod.[66] Stone was nurturing expertise in areas in which Chicago was already strong—algebra—and adding expertise in areas in which it was not—probability, "modern" analysis, and algebraic topology.

By the end of Stone's first year, Hestenes, Kelley, and Zelinsky had all left, the first for a professorship at UCLA, the second for a position at Berkeley, and the third for a fellowship at the Institute. Whitney had declined Chicago's offer, and Stone had been authorized by February 1947 to move on to Weil and Mac Lane.[67] Both accepted—although, in Weil's case, not without much negotiation and some trepidation on the part of the higher administration—and arrived in the fall of 1947.[68] So, too, did Polish émigré

63. As noted in chapter four, Ivan Niven so characterized the 1930s in his article entitled, "The Threadbare Thirties." Interestingly, both Kaplansky and Halmos published their dissertation results in America's then newest general research journal, the *Duke Mathematical Journal*. See Irving Kaplansky, "Maximal Fields with Valuations," *DMJ* 9 (1942), 303–321 and Paul Halmos, "Invariants of Certain Stochastic Transformations: The Mathematical Theory of Gambling Systems," *DMJ* 5 (1939), 461–478.

64. Stone to Walter Bartky, 29 November, 1946, Series II, Division of Social Sciences, Box 308, Folder 5, Office of the President, Hutchins Administration, Records 1892–1951, University of Chicago.

65. Joan Baez et al., "Irving Ezra Segal (1918–1998)," *NAMS* 46 (1999), 659–668.

66. Morris Hirsch, "Edwin Henry Spanier (1921–1996)," *NAMS* 45 (1998), 704–705.

67. Lawrence Kimpton, Vice President and Dean of Faculties, to Walter Bartky, 21 February, 1947, Series II, Division of Social Sciences, Box 308, Folder 2, Office of the President, Hutchins Administration, Records 1892–1951, University of Chicago.

68. Kimpton sought information on Weil from various mathematicians—among them, E. T. Bell, Samuel Eilenberg, and Hermann Weyl—as well as from Haverford and Lehigh where Weil had taught before moving to Brazil (recall chapter seven). These sources acknowledged

and Penn harmonic analyst Antoni Zygmund. By the fall of 1949, Shiing-Shen Chern, had also joined them, having fled civil war–torn China and his position at the Institute for Mathematics of the Academia Sinica for a second sojourn at the IAS in Princeton. The Stone department—reflective of the new mathematical and geopolitical world order as well as of its chair's international outlook—was "complete" at least relative to the purer side of the field, although his broader aim of also building in applied mathematics, statistics, and electronic computing ultimately went unrealized.[69] Still, in three short years, he had succeeded brilliantly—although not without much effort, despite the university's initial assurances of total cooperation—in extending the department to the then cutting edges of purer mathematical research and in "transform[ing] a department of dwindling prestige and vitality once more into the strongest mathematics department in the U.S. (and at that point probably the world)."[70]

Chicago's metamorphosis in the immediately postwar years was unquestionably exceptional, but Chicago was by no means alone in recognizing and acting on the possibilities presented by the "booming" postwar market for mathematicians.[71] Some universities continued to build on the gains they had made in the 1930s. Others saw their opportunity actually to move into the research ranks. Among the latter were universities in the South, where, as noted, research-level mathematics had been comparatively slow to take hold.

In 1952, the National Research Council's Division of Mathematics, chaired by Marston Morse, issued a preliminary report on the regional development of mathematical research that, in providing a remarkable midcentury snapshot of the lay of the American mathematical landscape, highlighted those graduate programs that were then "regarded, in some sense, as emerging, or

---

Weil's mathematical gifts as well as his often difficult personality. See Series II, Division of Social Sciences, Box 308, Folder 2, Office of the President, Hutchins Administration, Records 1892–1951, University of Chicago.

69. Chicago did establish a Department of Statistics in 1949.

70. Felix Browder, "The Stone Age of Mathematics on the Midway," *The University of Chicago Magazine* 69 (Sept. 1976), 28–29; reprinted in *A Century of Mathematics in America*, ed. Peter Duren et al., 2: 191–193 on p. 191.

71. The characterization of the job market in mathematics is in Gerhard Kalisch to Herman Meyer, 17 May, 1946, Meyer Papers. Kalisch was a 1942 Ph.D. student of Adrian Albert and had taken short-term positions at the University of Kansas, at Cornell, and at the Institute before landing an assistant professorship at the University of Minnesota. Kalisch was by no means alone in his assessment. See also Halmos, *I Want To Be a Mathematician*, pp. 121–122.

new centers of serious and scholarly activities in mathematics."[72] Seventy-six departments, "selected by general reputation only, as those institutions where research in mathematics was being pursued," were sent a questionnaire soliciting such data as the number of undergraduate and graduate students in mathematics, the number of full- and part-time instructors and their respective numbers of teaching hours per week, the number of M.A.s and Ph.D.s awarded since January 1949, the number of papers published per faculty member since 1940, and the number of project grants awarded, etc. Of those, twenty-nine were in the category of "established" as opposed to "newly developing" (see fig. 9.2) and were invited not to "fill in and return the questionnaire unless they wished to."[73] Only eight or 28% did, but among the representatives of 83% or thirty-nine of the forty-seven "new centers," many "expressed interest and appreciation" on sending in their replies. Their inclusion in the survey marked, in their view, recognition of the mathematical developments that they had been effecting outside the country's traditional mathematical strongholds. Among those "new" centers were two—Tulane and Louisiana State University (LSU)—in Louisiana, a state that was part of a greater southern region where only two such centers—Texas at Austin and Rice—had been recognized before (see fig. 9.3). Both departments had leaders with deep ties to the South, who were intent on building mathematics there.[74]

At Tulane, Mississippi, native, William Duren, 1926 Tulane alumnus, and 1930 Chicago Ph.D. under Bliss, had taken a job at his undergraduate alma mater in 1931 and had risen through the ranks there to become chair of the mathematics department in 1947. Duren had been "faced with the young southerner's problem" after graduate school "of whether to stay North where conditions were obviously better or to return South."[75] Although he chose the latter, the question nagged.

72. "The Regional Development of Mathematical Research: A Preliminary Report on a Questionnaire Sent Out to Seventy Six Colleges and Universities by the Division of Mathematics," Box 2019-192/4, Duren Papers. This report is unpaginated, but the quotation is on p. 1. The quotations that follow in this paragraph are on p. 6.

73. Among the latter were a number of the large state universities—Michigan, Minnesota, Illinois, Indiana, Ohio State, Purdue, Wisconsin, UCLA, and Berkeley—which had continued to develop their programs throughout the 1940s and into the 1950s.

74. The only historically black institution polled, Howard University in Washington, D.C., did not return the survey.

75. William Duren, "Scientists for the South," p. 2, Box 2019-192/9, Duren Papers. The quotations that follow in the next paragraph are also from this text on p. 3.

SUMMARY OF QUESTIONNAIRE AND REPLIES, ARRANGED GEOGRAPHICALLY

New England	Middle Eastern	South Atlantic	North Central	South Central	Far Western
*Brown (N)	*Columbia (N)	*Duke	*Chicago	*Rice	*Calif. (N)
*Brown (Ap.M) (N)	*Cornell (N)	*U.No.Car.	*Illinois	*Texas (N)	*U.C.L.A. (N)
*Harvard (N)	*Johns Hopkins (N)		*Michigan (N)		*Cal. Tech. (N)
*M.I.T. (N)	*N.Y.U (N)		*Minnesota (N)		*U.S.C.
*Yale (N)	*Ohio State (N)		*Northwestern (N)		
	*Penn (N)		*Wisconsin		
	*Princeton (N)				
	*Purdue				
	*Virginia (N)				
Boston U.	Carnegie T. (N)	U. Fla. (N)	Ill.Inst.Tech.	La.St.U. (N)	Colorado (N)
Smith	Catholic U.	U. Georgia	Iowa St. C.	Oklahoma	Idaho
Wellesley	Cincinnati	U. S. Car.	U. Iowa	Tulane	Montana
	Howard (N)	U. Tenn.	Kansas St. C.		New Mexico
	Kentucky	Vanderbilt	U. Kansas		U. Oregon
	Lehigh		Michigan St. C.		Oregon St.C.
	Maryland		Missouri		U. Washington
	Notre Dame (N)		Nebraska		St.Col.Wash.
	Penn State		Washington (St.L.)		U. Wyoming
	Pittsburgh (N)		Wisconsin (Mil.)		
	Rensselaer (N)		Wayne		
	Rochester		Rutgers		
	Syracuse				
	Vassar				
	W. Virginia				

Footnote: The notation (N) means that no reply was received. The starred colleges were not urged to return the questionnaire unless they desired to.

FIGURE 9.2. "Summary of Questionnaire and Replies, Arranged Geographically" from Preliminary Report on a Questionnaire Sent Out to Seventy-Six Colleges and Universities by the Division of Mathematics. (Typed facsimile of the document in William L. Duren Papers, Archives of American Mathematics, Dolph Briscoe Center for American History, University of Texas at Austin.)

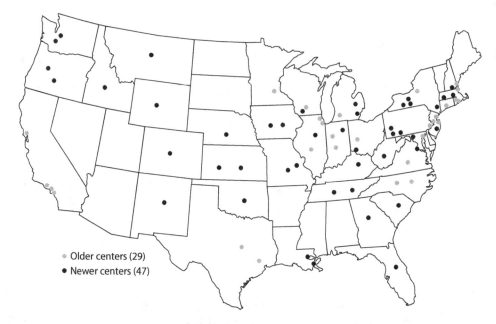

FIGURE 9.3. Map showing the distribution of major (in light gray) and secondary (in dark gray) centers of mathematics in the United States based on data collected in 1951. (Redrawn facsimile of the document in William L. Duren Papers, Archives of American Mathematics, Dolph Briscoe Center for American History, University of Texas at Austin.)

His war work with the Second Army Air Force in Colorado had exposed him to "the well established centers of research and study" around the country. He had thus seen firsthand how those institutions "were being vastly enriched with money, equipment and trained people by the government war research contracts while Tulane and other southern institutions taught trigonometry, which is not rewarding in either money or training." This led him to wonder "whether the South was ready for universities in this generation" and "whether [he] ought to seek other fields for [him]self, something frankly predatory instead of idealistic." Idealism won out.

In a lecture entitled "Scientists for the South" given before his fellow members of New Orleans's distinguished Round Table Club in October 1948, Duren made a compelling case for why the immediate postwar years were right for the promotion in the South of science in general and of mathematics in particular. First, he argued, "[f]loods of veterans would be coming back to study with government money for tuition," and "[a]s after World War I, there would be a new swing towards mathematics and the sciences."[76] Second, ongoing discussions in Washington made it clear that the government was "committed in principle to subsidies for scientific study and research as necessary for the national welfare and security." Since Federal dollars were likely to flow regionally, it would be "necessary to take the lead in one's own region to qualify for these funds." Third, "the South itself was stirring with a new spirit to develop its own economic resources for the good of its people," so "[s]cientists, good ones, would be needed." And, fourth and finally, "a dollar or a day spent in training young people in the South is more rewarding than it is in more highly developed sections of the country" because "there is a wealth of resources in undeveloped talent here" just waiting to be discovered and nurtured. The challenge was palpable, so he had resolved "to [come] back . . . and try to rebuild our own Department of Mathematics" at Tulane.

That process began when the department applied for and was awarded an ONR contract for "A Study of Functional Analysis in Topological Algebras" that included the explicit regional development goal of establishing a doctoral program in mathematics. By 1951, that program had turned out the first of its some two dozen 1950s, Ph.D.s, the earliest of which were in topology under Alexander Wallace, who had earned his 1939 doctorate under Whyburn at Virginia, and in analysis under Billy James Pettis, another Virginia Ph.D. but under Whyburn's colleague, McShane. Both of these men had joined the Tulane department in 1947, Duren's first year as chair. Like him and their most senior colleague, Herbert Buchanan, both had also been born in the South—Wallace in Virginia and Pettis in South Carolina—and both had been lured back to pursue their mathematical careers there.[77]

Duren's counterpart at LSU was Texan Luther Wade, who had completed his 1941 analytic-number-theoretic dissertation under the direction of Leonard Carlitz at Duke. After short stints at Hopkins and the Institute, Wade had returned to Duke before accepting LSU's offer to head its Department of Mathematics in 1948. His institutional mandate was explicit—build a research department—but, as at Tulane, that would be an uphill battle, if for different reasons.[78] Tulane was privately funded; LSU was financially

76. Ibid., p. 3. The quotations that follow in this paragraph are from pp. 3–4.

77. Buchanan had been born in Arkansas in 1881, earned his Ph.D. in 1909 at the University of Chicago under astronomer Forest R. Moulton, and, after teaching at the University of Tennessee, became chair of the department at Tulane in 1920. He retired from Tulane in 1949.

78. "Professor Luther Wade (1916–1996)" at https://www.math.lsu.edu/MathDepartmentHistory.

dependent on the vagaries of the Louisiana State Legislature. Tulane's department was small and successfully hired two researchers in Duren's first year as chair; LSU's was large and fully staffed at the higher ranks by midcareer Master's and/or Ph.D. holders who did little or no research and were effectively there for the long haul. Like Duren at Tulane, however, Wade began by trying to hire strong researchers in order to change the departmental dynamics.

In his first year, Wade approached Edwin Hewitt, the Harvard-trained harmonic analyst who had distinguished himself in operations research with the Eighth Army Air Force and who, in 1948, had just taken up an assistant professorship in the Northwest—at the University of Washington—another mathematically developing area of the country. Hewitt quite naturally sought the take on the situation at LSU of his Tulane friend, Duren, and received a very candid assessment in reply. "Let me preface the somewhat negative things I am going to say by revealing my prejudices in favor of LSU," Duren opened.[79] "I am very anxious to see the LSU department develop, for what we do here depends largely upon the general level of development in this region." After confessing to both his "missionary zeal for building up a center of mathematics" in Louisiana and in the South "where very little has existed before" and his hope that "a good mathematician" like Hewitt could be snared by LSU, Duren laid out his views on its mathematics department.

He started with the positives. LSU seemed to have resources, a relatively large pool of potential graduate students, an acceptable mathematics library, and a new president "who is determined to do things and seems to know how."[80] The negatives, though, were formidable. The department, comprised of some "thirty-five to forty full-time people," had a large "collection of dead wood" with "former high school teachers, junior college teachers, etc., at lower ranks." "I think," Duren explained, "all the main people have ranks of Associate Professor or higher, and not one has any chance of developing as

79. Duren to Hewitt, 25 January, 1949, Box 2019-192/6, Folder: Correspondence regarding employment 1948–1949, Duren Papers. The quotations that follow in this and the next two paragraphs are also from this letter (with his emphasis).

80. Harold Stoke was an historian and political scientist who had earned his Ph.D. at Johns Hopkins in 1930 and who, after a two-year teaching stint at Berea College in Kentucky and a position at the University of Nebraska, became a professional college administrator. He served as president of LSU from 1947 to 1951. See "Harold W. Stoke, College President," *New York Times,* 7 April, 1982.

a mathematician."[81] Still, "Wade evidently has a green light to bring some mathematicians in over their heads," and "[w]hile they undoubtedly accept this," Duren opined, "it does not make for 'sweetness and light.'"

Duren summed things up for Hewitt this way. "[I]f you went to LSU you would be going to a place with a fine long term prospect, a place where the president and the head of the department would be enlightened and congenial, and a place where you would have some very fine men personally as colleagues. But you would be going to a place where you would encounter much stupidity, many aggravations, where you would feel that academic freedoms were insecure, and where the quality of student is *much* lower than," say, at a Harvard or a Chicago. In all honesty, Duren concluded, "[m]any of these same things are also true of Tulane. . . . I would have great hesitation in trying to bring in a man in a professorial rank unless he knows pretty well what he will encounter in the South." In the end, Hewitt stayed in Seattle, but Wade succeeded in making a number of research-active hires into the 1950s and in growing the graduate program at LSU.[82] Indeed, as Duren had predicted, the department's long-term prospects did prove "fine."

Although the stories of Chicago, Tulane, and LSU in the immediately postwar period are very different, they underscore the possibilities for mathematics that those years presented for departments in a position to capitalize on them. As the preliminary report on the regional development of mathematics documented, such programs had emerged in the Midwest, West, and Northwest, in addition to the South, while the Northeast also continued to grow. Despite the worries of Kline and others of a severe postwar shortage of mathematicians, it turned out to be a bull market for those mathematicians, like Halmos, who had been underemployed during the Depression and war years, as well as for more recent Ph.D.s like Hewitt and Spanier.

81. The one notable member of the department was Herman Smith. A 1926 Chicago Ph.D. under E. H. Moore known for Moore-Smith sequences and Moore-Smith convergence in functional analysis (recall chapter one), Smith, according to Duren, was "so peculiarly adjusted that he will not publish his work."

82. Among the research-active faculty Wade hired were: Newton Hawley and Eugenio Calabi, 1949 and 1950 Ph.D.s, respectively, under Salomon Bochner at Princeton; Heron Collins, a 1952 Tulane Ph.D. under Billy James Pettis; Robert Koch, another 1952 Tulane Ph.D. but under Alexander Wallace; and Eugene Schenkman, a 1950 Yale Ph.D. under Nathan Jacobson. The presence of two Tulane Ph.D.s in this list is further reflective of that program's success in raising the level of mathematical research in the South in the 1950s. I thank Michelle Melancon, Assistant University Archivist at LSU, for unearthing all of the members of the LSU mathematics faculty from 1947 to 1957 for me.

	1945	1946	1947	1948	1949	1950
AJM	616	688	872	908	975	867
AM	718	832	1096	1010	991	1520
BAMS	1008	1099	1204	1200	1200	704
AMM	610	632	716	672	724	720
TAMS	964	1106	1104	1196	1463	1068
DMJ	722	621	1140	1132	636	510
JSL	158	142	162	236	284	312
MR	334	621	707	735	857	873
QAM	384	417	459	435	458	439
MTOAC	143	196	207	332	263	259

FIGURE 9.4. Pages published per year per journal, 1945–1950.[85]

Disseminating Postwar Mathematical Results

That there was no postwar shortage of mathematicians—or of new mathematical results to be published—was almost immediately reflected in the
avalanche of submissions under which American journal editors found themselves buried. All of the journals increased their number of pages published
after 1945 (see fig. 9.4) coincident, unfortunately, with yet another period
of soaring publication costs.[83] By 1948, all were bursting at the seams, and
the editors of some like the *Duke Mathematical Journal* and the AMS's two
journals, the *Bulletin* and the *Transactions*, were particularly concerned. In
Durham, Leonard Carlitz, editor of the former, found himself in the unenvi-
able position of having to explain in 1949 that his journal was "at present very
crowded and at the same time [had] been forced, for financial reasons" to cut
the number of pages published. Consequently, he and his editorial board had
"decided some time ago not to accept any new papers until we have been able
to catch up with the backlog."[84]

83. G. Bailey Price to Gordon Whyburn, 1 January, 1947, Box 2, Folder 40: Price, G. Bailey,
Whyburn Papers.

84. See, for example, Carlitz to Harry Vandiver, 15 December, 1949, Box 4RM156, Folder:
Correspondence: Carlitz, 1949–1961, Vandiver Papers. This explains the drop in the number
of pages published in 1949 recorded in fig 9.4 after the meteoric rise of the preceding four years.

85. The numbers for the *Mathematical Reviews* include the indices. In the reports they
wrote for the *Bulletin*, the *Mathematical Reviews* editors gave only the number of pages of
actual reviews printed, that is, in chronological order, 284, 540, 616, 636, 766. Printing costs,

Backlog was also a pressing issue at the AMS. In a letter to its Council in February 1948, Adrian Albert, managing editor of the *Transactions*, reported a "large" and "steadily" increasing backlog that "will continue to mount."[86] "No doubt," he reasoned, the situation at least partly owed "to the completion of work delayed by the war rather than to current production" so that "[i]t may then not be true that the present increased volume represents a continuing permanent increase." Still, it was imperative that the AMS "restore a reasonable situation" by authorizing "an accelerated rate of publication, if only a temporary one." Albert and his fellow *Transactions* editors thus "recommend[ed] that the limit for the number of pages be increased to 1600 pages for the current year, and that the budget for the *Transactions* be increased proportionately."

Although that proved financially unfeasible despite the fact that the Council approved the move,[87] the AMS turned the matter over to its Committee on the Role of the Society in Mathematical Publication which made the more modest but still substantial recommendation that three, instead of two, volumes of the *Transactions* be published in 1949 at around 500 pages per volume.[88] Even with that increase, the editors were forced in 1949 to issue a "statement of editorial policy" to the effect that "the large accumulation of manuscripts and the heavy demand for space" were forcing them to reserve the right to "require substantial condensation of manuscripts before they are accepted" "in order to avoid long delay in publication."[89]

The story was the same at the *Bulletin*, which, tellingly, also reported an increase in the number of submissions from abroad.[90] Indeed, the AMS Council came to view its overall publication situation as so dire that it appointed an Emergency Publication Committee to make further recommendations. By April 1949, it had been decided to establish two new periodicals.

---

of course, included the extensive indices. The *Annals of Mathematics* printed an extra issue in 1941 and what was essentially a double issue in 1950; in 1950, the *Bulletin* went from appearing twelve times a year to appearing six times a year.

86. Albert to the AMS Council, 24 February, 1948, Box 1, Folder 3: Albert, A. Adrian, Whyburn Papers. The quotations that follow in this paragraph are also from this letter.

87. John Kline, "The February Meeting in New York," *BAMS* 54 (1948), 461–479 on p. 469.

88. John Youngs, "The Annual Meeting of the Society," *BAMS* 55 (1949), 261–311 on p. 267.

89. "Notes," *BAMS* 55 (1949), 321–324 on p. 321.

90. John Kline, "The Annual Meeting of the Society," *BAMS* 54 (1948), 260–297 on p. 271.

The *Proceedings of the AMS* would relieve the *Bulletin* by publishing all submissions except meeting reports, other Society-related pieces, and texts of invited addresses given before the Society. In other words, the *Proceedings* would carry the short research papers that had, since 1930, been comprising the *Bulletin's* even-numbered issues, and the *Bulletin* would carry the specifically Society-related material that had made up the odd-numbered issues.[91] Although this ultimately helped reduce what was then an eighteen-month delay from time of acceptance to time of publication for *Bulletin* articles, a sizable backlog persisted into the 1950s even following increases in the total number of pages published by the *Bulletin* and *Proceedings* combined.[92] The editors of the *Transactions*, on the other hand, were empowered to select "longer papers and groups of cognate papers" submitted for their review for inclusion in a "non-periodical but serial publication" to be called the *Memoirs of the AMS* and "to be issued and sold in separate volumes."[93] By 1951, the *Transactions's* backlog problem had been relieved by the publication of the first nine *Memoirs*.

The situation was further improved in 1951 by the creation of a brand-new journal, the *Pacific Journal of Mathematics*, yet another new model for the publication of mathematical research. The *Pacific Journal*, unlike the other extant mathematical periodicals, was not associated with a single university or with a society but rather represented the editorial and financial collaboration of twelve West Coast colleges and universities. Conceived by Berkeley's František Wolf and UCLA's Edwin Beckenbach as "a research journal open to mathematicians all over the country" just like its competitors, it nevertheless aimed to highlight the many institutional advances on the West Coast that had been made relative to mathematics.[94] Wolf was a Czech émigré who had come to the United States from Sweden in 1941 before being hired by Griffith Evans at Berkeley in 1942; Beckenbach, a Texas native who had earned his 1931 Rice Ph.D. under Lester Ford, had, after a series of intermediate positions, landed his UCLA professorship in 1945. Both felt the time was right for a West

91. Temple Hollcroft, "The April Meeting in Philadelphia," *BAMS* 55 (1949), 678–705 on p. 684. Pitcher also details these developments in *A History of the Second Fifty Years*, pp. 14–15.

92. Temple Hollcroft, "The Annual Meeting of the Society," *BAMS* 56 (1950), 140–190 on p. 152 and William Whyburn, "The Annual Meeting of the Society," p. 117.

93. Hollcroft, "The April Meeting in Philadelphia," pp. 684–685.

94. Charles Morrey to Saunders Mac Lane, 1 February, 1950, Box 2, Folder 22: Morrey, Charles B., Whyburn Papers. The quotation that follows in this paragraph is also from this letter.

Coast-based journal. The mathematics department chairs "of all the principal institutions on the West Coast"—at the University of British Columbia, Caltech, the Universities of California in Berkeley, Davis, Los Angeles, and Santa Barbara, the University of Southern California, Stanford, Washington State College, and the University of Washington—concurred and secured financial commitments ranging from $250 to $500 from their respective schools in support of the new publication venture as well as from the AMS and the newly founded Institute for Numerical Analysis (see the next chapter).[95] At the end of its first year of publication, the *Pacific Journal* had printed some 602 pages of new research produced by over fifty mathematicians on the West Coast, around the country, and from abroad.

The issue of backlog was perhaps even more overwhelming for the *Mathematical Reviews* than for any of the research journals. Because of both the greater availability of foreign journals after the war and what was perceived as "the increased production in most countries," material to review was ever-mounting.[96] Another factor was at play by 1950: "the policy of encouraging longer reviews for Russian papers."[97]

In the immediate aftermath of World War II, Stalin's Soviet Russia came quickly to dominate Eastern Europe thereby establishing, what Winston Churchill famously termed in a 1946 speech, an "Iron Curtain" separating Eastern from Western Europe.[98] In mathematics, this new geopolitical reality manifested itself in a decision that Russian mathematicians would publish solely in Russian.[99] This would make it increasingly difficult for American mathematicians to keep up with new Russian results at a time when that seemed particularly critical. The new policy that the *Mathematical Reviews* adopted thus aimed to make it easier for English-speaking mathematicians to get more than just the gist of the results that a given Russian paper contained.

95. In all, the contributions to the costs of the journal totalled $5,450, with $1,300 of that amount coming from the AMS. See Minutes of the Annual Meeting of the Board of Governors of the *Pacific Journal of Mathematics*, undated (but in a folder dated 28 December, 1951), Minutes of the Board of Trustees, 1923–1960, AMS Records, Ms.75.1, AMS Papers. The "real wealth" of $5,450 in 1951 dollars is equivalent to $56,000 in 2021 dollars (see https://www.measuringworth.com).

96. Kline, "The Annual Meeting of the Society [1948]," p. 271.

97. Hollcroft, "The Annual Meeting of the Society [1950]," p. 152.

98. For an excellent account of the postwar period, see Tony Judt, *Postwar: A History of Europe Since 1945* (London: Penguin Books, 2005). Judt briefly mentions Churchill's speech on p. 110.

99. Pitcher, *A History of the Second Fifty Years*, p. 135.

The AMS's Committee on the Role of the Society in Mathematical Publication also seemed to sense the beginning of the Cold War and its possible implications for mathematics. In February 1948, it proposed the launch of a project for the "publication of translations of important Russian mathematical articles" as well as for the preparation of a Russian-English dictionary "of important mathematical terms, supplemented by a pamphlet to include the Russian alphabet, rules of syntax, and short bibliography of works available in parallel Russian and English versions."[100] By April, not only had Samuel Eilenberg, Lipman Bers, Paul Halmos, George Rainich, and Arnold Ross agreed to prepare the pamphlet, but Mina Rees at the ONR had also allocated $25,000 to pay for both it and the translations themselves.[101] Yet another committee—this one comprised of Ralph Boas as chair, Derrick Lehmer, William Prager, Eilenberg, and Rainich—had been appointed by the end of the year actually to select articles for translation. Two years later in 1950, some forty-two papers were in various stages of translation and the dictionary, short grammar, and fifteen translations of Russian mathematical articles were complete. Overly high distribution costs had given the AMS pause, however. By 1953, when the ONR could no longer afford to support the project, 105 papers—by Andrei Kolmogoroff and Ivan Vinogradov, among many others—had been translated and published.[102]

The AMS had also begun publication of another new postwar series, the *Proceedings of Symposia in Applied Mathematics*. While it is true that the AMS had instituted its Gibbs Lectures in 1924 to try to spark interest in applied mathematics among the members of that predominantly purist community, experiences during the war—like the creation of Brown's Advanced Research and Instruction program and the success of the *Quarterly Journal of Applied Mathematics*—had finally spurred it to play a more proactive role. The AMS supplemented the advances made at Brown, NYU, and elsewhere, with the organization of annual short courses on applied mathematics in the spirit of its Colloquium Lectures.

100. Kline, "The February Meeting of the Society [1948]," p. 469.

101. John Kline, "The April Meeting of the Society," *BAMS* 54 (1948), 622–647 on p. 625 and Richard Bruck, "The Summer Meeting in Madison," *BAMS* 54 (1948), 1042–1086 on p. 1048. In 2021 dollars, the ONR's allocation amounts to $277,200 in "real wealth" (see https://www.measuringworth.com).

102. Pitcher, *A History of the Second Fifty Years*, pp. 136–137. A grant from the National Science Foundation allowed the AMS to begin a second series of translations in 1955 that continued with the NSF's support until 1966. Indeed, the now self-sustaining project continues to the present.

The first such symposium, on "Non-linear Problems in Mechanics of Continua," was held, fittingly, at Brown in August 1947 and drew 265 attendees, among whom were 143 members of the AMS.[103] Over the course of three days, those assembled heard four main lecturers—Hopkins's Murnaghan, Stoker and Friedrichs of NYU, and Alexander Weinstein of the Carnegie Institute of Technology—and participated in break-out sessions consisting of shorter talks grouped thematically and followed by half hour–long discussion periods. The printed conference volume appeared two years later. Two more volumes in the series—based on symposia on "Electromagnetic Theory" and "Elasticity" held in 1948 and 1949, respectively—came out in 1950.[104]

In addition to these new publications, the AMS also renewed its Colloquium Publication series and the series of "Mathematical Surveys" that it had inaugurated with two volumes in 1943 but then effectively suspended. Between 1946 and 1950, seven new Colloquium volumes appeared—by Weil, Radó, Hille, Wilder, Ritt, Walsh, and Albert Schaeffer and Donald Spencer—while the years 1949 and 1950 witnessed the publication of three new Surveys—by Morris Marden, Otto Schilling, and Stefan Bergman.[105] (See the next chapter for a discussion of some of these works.) The authors

103. William Prager and Temple Hollcroft, "First Symposium in Applied Mathematics," *BAMS* 53 (1947), 878–881. Interestingly, despite this symposium's designation, symposia on applied mathematics had, in fact, been hosted by the AMS at least since 1941, with one as late as 1944 in conjunction with its April meeting in Berkeley. The latter had been comprised of talks by Chia-Chiao Lin (Lin Jiaqiao) and Wei-Zang Chien (Qian Weichang), both of Caltech (recall the previous chapter), on "Hydrodynamical Stability" and "The Intrinsic Theory of Elastic Plates and Shells," respectively. See Aristotle Michal, "The April Meeting in Berkeley," *BAMS* 50 (1944), 476–477 on p. 476 as well as Temple Hollcroft, "The February Meeting in New York," *BAMS* 47 (1941), 333–342 on p. 334; Temple Hollcroft, "The May Meeting in Washington," *BAMS* 47 (1941), 523–531 on p. 524; and Temple Hollcroft, "The Annual Meeting of the Society," *BAMS* 48 (1942), 183–198 on p. 184. Pitcher mentions the latter three symposia in *A History of the Second Fifty Years* (on p. 91) but not the one in 1944.

104. See Eric Reissner, William Prager, and James Stoker, ed., *Nonlinear Problems in Mechanics of Continua*, Proceedings of Symposia in Applied Mathematics (PSAM), vol. 1 (New York: American Mathematical Society, 1949); Abraham Taub, Eric Reissner, and Ruel Churchill, ed., *Electromagnetic Theory*, PSAM, vol. 2 (New York: American Mathematical Society, 1950); and Ruel Churchill, Eric Reissner, and Abraham Taub, ed. *Elasticity*, PSAM, vol. 3 (New York: American Mathematical Society, 1950).

105. For the volumes by Weil, Radó, Hille, and Wilder, see the previous chapter. The others, all published by the AMS in New York City, were: Joseph Ritt, *Differential Algebra*, vol. 33 (1950); Joseph Walsh, *The Location of Critical Points of Analytic and Harmonic Functions*, vol. 34 (1950); and Albert Schaeffer and Donald Spencer, *Coefficient Regions for Schlicht Functions*, vol. 35 (1950). The Surveys, likewise all published by the AMS in New York, were: Morris

of these volumes reflected well the demographics of the postwar American mathematical research community with émigrés Weil, Radó, Schilling, and Bergman making fundamental contributions alongside American counterparts home-based in the Northeast, Midwest, and West at established as well as developing programs.

––––––––

The American mathematical community was in high gear in the immediately postwar years. Internationally, it endeavored to reestablish mathematical ties that the war had broken and to help its foreign colleagues regain their mathematical footing. In a national context, it worked to position itself effectively within emerging bureaucracies like the ultimately ill-fated Research Board for National Security and the highly successful Office of Naval Research, fully aware of the importance that external funding would likely have for future research productivity. It also sought to exercise its authority over matters, such as secondary mathematics education, in which it had failed to engage with sufficient energy and innovation before the war.

And, then, there was research. Departments focused tightly on their graduate programs, both in training future researchers and in boosting their faculties with the best talent they could acquire. The talent was out there. The postwar hiring boom had meant that the roughly thirty programs that had been considered "established" before the war were, in the course of just a few years, augmented by some one-and-a-half times that number of research-oriented programs across the country, including areas like the South where research-level mathematics had traditionally been less strong. This growth was reflected in a volume of research that, in swamping American publication outlets, prompted the creation of new ones by a community that was agile and mature enough by the 1940s to adapt by creatively marshalling both new and extant resources.

––––––––

Marden, *Geometry of Polynomials*, vol. 3 (1949); Otto Schilling, *The Theory of Valuations*, vol. 4 (1950); and Stefan Bergman, *The Kernel Function and Conformal Mapping*, vol. 5 (1950).

10

Sustaining and Building
Research Agendas

If the immediately postwar years—characterized by their "rapid progress of military demobilization; rapidly changing picture of world affairs; . . . uncertainties as to the future of peacetime structures of the Army and Navy, and . . . prolonged discussions in Congress on science legislation"—had "combined to create a confused picture," mathematicians, at least, had managed to see through it all fairly successfully.[1] Their efforts, however, fundamentally aimed at one thing: to foster an ever-better research environment. After all, if their community had indeed "assumed world leadership," then, at the very least, that new stature would have to be reflected in the quality of its research output.[2] But can "world leadership" really be measured? Is it possible to "prove" that such dominance has been attained? Does the concept of *the* "world leader" even make sense?

While answers to these questions are elusive, several things are clear. As at the turn of the twentieth century when Germany was widely considered *the* world leader in mathematics, Americans of the 1940s continued to recognize the validity of that concept. But even more was true. They sensed that *they* had achieved mathematical hegemony.[3]

1. Frank Jewett to Karl Compton, 25 March, 1946, Box 34, Folder: Research Board for National Security 1945–1946, Veblen Papers.

2. Kline, "Rehabilitation of Graduate Work," p. 131.

3. John Krige explores the broader notion of the postwar hegenomy of American science, with a focus on the physical sciences and American foreign policy, in his book, *American Hegemony and the Postwar Reconstruction of Science in Europe* (Cambridge: The MIT Press, 2006).

What, in their view, had changed? After all, they had been producing much mathematical research in the two decades immediately following the close of World War I, yet, while the leaders among them believed that the community was pulling even with Germany, they continued to sense both that they had not quite pulled ahead and that their work was not as appreciated abroad as it should be. As their efforts in, for example, classical algebraic geometry and the calculus of variations exemplified, it is quite possible to generate a prodigious number of new results in the context of old, largely spent paradigms and, in so doing, to inspire research followings within national boundaries, but that does not necessarily redound positively to the perception of the community beyond those boundaries. They had also clearly absorbed mathematicians of all calibers from abroad, but those mathematicians—even taking into account the more applied bent of some of them—had not fundamentally changed what was already a well-developed and ever-strengthening community.

As least one thing *was* different by the 1940s: the conviction of the American community's leaders that *its* members were in the position of actually setting new research agendas that would then define the mathematical problems that others both nationally and internationally would tackle. If they were right, then *that* would be key in shaping the perception abroad of their community as a or even *the* mathematical leader. How, though, could they realize their conviction? Explicitly, as David Hilbert had shown in the context of the twenty-three "Mathematical Problems" he had brashly articulated at the Paris ICM in 1900? Implicitly, by publishing new results in directions that would then capture the interest of those within the broader mathematical community? Mathematicians in the American research community of the 1940s did both.

The Princeton Bicentennial Conference on
"Problems of Mathematics"

In November 1945, just two months after the war ended in the Pacific, Princeton President Harold Dobbs called a press conference in New York City to make a major announcement. His university, founded as the College of New Jersey in 1746, would fête its bicentennial in a celebration over the course of the 1946–1947 academic year.[4] In so doing, it would follow the example set a

---

4. "Princeton Plans for 200th Year," *New York Times*, 15 November, 1945, pp. 21 and 38. The quotations that follow are on p. 21.

decade earlier by Harvard on the occasion of its tercentenary (recall chapter seven), although the two events, punctuated by a world war, could not have been more dissimilar.

Although social events—class reunions, a commemorative baseball game, maybe even a regatta—would be part of Princeton's celebration as they had been for Harvard's, Princeton's would be independent of other meetings. Its focal point would be a series of so-called bicentennial conferences with a solemn postwar aim: "consider[ing] the problems of the reestablishment and strengthening of liberal scholarship and of the instruction of men in the ways of peace" by "provid[ing] . . . an occasion when emerging problems can be assessed and lines of attack" laid out in all areas: the sciences, the social sciences, and the humanities.[5] Each carefully thought-out conference would, in bringing together no more than a hundred "scholars of world distinction in their fields" from the United States and abroad, provide a medium for the free exchange of ideas by re-creating, albeit on a larger scale, the conversational atmosphere of Princeton's preceptorial program. It was not lost on the organizers, moreover, that they would also highlight the quality of the Princeton faculty, unlike the protocol adopted by Harvard a decade earlier.

The first six of the three-day conferences—among them, one on the pressing issue of the future of nuclear physics that, like many of the conferences, received coverage in the *New York Times*[6]—were held over the course of the first half of the fall semester prior to a convocation on 19 October in celebration of the granting of the university's founding charter. Chronicled in a newsreel and shown in theaters around the country, that event served as the occasion for a majestic academic procession numbering some 500 strong as well as the conferral of twenty-three honorary degrees to luminaries like the Danish physicist Niels Bohr.[7] Nine more conferences rounded out the year-long program that closed with a final convocation in June 1947. The latter event brought President Harry Truman, General Dwight Eisenhower, and Admiral Chester Nimitz, in addition to other dignitaries and an audience of 6,500, to the Princeton campus for graduation ceremonies and yet another round of honorary degrees. Paramount News's cameras once again

5. "Princeton Plans for 200th Year," *New York Times*, 15 November, 1945, p. 21.

6. William L. Laurence, "New 10-Billion Volt Atom Smasher Is Planned by California Scientists," *New York Times*, 25 September, 1946, 46.

7. See the newsreel posted at https://blogs.princeton.edu/mudd/2011/03/princetons -bicentennial-charter-day-october-19-1946/ .

FIGURE 10.1. Princeton Bicentennial Conference on "Problems of Mathematics" (1946).
(Photo courtesy of the Institute for Advanced Study, Princeton.)

rolled to capture not only the pomp and circumstance but also the policy speech in which Truman renewed his 1945 call for universal military training as opposed to the "maintenance of a large standing army."[8] It was as part of these rarefied intellectual and politicized year-long events that ninety-three mathematicians from ten countries and some thirty-seven different institutions of higher education had been convened at the close of 1946 to discuss "Problems of Mathematics." They aimed to realize, as best they could given their field, the bicentennial year's lofty goals.

Princeton had tapped its best in mathematics, mathematical physics, statistics, and logic for the organizing committee of its mathematical event. Under Lefschetz as chair, the twelve-man committee (indeed, there were neither

8. See the newsreel posted at https://blogs.princeton.edu/reelmudd/2010/07/past-present-and-future-us-presidents-at-bicentennial-celebration-1947/ as well as Frank Cunningham, "Harry S. Truman and Universal Military Training, 1945," *The Historian* 46 (1984), 397–415.

women organizers nor, judging from the conference photo, women participants) arranged sessions in the nine "areas" of algebra, geometry both algebraic and differential, mathematical logic, topology, so-called "new fields,"
mathematical probability, analysis, and analysis in the large. These, after all,
were areas in which mathematicians home-based in Princeton, whether at
the university or at the Institute, excelled. Yet, as the organizers had to admit,
they had been "forced to some omissions."[9] In particular, they determined that
since "[a]pplied mathematics, because of its wide ramifications into many sciences, could not . . . be treated as one field," it would have to be left to "other
groups"—Princeton expertise lacking—to consider at some future date. The
Princeton conference would thus concentrate on what its organizers characterized as applied mathematics' "unifying spirit, pure mathematics." Once
again, then, fundamentally purist American mathematicians chose to follow
their natural predilections rather than focus on matters applied. At the same
time, they acknowledged those matters—and perhaps with somewhat greater
conviction given their wartime experiences and the new postwar context—
as worthy of separate, detailed consideration. Their choice of focus, however,
prompted one journalist to remark that "[t]he mathematicians were, apparently, the only group that remained happily insulated from the cares and
tribulations of life on this planet," although "they laid out the most challenging
program for future research."[10]

The mathematics conference was held in Fine Hall from Tuesday through
Thursday, 17–19 December, 1946. By design, each of its sessions had a chair
who was a member of either the Princeton or the IAS faculty, a discussion
leader or leaders drawn from experts at other institutions nationally and
internationally, and a generally more junior reporter who was nevertheless
well familiar with the given mathematical area. As Lefschetz explained in

9. *Problems of Mathematics* (Princeton: Princeton University Press, 1946), reprinted (with
some omissions) as "The Princeton University Bicentennial Conference on the Problems of
Mathematics," in *A Century of Mathematics in America*, ed. Peter Duren et al., 2: 309–333. In
what follows, I will cite the page numbers from both versions. For this and the quotations in
this and the next paragraph (unless otherwise marked), see pp. 1–2 (or pp. 309–310). The
other organizing committee members were: mathematicians Emil Artin, Salomon Bochner,
Claude Chevalley, Ralph Fox, and Albert Tucker; mathematical physicists Valya Bargmann,
Bob Robertson, and Eugene Wigner; statisticians Samuel Wilks and John Tukey; and logician
Alonzo Church.

10. J. Douglas Brown, "The Princeton Bicentennial Conferences: A Retrospect," *The Journal
of Higher Education* 19 (1948), 55–59 on p. 58.

the foreword to the printed version of the proceedings, those assembled had a historic sense of purpose. "It has been nearly fifty years," he noted, since Hilbert put forth his conception of "a unified viewpoint in mathematics" before his audience at the Paris ICM. "It has seemed to us that our conference offered a unique opportunity to help mathematics swing again for a time toward unification." The organizers thus saw an opportunity for Princeton, and more broadly American, mathematics to achieve at midcentury the notoriety that Hilbert had gained at its start. They sought nothing less than to define mathematics' research paths in the century's second half.[11] They opened with algebra.

Emil Artin, a member of the American mathematical community since his emigration in 1937 and in his first year on the Princeton faculty in 1946, chaired the algebra session. Harvard's Garrett Birkhoff, émigré Richard Brauer then at the University of Toronto, and Nathan Jacobson then of the Johns Hopkins served as discussion leaders. Gerhard Hochschild, a 1941 Princeton Ph.D. under Claude Chevalley and, in 1946, a Harvard postdoc, did the reporting. In addition to these men, Adrian Albert, Nelson Dunford, Saunders Mac Lane, Tibor Radó, and Marshall Stone, among others, participated in a give-and-take that revealed "[t]wo main lines" for future algebraic research: "one the generalization of known results with an eye toward increasing their scope and learning more of their inner meaning . . . and the other the continuation along classical lines, represented by Brauer's imposing advance," namely, his proof of Artin's conjecture that "characters known to be rational combinations of certain special characters are in fact integral rational combinations" (see the next section).[12] Generalizations, in the forms, for example, of lattice theory and Boolean algebras, were already well underway and fit with Lefschetz's notion for the conference of directing a new "swing . . . toward unification" in mathematics. Brauer's result represented generalization of a different kind. The session participants deemed it "a decisive step in the generalization of class-field theory to the non-Abelian case, which is commonly regarded as one of the most difficult and important problems in modern algebra."[13]

11. I discussed the import of this conference in "Perspectives on American Mathematics," pp. 396–400. David Rowe also considered it in "On Projecting the Future and Assessing the Past: The 1946 Princeton Bicentennial Conference," *MI* 25 (2003), 8–15.

12. *Problems of Mathematics*, pp. 4–5 (or pp. 310–311). Compare Richard Brauer, "On Artin's *L*-Series with General Group Characters," *AM* 48 (1947), 502–514.

13. *Problems of Mathematics*, p. 5 (or p. 311). As Benedict Gross and John Tate note in their 1989 commentary on this session (see *A Century of Mathematics in America*, ed. Peter

Two sessions on geometry—algebraic, chaired by Lefschetz, and differential, chaired by Veblen—followed. As in the algebra session, and indeed in all of the conference's sessions, American and foreign talent, among the latter, émigrés newly absorbed into the American mathematical community, had been tapped for the various session roles and participated actively in the discussions. The sessions were thus another manifestation of the American mathematical community's new look and of the full integration into it of its newest members.

Like their more purely algebraic brethren, the algebraic geometers also saw two main paths for future research: "new, deeper problems for the classical algebraic geometry over the field of complex numbers contrasted with new methods for developing algebraic geometry over abstract fields."[14] Following Hilbert, Lefschetz laid out four specific problems—three in the first vein and one in the second—that he thought merited investigation. One in particular—extending the Riemann-Roch theorem to higher dimensions—proved particularly fruitful.[15] The discussion leaders—Cambridge's William Hodge and Oscar Zariski then at Illinois—also generated sustained conversation that, in retrospect, proved prescient. Hodge discussed the conjecture he had recently made relating the algebraic topology of a non-singular complex algebraic variety to its subvarieties, while Zariski described his work on the problem of minimal models in the classification theory of birational transformations.[16]

Veblen oversaw the differential geometers. Václav Hlavatý from the University of Prague and Tracy Thomas, as noted, a 1923 Princeton Ph.D. under Veblen who was on the faculty at Indiana in 1946, served as discussion leaders. Haverford's Carl Allendoerfer, a 1937 Ph.D. under a Thomas then still at Princeton, took notes. Hlavatý aside, this was the only fundamentally inbred session. Differential geometry in the United States had been a largely Princeton-centric affair that, by 1946, had played itself out there. Still, the Pasadena school (for one reason or another not represented at the symposium) continued into the 1940s and Princeton's Salomon Bochner had then

---

Duren et al., 2: 335–336), the result, although perhaps not as "decisive" as was thought in 1946, was nevertheless a harbinger of Robert Langlands's later linkage between non-abelian Galois representations and Harish-Chandra's theory of automorphic forms on reductive groups.

14. *Problems of Mathematics*, p. 6 (or p. 312).

15. Loosely speaking, given a space of meromorphic functions with a certain number of zeroes and poles, the Riemann-Roch theorem gives a way to compute the space's dimension.

16. Compare Herbert Clemens's commentary on this session in *A Century of Mathematics in America*, ed. Peter Duren et al., 2: 337–338.

recently linked curvature to topology by deriving "new relations between the Ricci curvature of a compact Riemann space and the characteristics of vector fields defined over the space."[17] Veblen presciently saw in the latter linkage "the direction in which we want to see differential geometry go."[18] This insight aside, the session on differential geometry was the one, in retrospect, that was "more notable for what (and who) [wa]s left out than for what [wa]s included."[19]

The first day closed with a session on mathematical logic. Alonzo Church was in the chair; émigré and Berkeley professor Alfred Tarski led the discussion; J.C.C. "Chen" McKinsey, then on the faculty at the Oklahoma Agricultural and Mechanical College (now Oklahoma State) but a 1936 Berkeley Ph.D. under Benjamin Bernstein, served as reporter. At issue were *Entscheidungsprobleme* or decision problems, that is, given an object in some set, is it or is it not possible to construct an algorithm for deciding if the object has a particular property? Apparently a "spirited discussion" ensued, in which the Institute's Kurt Gödel, Stephen Kleene from Wisconsin, and Harvard's Willard Quine, among others, took part.[20] Church, in particular, generated quite a bit of research into the 1950s in this vein (see the next section) with, for example, his symposium conjectures that "the word problem for groups" and the problem "of giving a complete set of topological invariants . . . for closed simplicial manifolds of dimension n" were unsolvable decision problems. Both were ultimately proven true: the first independently by Soviet mathematician Pyotr Novikov and the University of Illinois's William Boone and the second by Andrei Markov, another Soviet mathematician and son of the noted probabilist.[21] In mathematical logic, as in other areas, mathematicians

17. Salomon Bochner, "Vector Fields and Ricci Curvature," *BAMS* 52 (1946), 776–797. For the quotation, see *Problems of Mathematics*, p. 9 (or p. 314). In his commentary on this session in *A Century of Mathematics in America*, ed. Peter Duren et al., 2: 339–341, Robert Osserman sketches the path from Bochner's work to Kodaira's vanishing theorems.

18. *Problems of Mathematics*, p. 9 (or p. 314).

19. Osserman, p. 339. The quotations that follow are also on this page. In particular, Osserman notes that, "at least as reported," "the names of [Élie] Cartan, Chern, [Herman] Weyl, and [Sumner] Myers are nowhere mentioned."

20. *Problems of Mathematics*, p. 11 (or p. 316). The quotations that follow are on p. 10 (or p. 315). Compare Yiannnis Moschovakis's commentary in *A Century of Mathematics in America*, ed. Peter Duren et al., 2: 343–346.

21. The papers are Pyotr Novikov, "On the Algorithmic Unsolvability of the Word Problem in Group Theory" (in Russian), *Proceedings of the Steklov Institute of Mathematics* 44 (1955), 1–143; William Boone, "The Word Problem," *PNAS* 44 (1958), 1061–1065; and Andrei

internationally worked on problems identified during the Princeton conference, but after the 1950s, the logicians moved in directions fundamentally different from those discussed in 1946.[22]

Following an evening spent together in the clubby atmosphere of the Princeton Inn, the mathematicians reconvened on Wednesday morning to consider topology. "The discussion," reported émigré Samuel Eilenberg then newly hired by the University of Indiana, "centered around two main topics: (a) groups of transformations, [and] (b) classification of homotopy classes of maps, fibre bundles and related questions."[23] Perhaps not surprisingly, given the Princeton venue and Lefschetz's role as chair of the planning committee, "topology" at the "Problems of Mathematics" conference did not come in the two varieties that had evolved especially in the 1920s and 1930s. "Topology" there meant Lefschetz's algebraically oriented version not the point-set approach pursued by R. L. Moore and his followers. Although point-set stalwarts John Kline and Gordon Whyburn as well as reconciler Raymond Wilder were among the conference participants and although the "detailed discussion" in the session "was extremely active" and "wide-ranging," no point-set topological notions made it into the printed proceedings. Instead, questions of the future involved the "central conjecture . . . that any compact group which acts effectively on a manifold must be a Lie group,"[24] "finding algebraic methods for the numeration of the homotopy classes of continuous maps from a polyhedron of dimension n to one of dimension m," and others. More broadly, the topology session—chaired by Princeton's Albert Tucker and informed by discussion leaders Heinz Hopf from the ETH in Zürich, Yale's Deane Montgomery, Michigan's Norman Steenrod, and Henry Whitehead of Cambridge University—focused on techniques of "pressing up from

---

Markov, "The Insolubility of the Problem of Homeomorphy" (in Russian), *Proceedings of the USSR Academy of Sciences* 121 (1958), 218–220, respectively.

22. See Moschovakis's commentary, p. 346.

23. *Problems of Mathematics*, p. 13 (or p. 317). The otherwise unattributed quotations that follow in this paragraph may also be found here.

24. This unsolved conjecture is related to Hilbert's fifth problem, namely, as Hilbert expressed it, "[h]ow far Lie's concept of continuous groups of transformations is approachable in our investigations without the assumption of the differentiability of the functions." Hilbert, "Mathematical Problems," p. 451. The crisper formulation—is every locally Euclidean group a Lie group?—was established in 1952 by Andrew Gleason and by Dean Montgomery and Leo Zippin. See Andrew Gleason, "Groups without Small Subgroups," *AM* 56 (1952), 193–212 and Dean Montgomery and Leo Zippin, "Small Groups of Finite-Dimensional Groups," *AM* (1952), 213–241.

the homology to the homotopy region, partly by the application of the former to the latter."[25]

Conference participants moved from topology to consider what the organizers termed "new fields," although, Norbert Wiener's focus on the problems involved with the communication of information aside, the discussion was not of new fields per se but rather of "classical problems *related* to application, and of the need and feasibility of revitalizing work in these fields."[26] This kind of applications orientation was perhaps not surprising given that von Neumann chaired the session, and Griffith Evans, Francis Murnaghan, John Synge, and Norbert Wiener led the discussion. In reflecting and focusing on the underlying pure mathematics, it emphasized areas like potential theory, multivalued harmonic functions in three variables, and linear and nonlinear differential equations. In particular, von Neumann "pointed out that the success of mathematics with the linear differential equations of electrodynamics and quantum mechanics had concealed its failure with the nonlinear differential equations of hydrodynamics, elasticity, and general relativity." More work, he clearly inferred, needed to be done.

In a sense, the session that ended the conference's second day actually did deal with a "new" field of mathematics: mathematical probability. Andrei Kolmogoroff's 1933 *Grundbegriffe der Wahrscheinlichkeitsrechnung* had provided the first rigorous, measure-theoretic foundation for the theory of probability, but thirteen years later in 1946, it was still perceived that "only a few mathematicians were taking mathematical probability seriously as mathematics."[27] With, however, Samuel Wilks in the chair, with discussion leaders Harald Cramér of the University of Stockholm, Illinois's Joseph Doob, and William Feller (who had recently left Brown for Cornell), and with reporter, the Princeton and Bell Labs statistician, John Tukey, the area not only received careful consideration but also generated much discussion. For example, Cramér highlighted problems related to random variables and pointed out various open problems; all three discussion leaders emphasized open questions about stochastic processes; and the University of North Carolina's

25. *Problems of Mathematics*, p. 15 (or p. 318). It should be noted that Tucker, Steenrod, and Whitehead were all Princeton-trained.

26. Ibid. (my emphasis). For the quotation that follows, see ibid., p. 26 (or p. 319).

27. Andrei Kolmogoroff, *Grundbegriffe der Wahrscheinlichkeitsrechnung*, Ergebnisse der Mathematik und ihrer Grenzgebiete, vol. 2, no. 3 (Berlin: Julius Springer, 1933). Joseph Doob gives the characterization in his commentary in *A Century of Mathematics in America*, ed. Peter Duren et al., 2: 353–354 on p. 354.

Harold Hotelling cautioned the assemblage not to forget the related area of statistics. In particular, Hotelling stressed "the need for a complete and careful study of the estimation problem," that is, how to use information from a sample to make inferences about a population.[28] Mathematical probability was a well-defined "new" field that, together with statistics, held much promise for the future.

The third and final day of the conference opened with a consideration of analysis, the last of the broad areas of mathematics to be treated and one that had been a particular American focal point. With Salomon Bochner in the chair and Ralph Boas as the reporter, the four discussion leaders— Harvard's Lars Ahlfors, Einar Hille of Yale, Antoni Zygmund at Penn but soon en route to Chicago, and the University of Lund's Marcel Riesz—isolated three broad classes of problems for future exploration: the generalization of classical potential theory to both Euclidean and Lorentz space, the problem of representing functions by series, and "the distribution of the values of analytic functions of a complex variable."[29] As noted, Americans had long engaged in potential-theoretic research; much work had been done, especially in the 1930s, in harmonic analysis; and complex analysis had been developed at the hands of Paley, Wiener, Walsh, and Szegő, among others. The American mathematical community thus seemed well-positioned to push its analytic researches along these (and likely other) lines into the second half of the twentieth century.

A session on "analysis in the large" brought the thematic part of the conference to a close. Marston Morse chaired it, while Richard Courant and Heinz Hopf led the discussion. "[F]or convenience" and not to reflect "essential differences," it was subdivided into what Morse styled "differential geometry in the large, applied mathematics in the large, and equilibrium analysis in the large."[30] The principal mathematical tools used were the calculus of variations and fixed point theorems, but given a particular question, it was deemed "of central importance to choose the approach which will give the best results with the least difficulty." The questions ranged widely. Some came from differential geometry and topology, such as is it possible to develop a program for connecting "differential invariants with topological invariants"?[31] Some

28. *Problems of Mathematics*, pp. 17–18 (or p. 320).
29. Ibid., p. 20 (or p. 322).
30. Ibid., p. 21 (or p. 323). The quotation that follows may also be found here.
31. Ibid., p. 22 (or p. 323). The next quotation may also be found here.

came from dynamical systems, such as "[i]s a metric on the torus either flat or such that each geodesic has two conjugate points?" In applied mathematics, regardless of the tools used, existence and uniqueness theorems were deemed essential, while in what Morse termed equilibrium theory, "[t]he topological foundation of planetary orbits," that is, "the three body problem with one infinitesimal body," represented an open question in analysis in the large.[32] This grouping of such disparate questions under the single rubric of "analysis in the large" may strike the modern mathematician as more than "slightly puzzling."[33] After all, it reflects what became an explosion of subfields—"dynamical systems, differential topology, minimal surface theory, global differential geometry, analysis on Riemannian manifolds, complex geometry and several complex variables, nonlinear elliptic and parabolic equations, nonlinear hyperbolic equations and a whole realm of various subjects in mathematical physics and applied mathematics"—that, in 1946, had yet to occur.

The three-day affair ended over dinner and further discussion of "the state of mathematics as a whole and its prospects in the century ahead."[34] Mac Lane, Weyl, and Stone had been tapped to speak, but Mac Lane made perhaps the most perspicuous observation. "In almost all the conferences," he said, "we ran across the phenomenon of someone else moving in. . . .. The logicians moved in on the algebraists, the topologists moved in on the differential geometers (and vice versa), and the analysts moved in on the statisticians." He may well have added that the probabilists moved in on the analysts, the algebraists moved in on the geometers (and vice versa), and those working in "analysis in the large" moved in on the topologists and geometers. "When you set out to solve problems in mathematics, even in nicely labelled fields," Mac Lane concluded, "they may well lead you into some other field."[35] Indeed, in what follows, the blurring of subfield lines that Mac Lane underscored— already visible in the 1930s—will become even more apparent. Whereas George Birkhoff and his colleagues at Harvard in 1936 had "laid emphasis on exhibiting the resources of the University of today," Lefschetz and his

32. Ibid., p. 24 (or p. 325).

33. See p. 357 of Karen Uhlenbeck's commentary in *A Century of Mathematics in America*, ed. Peter Duren et al., 2: 357–359 for this characterization and the quotation that follows. That Morse himself had trouble articulating exactly what "analysis in the large" was is reflected, ironically, in his article "What Is Analysis in the Large?" *AMM* 49 (1942), 358–364.

34. *Problems of Mathematics*, p. 24 (or p. 325).

35. Ibid., p. 25 (or p. 325).

colleagues at Princeton in 1946 looked to the mathematics of the future and to the leadership role that they and American mathematics as a whole expected to play in it.[36]

Sustaining American Research Agendas

In opening with algebra, Princeton's "Problems of Mathematics" conference tacitly highlighted one of the areas of American mathematical expertise that had the deepest roots in the United States. Benjamin Peirce's work on linear associative algebras at Harvard in the 1870s had been reinforced in the 1880s by James Joseph Sylvester and Peirce's son, Charles, at Hopkins and had become firmly institutionalized at Chicago from the turn of the twentieth century onward. That Chicago continued to assume a leadership role in algebra into the 1940s is clear from the fact that it followed its 1938 conference on algebra with two more: one in 1941 on the occasion of the university's semicentennial and another just after the war in 1946.[37] In the early 1940s, Chicago's algebraic torch was carried primarily by Adrian Albert, as his émigré colleague, Otto Schilling, acculturated himself to the American academic scene. By the end of the decade, Schilling had hit his stride, completing, among other things, the book on the theory of valuations that appeared in the AMS's series of Mathematical Surveys in 1950,[38] and he and Albert had been joined by both Kaplansky and Mac Lane.

For his part, Albert followed the work he had done in the 1930s on associative algebras as well as his Cole Prize–winning research on Riemannian matrices with more seminal work, this time on the structure theory of non-associative algebras.[39] Around 1942, he began to focus on a special kind of non-associative algebra, what would come to be known as a Jordan algebra. This structure had arisen in the quantum-theoretic work of German

36. Elliott, p. 156, quoting *The Tercentenary of Harvard College: A Chronicle of the Tercentenary Year 1935–1936* (Cambridge: Harvard University Press, 1937), p. 6.

37. William Ayres, "The Summer Meeting in Chicago," *BAMS* 47 (1941), 832–845 on p. 834 and Nathan Jacobson, "A Personal History and Commentary, 1943–1946," in *Nathan Jacobson: Collected Mathematical Papers*, ed. Nathan Jacobson, 1: 276–284 on p. 276.

38. There, Schilling focused on the theory's significance "for the algebraic and arithmetic structure of fields, division rings, and simple algebras, including topics such as the local class field theory and non-commutative ideal theory." See Gerhard Hochschild, Review of "*The Theory of Valuations*. By O.F.G. Schilling," *BAMS* 57 (1951), 91–94 on p. 92.

39. Jacobson, "Abraham Adrian Albert 1905–1972," p. 1085.

physicist Pascual Jordan in the early 1930s and had been pursued from a purely mathematical point of view in 1934 by Jordan together with von Neumann and Eugene Wigner.[40] Von Neumann had then introduced Albert to the concept during the latter's 1933–1934 year at the Institute. As Albert codified it, an algebra A over a field F is called a Jordan algebra if, for all $a, b \in A$, $ab = ba$ and $a^2(ba) = (a^2b)a$.[41] Particularly from 1946 to 1950, he fleshed out the structure theory of such algebras over fields of characteristic unequal to 2 and, in so doing, sparked the interest of Nathan Jacobson, among others.[42]

Although Jacobson had written his doctoral dissertation on the theory of associative algebras under Wedderburn at Princeton, he had been exposed to (non-associative) Lie algebras in Weyl's 1933–1934 course at the Institute and had subsequently turned his research focus to the latter. By the mid 1940s, he had made significant progress in what would become yet another major interest, the general structure theory of rings, developing in particular the key notion of what others dubbed the Jacobson radical of a ring.[43] Jacobson lectured on these ideas at Chicago in the summer of 1947 while "en route" from Hopkins to Yale and the position he would hold for the rest of his

40. Pascual Jordan, John von Neumann, and Eugene Wigner, "On an Algebraic Generalization of the Quantum Mechanical Formalism," *AM* 35 (1934), 29–64. This paper was immediately followed in the *Annals* by Albert's paper "On a Certain Algebra of Quantum Mechanics" (see pp. 65–73), in which he showed that one of the three classes of algebras that Jordan, von Neumann, and Wigner had isolated was an exceptional, as opposed to a special, Jordan algebra. For the definitions, see, for example, Jacobson, "Abraham Adrian Albert 1905–1972," p. 1086.

41. Albert opened "A Structure Theory of Jordan Algebras," *AM* 48 (1947), 546–567 with this definition.

42. See ibid. as well as "On Jordan Algebras of Linear Transformations," *TAMS* 59 (1946), 524–555 and "A Theory of Power-associative Commutative Algebras," *TAMS* 69 (1950), 503–527. For commentary on this work, compare Jacobson, "Abraham Adrian Albert 1905–1972," pp. 1087–1088.

Jacobson's first paper on Jordan algebras—"Isomorphisms of Jordan Rings," *AJM* 70 (1948), 317–326—led to much more work in the area, including his later codification, *Structure Theory of Jordan Algebras*, Lecture Notes in Mathematics, University of Arkansas, 1981. Kevin McCrimmon gave a nice overview of Jacobson's work in this subfield in Georgia Benkart, Irving Kaplansky, Kevin McCrimmon, David Saltman, and George Seligman, "Nathan Jacobson (1910–1999)," *NAMS* 47 (2000), 1061–1071 on pp. 1066–1070.

43. See Jacobson, *The Theory of Rings* as well as "Structure Theory of Simple Rings Without Finiteness Assumptions," *TAMS* 57 (1945), 228–245 and "The Radical and Semi-simplicity for Arbitrary Rings," *AJM* 67 (1945), 300–320.

career.[44] One member of his audience, the thirty-year-old Irving Kaplansky, was hooked.

Conversations between the two young algebraists that summer led them to the work of Marshall Hall, a 1936 Ph.D. student of Ore's at Yale. In 1943, Hall had shown that for a division ring K with center F, if the polynomial identity—PI—"$z(xy - yx)^2 = (xy - yx)^2 z$ [holds] for three arbitrary elements of K, then either $K = F$ or K is a quaternion algebra over F."[45] As Jacobson explained, this had suggested the question: "Does the existence of a polynomial identity for a division ring imply finiteness of dimensionality over the [division ring's] center?"[46] In what has been described as "[o]ne of his most influential papers," Kaplansky, drawing directly on Jacobson's structure theory, in fact showed more: "a primitive algebra," that is, an algebra A that contains a maximal right ideal J such that the quotient $(J : A) = 0$, "with polynomial identity is finite dimensional over its center."[47] This discovery both revivified the study of PI-algebras—isolated by Max Dehn in 1922—and marked one of the many fundamental contributions to algebra that Kaplansky would make while at Chicago.[48]

The addition of Mac Lane to the Chicago group in 1948 brought algebra of a very different kind. Mac Lane and Eilenberg had inaugurated the abstract theory of categories in 1945 when they were together in Columbia's Applied Mathematics Group; in the late 1940s, they developed the cohomology of groups, Mac Lane from his new home base at Chicago and Eilenberg at Columbia following his move there from Indiana in 1947.[49] Theirs was

---

44. Nathan Jacobson, "A Personal History and Commentary 1947–1955," in *Nathan Jacobson: Collected Mathematical Papers*, 2: 1–16 on p. 1.

45. Marshall Hall, "Projective Planes," *TAMS* 54 (1943), 229–277 on p. 262. Albert had defined and explored the notion of a quaternion algebra in his 1939 Colloquium volume, *Structure of Algebras*, pp. 145–147.

46. Jacobson, "A Personal History and Commentary 1947–1955," p. 1.

47. For the quotations, see Hyman Bass and T. Y. Lam, "Irving Kaplansky (1917–2007)," *NAMS* 54 (2007), 1477–1493 on p. 1479. The paper in question is Irving Kaplansky, "Rings with a Polynomial Identity," *BAMS* 54 (1948), 575–580 on p. 577. Jacobson defined a primitive algebra in "Structure Theory for Algebraic Algebras of Bounded Degree," *AM* 46 (1945), 695–707 on p. 696.

48. Max Dehn, "Über die Grundlagen der projektiven Geometrie und allgemeine Zahlsysteme," *Mathematische Annalen* 85 (1922), 184–194.

49. See their paper "General Theory of Natural Equivalences," *TAMS* 58 (1945), 231–294 as well as, for example, their "Cohomology Theory in Abstract Groups I and II," *AM* 48 (1947), 51–78 and 326–341, resp. and Mac Lane's "Cohomology Theory in Abstract Groups III," *AM*

precisely the kind of generalization "of known results with an eye toward increasing their scope and learning more of their inner meaning" that the algebraists assembled at the "Problems of Mathematics" conference foresaw as the future of their field.[50]

Yet, as that conference also underscored, Chicago, although strong in algebra, was not, in the 1940s, the only bastion of important algebraic research in North America. Brauer had been at the University of Toronto when he proved the stunning results on induced characters that he announced at the Princeton event, namely, "[e]very character χ of a finite group G is an integral linear combination of characters induced from linear characters of subgroups" and "[i]f q is the exponent of G, then every representation of G can be written in the field of qth roots of unity."[51] When he won the AMS's Cole Prize in Algebra in 1949 for this work, however, it was from the position at the University of Michigan that he had taken up in 1948 and that he would hold until his move to Harvard four years later. Another key émigré, Artin, continued work—from his position at Indiana University just prior to his 1946 move to Princeton—on the theory of braids that he had begun in Germany as early as 1925–1926.[52]

If Chicago had been an American powerhouse in algebraic research over the course of the first half of the twentieth century, Princeton had had that distinction in geometry at least since the 1920s. Veblen and Eisenhart led the vibrant group in differential geometry; Lefschetz did the same in algebraic geometry (as well as in closely related algebraic topology). As Princeton's program in differential geometry waned in the 1940s, new loci of algebraic geometry coalesced. One of those centered on Zariski, at Hopkins from 1927 to 1946 and then at Harvard following one year at Illinois, another on Weil at Chicago until his move to the Institute in 1958. Zariski and Weil, moreover, found in each other kindred algebraic geometric spirits as they pursued their respective research.

Zariski's close examination of work of the Italian school of algebraic geometry in the first half of the 1930s had led him, by the decade's close, to see that

50 (1949), 736–761. Mac Lane discusses his collaboration with Eilenberg in *A Mathematical Autobiography*, pp. 125–132 as well as in Bass, Cartan, Freyd, Heller, and Mac Lane, pp. 1346–1348.

50. *Problems of Mathematics*, p. 4 (or p. 310).

51. Jacobson, "A Personal History and Commentary 1943–1946," p. 277.

52. See Emil Artin, "Theorie der Zöpfe," *Abhandlungen aus dem Mathematischen Seminar der Universität Hamburg* 4 (1926), 47–72 as well as "Theory of Braids," *AM* 48 (1947), 101–126 and "Braids and Permutations, *AM* 48 (1947), 643–649.

the Italian approach, while fruitfully intuitive, could be made rigorous only through adequately developed algebraic underpinnings. In particular, he realized that ring theory, especially the notions of the integral closure of a ring and of a valuation ring, "could be extremely useful for the analysis of singularities and for the problem of reduction of singularities" of an algebraic surface.[53] He proceeded to develop such methods in the four papers in 1939 and 1940 for which he was awarded the AMS's 1944 Cole Prize in Algebra.[54] The geometer had sufficiently blurred subdisciplinary lines to win the nation's major prize in algebraic research.

Zariski followed this work with what he considered to be its "natural outgrowth," namely, the extension "to varieties V over arbitrary ground fields [of] the classical notion of [the] analytic continuation of a holomorphic function defined in the neighborhood of a point P of V."[55] As he explained, he "sensed the probable existence of such an extension provided the analytic continuation were carried out along an algebraic subvariety W of V passing through P." Unfortunately, the outbreak of the war and the wartime increase in his teaching load at Hopkins left him little time for such pursuits. That changed in the 1945 calendar year when he, like George Birkhoff and Marshall Stone before him, served as a "Good Neighbor" exchange professor in Brazil. The timing of Zariski's sojourn—in his case under State Department auspices at the University of São Paulo—was even more fortuitous for his evolving algebraic geometric ideas, for it coincided with Weil's first year on the faculty there. Zariski had what he described as "a superlative audience consisting of one person—André Weil," and the two men discussed their mutual mathematical interests during their "frequent walks" around the city.

Zariski and Weil had known each other at least since 1925 when both had been students in Rome, but the two had become close friends in the spring of 1937 when Weil was at the Institute and Zariski was nearby in Baltimore.[56]

53. Zariski, Response in "1981 Steele Prizes," p. 505.

54. Oscar Zariski, "Some Results in the Arithmetic Theory of Algebraic Varieties," *AJM* 61 (1939), 249–294; "The Reduction of the Singularities of an Algebraic Surface," *AM* 40 (1939), 639–689; "Algebraic Varieties over Ground Fields of Characteristic Zero," *AJM* 62 (1940), 187–221; and "Local Uniformization on Algebraic Varieties," *AM* 41 (1940), 852–896. This work is discussed in historical and mathematical context by Slembek as well as by David Mumford in Appendix A of Parikh, pp. 204–214 and by Dieudonné in *History of Algebraic Geometry*, pp. 77–80.

55. Zariski, Response in "1981 Steele Prizes," p. 506. The quotations that follow in this paragraph may also be found here.

56. Parikh, pp. 40–41 and 84.

They shared the conviction that the work of the Italians needed to be placed on a firm algebraic footing, but Weil, for his part, was also motivated by finding an analogy to "the Riemann hypothesis for the zeta function of curves over finite fields," the proof of which he had announced in 1940.[57] As he had explained in a lengthy letter to Artin in 1942, his proof hinged "on the establishment of an intersection theory for surfaces," but he would only succeed in hammering that out in 1944 in the manuscript of what would become his 1946 Colloquium volume, *Foundations of Algebraic Geometry*. Weil was thus as perfect a sounding board for Zariski's new ideas as Zariski was for Weil's. By the end of the 1940s, both men had pushed the agenda articulated at the Princeton conference of developing algebraic geometry over abstract fields: Weil had made his famous conjectures on smooth complete varieties over finite fields, and Zariski had completed the 1951 AMS Memoir on the *Theory and Applications of Holomorphic Functions on Algebraic Varieties over Arbitrary Ground Fields* in which he exploited, among other things, what would come to be called the Zariski ring.[58] Both men, Zariski through his students at Harvard, such as Michael Artin and David Mumford, and Weil primarily through his participation in Bourbaki, also introduced a next generation of algebraic geometers to their new approaches.

Similarly, given its equally firm institutionalization in American mathematics, topology—both point-set and algebraic—also continued to flourish in the 1940s as seasoned mathematicians pursued established research programs and new talent entered the fold. Although it had largely been ignored at the 1946 Princeton conference, point-set topology had strong and vocal senior proponents in, among others, R. L. Moore and his mathematical "son," Whyburn, with Moore turning out eight new point-set theorists in the forties and Whyburn firmly establishing the subfield at the University of Virginia especially after the 1942 publication of his AMS Colloquium volume, *Analytic Topology*. Among Moore's new crop of students, moreover, R. H. Bing created yet another major point-set-topological node at the University of Wisconsin beginning in 1947 that had attracted his mathematical "sister," Mary Ellen Rudin, by 1959.

---

57. Michel Raynoud, "André Weil and the Foundations of Algebraic Geometry," *NAMS* 46 (1999), 864–867 on p. 864. The quotation that follows may also be found here.

58. André Weil, "Numbers and Solutions of Equations in Finite Fields," *BAMS* 55 (1949), 497–508 and Oscar Zariski, *Theory and Applications of Holomorphic Functions on Algebraic Varieties over Arbitrary Ground Fields*, Memoirs of the American Mathematical Society, no. 5 (New York: American Mathematical Society, 1951).

Bing had made an early mathematical name for himself with his answer, in the affirmative, of a question that had been posed by Moore's first student, John Kline, namely, "[i]s a nondegenerate, locally connected, compact continuum which is separated by each of its simple closed curves but by no pair of its points homeomorphic with the surface of a sphere?"[59] As Bing explained, his result gave a characterization of a simple closed curve different, notably, from those that Kuratowski and Claytor had formulated in the 1930s and that had attracted the favorable attention of the leading proponent of the then archrival, algebraic approach to topology, Lefschetz.

By the 1940s, that rivalry was largely a thing of the past. Lefschetz, following the 1942 publication of the AMS Colloquium volume, *Algebraic Topology*, in which he recast in group-theoretic terms the theory of complexes (defined in chapter one), left the algebraic topological (as well as the geometric) fray, shifting his mathematical focus to differential equations for the remainder of his long career.[60] Reconciler Wilder nevertheless continued to serve as a harmonizing force. In 1949, his AMS Colloquium volume on the *Topology of Manifolds* carefully laid out how, in his own development as a topologist, both the point-set and the algebraic topological approaches had played formative roles. He credited his advisor, R. L. Moore, not only for the "thorough grounding in point set theory" that he had gotten under his tutelage but also for the early realization of "the vacuum in our knowledge of the set-theoretic structure of the n-cell, particularly the lack of a topological characterization."[61] It was that vacuum that his book aimed to fill, using "new tools, especially the extension to general spaces of the theory of connectivity (homology)" developed in the 1930s by algebraic topologists Paul Alexandroff and Eduard Čech (see later in this section) as well as even newer "and more powerful tools . . . such as the theory of cohomology and chain products."

At Michigan, moreover, the "policy" was adopted "of bringing in new material and making sure that *all* aspects of [the] rapidly growing field [of topology] . . . were represented."[62] Wilder thus succeeded in building a group

59. R. H. Bing, "The Kline Sphere Characterization Problem," *BAMS* 52 (1946), 644–653 on p. 644.

60. Griffiths, Spencer, and Whitehead, pp. 300–302.

61. Wilder, *Topology of Manifolds*, p. iii. The quotations that follow are on pages iii and iv, respectively. For the definition of an n-cell, recall chapter one.

62. Typed transcript of a recording made in 1976 of Wilder's impressions of mathematics at the University of Michigan, p. 12 (my emphasis), Box 86-36/42: Folder 7, Faculty and Administrative Activities, University of Michigan, Wilder Papers.

in the 1940s that included, in addition to himself, point-set theorist William Ayres until his 1941 departure to chair the mathematics department at Purdue University and two algebraic topologists, Eilenberg, from 1940 until his departure for Indiana in 1946,[63] and former Michigan undergraduate and Wilder protégé, Norman Steenrod, from 1942 until his move to Princeton in 1947. It was, at least according to John Kline, "one of the outstanding groups in topology in the world."[64] As a fellow topologist and with his bird's-eye view of the mathematical scene as AMS Secretary, Kline was certainly in a position to judge.

The two young Turks, Eilenberg and Steenrod, unquestionably thrived at Michigan. Eilenberg, in addition to the collaboration with Mac Lane that linked their respective expertise in topology and algebra in the development of both category theory and the cohomology of groups, established a collaboration with Steenrod as early as 1945 that had resulted by 1952 in their book-length axiomatization of a theory of homology.[65] As they explained, "[h]omology is a transition (or function) from topology to algebra" that "assigns groups to topological spaces and homomorphisms to continuous maps of one space into another. To each array of spaces and maps is assigned an array of groups and homomorphisms. In this way, a homology theory is an *algebraic image* of topology. The *domain* of a homology theory is the topologist's field of study. Its *range* is the field of study of the algebraist. Topological problems are converted into algebraic problems." Homology was, in short, yet another prime example of subdisciplinary cross-fertilization.

The problem that Eilenberg and Steenrod confronted was that many homology theor*ies* had been developed especially over the course of the 1930s—among them, "the singular homology groups of Veblen, Alexander, and Lefschetz, the relative homology groups of Lefschetz, the Vietoris homology groups, the Čech homology groups, the Alexander cohomology groups, etc."—and it was not clear whether a proof in the context of one homology theory would "still hold if another homology theory replace[d]" it. Their axiomatization was viewed as having "drained the Pontine Marshes

63. Eilenberg left Indiana for Columbia after only one year.

64. Kline to Raymond Wilder, 1 October, 1942, Box 86-36/7, Folder 2: General Correspondence J. R. Kline, Wilder Papers.

65. Samuel Eilenberg and Norman Steenrod, "Axiomatic Approach to Homology Theory," *PNAS* 31 (1945), 117–120 and *Foundations of Algebraic Topology* (Princeton: Princeton University Press, 1952). The quotations that follow in this and the next paragraph are on pp. vii, viii, and x of the latter, respectively (with their emphases).

of homology theory, turning an ugly morass of variously motivated constructions into a simple and elegant system of axioms applied, for the first time, to functors."[66] Their "radical innovation" meant that homology theories would henceforth "be mathematical objects in their own right." In other words, out of many homology theories there had arisen a theory of homology.

Like that theory, differential topology was located at the interface of two mathematical subdisciplines, but in its case topology and analysis. Emerging in the 1930s in work particularly of Morse and Whitney, its development continued at their hands and from their respective points of view through the 1940s. For his part, Morse turned to a development in more abstract settings of the Morse theory that he had been exploring since the 1920s. He also did fruitful collaborative work: on geodesic flow with his former student, Hedlund; on minimal surfaces with Charles Tompkins, a student of George Rainich at the University of Michigan who had spent the two years from 1936 to 1938 working with Morse at the Institute; and on topological methods in a single complex variable with Maurice Heins, a former Harvard undergraduate under Morse who had worked with him in Washington during World War II.[67] Morse's work in the 1940s, like his earlier research, reflected his underlying interest in analysis, mechanics, and differential geometry. His approach to differential topology was fundamentally geometrical as opposed to algebraic, and he actually rued the dominance algebraic topology had attained.

Whitney was more forgiving of algebra's encroachment. Although his differential topological work was profoundly guided by geometrical insights, he, more often than not, couched them in algebraic terms in order to establish them with what he viewed as an appropriate level of rigor.[68] In two papers in 1944, for example, he considered the behavior of smooth manifolds of dimension n, showing, first, that any such manifold can be differentially embedded in $\mathbb{E}^{2n}$, and, second, that it can actually always be immersed in $\mathbb{E}^{2n-1}$, where

66. This is Heller's characterization in Bass, Cartan, Freyd, Heller, and Mac Lane, p. 1348. The quotations that follow in the next sentence may also be found here. For a historical account of the various homology and cohomology theories, see Dieudonné, *A History of Algebraic and Differential Topology*, pp. 67–157.

67. Bott gives a self-contained overview of these deep and technical results on pp. 915–937 and provides a more than 180-entry bibliography of Morse's works. The following characterization of Morse's approach is also due to Bott, pp. 907–908.

68. Kendig, *Never a Dull Moment*, p. 144.

embedding implies the existence of a map (with certain properties) between the two spaces that is also one-to-one and immersion implies the existence of the same sort of map without the one-to-one requirement.[69] In the view of at least one of his younger contemporaries, "Whitney's embedding theorems changed the landscape of manifold theory," since they "assured that any manifold sits in some Euclidean space" and so "inherits a metric"—and all that implies—"from that surrounding space."

The work of Morse and Whitney—as well as that of Morse's students Sumner Myers and Stewart Cairns, Whitney's student Wilfred Kaplan, and others—demonstrated some of the advantages of looking at certain kinds of analytical questions through a topological lens. So did Tibor Radó's 1948 AMS Colloquium volume, *Length and Area*, but from a different topological perspective. In his book, Radó built on the foundation that Lebesgue had laid earlier in the century to provide what was described as an "outstanding application of *analytic* topology"—the same point-set-inspired "brand" of topology that Whyburn had detailed in his Colloquium volume—to the problem of defining and finding the area of a curve or surface.[70] Two other AMS Colloquium volumes written in the 1940s—one on nonlinear differential equations and the other on functional analysis—showed that algebra also continued meaningfully to inform analysis.

Columbia's Joseph Ritt published a second AMS Colloquium volume, *Differential Algebra*, in 1950 that represented the evolution of his ideas as well as those issuing from what Bartel van der Waerden described as Ritt's "whole school of able collaborators" on the "gigantic task" of giving "the classical theory of nonlinear differential equations a rigorous algebraic foundation."[71] By 1950, that approach was not only "written from a much higher point of view"

69. See Hassler Whitney, "The Self-Intersections of a Smooth *n*-Manifold in 2*n*-Space," *AM* 45 (1944), 220–246 and "The Singularities of a Smooth *n*-Manifold in (2*n* − 1)-Space," *AM* 45 (1944), 247–293. Whitney gives the precise definitions of embedding and immersion in the former paper on p. 221. See also the discussion in Kendig, *Never a Dull Moment*, pp. 175–186. The quotation that follows is from the latter source, p. 182.

70. Edward McShane, Review of "*Length and Area*. By Tibor Radó," *BAMS* 54 (1948), 861–863 on p. 861 (my emphasis). McShane made a contribution of a different sort in this vein with the graduate textbook, *Integration*, that he wrote based on courses he had given at the University of Virginia. See Edward McShane, *Integration*, Princeton Mathematical Series, vol. 7 (Princeton: Princeton University Press, 1944).

71. For the quotations here and in the next two sentences, see Bartel van der Waerden, Review of "*Differential Algebra*. By Joseph Fels Ritt," *BAMS* 56 (1950), 521–523 on pp. 521 and 521–522, respectively.

than it had been in Ritt's earlier Colloquium volume, but it was also "based upon new principles" established by Ritt in conjunction particularly with his student Henry Raudenbush. As van der Waerden saw things from the perspective of 1950, the first version looked "tentative," while the new one was "a classic." This may have been a natural conclusion coming as it did from one of the principal proponents of the modern algebra that had developed particularly in the 1920s and 1930s, but Ritt's book ultimately proved out of the mainstream.[72]

The same was not true of Einar Hille's *Functional Analysis and Semi-groups*. It was described as a veritable " 'cours d'analyse' for the study of vector-valued functions, Banach norms and spectral theory" that also "contain[ed] such 'classical' material as Laplace integrals, Fourier series, Hermite expansions, [and] real functions in Euclidean spaces."[73] Polish mathematician Stefan Banach had made major contributions to functional analysis beginning in the 1920s with his axiomatization of the concept of a complete normed vector space—subsequently called a Banach space—and had pushed his ideas further in the 1932 book-length treatise, *Théorie des opérations linéaires*.[74] By the 1940s, Banach spaces had become important objects of study as had Banach algebras, that is, associative algebras over the real or complex numbers that are also Banach spaces and such that the norm satisfies the relation $\|x \cdot y\| \leq \|x\| \cdot \|y\|$, for all elements x and y in the algebra.

From an algebraic point of view, Hille's book not only included "a most handy monograph" on Banach algebras, but also a development of semi-groups, sets of elements closed under an associative binary operation.[75] As Hille explained, although the notion of a semi-group is algebraic, its "main importance . . . does not seem to lie in the algebraic field, but rather in the applications to analysis where topological semi-groups and in particular one-parameter semi-groups of linear transformations on a function space to itself

72. Indeed, Ritt felt that, despite two AMS Colloquium volumes on it, his research had not gotten the reception it deserved. This was the epitaph he composed for himself: "Here at your feet J. F. Ritt lies; He never won the Bôcher prize." See Edgar Lorch, "Mathematics at Columbia during Adolescence," in *A Century of Mathematics in America*, ed. Peter L. Duren et al., 3: 149–161 on p. 158.

73. Salomon Bochner, Review of *"Functional Analysis and Semi-groups*. By Einar Hille," *BAMS* 55 (1949), 528–533 on p. 528.

74. Stefan Banach, *Théorie des opérations linéaires* (Warszawa: Z subwencji Funduszu kultury narodowej, 1932).

75. Bochner, Review of *"Functional Analysis and Semi-groups*. By Einar Hille," p. 528.

come up in the most diversified connections."[76] It was the latter application that Hille explored at length in his book, but developments later in the 1940s and in the early 1950s in the theory of semi-groups at the hands particularly of Ralph Phillips resulted in a co-authored revision of the volume in 1957 that became and remained a standard work.[77]

Another important contributor to functional analysis, John von Neumann, also fundamentally exploited its connections with algebraic structures. The Hungarian prodigy had begun developing a theory of rings of operators—now called von Neumann algebras—before his emigration to the United States in 1930. He had then pursued it in earnest in a series of substantial papers—solely and in collaboration with Francis Murray, a 1936 Columbia Ph.D. under Bernard Koopman—in the late 1930s and 1940s and had high-lighted it before his Institute audiences.[78] One young mathematician who not only heard these lectures in the 1939–1940 academic year but who also wrote up the lecture notes for them was Paul Halmos.[79] Still, despite the fact that he found the subject "awful good fun," Halmos did not actually shift his personal research focus to the theory of operators on Hilbert space until the completion of the work in measure theory that resulted in his acclaimed 1950 book by that same name.[80]

76. Hille, p. 146. On the history of semi-groups, see Christopher Hollings, "The Early Development of the Algebraic Theory of Semigroups," *Archive for History of Exact Sciences* 63 (2009), 497–536 and *Mathematics across the Iron Curtain: A History of the Algebraic Theory of Semigroups*, HMATH, vol. 41 (Providence: American Mathematical Society, 2014).

77. Einar Hille and Ralph Phillips, *Functional Analysis and Semi-groups*, AMS Colloquium Publications, vol. 31, rev. ed. (Providence: American Mathematical Society, 1957). Phillips had earned his Ph.D. under Theophil Hildebrandt at the University of Michigan in 1939. A series of short appointments and war work at the Radiation Laboratory at MIT preceded his appointment to the faculty at the University of Southern California in 1947. For more on Phillips's life and work, see Peter Sarnak, "Ralph Phillips (1913–1998)," *NAMS* 47 (2000), 561–563. Phillips and Bochner also collaborated on other algebraically informed work in functional analysis such as their "Absolutely Convergent Fourier Expansions for Non-commutative Normed Rings," *AM* 43 (1942), 409–418.

78. See, for example, Francis Murray and John von Neumann, "On Rings of Operators," *AM* 37 (1936), 116–229; "On Rings of Operators II," *TAMS* 41 (1937), 208–248; and "On Rings of Operators IV," *AM* 44 (1943), 716–808, in addition to John von Neumann, "On Rings of Operators III," *AM* 41 (1940), 94–161. Dieudonné discusses this work in *History of Functional Analysis*, pp. 183–188.

79. Halmos, *I Want To Be a Mathematician*, pp. 92–93.

80. See Paul Halmos, *Measure Theory* (New York: Van Nostrand, 1950) and Halmos, *I Want To Be a Mathematician*, pp. 140–166 (the quotation, however, is on p. 92).

By the end of the 1940s, new ideas about linear transformations in Hilbert space had developed from Stone's path-breaking 1932 treatise on the subject, and Stone's then colleague, Halmos, decided to learn more.[81] He published his first paper in the area in 1950, therein defining several concepts—dilation and extension of an operator and subnormal and hyponormal operators—that ultimately resulted "in the creation of a body of mathematics," distinct "subarea[s] of operator theory."[82] A year later in 1951, he had also published an *Introduction to Hilbert Space and the Theory of Spectral Multiplicity*, in which, unlike Stone in 1932, he worked without the underlying assumption of the separability of the Hilbert space.[83]

Halmos's Chicago colleague, Antoni Zygmund, also pursued research in analysis, but, in his case, the research in harmonic analysis, specifically one-dimensional Fourier analysis, that he had begun in Poland and that had resulted in his influential 1935 treatise, *Trigonometrical Series*.[84] There, in addition to codifying "practically all the important results that were [then] known" on the subject and establishing its "connections with other disciplines," Zygmund cast his results in terms of the late-breaking functional analytic work of his countryman, Banach. At the same time, he persisted in his own "concrete" techniques in Fourier analysis and largely eschewed Bourbaki's "algebraic approach."[85] Zygmund passed on his deep research insights and mathematical style to a distinguished string of students, among them, his 1950 Ph.D. and later colleague and collaborator, Alberto Calderón. Together, they created a research node in harmonic analysis distinct from the one Wiener defined at MIT and that included, among other students, Norman Levinson.

The work of Wiener and Levinson, moreover, reflected some of the research being done on American shores in the 1940s in complex analysis. For

81. Halmos, *I Want to Be a Mathematician*, p. 158.

82. John Conway, "Paul Halmos and the Progress of Operator Theory," in *Paul Halmos: Celebrating 50 Years of Mathematics*, ed. John Ewing and Fred Gehring (New York: Springer-Verlag, 1991), pp. 155–167 on p. 156. Conway gives the various definitions on pp. 156–157. The paper in question is Paul Halmos, "Normal Dilations and Extensions of Operators," *Summa Brasiliensis Mathematicae* 2 (1950), 125–134.

83. Paul Halmos, *Introduction to Hilbert Space and the Theory of Spectral Multiplicity* (New York: Chelsea Publishing Company, 1951).

84. Antoni Zygmund, *Trigonometrical Series* (Warszawa: Z subwencji Funduszu kultury narodowej, 1935).

85. Coifman, and Strichartz, pp. 345 (for the first two quotations) and 347 (for the latter two).

example, from the permanent position at MIT that he ultimately secured in 1940, Levinson followed up on Paley and Wiener's 1934 *Fourier Transforms in the Complex Domain* with his own AMS Colloquium volume, *Gap and Density Theorems*; Princeton's Bochner codified his evolving ideas in the book, *Several Complex Variables*, that he co-authored with William Ted Martin of MIT in 1948; and Albert Schaeffer and Donald Spencer, both at Stanford in the 1940s, completed the collaborative work on schlicht functions for which the AMS awarded them its 1948 Bôcher Memorial Prize.[86] Levinson's body of work, which highlighted the fundamental interconnectedness of the theories of linear, nonlinear, ordinary, and partial differential equations, snared the very next Bôcher Prize in 1953. Partial differential equations, in particular, engaged, among other contributors, Charles Morrey and Hans Lewy at Berkeley, Charles Loewner and Lipman Bers at (in the latter half of the 1940s) Syracuse University, and Stoker, Courant, and Friedrichs at NYU.[87]

If analysis, as pursued in the United States in the 1940s, thus ranged over several of the discrete topics discussed at Princeton's "Problems of Mathematics" conference in 1946, mathematical logic was perhaps the most focused and circumscribed of the areas represented there. The *Journal of Symbolic Logic* had, after all, been founded in 1936 to provide a targeted forum for

86. See Norman Levinson, *Gap and Density Theorems*, AMS Colloquium Publications, vol. 26 (New York: American Mathematical Society, 1940); Salomon Bochner and William Ted Martin, *Several Complex Variables* (Princeton: Princeton University Press, 1948); and Albert Schaeffer and Donald Spencer, "Coefficients of Schlicht Functions, I, II, III, IV," *DMJ* 10 (1943), 611–635; *DMJ* 12 (1945), 107–125; *PNAS* 23 (1946), 111–116; and *PNAS* 35 (1949), 143–150. Analytic functions f on the unit disk are said to be schlicht if they are one-to-one and satisfy the conditions $f(0) = 0$ and $f'(0) = 1$. Schaeffer and Spencer used schlicht functions to analyze various special cases of Ludwig Bieberbach's 1916 conjecture to the effect that the nth coefficient in the power series of a univalent function is less than or equal to n. Recall from the previous chapter that Schaeffer and Spencer also published a Colloquium volume on this work in 1950.

87. Louis Nirenberg provides a self-contained historical overview with an extensive bibliography in "Partial Differential Equations in the First Half of the Century," in *Development of Mathematics 1900–1950*, ed. Jean-Paul Pier (Basel: Birkhäuser Verlag, 1994), pp. 479–515. In particular, Morrey did work fundamental to the solution of Hilbert's nineteenth and twentieth problems, namely, "are the solutions of regular problems in the calculus of variations always necessarily analytic" and the related "question concerning the existence of solutions of partial differential equations when the values on the boundary of the region are prescribed," respectively. See Hilbert, "Mathematical Problems," pp. 469–470 and Charles Morrey, "Multiple Integral Problems in the Calculus of Variations and Related Topics," *University of California Publications in Mathematics*, new ser., 1 (1943), 1–130.

the nation's symbolic logicians, a group that had reached a critical mass in the 1930s. By the 1940s, there were key programs at Princeton advocated by Veblen, sustained by his student, Church, and reinforced by Gödel's presence at the Institute; at the University of Wisconsin in Madison and at Cornell fostered by Church's students, Stephen Kleene and Barkley Rosser, respectively; at Harvard (in philosophy) animated first by Alfred North Whitehead and then by his student, Willard Quine; and at Berkeley begun by Benjamin Bernstein and then shepherded by Alfred Tarski and Church's student Leon Henkin. These had generated a significant body of work on a range of fundamental questions, including the decision problem as highlighted during the Princeton bicentennial.[88]

In laying out a program for solving the continuum hypothesis in his 1926 paper "Über das Unendliche," Hilbert had launched a research agenda on general recursive functions, that is, functions from a subset of the natural numbers into the natural numbers that are "computable" or "calculable" in some intuitive sense.[89] Church, in cooperation with his former students Kleene and Rosser, made a major breakthrough in the spring of 1935 when he not only gave a definition of "effective calculability" in terms of what he called λ-definable functions but also showed that not every effectively calculable problem is solvable.[90] These three mathematicians continued to develop this theory into the 1940s with, for example, Church publishing in 1941 "a considerable improvement" with "simplified and improved" notation of his 1936 Princeton lectures.[91] They were joined in this work by, among others, Chen

88. For an idea of work being done in the 1940s in other areas of mathematical logic such as set theory and model theory, see Marcel Guillaume's detailed historical account, "La Logique mathématique en sa jeunesse," in *Development of Mathematics 1900–1950*, ed. Jean-Paul Pier, pp. 185–367.

89. David Hilbert, "Über das Unendliche," *Mathematische Annalen* 95 (1926), 161–190 and Guillaume, pp. 269–289.

90. Alonzo Church, "An Unsolvable Problem of Elementary Number Theory," *AJM* 58 (1936), 345–363. Independently and in work submitted in 1936 but published in 1937, Church's soon-to-be Ph.D. student at Princeton, Alan Turing, proved the same result, developing what came to be called the Turing machine. See Alan Turing, "On Computable Numbers, with an Application to the Entscheidungsproblem," *Proceedings of the London Mathematical Society* 42 (1937), 230–265.

91. Alonzo Church, *The Calculi of Lambda-Conversion*, Annals of Mathematics Studies, no. 6 (Princeton: Princeton University Press and London: Humphrey Milford and Oxford University Press, 1941). Orrin Frink so characterized the work in his review of "*The Calculi of Lambda-Conversion*. By Alonzo Church," *BAMS* 50 (1944), 169–172 on p. 169.

McKinsey, Emil Post, professor of mathematics at the City College of New York and 1920 Ph.D. under Cassius Keyser at Columbia, and Julia Robinson, a 1948 Berkeley Ph.D. under Tarski.[92]

The American mathematical community's robust cadre of mathematical logicians contrasted markedly with its representation in number theory, however. That area had apparently been one of the Princeton symposium's "forced ... omissions" in 1946, despite the facts that Hans Rademacher had, beginning in 1940, built a small but active program in the field at the University of Pennsylvania and émigré analytic number theorist Carl Siegel had become a member at the Institute in 1940 and a professor there in 1945.[93] True, the AMS had awarded its first Cole Prize in Number Theory in 1931 to Harry Vandiver for his work on Fermat's Last Theorem, but, in the absence of a suitable candidate in 1936, the next award was not made until 1941.[94] By that point, French mathematician, Claude Chevalley, who had been at the Institute at the outbreak of World War II, was on the faculty at Princeton and actively participating in the American mathematical community. It was for his 1940 *Annals* paper, "La Théorie du corps de classes," that he won the 1941 Cole Prize in Number Theory.[95] There, he began "the work," as he described it, "of simplification and especially of clarification" of class field theory via the theory of idèles that he had introduced in 1936. It was his hope that "new researchers" would be "induced" to follow this lead, and, indeed, class field theory was targeted as one of the "Problems of Mathematics" by the *algebraists* in 1946.

92. See, for example, J.C.C. "Chen" McKinsey, "A Solution of the Decision Problem for the Lewis systems S2 and S4, with an Application to Topology," *JSL* 6 (1941), 117–134 and "The Decision Problem for Some Classes of Sentences Without Quantifiers," *JSL* 8 (1943), 61–76; Emil Post, "Recursively Enumerable Sets of Positive Integers and Their Decision Problem," *BAMS* 50 (1944), 284–316 and "A Variant of a Recursively Unsolvable Problem," *BAMS* 52 (1946), 264–269; and Julia Robinson, "Definability and Decision Problems in Arithmetic," *JSL* 14 (1949), 98–114.

93. *Problems of Mathematics*, p. 1 (or p. 309). The Norwegian number theorist, Atle Selberg, became an Institute professor in 1949 and won the Fields Medal in 1950 for his work.

94. As noted in chapter two, however, it had also been deemed the case that there was no suitable candidate in algebra for what would have been the second Cole Prize in that field in 1934. This was despite the fact that it was much better represented than number theory within the American mathematical research community in the 1930s.

95. Claude Chevalley, "La Théorie du corps de classes," *AM* 41 (1940), 394–418. The quotations that follow are on p. 394 (my translation).

The next two Cole Prizes in Number Theory were also awarded for work in the 1940s by those trained outside the United States. The 1946 winner, Henry (although originally Heinrich) Mann, then an associate professor at the Ohio State University, had earned his Ph.D. under Philipp Furtwängler in Vienna in 1935 and had emigrated from Austria to New York City in 1938. After initially supporting himself through private tutoring, he was awarded a Carnegie Foundation fellowship to retool in statistics and to help the statistical team at Columbia during the war. Early in 1942, however, Mann submitted "A Proof of the Fundamental Theorem on the Density of Sums of Sets of Positive Integers" to the *Annals*, inspired by lectures he had heard fellow émigré and soon-to-be member of the faculty at the University of North Carolina, Alfred Brauer, give on his own then-recent work on the additive theory of numbers.[96] Mann's short paper not only settled but also gave a sharper version of a long-standing conjecture of Edmund Landau and others.[97] Hungarian-born and -educated Paul Erdős won the next prize in 1951 for his "many papers in the Theory of Numbers" but, in particular, for the elementary—that is, not using advanced arguments from complex analysis—proof of the prime number theorem that he gave in 1949, drawing from work Atle Selberg had published a year earlier.[98] The émigrés thus infused the American mathematical research community with number-theoretic research, even if that community did not especially nurture the area through faculty positions and graduate training.[99]

96. Henry Mann, "A Proof of the Fundamental Theorem on the Density of Sums of Sets of Positive Integers," *AM* 43 (1942), 523–527 and Alfred Brauer, "Über die Dichte der Summe von Mengen positiver ganzer Zahlen I," *AM* 39 (1938), 322–340 and "On the Density of the Sum of Sets of Positive Integers II," *AM* 42 (1941), 959–988.

97. For the interested reader, Mann sets up the necessary notation and states the conjecture on p. 523 of his paper.

98. Paul Erdős, "On a New Method in Elementary Number Theory Which Leads to An Elementary Proof of the Prime Number Theorem," *PNAS* 35 (1949), 374–384. For the quotation, see Leon Cohen, "The Annual Meeting of the Society," *BAMS* 58 (1952), 157–217 on p. 159. The two relevant papers of Selberg are "An Elementary Proof of the Prime-Number Theorem," *AM* 50 (1948), 305–313 and "An Elementary Proof of Dirichlet's Theorem About Primes in Arithmetic Progression," *AM* 50 (1948), 297–304. Recall that the prime number theorem states that $\lim_{x \to \infty} \frac{\log(x)\pi(x)}{x} = 1$, where $\pi(x)$ is the number of primes less than or equal to x.

99. Eric Temple Bell continued his rather idiosyncratic (recall his work on "arithmetical paraphrases" briefly discussed in chapter one) number-theoretic researches from his home base at Caltech, but he trained few students after 1940.

As Princeton's "Problems of Mathematics" conference underscored, the fundamental research in algebra, geometry, topology, analysis, and mathematical logic that had engaged Americans in the 1920s and 1930s had revealed new mathematical horizons in those fields. At the same time, the long-established but less vibrant—at least in the United States—area of number theory was represented at the Institute as well as at a few scattered programs around the country. Established members of the American mathematical community in all of these fields worked side by side with their newly emigrated colleagues—as well as with the steady stream of fresh Ph.D.s that they collectively produced—to swell the nation's journals and its mathematical bookshelves with the fruits of their labors in all of these areas. In so doing, they also defined new research foci—in, for example, ring theory, the theories of homology and cohomology, intersection theory, the theory of manifolds, decision theory, functional and harmonic analysis—that drew in mathematicians internationally.

Building "New" American Research Agendas

These, however, were not the only areas of active mathematical interest in the United States in the 1940s. Even if it did not explicitly acknowledge number theory with its own individual session, the "Problems of Mathematics" conference did single out another under-represented—again, within the American mathematical research community—area: probability. The field had gained a certain notoriety in 1900 when Hilbert targeted it as ripe for axiomatization.[100] That challenge had been taken up at least as early as 1919 by a Richard von Mises, then still in Germany, in the form of his so-called frequentist approach.

Von Mises defined probability in terms of infinite sequences of elements representing observable events and then defined those sequences by means of a set of articulated postulates.[101] An event's probability was interpreted as

---

100. See the sixth problem in Hilbert, "Mathematical Problems," p. 454. Jan von Plato gives a historical overview of developments in *Creating Modern Probability: Its Mathematics, Physics and Philosophy in Historical Perspective* (Cambridge: Cambridge University Press, 1994).

101. Richard von Mises, "Fundamentalsätze der Wahrscheinlichkeitsrechnung," *Mathematische Zeitschrift* 4 (1919), 1–97 and "Grundlagen der Wahrscheinlichkeitsrechnung," *Mathematische Zeitschrift* 5 (1919), 52–99. Reinhard Siegmund-Schultze discusses von Mises's work at this time in "Probability in 1919/20: The von Mises-Polya-Controversy," *Archive for History of Exact Sciences* 60 (2006), 431–515.

the limit of its relative frequency over the course of many trials. Von Mises's ideas sparked the interest—and critique—of some of his contemporaries, in particular the then ETH-based Hungarian mathematician George Pólya, and, somewhat later, the University of Michigan's Arthur Copeland and Austrian Abraham Wald.[102]

In contradistinction, Wiener's construction—over the course of a number of papers in the early 1920s—of a mathematical model for Brownian motion had, for the most part, fallen on deaf ears in the United States. For him, "the basic probabilities were values of a measure"—now called the Wiener measure—"defined on subsets of a space of continuous functions." Although it anticipated by a decade Kolmogoroff's axiomatic grounding of probability in measure theory, the "forbiddingly formal" nature of Wiener's work, together with the fact that it tended to appear in MIT's essentially in-house *Journal of Mathematics and Physics*, initially prevented it from having "more influence."[103] Similarly, isolated attempts—like Julian Coolidge's 1925 textbook, *An Introduction to Mathematical Probability*—to introduce into the American collegiate curriculum actual *mathematical* probability, as opposed to discussions of games of chance or observational errors, had largely failed.[104]

The situation was different after 1939. Copeland, Wiener, Doob, Stanford's James Uspensky, and others had been joined by émigrés von Mises at Harvard, Hilda Geiringer first at Bryn Mawr and then at Wheaton, Mark Kac at Cornell,

102. Arthur Copeland, "Consistency of the Conditions Determining Kollektivs," *TAMS* 42 (1937), 333–357 and Abraham Wald, "Die Widerspruchtsfreiheit des Kollectivbegriffes in der Wahrscheinlichkeitsrechnung," *Ergebnisse eines mathematischen Kolloquiums* 8 (1937), 38–72.

103. See, for example, Norbert Wiener, "Differential Space," *JMP* 2 (1923), 131–174 and Joseph Doob, "Wiener's Work in Probability Theory," *BAMS* 72 (1966), 69–72 (the quotations in this and the previous sentence are on p. 69).

104. See Julian Coolidge, *An Introduction to Mathematical Probability* (Oxford: Clarendon Press, 1925) and the review of it by Henry Rietz, "Coolidge on Probability," *BAMS* 32 (1926), 83–85. Copeland, at least, was aware of Coolidge's book from his doctoral studies at Harvard and actually cited it in "Admissible Numbers in the Theory of Probability," *AJM* 50 (1928), 535–552 on p. 535, note †.

In a 1926 study, it was reported that, of the 122 institutions of higher education responding to a survey on teaching in statistics and probability, only twenty-one courses in probability *total* were being taught by mathematics departments nationwide during the 1924–1925 academic year. Of those courses, only thirteen required calculus. Ninety-seven reported offering no courses in probability at all. See James Glover, "Statistical Teaching in American Colleges and Universities," *JASA* 21 (1926), 419–446 on p. 420.

and William Feller at Brown from 1939 to 1945, then at Cornell, and finally at Princeton starting in 1950. In the 1940s, these mathematicians continued to elaborate the different strains of mathematical probability at the same time that some of them actively trained students in the field.

Among the Americans, Wiener developed linear prediction theory in the early 1940s and, when his work was finally declassified after the war, produced his 1949 book, *Extrapolation, Interpolation and Smoothing of Stationary Time Series with Engineering Applications*.[105] For his part, Doob—among much other work, including his development beginning in 1936 of martingales—applied "the methods and results of modern probability theory," that is, measure theory, to work on Brownian motion that had been advanced independently of Wiener's approach by physicists Leonard Ornstein and George Uhlenbeck.[106] Doob's 1941 Ph.D. student, David Blackwell, moreover, not only extended in his doctoral thesis his advisor's work on Markov chains but also became, during the 1941–1942 academic year, the first African-American member of the Institute for Advanced Study.[107] After a series of teaching positions at historically black colleges, Blackwell began the distinguished career at Berkeley in 1954 that also gave him the opportunity to advise an impressive number of students in probability.

Among the émigrés, Kac and Feller also animated a small but vibrant graduate program at Cornell that reinforced their respective research programs. For example, Kac, like Doob, brought work on Brownian motion more to the fore in his Chauvenet Prize–winning paper, "Random Walks and the Theory of Brownian Motion," while Feller produced the first of his ultimately

---

105. Norbert Wiener, *Extrapolation, Interpolation and Smoothing of Stationary Time Series with Engineering Applications* (Cambridge: The MIT Press; New York: Wiley; and London: Chapman and Hall, 1949). Pesi Masani, who joined Wiener in his research on prediction theory, gives a technical discussion of it in "Wiener's Contributions to Generalized Harmonic Analysis, Prediction Theory and Filter Theory," especially pp. 94–105. Wiener also discusses his work in this area in *I Am a Mathematician*, pp. 241–255.

106. See Joseph Doob, "Note on Probability," *AM* 37 (1936), 363–367 and "The Brownian Movement and Stochastic Equations," *AM* 43 (1942), 351–369 (the quotation is on p. 352). For the interested reader, Kai-Lai Chung gives an overview of Doob's work on martingales in his article, "Probability and Doob," *AMM* 105 (1998), 28–35 on pp. 32–34.

107. David Blackwell, "Idempotent Markoff Chains," *AM* 43 (1942), 560–567 and Joseph Doob, "Topics in the Theory of Markoff Chains," *TAMS* 52 (1942), 37–64. Linda Kirby provides Blackwell's biography as well as his extensive bibliography in "David H. Blackwell: Biography," *Celebratio Mathematica* (2011) at https://celebratio.org/Blackwell_DH/article/27/ and https://celebratio.org/Blackwell_DH/article/26/, respectively.

FIGURE 10.2. David Blackwell (1919–2010) in 1956. (Photo courtesy of Paul R. Halmos
Photograph Collection, di_07305, Archives of American Mathematics,
Dolph Briscoe Center for American History, University of Texas at Austin.)

two-volume *An Introduction to Probability Theory and Its Applications*.[108] In
reviewing the latter in 1951, Polish-born but NYU-trained statistician Jacob
Wolfowitz categorically asserted what the "Problems of Mathematics" symposium had implied, namely, that "[p]robability theory is now a rigorous
and flourishing branch of analysis, distinguished from, say, measure theory,

108. Mark Kac, "Random Walks and the Theory of Brownian Motion," *AMM* 54 (1947),
369–391 (which won the 1950 Chauvenet Prize) and William Feller, *An Introduction to
Probability Theory and Its Applications* (New York: John Wiley, 1950).

by the character and interest of its problems." In his book, Feller intimately linked that theory and its applications, providing "a huge number of examples" that not only "illustrat[ed] almost every aspect of the theory developed" but also "enhance[d] the interest of the theory even for the pure mathematician, except perhaps for the extreme diehard of the 'God save mathematics from its applications' school."[109]

This review of probabilist Feller's book by statistician Wolfowitz also underscores the interrelation of probability and statistics to which Hotelling had called attention in a "Problems of Mathematics" session on mathematical probability that was chaired and reported on, respectively, by statisticians Wilks and Tukey. Indeed, as Hotelling had insisted, "[w]ithout probability theory, statistical methods are of minor value, for although they may put data into forms from which intuitive inferences are easy, such inferences are very likely to be incorrect. The objective weighing of the degree of confidence to be placed in inductive conclusions is necessary to avoid fallacies. Indeed, the whole foundation of descriptive statistical methods, of inductive inference, and of the design of experiments, rests upon probability theory."[110]

It had, to some extent, been this critical dependence on probability that had sparked, in the 1930s, the professional delineation of *mathematical* statisticians, that is, those who focused on the theoretical underpinnings of statistics (which intimately involved the theoretical underpinnings of probability) from those social scientists who used statistics as a tool and viewed it as the gathering and tabulating of data. This process had manifested itself in at least two key ways: the publication of the first volume of *The Annals of Mathematical Statistics* in 1930, as distinct from the more social science–, less mathematically-oriented *Journal of the American Statistical Association* and the subsequent formation of the Institute of Mathematical Statistics (IMS) in 1935, as distinct from the American Statistical Association that had been founded in 1839 to foster descriptive statistics in the service of society. These accoutrements of professionalization likewise formed a certain

109. Jacob Wolfowitz, Review of "*An Introduction to Probability Theory and Its Applications*. Vol. 1. By William Feller," *BAMS* 57 (1951), 156–159 on pp. 156–157.

110. Harold Hotelling, "The Place of Statistics in the University," in *Proceedings of the Berkeley Symposium on Mathematical Statistics and Probability*, ed. Jerzy Neyman (Berkeley: University of California Press, 1949), pp. 21–40 on p. 25. Although its proceedings were only published in 1949, the symposium took place 13–18 August, 1945, and 27–29 January, 1946. For more on it, see later in this section.

boundary between mathematical statisticians and pure mathematicians that was nevertheless permeable. Disciplinary lines were blurred.[111]

Statisticians Wilks and Neyman, who recognized the importance of providing appropriate textbooks for students' entry into their special field,[112] were professors of mathematics at Princeton and Berkeley, respectively. Probabilist Doob was both a founding member of the IMS and an active participant in the AMS. Doob and others—like von Mises and Columbia statistician Abraham Wald—also published in the journals of both, while Wald and others—like Neyman and Geiringer—consciously crossed the constructed boundary between the mathematically- and social science–oriented statisticians.

Doob and von Mises, for example, debated the merits of their respective interpretations of the foundations of probability in invited addresses that were delivered at a meeting of the Institute of Mathematical Statistics in September 1940 and subsequently published in its journal, which had been under Wilks's editorship since 1939.[113] Doob and von Mises each aimed not only to convince mathematical statisticians that his own approach to probability was the better on which to ground mathematical statistics but also to underscore what each viewed as the shortcomings of the rival point of view. Wald, on the other hand, first introduced his work on decision theory—that is, in the statistical context, a mathematical generalization of the problems of hypothesis testing and constructing confidence intervals that includes both as special cases—in *The Annals of Mathematical Statistics* in 1939 but presented it in

111. Patti Hunter explores the delineation of an American community of mathematical statisticians from both the mathematical research community represented by the AMS and various of the social science communities in "An Unofficial Community: American Mathematical Statisticians before 1935," *Annals of Science* 56 (1999), 47–68 and "Drawing the Boundaries: Mathematical Statistics in 20th-Century America," *Historia Mathematica* 23 (1996), 7–30.

112. Samuel Wilks, *Mathematical Statistics* (Princeton: Princeton University Press, 1943) and Jerzy Neyman, *First Course in Probability and Statistics* (New York: Henry Holt and Co., 1950). Patti Hunter considers the key role of textbooks in defining mathematical statistics in the United States in "Foundations of Statistics in American Textbooks: Probability and Pedagogy in Historical Context," in *From Calculus to Computers: Using the Last 200 Years of Mathematics History in the Classroom*, ed. Amy Shell-Gellasch and Richard Jardine (Washington, D.C.: Mathematical Association of America, 2005), pp. 165–180.

113. Richard von Mises, "On the Foundations of Probability and Statistics," *AMS* 12 (1941), 191–205; Joseph Doob, "Probability as Measure," *AMS* 12 (1941), 206–214; and Richard von Mises and Joseph Doob, "Discussion of Papers on Probability Theory," *AMS* 12 (1941), 215–217.

1945 to the more purely mathematical audience targeted by the *Annals of Mathematics*. He also published, in 1945, the not-unrelated work he had done during the war on the sequential probability ratio test in versions intended for two audiences: an exposition "accessible to statisticians with little mathematical background" in the *Journal of the American Statistical Association* and a massive technical paper in the IMS's *Annals*.[114] Neyman and Geiringer, too, targeted both mathematically- and non-mathematically-oriented statisticians with their work, Neyman on the theory of sampling human populations as well as on industrial sampling and Geiringer on Mendelian genetics.[115]

This intertwining of probability and statistics—in their mathematical as well as applications orientations—was further underscored in 1945 and 1946 when Neyman hosted the first of what would ultimately be six Berkeley Symposia on Mathematical Statistics and Probability. In so doing, he united an eclectic selection of statisticians, among them, Hotelling and Wolfowitz; probabilists, among them, Doob and Feller; and scientists who used statistics in their work, among them, Berkeley astronomer Robert Trumpler and his colleague, physicist Victor Lenzen. As even Neyman had to admit, "[t]he subjects of the papers vary considerably, illustrating the range of problems which in one way or another tie up with mathematical statistics and probability."[116] In trying to review the volume, Mark Kac was even more categorical. "The thirty papers which are collected in the volume," he confessed, "cover such a wide range of subjects that a comprehensive review is impossible."[117] Still, he

114. Abraham Wald, "Contributions to the Theory of Statistical Estimation and Testing Hypotheses," *AMS* 10 (1939), 299–326; "Statistical Decision Functions Which Minimize the Maximum Risk," *AM* 46 (1945), 265–280; "Sequential Method of Sampling for Deciding between Two Courses of Action," *JASA* 40 (1945), 277–306 (for the quotation see note * by the journal's editor on p. 277); and "Sequential Tests of Statistical Hypotheses," *AMS* 16 (1945), 117–186. Patti Hunter discusses all of this work in "Connections, Context, and Community," especially pp. 29–31.

115. Jerzy Neyman, "Contribution to the Theory of Sampling Human Populations," *JASA* 33 (1938), 101–116 and "On a Statistical Problem Arising in Routine Analyses and in Sampling Inspections of Mass Production," *AMS* 12 (1941), 46–76 as well as Hilda Geiringer, "On the Probability Theory of Linkage in Mendelian Heredity," *AMS* 15 (1944), 25–57 and "On Some Mathematical Problems Arising in the Development of Mendelian Genetics," *JASA* 44 (1949), 526–547.

116. Jerzy Neyman, "Foreword," *Proceedings of the Berkeley Symposium on Mathematical Statistics and Probability*, p. vi.

117. Mark Kac, Review of "*Proceedings of the Berkeley Symposium on Mathematical Statistics and Probability*. Ed. by J. Neyman," *BAMS* 56 (1950), 267–268 on pp. 267–268. The quotation that follows is on p. 267.

added, "[i]n spite of the somewhat chaotic arrangement and an overwhelming battery of topics this volume is a tribute to the great vitality of statistics and statistical methods." Indeed, Neyman, through his Statistical Laboratory as well as through the symposia, had transformed Berkeley into what was perceived as "the world headquarters" of mathematical statistics by the 1950s.[118]

Like probability and statistics, applied mathematics had also become an area of heightened activity within the American mathematical research community of the 1940s. This activity, spurred, in part, by American mathematicians' wartime experiences, well reflected the concerns expressed during the discussion of "New Fields" at the Princeton bicentennial conference, namely, that various areas—the theory of diffraction, the theory of plasticity, control and communications engineering, neurophysiology, modern computing, hydrodynamics, and others—would all benefit from the concerted efforts of mathematicians. As von Neumann exclaimed in exasperation, "we [do not even have an] adequate theory of the interactions of two shock waves!"[119]

The mathematical behavior of shock waves had, in fact, defined some of the classified wartime work of Richard Courant and Kurt Friedrichs. On its declassification in 1946, the two men—with the editorial help of native English-speaker Cathleen Morawetz, a then-recent mathematics M.A. recipient from MIT and the daughter of applied mathematician John Synge— began work on a book on the subject.[120] The result, their 1948 *Supersonic Flow and Shock Waves*, was described as "an excellent and up-to-date account of the related problems of supersonic flow and non-linear wave propagation" that not only laid out "the underlying theory of hyperbolic partial differential equations" but also discussed "the practical problems of flow in nozzles and jets."[121] The first book published in a new "Monographs in Pure and Applied Mathematics" series launched by New York–based publisher Interscience, it reflected Courant's powers of persuasion as well as commercial faith in

118. David Kendall, Maurice Bartlett, and Thornton Page, "Jerzy Neyman, 16 April, 1894– 5 August, 1981," *Biographical Memoirs of Fellows of the Royal Society* 28 (1982), 379–412 on p. 385.

119. *Problems of Mathematics*, p. 16 (or p. 320).

120. Reid, *Courant*, pp. 255. Morawetz put her editorial and mathematical experiences to good use, staying on at NYU and earning her 1951 Ph.D. on "Contracting Spherical Shocks Treated by a Perturbation Method" under Friedrichs's supervision.

121. Richard Courant and Kurt Friedrichs, *Supersonic Flow and Shock Waves* (New York: Interscience Publishers, 1948) as well as Chia-Chiao Lin, Review of "*Supersonic Flow and Shock Waves*. By R. Courant and K. O. Friedrichs," *BAMS* 57 (1951), 85–87 (the quotations are on p. 85).

growing American productivity in, especially, applied mathematics. Courant and Friedrichs's book was later credited for the fact that "fluid dynamics was one of the first fields [of applied mathematics] to undergo a renaissance" in the United States beginning in the 1950s.[122]

Courant and Friedrichs's colleague James Stoker, the third member of the NYU triumvirate, produced the second book in the Interscience series in 1950, also on the thorny problem of nonlinear phenomena. Like his colleagues' volume, Stoker's *Nonlinear Vibrations in Mechanical and Electrical Systems* motivated the mathematical concepts involved through numerous concrete examples of intrinsic appeal to physicists and engineers. After thus rendering those mathematical concepts intuitively clear, rigorous proofs, involving in many places "a considerable amount of original work," were then provided of their existence theorems. According to Nicolas Minorsky, by this point a member of the engineering faculty at Stanford, Stoker's book marked "an important landmark in the first decade of the studies of nonlinear problems in the United States."[123]

Although qther applied mathematical research in physical science settings could be cited here—for example, the advances in aerodynamics made by von Kármán and his colleagues and students at Caltech—applications, as suggested by the statistical work of Neyman and Geiringer mentioned previously, were by no means limited to the physical sciences. In the 1940s, one of the most striking examples of a non-physical-science application of mathematics issued from the collaboration of John von Neumann and Princeton economist Oskar Morgenstern. Their 1944 book, *Theory of Games and Economic Behavior*, exploited the novel analogy between how businesses operate and how games are played. In his overview of their ideas, probabilist Arthur Copeland explained that von Neumann and Morgenstern observed "that in the game of life as well as in social games the players are frequently called upon to choose between alternatives to which probabilities rather than certainties are attached." They "show that if a player can always arrange such fortuitous alternatives in the order of his preferences, then it is possible to assign to each alternative a number ... expressing the degree of the player's

122. Peter Lax, "The Flowering of Applied Mathematics in America," in *A Century of Mathematics in America*, ed. Peter L. Duren et al., 2: 455–466 on p. 456.

123. James Stoker, *Nonlinear Vibrations in Mechanical and Electrical Systems* (New York: Interscience Publishers, 1950) and Nicolas Minorsky, Review of "*Nonlinear Vibrations in Mechanical and Electrical Systems*. By J. J. Stoker," *BAMS* 56 (1950), 519–521. For the quotations in this and the previous sentence, see pp. 519 and 520, respectively.

preference for that alternative. The assignment," however, "is not unique but two such assignments must be related by a linear transformation."[124] With that as their starting point, von Neumann and Morgenstern proceeded to formalize the concept of a game in terms of a set of function-theoretic postulates and to derive mathematical consequences from them. "Posterity," Copeland hazarded in 1945, "may regard this book as one of the major scientific achievements of the first half of the twentieth century."

Von Neumann's is also one of the many mathematical names—among them, Alan Turing in England—associated with another of the "major scientific achievements of the first half of the twentieth century," that is, the development of electronic computers and the mathematical exploration of their myriad implications. Von Neumann's war work had convinced him of the potential of and need for high-speed computing, and, to that end, he set up the Electronic Computer Project (ECP) at the Institute for Advanced Study.[125] "[A] multifaceted [project] embracing engineering, formal logic, logical design, programming, [and] mathematics," the ECP aimed to demonstrate the power of the new theories and technology via "some significant revolutionary application."[126] One of the first mathematical problems that the project explored involved the cumulative effect of round-off errors when performing large numbers of iterative computations. Von Neumann and his associate on the ECP, Herman Goldstine, who, at the Ballistic Research Laboratory associated with the Aberdeen Proving Ground had, as noted, been a principal on the ENIAC project, approached it head on in what Goldstine later described as "the first modern paper on numerical analysis ever written."

Von Neumann and Goldstine focused on determining "the stability of a computational procedure ... with respect to ... round off errors," since "[e]rrors committed (that is, noise introduced) in an earlier stage of the

---

124. Arthur Copeland, Review of "Theory of Games and Economic Behavior. By John von Neumann and Oskar Morgenstern," BAMS 51 (1945), 498–504. The quotations are on p. 498.

125. Much has been written about this aspect of von Neumann's career in, for example, two biographical studies—one by Norman Macrae (cited in chapter eight) and another by Steve Heims, John von Neumann and Norbert Wiener: From Mathematics to the Technologies of Life and Death (Cambridge: The MIT Press, 1980)—as well as in Goldstine, The Computer from Pascal to von Neumann and William Aspray, John von Neumann and the Origins of Modern Computing (Cambridge: The MIT Press, 1990).

126. Goldstine, The Computer from Pascal to von Neumann, p. 209. The "revolutionary application" on which von Neumann focused was meteorology. The quotation that follows is on p. 113.

computation should be exposed to a possibility of considerable amplification by the subsequent operations of the computation."[127] To analyze the problem, they considered, in a massive paper that appeared in the AMS's *Bulletin* in 1947, the example of digitally solving n simultaneous linear equations in n unknowns, for n large, and developed an elimination algorithm that allowed them to get a theoretical handle on the rounding errors inherent in the process of matrix inversion. As they explained, it was this problem that would become "very important . . . when the fast digital machines . . . become available," since "those machines will create a prima facie possibility to attack a wide variety of important problems that require matrix manipulations, and in particular inversions, for unusually large values of n."

Von Neumann and Goldstine were by no means alone in recognizing this fact. Early in 1946, the ONR's Mina Rees and Edward Condon, nuclear physicist on the Manhattan Project and then head of the National Bureau of Standards (NBS), became involved in a Navy initiative to establish a national center for mathematical computation.[128] It had become clear that high-speed computing machinery would "not only greatly increase the effectiveness of the mathematical approach to standard problems, but [would] also permit successful attacks . . . on problems which ha[d] hitherto been considered . . . inaccessible by mathematical methods," such as "weather forecasting," "the rational planning of military and business operations," that is, operations research, and "the automatic control of industrial processes."

By 1947, the National Applied Mathematics Laboratories of the NBS had been founded with four components: the Institute for Numerical Analysis (INA) located at UCLA; the Computation Laboratory initially in New York City but, after 1948, in Washington, D.C.; and the Statistical Engineering and Machine Development Laboratories, both also located in Washington. The INA, in particular, served as "the focal point . . . for basic research and *training* in the types of mathematics," namely, numerical analysis, "which are pertinent to the efficient exploitation and further development of high-speed automatic digital computer equipment."[129] Initially headed by John Curtiss,

127. Herman Goldstine and John von Neumann, "Numerical Inverting of Matrices of High Order," *BAMS* 53 (1947), 1021–1099 on pp. 1030–1031. The quotations that follow are on p. 1031.

128. John Curtiss, "A Federal Program in Applied Mathematics," *Science* 107 (1948), 257–262. The quotations that follow in the next sentence are on p. 257.

129. John Curtiss, "The Institute for Numerical Analysis of the National Bureau of Standards," *AMM* 58 (1951), 372–379 on p. 376 (my emphasis).

a 1935 Harvard Ph.D. under Walsh, with Gertrude Blanch from the Mathematical Tables Project as assistant director for computation, the INA was to applied mathematics and the West Coast what the IAS was to pure mathematics and the East Coast, attracting a steady stream of year-long or shorter-term visitors. Its first senior researchers in the 1947–1948 academic year were Olga Taussky-Todd and her husband, John Todd, then both home-based in England, and Cincinnati's Otto Szász. In the summer of 1948, they were not only joined by eight visiting researchers, among them, Penn's Hans Rademacher, but they also participated in the first of what would become regular symposia jointly sponsored by the INA and UCLA.[130] On "Modern Calculating Machinery and Numerical Methods," the 1948 summer symposium included talks by, among others, Lefschetz on "Numerical Calculations in Nonlinear Mechanics" and 1946 Berkeley Ph.D. (under Neymann) George Dantzig on "Linear Programming." The collection of talks was subsequently published under the title *Problems for the Numerical Analysis of the Future* in what had become an Applied Mathematics Series published under NBS auspices.[131]

Advances in modern numerical analysis, then, were intimately intertwined with such mathematically informed, World War II–fostered areas as computer science, linear programming, of which Dantzig along with von Neumann and the Russian Leonid Kantorovich were pioneers, and operations research, which was developed after the war in the United States by, among others, Dantzig and physicist Philip Morse.[132] Although for many of the Princeton bicentennial discussants in 1946, "new fields" had meant a revitalization of various aspects of classical analysis and their applications, the areas stemming from modern numerical analysis really were "new fields" of *applied* mathematics that were ripe for cultivation beginning in the late 1940s.

As the formation of the INA within the Federal government suggests, moreover, these and other technical advances were accompanied by

130. Magnus Hestenes and John Todd give a fact-filled history of the INA in their *NBS-INA–The Institute for Numerical Analysis–UCLA 1947–1954*, NIST Special Publication 730 (Washington, D.C.: U.S. Government Printing Office, 1991). In particular, they discuss in some detail the work that Taussky-Todd did at the INA in 1948 on Hilbert matrices. See pp. 45–47. Judith Goodstein gives an overview of Taussky-Todd's life in "Olga Taussky-Todd," *NAMS* 67 (2020), 345–353.

131. *Problems for the Numerical Analysis of the Future* appeared in 1951 as volume 15 in the series.

132. See, for example, George Dantzig, "Origins of the Simplex Method" and Philip Morse, "Mathematical Problems in Operations Research," *BAMS* 54 (1948), 602–620.

significant changes in the professional development of applied mathematics within both academe and an expanding American mathematical research community. Brown University had founded its Graduate Division of Applied Mathematics in 1946, a unit that had grown from its wartime Program of Advanced Instruction and Research in Mechanics. In so doing, it joined von Kármán's more focused program in aeronautics at Caltech, Courant's program at NYU, and recently formed programs at the Carnegie Institute of Technology, MIT, and elsewhere—as well as more scattered practitioners around the country, like Ivan Sokolnikoff at Wisconsin until 1946 and then at UCLA—in actively training students in the area.

Such training required new textbooks in the various areas subsumed under the "applied mathematics" rubric. In this spirit, von Kármán and Maurice Biot published their text, *Mathematical Methods in Engineering: An Introduction to the Mathematical Treatment of Engineering Problems*, in 1940, while Sokolnikoff and his wife Elizabeth produced a second edition of their text, *Higher Mathematics for Engineers and Physicists* in 1941, a book on which "practically an entire generation of young engineers and physicists was mathematically raised."[133] Similarly, von Mises, in collaboration with William Prager as well as with Gustav Kuerti, von Mises's assistant in Harvard's School of Engineering, published the *Theory of Flight*, a 1945 upgrade for an English-reading audience of von Mises's more elementary 1918 *Fluglehre: Vorträge über Theorie und Berechnung der Flugzeuge in elementarer Darstellung* and based, at least partly, on lectures he had given in Brown's summer program.[134] As von Mises explained in the preface, since there seemed to be "[a]n actual demand . . . for a book at an intermediate level," the new text targeted the advanced undergraduate or beginning graduate student proficient in the calculus and who

133. Theodore von Kármán and Maurice Biot, *Mathematical Methods in Engineering: An Introduction to the Mathematical Treatment of Engineering Problems* (New York: McGraw-Hill, 1940) and Ivan and Elizabeth Sokolnikoff, *Higher Mathematics for Engineers and Physicists* (New York and London: McGraw-Hill, 1941 (the first edition, also published by McGraw-Hill, appeared in 1934) and, for the quotation, Magnus Hestenes, Raymond Redheffer, and John Green, "Ivan Stephan Sokolnikoff, Mathematics: Los Angeles (1901–1976), Professor Emeritus," posted in the California Digital Library at http://texts.cdlib.org.

134. Richard von Mises (with William Prager and Gustav Kuerti), *Theory of Flight* (New York: McGraw-Hill, 1945; reprint ed., New York: Dover Publications, 1959) (the Dover edition is essentially a facsimile of the 1945 edition; the quotations that follow are from the Dover edition, p. xi) and Richard von Mises, *Fluglehre: Vorträge über Theorie und Berechnung der Flugzeuge in elementarer Darstellung* (Berlin: Verlag von Julius Springer, 1918).

had "some training in general mechanics insofar as the standard college education provides for these things." It therefore did not assume knowledge of the mathematical theory of fluid mechanics. The reader desirous of a treatment at that level was referred to the mimeographed notes of the Brown lectures on advanced fluid mechanics that he, Friedrichs, and Stefan Bergmann had given. The first half of Sokolnikoff's Brown lectures on the *Mathematical Theory of Elasticity*, on the other hand, were actually published in 1946, long remaining "the leading book on the subject,"[135] while Philip Morse and chemist George Kimball's *Methods of Operations Research* codified, on its declassification in 1945, wartime work in the new field and, in so doing, became a resource for courses that had been developed at "several American Universities" by 1951.[136]

As these and other educational efforts began slowly to bear fruit in the late 1940s and into the 1950s, applied mathematicians increasingly came together to develop the same kind of professional accoutrements—journals and a society—that the mathematical statisticians had created a decade earlier. Brown's Prager brought out the first American journal specifically dedicated to the field, the *Quarterly of Applied Mathematics*, in 1943; his colleague, Raymond Archibald, together with Berkeley number theorist Derrick H. Lehmer (not to be confused with his father, Derrick N. Lehmer, the Berkeley number theorist who had died in 1938), started the quarterly journal, *Mathematical Tables and Other Aids to Computation*, that same year under the auspices of the National Research Council.[137] Many of the fruits of war research in applied mathematics subsequently appeared in the former, while in the latter, Herman and Adele Goldstine first described the ENIAC and how it worked to a broader public.[138] Five years later, Courant and his NYU colleagues built on these initial efforts with the publication of the first issue of *Communications on Applied Mathematics*. Initially intended, like MIT's *Journal of Mathematics and Physics*, as an in-house journal, it was selectively opened to non-NYU contributors as of its fifth issue and renamed *Communications*

135. Ivan Sokolnikoff, *Mathematical Theory of Elasticity* (New York: McGraw-Hill, 1946) and, for the quotation, Hestenes, Redheffer, and Green.

136. Philip Morse and George Kimball, *Methods of Operations Research* (Cambridge: Technology Press of the Massachusetts Institute of Technology and New York: John Wiley & Sons, Inc., 1945; rev. first ed., 1951). The quotation is from the preface (p. i) of the 1951 edition.

137. The latter journal was renamed *Mathematics of Computation* in 1960.

138. Herman and Adele Goldstine, "The Electronic Numerical Integrator and Computer (ENIAC)," *Mathematical Tables and Other Aids to Computation* 2 (1946), 97–110.

on Pure and Applied Mathematics better to reflect the full range of the NYU program.[139]

The Society for Industrial and Applied Mathematics (SIAM) followed in 1951, founded with the explicit aims of "further[ing] the application of mathematics to industry and science," "promot[ing] basic research in mathematics leading to new methods and techniques useful to industry and science," and "provid[ing] media for the exchange of information and ideas between mathematicians and other technical and scientific personnel."[140] By the end of 1952, SIAM had 130 members; by September 1953, it had brought out the first volume of its *Journal of the Society for Industrial and Applied Mathematics*; by 1955, its membership had grown to 1,000. In the immediate aftermath of World War II, then, the explicit calls from within the American mathematical research community—that had been repeatedly sounding for greater emphasis on and support of applied mathematics—had finally been answered, and American programs in, for example, mathematical statistics, had come to lead internationally.

———

The American mathematical research community had emerged from World War II with a renewed sense of possibility and purpose as symbolized by Princeton's 1946 "Problems of Mathematics" conference. As the hosts of "the first international gathering of mathematicians in a long and terrible decade,"[141] Princeton mathematicians and, by extension, the American community as a whole began to assert what they collectively viewed as their newly attained international leadership role. At their invitation, key world mathematical leaders had joined them in their quest to define postwar mathematical research in what they deemed some of the principal fields of pure mathematics: algebra, geometry, topology, analysis, mathematical logic, probability.

139. Reid, *Courant*, pp. 269–270.

140. Front matter, *Journal of the Society for Industrial and Applied Mathematics* 1 (1953). For this and the membership figures that follow, see http://mathshistory.st-andrews.ac.uk/Societies/SIAM.html. Some within the AMS had lobbied—unsuccessfully—at least as early as the summer of 1946 for a division of applied mathematics within the AMS as opposed to an independent and separate society. See the open letter dated 21 June, 1946, signed by Coble, Eisenhart, Lefschetz, Mac Lane, Morse, Tukey, Veblen, and Wilks, Box 2, Folder 74: Tukey, John, Whyburn Papers.

141. *Problems of Mathematics*, p. 2 (or p. 309).

Although they had conspicuously left aside applied mathematics, the war, especially in the United States, had served as a catalyst for research into the applications of mathematics to the physical as well as to the biological and social sciences and had significantly extended the boundaries of American mathematics. Whether pure or applied, mathematical research issuing from the United States in the 1940s represented the concerted effort of a burgeoning community from coast to coast and north to south comprised of American-born and émigré mathematicians who self-confidently pursued research interests they knew to be important. At the same time, they trained future researchers in and strengthened the professional infrastructure of their various fields. Whatever other characteristics *a* or *the* world leader in mathematics might have, such self-confidence and energy must certainly be among them.

CODA

Mathematicians in the postwar United States not only returned full time to their various prewar occupations—teaching, research, publication, self-governance, program-building—but also found themselves in changed political environments both at home and abroad. Domestically, the war had resulted in the creation of new bureaucratic models—like the ad hoc Office of Scientific Research and Development and the permanent Office of Naval Research—for science in general and for mathematics in particular. It had also spurred discussion—for example, of the Kilgore Bill as early as 1942 and of the aborted Research Board for National Security in 1945 and 1946—about the possibility, desirability, and even necessity of the long-term Federal support of science. Although much was in flux, one thing seemed clear. The postwar world would require even more active engagement in science politics on a national level. For mathematicians, that would mean making sure that their voices were heard—and heard compellingly—as new decisions were made and policies shaped.

Abroad, the war had left much of Europe ravaged and many of its institutions crippled. Rebuilding there would take time. The war had also redrawn political boundaries. An Iron Curtain had descended that separated West from East. Germany was divided and occupied by the Americans, the Russians, the British, and the French. Countries like Poland, Hungary, and Czechoslovakia had come under the sway of the Soviet Union. By the end of the 1940s, a Cold War between the United States, the Soviet Union, and their respective allies had begun to set in that made communications between the Western and Eastern Blocs increasingly difficult. Elsewhere, the United States occupied Japan. China was plunged once again into civil war, resulting in the ascendance there in 1949 of Mao Zedong and the Chinese Communist Party. New political alliances, like the Organization of American States and the North Atlantic Treaty Organization, formed. All of these developments presented challenges for an American mathematical community intent on demonstrating what it viewed as its new international leadership role as host of a postwar International Congress of Mathematicians.

The five years from 1945 to 1950 found American mathematicians engaged in both of these political arenas. New policies and legislation were being debated in Washington that had the potential to impact mathematics, and, as discussions resumed in earnest about when to hold the first postwar ICM, the prospect of revivifying the International Mathematical Union also emerged. The time had come for the putative world leader in mathematics to test its political mettle.

The New Domestic Politics of Mathematics

As noted in chapter nine, the joint AMS/MAA War Policy Committee evolved in 1946 into what was initially a six-person Policy Committee for Mathematics. Four represented the AMS—Griffith Evans (for a year-long term), Theophil Hildebrandt (for a two-year term), Marston Morse (for a three-year term), and Marshall Stone as chair (for a four-year term)—while one each—Alonzo Church and William Feller—represented the Association for Symbolic Logic and the Institute of Mathematical Statistics, respectively. John Kline, Secretary of the AMS, served ex officio as secretary of the committee as well. The MAA's participation beginning in 1948 increased the membership by three: former Oberlin professor of mathematics and then Assistant Commissioner for Higher Education for the State of New York, Carroll Newsom; immediate MAA Past President and Illinois Institute of Technology professor, Lester Ford; and then MAA Secretary-Treasurer, Harry Gehman of the University of Buffalo (now the State University of New York at Buffalo).[1] The committee's charge reflected new postwar priorities. Specifically, it was "empowered to speak for the constituent organizations on matters which concern the position of mathematics" such as "proposed or enacted legislation concerning science, problems concerning the effective use of mathematicians or potential members of the mathematical profession, and other questions

1. See Harry Gehman, "The Thirtieth Summer Meeting of the Association," *AMM* 55 (1948), 605–611 on p. 611 and "The Thirty-Second Annual Meeting of the Association," *AMM* 56 (1949), 200–207 on p. 206. The National Council of Teachers of Mathematics joined with one member in 1951.

which tend to affect the dignity and effective position of mathematics among related sciences, both nationally and internationally."[2] Funded by a modest grant from the Rockefeller Foundation, the committee thus had a wide remit which was fully reflected in its initial discussions.[3]

Indeed, "a number of problems" were seen to "face the mathematical profession in this post-war period." Among the most pressing in 1946 were: "[t]he various science bills before Congress," "[a]tomic research, particularly as it affects the freedom of scientific investigation," "[t]he various changes in Selective Service regulations," and "[i]nternational cooperation problems raised by the United Nations Educational, Scientific and Cultural Organization" (UNESCO). The latter issue, it seemed, would "probably be before the committee in more concrete form within a short time, together with the associated problem of the revival of the International Mathematical Union" and primarily concerned the international front (see the next section). The former, however, were of immediate concern domestically.

In the war's aftermath, the United States began to demobilize its large armed forces. This rather quickly prompted the question of how the country would defend itself in the future. Would it maintain a standing army? Would it continue to rely on the draft? If the latter, would there be occupational deferments, and, if so, under what conditions? In 1946, the Selective Training and Service Act of 1940 was still in effect, but as early as October 1945, Truman had gone before a joint session of Congress to argue for the establishment of a system of universal military training that "would require every young man between the ages of eighteen and twenty to take a year of military training and afterwards to serve six years in a General Reserve system."[4] As noted,

2. J. R. Kline to the Rockefeller Foundation, 5 August, 1946, Ms.75.2, Box 15, Folder 100, AMS Papers. The quotations that follow in the next paragraph are also from this letter. See also Temple Hollcroft, "The Summer Meeting in Ithaca," *BAMS* 52 (1946), 964–975 on p. 969 and Pitcher, *A History of the Second Fifty Years*, p. 277.

3. Pitcher details this committee's activities in the 1940s and 1950s in *A History of the Second Fifty Years*, pp. 276–280 as does G. Baley Price in "The Mathematical Scene, 1940–1965," in *A Century of Mathematics in America*, ed. Peter Duren et al., 2: 379–404 on pp. 392–397. In some sense, its work continues to the present in the form of the Joint Policy Board for Mathematics that links the AMS, the MAA, the American Statistical Association, and the Society for Industrial and Applied Mathematics (founded, as noted in the previous chapter, in 1951).

4. Cunningham, p. 397. Even earlier, the War Policy Committee had drafted the text of a bill "[t]o authorize the assignment of persons to a national scientific education program instead of to service in the armed forces, in order to make possible the education and training of scientists to meet essential needs." See John Kline, Report of the War Policy Committee, 28 April, 1945, Ms.75.2, Box 15, Folder 88, AMS Papers.

the President had chosen to highlight precisely this initiative in his nationally publicized speech in Princeton in 1947.

Like the War Policy Committee during the war, the postwar Policy Committee for Mathematics closely monitored the ensuing debate, ready to act when appropriate. For example, it engaged directly with those in Washington on behalf of the American mathematical community when, in June 1948, a new Selective Service Act was passed in response to the escalating Cold War.[5] Less than a year later it was also actively involved in discussions initiated by the Army and Air Force "concerning the possibility of an arrangement for the exchange of ideas and various forms of mutual assistance" and was "planning to establish a liaison committee which may be consulted by the various branches of the armed services on problems of research and on questions of personnel."[6] This sort of direct connection between the AMS and the military had been exactly what Morse, Stone, and others on the War Preparedness Committee had advocated as early as 1940, but their initiatives then had been preempted by the creation of the NDRC.

As for the question of atomic research and its effects on free scientific inquiry, that certainly affected mathematicians and engaged some among them like von Neumann. The issue, however, that dominated the Policy Committee's domestic agenda was that of Federal support for research. From 1946 until the passage in 1950 of the bill that created the National Science Foundation, its members made sure that mathematics remained integral to the discussions of whatever Federal funding scheme might emerge.

The same Harley Kilgore who had introduced legislation in the Senate in 1942 aimed, among other things, at financing scientific research during wartime, and who had exerted legislative pressure in favor of Federal support of science throughout the war, continued to press his position. Early in 1946, he joined forces, although only after heated wrangling, with Warren Magnuson, Democratic Senator from the State of Washington, to sponsor a bill that finally emerged in February 1946 as Senate Bill 1850 (S. 1850).[7]

5. The AMS's Policy Committee actually formed a separate Subcommittee on Selective Service Regulations with Gordon Whyburn as chair and, as members, Carroll Newsom and Haverford College's Carl Allendoerffer. On its activities, see Box 7, Folder 19, Whyburn Papers.

6. Kline to (U. S. Air Force) Brigadier General Donald Putt, 26 January, 1949, Box 7, Folder 18, Whyburn Papers. See also Pitcher, A History of the Second Fifty Years, p. 279.

7. Robert Maddox details Kilgore's efforts between 1942 and 1950, when a version descended from the 1946 bill finally passed both houses of Congress, in "The Politics of World War II Science: Senator Harley M. Kilgore and the Legislative Origins of the National Science Foundation," West Virginia History 41 (1979), 20–39. On the dizzying proliferation of bills in

This proposed legislation aimed to "provide support for scientific research and development, to enable young men and women of ability to receive scientific training, to promote the conservation and use of the natural resources of the Nation, to correlate the scientific research and development programs of the several Government agencies, to achieve a full dissemination of scientific and technical information to the public, and to foster the interchange of scientific and technical information in this country and abroad."[8] In order to meet these objectives, "it [would be] essential," the bill posited, "to create a central scientific agency within the Federal Government," specifically, a "National Science Foundation" (NSF).

The Policy Committee agreed, sending two telegrams in support of the bill, one "to influential Senators" and one "to appropriate Congressmen after the Bill was sent to the House and referred to the Interstate Commerce Committee" (ICC).[9] So far, so good. When it was learned from a Washington insider, however, that the ICC had voted not to report the bill in the then current session owing to the fact that the Congressional recess was imminent, that insider suggested that a telegram to the chair "might reverse" the decision.[10] Over Stone's signature as Policy Committee chair, a letter was duly sent "urging reconsideration of the Committee's vote against reporting out the Bill."[11] In it, Stone reemphasized the mathematical community's "support of the Kilgore-Magnuson Bill in essentially the same terms as those used in the two telegrams, with the addition of a few embellishments." Although likely not the only entreaty from a scientific body he received, the ICC chair was unmoved. As one historian put it, whereas the atomic scientists had been very effective in getting the legislation passed that resulted in the creation in 1946

1945 and 1946 that ultimately resulted in S. 1850, see Howard Meyerhoff, "The National Science Foundation: S. 1850, Final Senate Bill," *Science* 103 (1946), 270–273. J. Merton England discusses the tensions between Kilgore and Magnuson that finally ended in compromise in *A Patron of Pure Science: The National Science Foundation's Formative Years, 1945–57* (Washington, D.C.: National Science Foundation, 1982), pp. 25–43.

8. "Text of the New Kilgore-Magnuson Bill," *Science* 103 (1946), 225–230 on p. 225. The quotations in the next sentence are also on this page.

9. Stone to Kline, 26 July, 1946, Ms.75.5, Box 31, Folder 120, AMS Papers.

10. Kline to Stone, 23 July, 1946, Ms.75.5, Box 31, Folder 120, AMS Papers. That insider was M(erriam) H. Trytten, the Director of the National Research Council's Office of Scientific Personnel.

11. Stone to Kline, 26 July, 1946, Ms.75.5, Box 31, Folder 120, AMS Papers. The quotation that follows is also from this letter.

of the Atomic Energy Commission, "the summer of '46 showed that the scientists advocating a national science foundation had not yet mastered the art of politics."[12]

Of course, mastering that art was complicated, for the mathematicians at least, by the fact that decisions on how best to proceed did not rest solely in the hands of the Policy Committee for Mathematics. That body, for example, had urged that a circular be sent early on in the legislative process to the memberships of the AMS, the Association for Symbolic Logic, and the Institute of Mathematical Statistics that would have not only laid out the main features of S. 1850 but also suggested that individual "letters and telegrams" be sent in support of it "to the Congressmen and Senators."[13] When this proposal was brought before the AMS Council at its meeting in April 1946, the committee was authorized merely "to give public expression to opinion in support of the Kilgore-Magnuson Bill."[14] The organization of a write-in campaign from the mathematical community would be, in the Council's view, "to indulge in . . . 'mixing in political affairs,'" and it reminded the Policy Committee that its "function was to speak *for* the Constituent bodies" as opposed, it would seem, to encouraging their members to strengthen the position with their collective voice.[15] It would be hard to play politics, especially in Washington, without being allowed actually to *play* politics.

Still, the Policy Committee repeatedly engaged over the course of the four years from 1946 to 1950 in the tortured legislative process that brought fashioned and refashioned versions of the original S. 1850 up for debate in both the House and Senate.[16] In particular, in July 1949 as that debate became increasingly entangled with the issue of loyalty oaths sparked by the Cold War, the Policy Committee, by then under Morse as chair, formed a committee, first, "to study the way in which the Science Foundation if established could serve science and mathematics" and, then, to draw up a report for the Policy Committee's consideration.[17] Under MIT's Ted Martin as chair, Griffith

12. England, p. 59.

13. Kline to Stone, 14 August, 1946, Ms.75.5, Box 31, Folder 121, AMS Papers.

14. Temple Hollcroft, "The April Meeting in New York," *BAMS* 52 (1946), 580–593 on p. 589.

15. Kline to Stone, 14 August, 1946, Ms.75.5, Box 31, Folder 121, AMS Papers (my emphasis).

16. For more on this saga, see England, especially pp. 45–110.

17. Marston Morse and John Kline, "Report of Policy Committee for Mathematics," August 1949, p. 4, Ms.75.2, Box 15, Folder 136, AMS Papers. The quotations that follow in the next

Evans, Saunders Mac Lane, and Mina Rees considered such questions as "[s]hould the selection of recipients of grants be in the Federal Government or outside?" "[i]f outside, is the N.R.C. the best agency?" and "[i]f a Foundation is established, would the interests of mathematics be better served by a separate Division of Mathematics in the N.R.C. . . . ?" The former questions were partially addressed in 1950 by the creation of a twenty-four-person National Science Board under a Director, all nominated by the President and approved by the Senate.[18] The latter question centered on the move then afoot to create not only a "Mathematical Foundation" that would have actively sought financial support for mathematics from industrial concerns but also a division for mathematics, separate from physics, within the National Research Council. While the first initiative was not realized, the second was in 1951 under Morse as chair and Stone as vice chair.[19] It, like the facts that mathematics had always been part of the discussions of both the abortive Research Board for National Security and the NSF, represented a clear mark of the national recognition that the mathematical community had achieved since the 1920s when Veblen had had to fight for mathematics to be included in the NRC postdoctoral fellowship program. By midcentury, mathematics was on a more equal footing with the other sciences.

Perhaps of even more immediate concern to Martin's committee in 1949, however, were recent amendments to a bill then in the House. These included a provision that would prohibit the Foundation from financially supporting those who had failed to submit "an affidavit" to the effect that "he [or she] does not believe in, and is not a member of and does not support any organization that believes in or teaches the overthrow of the United States Government

---

sentence are also from p. 4 of this report. By this point, the committee was comprised of Morse, together with Einar Hille, Rudolph Langer, and Stone representing the AMS, Albert Meder of Rutgers representing the Association for Symbolic Logic, Samuel Wilks for the Institute of Mathematical Statistics, and Ford, Gehman, and Newsom for the MAA.

18. Yale physicist Alan Waterman, then Deputy Chief of the Office of Naval Research, served as the NSF's first Director. Marston Morse was the Board's sole mathematician. England discusses the Board and its formation on pp. 113–140.

19. As Morse explained in a circular to "Dear Colleagues" on 29 September, 1951, the new Division of Mathematics was concerned specifically "with advising government agencies in matters having to do with mathematical science" (Box 8, Folder 39, Whyburn Papers). Of critical importance would be advising a new NSF on "a proper budget for mathematics" ("Mathematics and the National Science Foundation," *BAMS* 50 (1950), 285–287 on p. 285). Pitcher briefly discusses both the Mathematical Foundation and the new division within the NRC in *A History of the Second Fifty Years*, p. 279.

by force or violence or by any illegal or unconstitutional methods."[20] By November 1949, a memorandum had been readied that contained specific questions, observations, and tentative recommendations regarding a possible NSF and that was directed to the Presidents and Secretaries of the mathematical societies represented by the Policy Committee, the chairs of fifty-two mathematics departments from coast to coast, the members in mathematics of the National Research Council, and the editors of nine mathematical journals. It asked specifically for comments and further suggestions so as to make its final report as comprehensive as possible.[21] In particular, it "recommended that action opposing the inclusion of a loyalty oath be recommended to the several mathematical societies; in particular that the phrase '*does not believe in* . . . *any organization that believes in*' be opposed."

Despite such opposition, various loyalty and security provisions survived in the bill that Truman finally signed in May 1950, among them, section 15.d. In addition to the precise language about affidavits that Martin's committee found abhorrent, that section also included the condition that NSF funding recipients must have "taken and subscribed to" yet another "oath or affirmation in the following form: 'I do solemnly swear (or affirm) that I will bear true faith and allegiance to the United States of America and will support and defend the Constitution and laws of the United States against all its enemies, foreign and domestic.' "[22] As the Cold War slid precipitously into the deep freeze and as faculties at Berkeley and elsewhere in California famously challenged an oath even more pointed than those in section 15.d in that it specifically referred to membership in the Communist Party, it was likely inevitable that no amount of external politicking from academic groups would have prevailed in removing such a clause once it had survived debate in Congress.[23] Indeed, mathematicians—like topologist John Kelley at Berkeley, Chandler Davis then at the University of Michigan, and Dirk

20. Martin to Kline, 15 November, 1949, Ms.75.5, Box 35, Folder 15, AMS Papers.

21. Ibid. The quotation that follows is from the materials enclosed with this letter (with the committee's emphasis). For the statement the Policy Committee issued based on the committee's report, see "Mathematics and the National Science Foundation."

22. See Public Law 507–81st Congress, Chapter 171-2D Session at the URL given in the references.

23. The situation in California in 1949–1950 is detailed in, for example, George Stewart, *The Year of the Oath: The Fight for Academic Freedom at the University of California* (Garden City: Doubleday, 1950) and David Gardner, *The California Oath Controversy* (Berkeley: University of California Press, 1967).

Struik of MIT, among others—soon ran afoul of these and similar oaths. As all of this makes clear, the founding of the National Science Foundation was by no means without controversy. Still, it represented a major political coup on the domestic front for science in general and for mathematics in particular.

The New Geopolitics of Mathematics

As leaders within America's immediately postwar mathematical community focused on domestic issues so, too, did they return their attention to the international scene. Even before the end of the war in the Pacific, some among them were already intent on renewing international ties especially in the form of the International Congress of Mathematicians that was to have taken place in the United States in 1940. When plans for that event were halted in 1939, an Emergency Executive Committee comprised of George Birkhoff, William Graustein, Einar Hille, Mark Ingraham, John Kline, Marston Morse, Roland Richardson, and Marshall Stone had been "appointed to act during the interim," that is, to be responsible for overseeing the funds that had been raised for the 1940 event and for reviving the initiative whenever that was deemed feasible.[24] By the time the war was over, Birkhoff and Graustein were dead, and Theophil Hildebrandt and Harvard's David Widder had replaced them on the committee.

In July 1945, Hildebrandt, who was then the AMS's President, wrote to Richardson with procedures for reanimating the ICM already on his mind. In reply, Richardson counseled that, first, "there should be some preliminary discussion of the date of the Congress" on the part of the Emergency Committee so as to avoid "getting an organization set up too many years in advance."[25] In his view, "the most difficult question" would concern "the scope of the Congress." Would, for example, the "Poles, Dutch and French . . . accept invitations to a Congress to which Germans were invited"? As he saw it, "the crux of the problem" was that "if we await the time when Germany's neighbors will be willing to sit down in Congress with Germans, the Congress will be postponed many years. On the other hand, there are those Americans who would not want to see a Congress until it was *truly international.*" Three

24. Ayers, "The September Meeting in Madison," pp. 804–805.

25. Richardson to Hildebrandt, 20 July, 1945, Box 15, Folder: H, Richardson Papers. Unless otherwise indicated, the quotations that follow in this paragraph are also from this letter (with my emphasis).

months later in October, Morse, in his capacity as Emergency Committee chair, engaged his group in a discussion of precisely these issues.[26] By February 1946, they had concluded that, while "it would be too soon to hold the International Congress in 1948," "the possibility of holding an *open* Congress in 1950 should be explored."[27]

Explored it was, with consensus settling on 1950. That timing, however, was further reinforced when the AMS learned in 1946 that the Rockefeller Foundation—which had not only agreed to underwrite the 1940 Congress to the tune of $7,500 (recall chapter five) but had also pledged to sustain that support following the event's postponement in 1939—let it be known that the latter offer would lapse were the congress not held before 1950's close.[28] It was time to get busy.

By April 1947, the AMS Council had accepted Harvard University's renewed invitation to host an ICM, this time in 1950. Eight months later in December, it had dissolved the Emergency Committee and instructed a nominating committee to draw up a slate of names to fill out the committees requisite for an ICM's organization. Building on the foundations laid in 1937 and 1938, new committees were in place in the winter of 1948, and the organization of an ICM in the United States began anew.[29]

There were three main committees: Editorial, Financial, and Organizing chaired by Chicago's Lawrence Graves, the Institute's John von Neumann, and Harvard's Garrett Birkhoff, respectively.[30] Scores of mathematicians, most of whom were among the American mathematical community's most

26. Michael Barany details the activities of the Emergency Committee in advance of the formation of an actual Congress Organizing Committee in "Distributions in Postwar Mathematics" (Doctoral dissertation, Princeton University, 2016), pp. 150–161.

27. "Memorandum RE Meeting of Emergency Committee of International Congress of Mathematicians on 2/16/46," Ms.75.2, Box 15, Folder 97, AMS Papers, as quoted (with my emphasis) in my article "Marshall Stone and the Internationalization of the American Mathematical Research Community," *BAMS* 46 (2009), 459–482 on p. 470.

28. Warren Weaver to Theophil Hildebrandt, 22 October, 1946, as discussed in Barany, "Distributions in Postwar Mathematics," pp. 156–157.

29. See John Kline, "Secretary's Report: The International Congress of Mathematicians," in *Proceedings of the International Congress of Mathematicians: Cambridge, Massachusetts, U.S.A., August 30–September 6, 1950*, ed. Lawrence Graves, Einar Hille, Paul Smith, and Oscar Zariski, 2 vols. (Providence: American Mathematical Society, 1952), 1: 121–145 on p. 121 and Pitcher, *A History of the Second Fifty Years*, p. 148–149.

30. For the full list of committees and their members, see "The International Congress of Mathematics: Committees of the Congress," in *Proceedings*, 1: 1–4.

research-productive yet only several of whom were women, participated in the planning, with the Organizing Committee creating a proliferation of sub-committees, some, of course, dealing with the program.[31] Like the planners of the aborted 1940 ICM, Birkhoff's group incorporated the "conference" model into the "usual" model of invited addresses and contributed papers. In 1950, as had been the plan for 1940, there would be four targeted conference topics. Three were predictable and essentially the same as in 1940: algebra, analysis, and topology. The fourth, however, reflected that half-century-long, American discussion that, likely owing to the war, had finally come to a resolution: applied mathematics.

Seven sections—only six had been planned for 1940—complemented these conferences: algebra and the theory of numbers; analysis; geometry and topology; probability and statistics, actuarial science and economics; mathematical physics and applied mathematics; logic and philosophy; and history and education. The 1950 planners thus explicitly highlighted number theory and divided in two what, in 1940, would have been a single section on logic, philosophy, history, and didactics. With these decisions made, speakers had to be chosen, invitations extended, and, for those foreign mathematicians who would make the trip to Cambridge, travel and visa issues addressed. It would be in dealing with the latter that American mathematicians would confront the new geopolitics of their field, a geopolitics further complicated by the contemporaneous initiative to re-found an International Mathematical Union.

Just as some in the American mathematical community turned their thoughts to the postponed ICM in the immediate aftermath of the Victory-in-Europe celebrations in May 1945, others focused on the resumption of international connections more generally. Among the latter, Stone, chair of the joint AMS/MAA War Policy Committee and newly returned to Washington from his wartime reconnaissance mission to South Asia for the War Department's Military Intelligence Service, queried Kline on "what steps, whether formal and official or not, have been taken towards renewing our

---

31. Mina Rees, a member of the committee on the Conference in Applied Mathematics, was the only woman on a scientific committee. Mary Graustein, wife of Harvard professor William Graustein and assistant professor of mathematics at Tufts University, and Wellesley College's Helen Russell, a 1932 Radcliffe College Ph.D. under Harvard's Joseph Walsh, served with a number of the wives of Cambridge-area mathematicians and others on the Entertainment Committee.

scientific ties with mathematicians in previously occupied territories (including those now occupied by our side)."[32] Stone clearly thought that such international outreach was important. He also hoped that, given his political connections, he might be able to "do a little exploration of the possibilities of government assistance," even though "at the moment the State Department appears to be on the verge of important changes in personnel if not in organization." Truman was, after all, settling, into the Presidency he had assumed less than a year earlier on Roosevelt's death. Kline recognized the import of Stone's question not only in the abstract but also for any ICM planning that might eventually get under way and encouraged him to check around. As Kline saw it, "[t]his is an important item and should be carefully discussed by the War Policy Committee soon."[33]

Although that committee was discharged of its duties before it could take up the issue, it was on the newly created Policy Committee's agenda in the spring of 1946 in the form of both "[i]nternational cooperation problems" and the possible re-founding of the IMU.[34] One of the "problems" specifically concerned the Russians, whom Stone was intent on trying to engage, especially given the mathematical connections he had made with them as early as 1935 when he attended the international conference on topology in Moscow. In February 1946, in fact, he had suggested that the AMS "arrange an exchange of members with some of the Russian Mathematical Societies," enlisting Kolmogoroff and Alexandroff as intermediaries.[35] As for the matter of a new IMU, English biochemist turned sinologist and historian of science, Joseph Needham, in his capacity as the first head of the Natural Sciences Section of newly formed UNESCO in Paris, had written to Richardson in May 1946 suggesting that the Americans "take the lead in reconstituting the International Mathematical Union."[36] Given that UNESCO sought to establish a working relationship with the International Council of Scientific Unions,

32. Stone to Kline, 17 June, 1945, Ms.75.5, Box 30, Folder 112, AMS Papers. The quotation that follows is also from this letter.

33. Kline to Stone, 23 June, 1945, Ms.75.5, Box 30, Folder 112, AMS Papers.

34. John Kline to the Rockefeller Foundation, 5 August, 1946, Ms.75.2, Box 15, Folder 100, AMS Papers.

35. Stone to Kline, 27 February, 1946 (for the quotation), and Kline to Stone, 5 March, 1946, Ms.75.5, Box 31, Folder 116, AMS Papers.

36. John Kline to Marshall Stone, 12 July, 1946, Ms.75.5, Box 31, Folder 120, AMS Papers. Compare Barany, "Distributions in Postwar Mathematics," p. 179. Indeed, it was Needham who put the "S" in UNESCO.

the natural home of a re-founded IMU, the suggestion, coming as it did from Needham, was not unnatural.[37] Moreover, as noted in chapter three, the original IMU had coordinated the organization of the ICMs in 1920 in Strasbourg and in 1924 in Toronto before it was repudiated by the Italian organizers of the Bologna ICM in 1928. It effectively ceased to exist at the Zürich ICM in 1932, owing to what many viewed as its exclusionary policies. Who better than the American organizers of what would be the first postwar ICM to establish a healthy, postwar ICM-IMU connection?

Needham's suggestion thus (almost) inevitably became entangled with plans for the ICM. Kline, for one, felt "strongly that we should not be a party to a Union which excludes mathematicians because of national and racial ties, just as we feel that the Congress must be an open Congress. There will surely be some knotty problems," he continued, "because of the connection with the State Department and UNESCO. What would be the relation between the Congress and the International Mathematical Union?"[38] In other words, were a new IMU to be created before the 1950 ICM, the planning for which was on the verge of gearing up, how would that affect the Congress? Such complexities led Stone to conclude that "[t]he question of an International Mathematical Union should undoubtedly be taken up by the Policy Committee."[39] When, in August 1946, the AMS Council explicitly asked that committee to do just that, Stone, as its committee's chair, was thrust into what quickly developed into a challenging geopolitical situation.[40]

Three months later, and owing largely to "[t]he move to communicate with the Russian mathematicians about the Congress and the discussion concerning it," Stone had yet to do "much with the matter of a statement concerning a Mathematical Union."[41] He had had time to consider a related question, though. How, he had been asked, could UNESCO "promote the development

37. Lehto, p. 73. On the relationship of UNESCO to scientific unions, see John Fleming, "The International Scientific Unions," *Proceedings of the American Philosophical Society* 91 (1947), 121–125.

38. Kline to Stone, 12 July, 1946, Ms.75.5, Box 31, Folder 120, AMS Papers. Since governments adhere to the United Nations and, so, to its associated organizations like UNESCO, UNESCO interfaces with the United States through the State Department.

39. Stone to Kline, 26 July, 1946, Ms.75.5, Box 31, Folder 120, AMS Papers.

40. Stone and the members of the Policy Committee, Alonzo Church, William Feller, Theophil Hildebrandt, Rudolph Langer, and Marston Morse, "Report to Council of the American Mathematical Society," 5 May, 1947, Ms.75.5, Box 32, Folder 60, AMS Papers.

41. Stone to Kline, 25 November, 1946, Ms.75.5, Box 31, Folder 122, AMS Papers.

of mathematics on the international front"?[42] In his view, the new organization could "help ... most effectively by taking whatever steps it can to facilitate the preparation and distribution of mathematical literature embodying the important new ideas in mathematics and by eliminating or minimizing the various obstacles, political and economic, which threaten to make the travel of scholars and students between the nations of the world extremely difficult even after the conditions of peace are established."[43] Six months later, when the Policy Committee reported back to the AMS Council on the IMU question, not only had what Stone sketched as UNESCO's possible benefits for mathematics been transferred to an IMU but also the whole IMU question had been complicated by concurrent French efforts to shepherd a new IMU into existence.

The French had informally invited a number of mathematicians (who would be in France for two mathematical meetings in June 1947) to come together to discuss the issue.[44] The Americans would apparently have competition relative to setting up a new IMU. Be that as it may, "[i]n the opinion of the Policy Committee the strong support which is emerging in favor of a new union and the need for endowing such a union at the very outset with the maximum potentialities for service to the entire world-community of mathematicians are compelling reasons for *the American group* to participate actively and officially" in laying such groundwork. Of course, there were at least two other "compelling reasons": the Americans' desire to assert what they viewed as their new standing as the world's mathematical leader and the negative role that the French, but particularly Émile Picard and Gabriel Koenigs (both deceased by 1947), were viewed—by G. H. Hardy, Richardson, Stone, Veblen, and others—as having played in the exclusionary policies of the original IMU. The members of the Policy Committee in general, and Stone in particular, were adamant that a new IMU would have to be founded

42. Stone to the members of the Policy Committee, 31 October, 1946, Ms.75.5, Box 31, Folder 122, AMS Papers.

43. Stone to Arthur Compton and William Noyes, 30 October, 1946, Ms.75.5, Box 31, Folder 121, AMS Papers.

44. Stone et al., "Report to Council of the American Mathematical Society," 5 May, 1947. Unless otherwise indicated, the quotations that follow in this and the next paragraph are also from this report (with my emphases). The mathematical meetings, on algebraic topology and harmonic analysis, were co-sponsored by the Rockefeller Foundation and the Centre National de la Recherche Scientifique. Barany treats the give-and-take at the informal meeting on a possible IMU in some detail in "Distributions in Postwar Mathematics," pp. 168–179.

on "the principle of representation for *all* national and geographical groups of mathematicians" be they German, Japanese, Russian,. . .

Consonant with the spirit of Needham's initial approach to "the American group," the Policy Committee had determined that the best way to proceed would be within established, as opposed to ad hoc, channels. It had thus proposed "a meeting held under the sponsorship of the International Council of Scientific Unions (ICSU) and in connection with the next General Assembly of UNESCO at Mexico City in the autumn of 1947."[45] The President of ICSU had affirmed, moreover, that "ICSU is the natural body to convoke such a meeting since the existing scientific unions, while autonomous, function under its aegis and their constitutions bear the stamp of its approval." At the same time, in the Policy Committee's view, "[t]he agreement between ICSU and UNESCO would furnish grounds for official aid from the latter body which would add substantially to the prospects for success."

The Policy Committee had thus already laid a fair amount of the groundwork for the discussion of a new IMU in the months before May 1947 when it reported on its activities to the AMS Council. Stone, in particular, did not want those efforts scuttled by an informal meeting in France and feared "that the motive back of the great activity of the French toward the formation of a Union is political, to promote French cultural domination over the satellite nations of Europe."[46] He did not want a new IMU to become mired in the kind of exclusionary politics that had been the downfall of the earlier incarnation, and he felt, as apparently did Needham, that the Americans were best suited to effect a satisfactory result.

The meeting in France took place in early July, and many views were aired.[47] While there was a general consensus—although not unanimity—that the re-founding of an IMU was desirable, there were many differences of opinion on issues like timing and inclusion vs. exclusion. It became clear

45. In "Distributions in Postwar Mathematics," p. 171 ff, Barany sees this as a deliberate "end-run" around the French. The read of the evidence that I give here differs, seeing it as part of deliberations that had begun with Needham's first approach in 1946.

46. Stone in remarks given on 27 April, 1947, as quoted in Lehto, pp. 76–77. Stone's skepticism was not unwarranted. The French government's immediately postwar position relative to Germany was far from conciliatory and only began to change later in 1947. See Judt, pp. 114–117. Stone's reference to "the satellite nations of Europe" likely refers to French diplomatic efforts, especially in the 1930s, to exert influence in Eastern Europe.

47. Barany details the various positions in "Distributions in Postwar Mathematics," pp. 174–179.

both to some in the ICSU as well as ultimately to Stone that more ground-work still needed to be laid. Accordingly, Stone recommended to the AMS Council, and it agreed, that "further action looking towards the establishment of an International Mathematical Union had best be postponed until the time of the International Congress contemplated for 1950."[48]

Following the events of the summer of 1947, the American Policy Committee took the lead internationally on the matter of an IMU. Within it, a three-person subcommittee comprised of Stone, Morse, and Kline coordinated the effort, with Stone the animating force. By July 1948, Stone had laid out a strategy for how to proceed that would involve much correspondence with national mathematics societies, with individual mathematicians, and ultimately with duly formed national IMU-focused committees around the world. The idea was that, working collaboratively and addressing issues as they arose, a set of statutes and bylaws for an IMU could first be hammered out. The draft could then be discussed collectively by representatives of those nations, the national committees of which had participated in crafting it, in a face-to-face meeting to be held just before the 1950 ICM. To underscore the independence of a new IMU from the ICM, moreover, that meeting would take place in a city other than Cambridge.[49] A reasonable plan, it was threatened at its outset.

In the summer of 1948, Garrett Birkhoff, in his role as ICM Organizing Committee chair, not only got into independent discussions about an IMU with representatives of UNESCO on approaching them for financial support for the ICM but also assumed that discussion of an IMU would be part of the ICM's program. As Morse saw it, "G. B. is making an intolerable mess" that was only cleared up when Stone got the AMS Council involved in September.[50] The Council made it crystal clear that the Policy Committee was "the sole agent to carry on negotiations and bring recommendations to the governing boards leading to the formation of a Union."[51]

That snafu dealt with, Stone worked tirelessly, studying the statutes and bylaws of other organizations, drafting and redrafting text based on feedback from myriad quarters, communicating with interested parties.[52] He also

48. Minutes of Council, 2 September, 1947, Ms.75.2, Box 15, Folder 111, AMS Papers.
49. Compare Lehto, pp. 77–79 and Kline to Morse and Stone, 6 August, 1948, Box 37, Folder 5, Stone Papers.
50. For the quotation, see Morse to Stone, 26 August, 1948, Box 37, Folder 5, Stone Papers.
51. Minutes of Council, 7 September, 1948, Ms.75.2, Box 15, Folder 124, AMS Papers.
52. Lehto details these efforts on pp. 79–82.

used the occasion of an around-the-world trip—from August 1949 through May 1950 that took him to Occupied Japan, Thailand, Vietnam, Java, Singapore, Ceylon (modern-day Sri Lanka), India, Egypt, France, and elsewhere in Europe—personally to conduct "very delicate negotiations with mathematicians throughout the world on the problem of the formation of a new International Mathematical Union."[53] The stop in Occupied Japan was particularly significant. As Stone later reported to the American authorities under General Douglas MacArthur, he "came as the first foreign academic visitor in the field of mathematics for many years" and capitalized on the visit "to arrange for the participation of Japanese Mathematicians in the formation of the proposed new International Mathematical Union."[54] "My visit," he continued, "can thus be said to have contributed in a quite specific way to the restoration of professional bonds between the mathematicians of Japan and those of other countries."

While Stone was in India, the related issue of Germany also came to the fore when he received a letter from Erich Kampke, President of the newly re-established Deutsche Mathematiker-Vereinigung. Kampke let Stone know that he had heard about the "preparations for an International Mathematical Union" and that he "would be very happy . . . if we could initiate an exchange of ideas about your plan."[55] Stone had been treading carefully relative to the question of when to approach the Germans, so Kampke's was a welcome overture. On consulting the national IMU-focused committees then in existence, Stone was both gratified and relieved to find no country opposed to inviting Germany to participate in setting up an IMU.

The Russians also presented a problem, but for different reasons. Stone, who, as noted, had been working to engage them at least as early as 1946, had yet to receive a response to his approach regarding the IMU (or the ICM). As the 1940s drew to a close, it became ever clearer that domestic policies within the Soviet Union were unlikely to permit Soviet participation in international mathematical affairs in 1950.

By the time the constitutive meeting of a new IMU took place at Columbia University in New York City on 27–29 August, 1950, some twenty-two

53. Kline to Rudolph Langer, Edwin Beckenbach, and Samuel Wilks, 4 October, 1949, Ms.75.5, Box 35, Folder 32, AMS Papers.

54. Stone to Bowen Dees, Scientific and Technical Branch, Economic and Scientific Section, Tokyo, Japan, 24 February, 1950, Box 38, Folder 8, Stone Papers. The quotation that follows is also from this report.

55. Kampke to Stone, 3 November, 1948, as quoted in Lehto, p. 83 (Lehto's translation).

countries had formed national IMU-focused committees. Although Russia was not one of them, Austria, Germany, Italy, and Japan were. Representatives from these and other countries were able to come together thanks to a $10,000 grant allocated by UNESCO to help defray the travel costs of one delegate per committee.[56] UNESCO was, as Stone had hoped, facilitating "the travel of scholars and students between nations of the world."[57]

The Paris-based organization earmarked an equal amount for travel grants to allow those from abroad to attend the ICM. Either at the invitation of the Congress, or as part of the sections, or as interested participants, mathematicians from abroad would be able to come to speak on their latest work, to learn of research being done elsewhere, and to make what for some would be the first personal contact with their foreign colleagues since before the war. The various ICM committees had been hard at work to decide on speakers, both foreign and domestic, for the invited hour addresses and for the four conferences.[58] A committee had also been formed to select the second two Fields Medalists, the first two having been awarded in Oslo in 1936. Comprised of Harald Bohr as chair together with Lars Ahlfors by then back at Harvard, Warsaw's Karol Borsuk, the Sorbonne's Maurice Fréchet, William Hodge of Cambridge University, Andrei Kolmogoroff of Moscow's Steklov Institute as well as the Moscow State University, Damodar Kosambi of the Tata Institute in Bombay (present-day Mumbai), and Marston Morse of the IAS, this committee, like the others, had a task before it complicated by both mathematical and geopolitical concerns.[59]

Among the latter were visas for attendees—both speakers and otherwise—at the ICM. As they repeatedly emphasized, the Congress organizers were determined that theirs would be an "open" and "truly international" Congress,

---

56. The other eighteen countries were: Argentina, Belgium, Brazil, Cuba, Denmark, Finland, France, Great Britain, Greece, India, Yugoslavia, the Netherlands, Norway, Sweden, Switzerland, Turkey, the United States, and Uruguay. See Lehto, p.85. In 2021 dollars, the "real wealth" of $10,000 is $110,900 (see https://www.measuringworth.com).

57. Stone to Compton and Noyes, 30 October, 1946.

58. Barany focuses on the financing of the ICM in "Remunerative Combinatorics," pp. 340–343. See also the list of donors to the Congress given in *Proceedings*, 1: 5–6. Barany calculates the Congress budget at just under $93,000 with just over $24,000 going toward travel grants to foreign mathematicians. In 2021 dollars, the total budget represents "real wealth" of $1,031,000 (see https://www.measuringworth.com).

59. Barany considers some of the difficulties faced in "Distributions in Postwar Mathematics," pp. 195–210 and 237–238.

yet the worsening Cold War and rising anti-Communist sentiments in the United States risked undermining that objective. Russian mathematicians were ultimately prohibited by their own government from attending, so the Congress was not effectively "open" to them, although through no fault of the Congress's organizers and despite their efforts to smooth the way for them politically. In effect, no amount of effort on the Americans' part could have influenced the Soviet government to allow its mathematicians to attend. Yet, it was also the case that anyone who was thought to have or to have had Communist connections faced an increasingly uphill battle with the State Department as 1950 approached. In particular, the thirty-five-year-old analyst Laurent Schwartz of the University of Nancy, who had been selected as one of the plenary speakers and who, it later emerged, would also be one of the two Fields Medalists, was only granted a visa at the eleventh hour thanks to the efforts of John Kline and the "high priced legal talent" he had engaged in anticipation of precisely this sort of problem.[60]

As is well-known, Schwartz's difficulties with the American authorities had begun a year earlier in 1949 when he had hoped to spend time at both the Institute for Advanced Study and the University of Chicago in conjunction with a trip to attend a meeting in Canada.[61] Owing to Trotskyist connections (from which he had subsequently distanced himself), his entry visa was denied. This resulted in a geopolitical firestorm. Outside the United States, the French threatened to boycott the 1950 Congress. Domestically, a so-called "Committee of Five"—William Eberlein of the University of Wisconsin, MIT's Witold Hurewicz, Paul Halmos of Chicago, Berkeley's John Kelley, and Maxwell Reade of Michigan—demanded at the summer meeting of the AMS in Boulder to meet with the Congress's Organizing Committee in order "to ascertain . . . the latest possible date for cancelling [sic] the congress in case difficulties become insurmountable."[62] The situation had, in fact, left such a bad taste in Marshall Stone's mouth—he had worked hard to secure Schwartz's visa and so to ensure his visit to Chicago—that he confessed in the summer of 1950 that "[s]ince a year ago I have believed that the Congress

60. Maxwell [Reade] to Raymond Wilder, 7 September, 1949, Box 86-36/8, Folder 1, Wilder Papers.

61. See, for example, Laurent Schwartz, *A Mathematician Grappling with His Century*, trans. Leila Schneps (Basel: Birkhäuser Verlag, 2001), pp. 311–316. Barany brings new archival evidence to bear on the story Schwartz tells in his autobiography in "Distributions in Postwar Mathematics," pp. 229–244.

62. Reade to Wilder, 7 September, 1949.

should *not* be held in the U.S. on account of the attitude of the government towards liberal and radical scientists from other countries."[63]

The year between the summer of 1949 and the summer of 1950 definitely presented a number of thorny visa problems, among them Schwartz's, over which John Kline and others lost sleep. Still, Kline was ultimately right in believing in December 1949 "that we shall be able to solve the major part of our visa difficulties."[64] Schwartz was in Cambridge in 1950 to give his invited address and to pick up—along with Norwegian analytic number theorist and, beginning in 1949, IAS professor Alte Selberg—his Fields Medal. So, too, were Schwartz's Communist-sympathizing great-uncle-in-law, Jacques Hadamard, and others, like the Uruguayan mathematician Rafael Laguardia, whose visas had been problematic.[65]

Consonant with his sense of both diplomacy and decorum, Kline only alluded vaguely in his official report on the Congress to the geopolitical struggles with which he and the other organizers had had to contend. "In attempting to maintain the non-political nature of the Congress," he wrote, "many serious difficulties had to be overcome."[66] Moreover, even in full knowledge of the often intense wrangling that had gone on with the State Department, he let the printed record show that the "officers of the Congress found the various officials of the Department of State most sympathetic and helpful." "As far as [the latter officials] know," he recounted, "only one mathematician from any independent nation was prevented from attending the Congress because he failed to pass a political test and this man did not notify the officers of the Congress about his difficulties. Only two mathematicians from occupied countries failed to secure visas."[67] The mathematicians had ultimately won (almost all of) the geopolitical battles associated with the 1950 ICM that it was within their power to fight. Theirs was perhaps as "open" and "truly international" an ICM as geopolitical realities in 1950 could have permitted.

---

63. Stone to Uruguayan mathematician Rafael Laguardia, 4 August, 1950 (Stone's emphasis), as quoted in Barany, "Distributions in Postwar Mathematics," p. 227.

64. Kline to Raymond Wilder, 16 December, 1949, Box 86-36/7, Folder 2, Wilder Papers.

65. On Laguardia, see Barany, "Distributions in Postwar Mathematics," pp. 223–228.

66. Kline, "Secretary's Report," p. 122. The quotations that follow in this paragraph are also on this page.

67. In "Distributions in Postwar Mathematics," pp. 215–228, Barany makes a convincing argument that the one mathematician who "failed the political test" was Uruguayan mathematician José Luis Massera.

FIGURE C.1. John Kline (1891–1955) in his office at Penn in 1954. (Photo courtesy of University Archives and Records Center, University of Pennsylvania.)

The International Congress of Mathematicians: Cambridge, MA, 1950

When the world's mathematicians finally came together in Cambridge, Massachusetts on Wednesday, 30 August, 1950, they hailed from forty-one countries representing every continent (save, of course, Antarctica). From the United States in particular, they represented the District of Columbia as well as every state in the union except South Dakota.[68] In all, just over 2,300 people attended with roughly 80% from an American mathematical community that had grown dramatically over the course especially of the 1930s and 1940s. (Another 255, of whom 160 were from either the United States or

---

68. Kline, "Secretary's Report," p. 135. The number of countries Kline lists here differs from those enumerated in *Proceedings*, 1: 7–20, since presumably delegates from some countries were finally unable to attend. The numbers in the next sentence are also on p. 135 of Kline's report. The count for women was drawn from the "List of Members," in *Proceedings*, 1: 36–87.

	Total Attendance	% from Host Country	% of Women Participants	# of Participating Countries
Strasbourg 1920	200	40%	3%	27
Toronto 1924	444	24%	7%	33
Bologna 1928	836	40%	8%	36
Zürich 1932	667	21%	5%	35
Oslo 1936	487	12%	12%	36
Cambridge 1950	2,302	80%	14%	41

FIGURE C.2. ICM attendance statistics 1920–1950.[69]

Canada, registered but, for one reason or another, did not attend.) Compared to the other post–World War I ICMs, this one was by far the largest, the most dominated by the host country, and yet the most internationally diverse (see Fig. C.2).

Additionally, three of the first African-Americans to earn a Ph.D. in mathematics in the United States—William Claytor, Walter Talbot, and Joseph Pierce—were among the Congress's registrants, with Claytor serving as an official delegate from Howard University.[70] Some 180 women also registered, most coming from North America but a few, like English analyst Mary Cartwright, traveling from abroad. No woman, however, gave an invited address; one, Dorothy Maharam Stone, did participate in the Panel on Measure Theory in the subconference on Algebraic Tendencies in Analysis; one, Mina Rees, served, as noted, on the selection committee for speakers in the Conference in Applied Mathematics; and only ten lectured in one of the sections. Although 14% of the total Congress registrants, women made up less than 3% of its speakers. The 1950 ICM was, perhaps not surprisingly, a North American, white-male-dominated affair, although the wives and children of

69. The European or North American host had the greatest percentage of participants except at the Toronto and Oslo Congresses, where the United States accounted for 43% and 17.6% of the participants, respectively. First names of participants were not given in the proceedings of the Strasbourg ICM, and only six participants were designated "Mlle," so the 3% figure may be an underestimate.

70. Talbot, the fourth, received his degree in 1935 at the University of Pittsburgh under the direction of Montgomery Culver in group theory and taught at historically black Lincoln University in Jefferson City, Missouri; Joseph Pierce, the sixth, earned his Ph.D. in 1938 at the University of Michigan under Harry Carver in statistics and taught at the Texas State University for Negroes (now Texas Southern University) in Houston.

FIGURE C.3. The 1950 International Congress of Mathematicians, Cambridge, MA. (Photo in the Harvard University Archives.)

some of the participants as well as some of the participants from abroad did bring greater diversity to at least the social aspects of the eight-day-long event.

The Congress opened in a plenary session that first Wednesday afternoon. As Organizing Committee chair, Garrett Birkhoff welcomed those assembled; Oswald Veblen was officially elected the Congress's President and gave a short opening address; a cablegram was read from physicist Sergei Vavilov, the President of the Soviet Academy of Sciences. Vavilov thanked the organizers for their invitation to Soviet mathematicians, explained that they were "very much occupied with their regular work" and so "unable to attend," and offered his hope "that [the] impending congress will be [a] significant event in mathematical science."[71] Bohr, as Fields Medal selection committee chair, closed the session with brief descriptions of the award-winning work of and with the presentation of medals to Selberg and Schwartz, the former for his recent findings on the Riemann zeta function as well as for his elementary proof of the prime number theorem, the latter for his work on distributions,

71. Kline, "Secretary's Report," p. 122.

FIGURE C.3. Continued.

"a generalization of the very notion of a function better adapted to the process of differentiation than the ordinary classical one."[72]

With that, the first four of twenty-two invited hour addresses opened the scientific part of the program. Arne Beurling of the University of Uppsala spoke "On Null-Sets in Harmonic Analysis and Function Theory"; the ETH's Heinz Hopf discussed "Die n-dimensionalen Sphären und projektiven Räume in der Topologie"; Henri Cartan of the Sorbonne in Paris treated "Problèmes globaux dans la théorie des fonctions analytiques de plusieurs variables complexes"; and the University of Michigan's Raymond Wilder shared his thoughts on "The Cultural Basis of Mathematics." A Swede, a German home-based in Switzerland, a Frenchman, and an American thus addressed the international audience in English, French, and German.

Over the course of the Congress, eighteen more leaders in the field spoke, all in English, on what the Congress organizers deemed among the most important work then under way. Belgium, Sweden, and Switzerland were each represented by one invited hour speaker: historian of science Adolphe Rome,

72. Harald Bohr, "Address," in Kline, "Secretary's Report," pp. 127–134 on p. 130.

	# of Plenary Addresses	% from Host Country	% from the United States
Strasbourg 1920	5	0%	20%
Toronto 1924	8	0%	20%
Bologna 1928	17	41%	12%
Zürich 1932	10	50%	10%
Oslo 1936	20	10%	20%
Cambridge 1950	22	68%	68%

FIGURE C.4. ICM plenary addresses 1920–1950.

Hopf, and Beurling, respectively. France had two: Cartan and Schwartz. England had two: Harold Davenport and William Hodge. The United States was, in 1950, home to the other fifteen or 68% of the invited hour speakers (see Fig. C.4). Six of these—Albert, Morse, Ritt, Whitney, Wiener, and Wilder—were American by birth. Seven—Bochner, Gödel, Hurewicz, von Neumann, Wald, Weil, and Zariski—had left Europe for the United States either just before or during the war. Two more—Shiing-Shen Chern and Shizuo Kakutani—emigrated from East Asia at the end of the 1940s. This line-up reflected the geographical shift in mathematical talent that had taken place not only as a result of the war but also in its aftermath as the United States was increasingly recognized as a mathematical leader. It also reflected the fact that the postwar American mathematical community, in hosting its first international congress, was introducing itself with éclat to the rest, at least symbolically, of the mathematical world. This was, however, not without some controversy.

As early as December 1948, Gordon Whyburn had written to Hassler Whitney, chair of the Committee to Select Foreign Hour Speakers in Topology and Geometry, suggesting for consideration of a number of names, among them, two point-set topologists, Sierpinski and Kuratowski.[73] Whitney's committee, which was rounded out by two émigrés—differential geometer Herbert Busemann, by then at the University of Southern California, and Columbia algebraic topologist Samuel Eilenberg—ultimately took neither suggestion. Whitney and Eilenberg, after all, were topologists in the rival algebraic topological camp. When Whyburn wrote Kline, fellow point-set topologist and ICM Secretary, to complain about the situation in August 1949, the

73. Whyburn to Whitney, 6 December, 1948, Box 5, Folder 27, Whyburn Papers.

latter acknowledged that "I also am not happy over the entire absence of point set topology." At the same time, he informed Whyburn that the invited hour speakers in topology would be Hopf, Hurewicz, and Whitney.[74] It was the second, high-profile snub of point-set topology in then recent memory, the other having been at the Princeton bicentennial "Problems in Mathematics" conference in 1946.

In addition to the invited hour addresses, the Congress, as noted, also incorporated the "conference" innovation, bringing together international leaders in algebra, analysis, applied mathematics, and topology to highlight those areas "in which vigorous advances have been made or are in progress."[75] Albert, as he had for the aborted 1940 Congress, chaired the organizing committee for the Conference in Algebra, which, on the morning of Thursday, 31 August, was the first of the conferences to convene. He was joined by two of his former committee members, Saunders Mac Lane and Richard Brauer, but two new members, Nathan Jacobson and Oscar Zariski, had replaced former members Garrett Birkhoff and Oystein Ore. They took as their "central theme" the theory of rings, which they conceived of in terms of four subsections: "modern aspects of lattice theory and group theory," "recent results on the structure and representation of various types of rings and algebras," "arithmetic aspects of ring and field theory," and algebraic geometry, viewed as "a continuation" of the previous subsection as opposed to as a branch of geometry.[76]

In all, they tapped sixteen speakers. Four were American-born. Six others had been born abroad but had moved to the United States during the war. Six—Hans Zassenhaus of Canada, Frenchman Jean Dieudonné, Tadashi Nakayama of Japan's Nagoya University, the Germans Wolfgang Krull and Max Deuring, and the Russian-born but Paris-based Marc Krasner—hailed from abroad. When Zassenhaus and Deuring were ultimately unable to attend, the Conference in Algebra necessarily became more "American" than its organizers had intended. Indeed, all of the conferences ultimately showcased mathematical work being done on American soil.

The Conference in Analysis was more complex in its organization than that in algebra. For one thing, its organizers—Morse as chair with Ahlfors, Bochner, Evans, Hille, and, initially, Stone—had hoped to engage "the

74. Kline to Whyburn, 26 August, 1949, Box 1, Folder 101, Whyburn Papers.

75. Kline, "Secretary's Report," p. 136.

76. *Proceedings*, 2: 3.

eminent Russian analysts Gelfand, Kolmogoroff, Vinogradow, Lusternik, S. Bernstein and others" in both the conference and as invited hour speakers, but to no avail.[77] As Congress planning proceeded, it became clear that the organizers would have to make their selections under the assumption that their Russian colleagues would not be in attendance.

Another challenge for the analysts was the fact that the Program Committee had specifically requested that they stress "the relations of algebra and topology to analysis." This trend (recall chapters six and ten) of what Saunders Mac Lane termed at the 1946 Princeton bicentennial conference the "moving in" of various mathematical subdisciplines to others was thus highlighted explicitly in 1950 at the Cambridge ICM. In that spirit, the conference was subdivided into three subconferences on "algebraic tendencies in analysis," "analysis and geometry in the large," and "extremal methods and geometric theory of functions of a complex variable" with the second of these further subdivided into "analysis in the large" and "analysis and geometry in the large." The session on the latter, moreover, which brought Jean Leray and André Lichnerowicz from France and Georges de Rham from Switzerland as the three speakers, was joint with the Topology Conference in explicit recognition of the cross-subdisciplinarity increasingly affecting analysis. With five sessions spread over the Thursday, Friday, Monday, and Tuesday of the Congress, the Conference in Analysis had more moving parts than any of the others.

A final complication proved to be the innovation, at Stone's suggestion, of organizing the first topic not as a series of individual talks but rather in terms of panels of experts in particular subareas. As chair of that subconference, Stone was thus in charge of the logistics. When he quit the committee in November 1949, frustrated by the State Department's stance on visas and so by the potential thwarting of an "open" Congress, Hille was left to pick up the pieces and to complete the plan's implementation.

Stone's idea had been that the members of each panel would collaborate in advance on surveying their subarea and on writing a report that would then be delivered by a designated panel spokesperson. The subareas chosen—group representations, topological algebra, spectral theory, applied functional analysis, and ergodic theory—had panels numbering between three and eight specialists from the United States and abroad who had to coordinate the

---

77. *Proceedings*, 2: 103. The quotations that follow in the next paragraph are also on this page.

drafting of reports "intended to give a thorough survey of the state of knowledge in the field under consideration with adequate bibliography, emphasis being placed on the interplay of algebra and analysis typical for the field."[78] When, as Hille explained, "[a] preliminary survey showed the task to be a staggering one" and "it became clear that publishing detailed reports with extensive bibliographies would call for much more space than was available in the Proceedings of the Congress," the subarea spokespersons "were invited to address the Congress on the subject matter of their reports so that their lectures [would] naturally find a place in [the] Proceedings."

For example, Shizuo Kakutani represented the panel on ergodic theory, giving a talk entitled simply "Ergodic Theory" that aimed both to highlight some of the area's fundamental theorems and to showcase those who had proved them. In particular, Kakutani indicated how the work that he and Nagoya University's Kōsaku Yoshida (who was also present in Cambridge) had done in Japan fit in with and complemented work done in the United States—by Halmos, Doob, Stone, von Neumann, Wiener, and Nelson Dunford, among others—as well as elsewhere internationally by analysts like Kolmogoroff in Russia.[79] The young Japanese mathematician was well-positioned to undertake this kind of survey.

Kakutani had earned an M.A. at Tohoku University in 1934 and had taken the teaching assistantship at Osaka University that same year that had allowed him both to continue his mathematical research and to begin his collaboration and ongoing mathematical conversation with Yoshida, who was then also at Osaka.[80] When some of that work caught Hermann Weyl's attention, an invitation was extended to Kakutani for a two-year stay at the Institute beginning in 1940. Although the declaration of war between the United States and Japan in December 1941 cut that visit short, mathematical connections were made not only with Weyl and von Neumann but also with fellow Institute visitors: Warren Ambrose, a 1939 Illinois Ph.D. under Doob and by 1947 a member of the Department of Mathematics at MIT; Paul Halmos, Ambrose's mathematical "brother"; and the peripatetic Paul Erdős. Kakutani returned to Japan to take his Ph.D. and an assistant professorship at Osaka, where he continued to

78. Einar Hille, "Algebraic Tendencies in Analysis," in *Proceedings*, 2: 104–105 on p. 104. The quotations that follow in this paragraph are on pp. 104–105.

79. Shizuo Kakutani, "Ergodic Theory," in *Proceedings*, 2: 128–142.

80. On Kakutani's life, see "In Memoriam: Yale Mathematician Shizuo Kakutani Known for His Work in Functional Analysis and Probability," *Yale Bulletin and Calendar* 33 (27 August, 2004).

produce and publish new results during the war. Invited back to the Institute in 1948, he decided to remain in the United States, accepting the position at Yale in 1949 that he would hold until his retirement in 1982. He thus knew well the movers and shakers in ergodic theory on both sides of the Pacific and had absorbed the ideas of others elsewhere through their published work.

The not unrelated Conference in Applied Mathematics, the smallest of the four with ten speakers, was chaired by von Neumann. He and his co-organizers—Courant, Evans, Prager, Mina Rees, Chicago astronomer and applied mathematician Walter Bartky, and Michigan analyst Ruel Churchill—chose to divide their program up into the three subsections of "random processes in physics and communications," "partial differential equations," and "statistical mechanics" with the second focusing on "problems in fluid dynamics, except for the lecture by W. Prager on the theory of plasticity."[81] Although perhaps not as well represented in Cambridge in 1950 as it had been in Toronto in 1924 or in Bologna in 1928, applied mathematics neverthe-less took its place alongside algebra, analysis, and topology as a mathematical equal for the first time in an American context. Of the ten speakers, two—Göttingen's Werner Heisenberg and Sydney Goldstein of the Israel Institute of Technology—were not members of the American mathematical commu-nity. The other eight—Courant, Feller, von Neumann, Prager, Stoker, Ulam, Wiener, and Claude Shannon then at Bell Labs—underscored the impact that the émigrés had had in establishing a viable community of applied mathemati-cians in the United States.

Of the four conferences, though, the one in topology left a vocal minority unhappy. Chaired by Whitney, its speaker selection committee was rounded out by Deane Montgomery and Norman Steenrod. Not surprisingly, given their individual interests, they chose algebraic topology as the theme of their conference "because of its great growth in recent years" and because of its "increasingly large contact with other fields of mathematics, in geometry, alge-bra, and analysis."[82] Like the analysts, moreover, the topologists highlighted the "moving in" or cross-subdisciplinarity of midcentury mathematics. The talks they listened to focused on the four subthemes of "homology and homo-topy theory," "fiber bundles and obstructions," "differentiable manifolds," and topological "group theory" by seventeen speakers, twelve from the United States and one each from Belgium, England, France, Japan, and Switzerland.

81. *Proceedings*, 2: 261.

82. *Proceedings*, 2: 343. The quotations that follow are also on this page.

This meant that, as with the invited hour addresses at the Congress, point-set topology was ignored in its Topology Conference despite the continued research activity in it in the United States, Poland, and elsewhere. F. Burton Jones, a 1935 Ph.D. under R. L. Moore at Texas who had taken a position at the University of North Carolina, put it this way to his elder academic "brother" and Texas faculty member Renke Lubben: "The Congress would have been a disappointment to you and Dr. Moore. What with all the good work that has been done in point-sets since the last Congress, there was not one conference on point-sets. In fact," he continued, "the only pure point-set papers were placed on the afternoon of the *last day* and poorly attended."[83] It was perhaps an early indication that algebraic topology was supplanting point-set topology as an internationally vibrant research area. If, as has been argued, the latter area's "golden age" was from 1920 to 1960, the 1950 Cambridge ICM may thus be seen, in retrospect, as a reflection that the sun was already setting on the field.[84]

All of this mathematics took place in a festive atmosphere of receptions, banquets, excursions, and other entertainments. On the afternoon of Friday, 1 September, Wellesley College hosted some members of the Congress for tea while the Harvard Observatory hosted others. Boston College arranged for an organ recital on Sunday, the Congress's official day of rest. The internationally celebrated Busch String Quartet, members of which had emigrated to the United States from Europe during World War II, had given a concert on Thursday, 31 August, the Congress's second night. Helen Traubel, American soprano, renowned interpreter of Wagner, and member of the company at the Metropolitan Opera, followed four days later on Monday with a concert that was "enthusiastically received by the audience" and that closed with "a tremendous ovation." For less high-brow tastes, Saturday night had brought "an informal dance" and a "beer party." In all of its facets, the Congress was an event that was "rather enjoyed" by those in attendance.[85]

Things came to a close on Wednesday, 6 September, a day that began with a final plenary session at which "the Congress unanimously voted to accept the gracious invitation of" the Dutch delegation to host the 1954 ICM in the Netherlands and at which Marshall Stone gave an update on the meeting that

83. Jones to Lubben, 19 September, [1950], Box 1, Folder: General Correspondence, Outgoing 1941–1971, Lubben Papers (his emphasis).

84. Koetsier and van Mill, p. 199.

85. Jones to Lubben, 19 September, [1950].

had been held at Columbia on the possible formation of an IMU.[86] Stone "reported that Statutes and By-Laws had been adopted and that these would be submitted to the proper scientific groups in the various national or geographic areas in which there was significant mathematical activity. When," he explained, "a specified number of groups have signified their acceptance of these Statutes and By-Laws, the Union will be declared in existence and a meeting of the General Assembly arranged." That moment came a year later in September 1951, and plans were laid for the first IMU General Assembly, ultimately held in Rome in March 1952. Perhaps not surprisingly given the role he had played in its re-formation, the American Stone was elected IMU President, with Émile Borel of France and Erich Kampke of Germany, the First and Second Vice Presidents, respectively, with Enrico Bompiani of Italy as Secretary, and with Cambridge University's William Hodge, the University of Tokyo's Shokichi Iyanaga, and the University of Copenhagen's Børge Jessen rounding out the Executive Committee. Stone's goal of uniting the mathematicians of the formerly warring nations had at least been partially realized.[87]

With the conclusion of the Cambridge ICM, the American mathematical community had also realized an ambition that some among them had had at least since the ICM met in Strasbourg in 1920. They had not only hosted an ICM, but, as Kline decorously put it, one that "was undoubtedly the largest gathering of persons ever assembled in the history of the world for the discussion of mathematical research."[88] Still, in his view, "the real measure of" the Congress's "success" lay "not in the large number of persons present, but in the excellence of the scientific program and in the contributions which it made to the cause of closer cooperation among scientists and to the cause of international good will." It was with this sense of gravitas that the midcentury leaders of the American mathematical community assumed what they and others perceived as their new international role.

————

The Congress President, seventy-year-old Oswald Veblen, had welcomed its members and guests to a United States, the "colonial period" of which he

86. Kline, "Secretary's Report," p. 145. The quotations that follow are also on this page.
87. Lehto gives a detailed account of the events from the 1950 constitutive meeting at Columbia to the first IMU General Assembly on pp. 86–88 and 91–103.
88. Kline, "Secretary's Report," p. 145. The quotation that follows is also on this page.

FIGURE C.5. Oswald Veblen (1880–1960) (ca. 1950). (Photographer: Kurt Reidemeister. Photo courtesy of the Archives of the Mathematisches Forschungsinstitut Oberwolfach.)

viewed as having ended for mathematics around 1940.[89] For him, 1940, when the first American ICM would have been held, "marked in rather a definitive sense the coming of age of mathematics in the United States." His reasons for this demarcation were several.

From the turn of the twentieth century to 1936 when the Oslo Congress took place, he cited the indisputable fact that "notable growth and transformation had taken place" in the country's mathematical community. In particular, as the first part of this book documented, the 1920s witnessed substantial mathematical output from that community's members as well

89. Oswald Veblen, "Opening Address," in *Proceedings*, 1: 124–125 on p. 124. The quotation that follows is also on this page.

as important advances in the financial support of their research endeavor. A Veblen then in his forties (recall Fig. 1.1) and in close concert with his slightly younger friend and colleague, George Birkhoff, had, with other like-minded mathematicians—Richardson, Bliss, Coolidge, Eisenhart, Hedrick, and others—succeeded in making the case that their field merited the support of individual private benefactors, foundations, and comprehensive scientific entities like the National Research Council. Moreover, they had recognized the importance of participating actively in the mathematical world beyond the United States, and especially in Europe, as well as of attracting mathematicians from abroad to American shores in order to drive home to them the point "that the study of mathematics in America is to be taken very seriously."[90] Their successes had led at least some of them to conclude that they were entering into "a new era in the development of our science."[91]

That new era's dawning, characterized in this book's second part, coincided inauspiciously with the crash of the American stock market in 1929 and the ensuing financial Depression. Although at the time it was by no means a foregone conclusion that the gains of the 1920s could or would be sustained into the 1930s, looking back from the perspective of 1950, Veblen recognized that "[i]mportant discoveries [were] made by American mathematicians. New branches of mathematics were being cultivated and new tendencies in research were showing themselves. Some American universities were receiving students and research workers from overseas, and interchanges of all sorts tended to be more and more on terms of equality." "At the same time," he continued, "mathematics had attained a small but growing amount of recognition from the rest of the American community."[92] It had, after all, witnessed the very public founding in 1933 of the Institute for Advanced Study with its initial emphasis on mathematics and had presented what it viewed as a compelling public face at the "Century of Progress" World's Fair in Chicago that same year.

But, the 1930s and 1940s had, of course, witnessed much more. From his perspective in 1950, Veblen claimed that the United States was "approaching the end of another epoch," namely, "the period during which North America

90. Earle Hedrick to George Birkhoff, 17 July, 1926, HUG 4213.2, Box 6, Folder: Correspondence 1926, H–J, Birkhoff Papers.

91. Roland Richardson to Oswald Veblen, 19 December, 1923, Box 10, Folder: Richardson, R.G.D. 1923, Veblen Papers.

92. Veblen, "Opening Address," p. 124. The quotations that follow in the next two paragraphs are also on this page.

... absorbed so many powerful mathematicians from all over the world." Of course, he, together with other mathematicians like von Neumann and Weyl, had been instrumental in that absorption process, but on the podium on the opening day of the 1950 ICM, he chose to emphasize the "powerful mathematicians" who had been absorbed and to leave unmentioned the majority of the mathematical émigrés—rated with Bs and Cs in an effort at objectivity in the face of adversity—who also started new mathematical lives on this side of the Atlantic. In 1950, Veblen was perhaps too close to events.

He did, however, recognize, as shown in this book's third part, that those émigrés had "enriched" what he termed "the indigenous traditions and tendencies of mathematical thought," even if he also held that the latter had "been radically changed," a position that the evidence presented in chapter ten does not support. In the 1940s, American mathematicians did engage actively in the war effort and, in so doing, did finally come to embrace more fully the applied aspects of the field that had intermittently caused angst among some of its members for a half-century. That certainly *was* a change, if perhaps not a "radical" one. The community also built on its wartime successes, at the same time that it (mostly) learned from its failures, to continue to develop itself in institutions around the country and to take its place as an equal on the postwar playing field of Federally funded science.

The strides made between 1920 and 1950 had in no way been inevitable. Much hard work had gone into building the American mathematical research community, after the close of the First World War, on the foundations that had been laid in the decades around the turn of the twentieth century. There had been successes when there could equally have been failures. There had been opportunities that could have been passed by but were not. Through all three of these decades, there had been a shared conviction that their community could and should be a mover and shaker internationally and not merely a provincial outpost, that it could and should assume world leadership. The 1950 Cambridge ICM symbolized for American mathematicians the latter conviction's realization, whatever they might have thought being the world leader meant and whatever implications they might have thought it had for their future. It marked the beginning of yet another "new era in the development" of their science.

REFERENCES

Archival Sources

American Mathematical Society Papers, John Hay Library, Brown University, Providence, RI.

George D. Birkhoff Papers, Harvard University Archives, Pusey Library, Cambridge, MA.

Fond Élie Cartan 38J, Archives de l'Académie des Sciences, Paris, France.

University of Chicago, Department of Mathematics Records, Special Collections Research Center, University of Chicago.

Max Dehn Papers, 93-476, Archives of American Mathematics, Dolph Briscoe Center for American History, The university of Texas at Austin.

Division of Applied Mathematics Papers, Office Files of the Division of Applied Mathematics, John Hay Library, Brown University.

Duke Mathematical Journal Records, 1927–1934, Archives of American Mathematics, Dolph Briscoe Center for American History, The University of Texas at Austin.

William L. Duren, Jr. Papers, Archives of American Mathematics, Dolph Briscoe Center for American History, The University of Texas at Austin.

Emergency Committee in Aid of Displaced German (later Foreign) Scholars, New York Public Library, New York, NY.

Robert Hutchins Administration, Records 1892–1951, Special Collections Research Center, University of Chicago.

R. G. Lubben Papers, 86-9, Archives of American Mathematics, Dolph Briscoe Center for American History, The University of Texas at Austin.

Mathematical Association of America Records, Archives of American Mathematics, Dolph Briscoe Center for American History, The University of Texas at Austin.

Edward J. "Jimmy" McShane Autobiographical Notes, Private Collection of his daughter and son-in-law, Jennifer and Thann Ward.

Herman Louis Meyer, Jr. Papers, Private Collection of his daughter, Elizabeth Meyer.

University of Michigan, Department of Mathematics Records: 1913–1981, Bentley Historical Library, University of Michigan.

Robert L. Moore Papers, 1875, 1891–1975, Archives of American Mathematics, Dolph Briscoe Center for American History, The University of Texas at Austin.

G. Baley Price Papers 1932–1993, Archives of American Mathematics, Dolph Briscoe Center for American History, The University of Texas at Austin.

Roland G. D. Richardson Papers, 1922–1942, John Hay Library, Brown University

Julius Rosenwald Fund Archives, Special Collections, John Hope and Aurelia E. Franklin Library, Fisk University.

J. Barkley Rosser Papers, 91-1, Archives of American Mathematics, Dolph Briscoe Center for American History, The University of Texas at Austin.

Marshall H. Stone Papers, John Hay Library, Brown University.

Harry S. Vandiver Papers, Archives of American Mathematics, Dolph Briscoe Center for American History, The University of Texas at Austin.

Papers of Oswald Veblen, Library of Congress Manuscripts Division, Washington, D.C.

Gordon T. Whyburn Papers, circa 1930–1965, Special Collections, University of Virginia Library, Charlottesville, VA.

Raymond L. Wilder Papers, 1914–1982, Archives of American Mathematics, Dolph Briscoe Center for American History, The University of Texas at Austin.

Web Sources and Databases

American Institute of Physics: https://www.aip.org/aip/history

Barber, Gordon, et al. *Ballisticians in War and Peace.* Vol. 1. *A History of the United States Army Ballistic Research Laboratories, 1914–1958.* Aberdeen, MD: Aberdeen Proving Ground, [1956]: https://apps.dtic.mil/dtic/tr/fulltext/u2/a300523.pdf

Bernat, Henry. "Overview of the Army Meteorological Program and the Program at Reed College": https://www.reed.edu/reed_magazine/autumn2007/features/reed_at_war/AMP%20memoir%20by%20Bernat.pdf

Biographies of Women Mathematicians: https://www.agnesscott.edu/lriddle/women/women.htm

California Digital Library: https://cdlib.org

Celebratio Mathematica: https://celebratio.org

Dantzig, George. "Impact of Linear Programming on Computer Development." Technical Report SOL 87-5. Systems Optimization Laboratory. Department of Operations Research. Stanford University. June 1985: https://apps.dtic.mil/dtic/tr/fulltext/u2/a157659.pdf

———. "Origins of the Simplex Method." Technical Report SOL 87-5. Systems Optimization Laboratory. Department of Operations Research. Stanford University, May 1987. https://apps.dtic.mil/dtic/tr/fulltext/u2/a182708.pdf

Dickinson, Elizabeth R. "The Production of Firing Tables for Cannon Artillery." Report No. 1371. U.S. Army Materiel Command. Ballistic Research Laboratories. Aberdeen Proving Ground. November 1967: https://apps.dtic.mil/dtic/tr/fulltext/u2/826735.pdf.

Garbe, Evelyn R. (obituary): https://www.legacy.com/obituaries/capitalgazette/obituary.aspx?n=evelyn-garbe&pid=145603992

Goldstine, Adele. "Report on the ENIAC (Electronic Numerical Integrator and Computer)." Technical Report 1. Moore School of Electrical Engineering. Philadelphia, Pennsylvania. 1 June, 1946: https://ftp.arl.army.mil/ mike/comphist/46eniac-report/index.html.

Green, Judy and LaDuke, Jeanne. *Supplementary Material for Pioneering Women in American Mathematics: The Pre-1940 PhD's* http://www.ams.org/publications/authors/books/postpub/hmath-34-PioneeringWomen.pdf

John Simon Guggenheim Foundation, especially http://www.gf.org/fellows

Hestenes, Magnus; Redheffer, Raymond; and Green, John W. "Ivan Stephan Sokolnikoff, Mathematics: Los Angeles (1901–1976), Professor Emeritus." http://texts.cdlib.org/view? docId=hb4q2nb2px&doc.view=frames&chunk.id=div00043&toc.depth=1&toc.id=.

History of Louisiana State University's Department of Mathematics: https://www.math.lsu.edu/MathDepartmentHistory

History of the Ohio State University's Department of Mathematics: https://math.osu.edu/about-us/history

History of the University of Notre Dame Department of Mathematics: https://math.nd.edu/about/history

Kim, Claire. "Math Derived, Math Applied: The Establishment of Brown University's Division of Applied Mathematics, 1940–1946": https://www.brown.edu/academics/applied-mathematics/sites/brown.edu.academics.applied-mathematics/files/uploads/Clare%20Kim%20-%20History%20Thesis.pdf

MacTutor History of Mathematics Archive: http://mathshistory.st-andrews.ac.uk

Mathematics Genealogy Project: http://www.genealogy.ams.org

MathSciNet, an online database of mathematical publications: https://mathscinet-ams-org

McShane, Edward J. "Corrections of Elements of Bombing Tables for Effects of Height of Target." Report No. 374. Aberdeen Proving Ground. 2 July, 1943.": https://apps.dtic.mil/dtic/tr/fulltext/u2/733887.pdf

Measuring Worth, a site for making historical monetary conversions: https://www.measuringworth.com

Merjos, Anna (obituary): at https://prabook.com/web.anna.merjos/198348

Miller, Herman Lunden. "First-Hand: My Experience in the Army Air Force, 1943 to 1946": https://ethw.org/First-Hand:My_Experience_in_the_Army_Air_Force,_1943_to_1946.

Princeton Mathematics Community in the 1930s: An Oral History Project: https://findingaids.princeton.edu/collections/AC057/c48

Princeton University Bicentennial Celebration: https://blogs.princeton.edu/mudd/2011/03/princetons-bicentennial-charter-day-october-19-1946/ https://blogs.princeton.edu/reel mudd/2010/07/past-present-and-future-us-presidents-at-bicentennial-celebration-1947/

ProQuest Historical Newspapers: *The New York Times* https://www.proquest.com/products-services/pq-hist-news.html

Public Law 507–81st Congress, Chapter 171-2D Session https://www.nsf.gov/about/history/legislation.pdf

Rice University Pamphlets–Rice University Studies: https://scholarship.rice.edu/handle/1911/8328

Rowland, R. E. "The Premeteorology Program of the Army Air Force, 1942–1945": http://www.rerowland.com/premet.html

Schiffer, Menahem M.; Osserman, Robert; and Samelson, Hans. "Memorial Resolution: Stefan Bergman (1895–1977)." Stanford University Faculty Senate Records. Stanford University Library: https://stacks.stanford.edu/file/druid:fq999qp2915/SC0193_Memorial Resolution_BergmanS.pdf.

Summary Technical Report of the Applied Mathematics Panel. Washington, D.C.: NDRC, 1946: Vol. 1: Ivan Sokolnikoff, Ed. *Mathematical Studies Relating to Military Physical*

Sciences: https://apps.dtic.mil/dtic/tr/fulltext/u2/221604.pdf Vol. 2: Mina Rees, Ed. *Analytical Studies in Aerial Warfare*: https://www.loc.gov/collections/selected-digitized-books/?dates=1940/1949&fa=segmentof:dcmsiabooks.analyticalstudie02bush/%7C subject: military+research&sb=shelf-id&st=gallery Vol. 3: Samuel Wilks, Ed. *Probability and Statistical Studies in Warfare Analysis*: https://apps.dtic.mil/dtic/tr/fulltext/u2/b809137.pdf

Printed Sources

Abir-Am, Pnina and Elliott, Clarke A., Ed. *Commemorative Practices in Science: Historical Perspectives on the Politics of Collective Memory*. Osiris 14 (1999).

"Academic Unemployment." *Bulletin of the American Association of University Professors (1915–1955)* 19 (October 1933), 354–355.

Aitchison, Beatrice. "Concerning Regular Accessibility." *Fundamenta Mathematicae* 20 (1933), 117–125.

Albers, Donald J.; Alexanderson, Gerald L.; and Reid, Constance. *International Mathematical Congresses: An Illustrated History*. New York: Springer-Verlag, 1987.

Albert, A. Adrian. "The Chicago Conference and Seminar on Algebra." *Bulletin of the American Mathematical Society* 44 (1938), 756–757.

―――――. "A Determination of All Normal Division Algebras in Sixteen Units." *Transactions of the American Mathematical Society* 31 (1929), 253–260.

―――――. "New Results in the Theory of Normal Division Algebras." *Transactions of the American Mathematical Society* 32 (1930), 171–195.

―――――. "The Non-Existence of Pure Riemann Matrices with Normal Multiplication Algebras of Order Sixteen." *Annals of Mathematics* 31 (1930), 375–380.

―――――. "Normal Division Algebras of Degree Four over an Algebraic Field." *Transactions of the American Mathematical Society* (1932), 363–372.

―――――. "On a Certain Algebra of Quantum Mechanics." *Annals of Mathematics* 35 (1934), 65–73.

―――――. "On the Construction of Riemann Matrices I." *Annals of Mathematics* 35 (1934), 1–28.

―――――. "On the Construction of Riemann Matrices II." *Annals of Mathematics* 36 (1935), 376–394.

―――――. "On Jordan Algebras of Linear Transformations." *Transactions of the American Mathematical Society* 59 (1946), 524–555.

―――――. "A Solution of the Principal Problem in the Theory of Riemann Matrices." *Annals of Mathematics* 35 (1934), 500–515.

―――――. *Structure of Algebras*. AMS Colloquium Publications. Vol. 24. New York: American Mathematical Society, 1939.

―――――. "The Structure of Pure Riemann Matrices with Non-Commutative Multiplication Algebras." *Rendiconti del Circolo matematico di Palermo* 55 (1931), 1–59.

―――――. "A Structure Theory of Jordan Algebras." *Annals of Mathematics* 48 (1947), 546–567.

―――――. "A Theory of Power-associative Commutative Algebras." *Transactions of the American Mathematical Society* 69 (1950), 503–527.

Albert, A. Adrian and Hasse, Helmut. "A Determination of All Normal Division Algebras over an Algebraic Number Field." *Transactions of the American Mathematical Society* 34 (1932), 722–726.

Albert, Nancy E. A^3 *& His Algebra: How a Boy from Chicago's West Side Became a Force in American Mathematics.* New York: iUniverse, 2005.

Alexander, James. "Combinatorial Analysis Situs." *Transactions of the American Mathematical Society* 28 (1926), 301–329.

──────. "A Combinatorial Theory of Complexes," *Annals of Mathematics* 31 (1930), 292–320.

──────. "An Example of a Simply Connected Surface Bounding a Region Which is Not Simply Connected," *Proceedings of the National Academy of Sciences of the United States of America* 10 (1924), 8–10.

──────. "On the Chains of a Complex and Their Duals." *Proceedings of the National Academy of Sciences of the United States of America* 21 (1935), 509–511.

──────. "On the Connectivity Ring of an Abstract Space." *Annals of Mathematics* 37 (1936), 698–708.

──────. "On the Ring of a Compact Metric Space." *Proceedings of the National Academy of Sciences of the United States of America* 21 (1935), 511–512.

──────. "A Proof and Extension of the Jordan-Brouwer Theorem." *Transactions of the American Mathematical Society* 23 (1922), 333–349.

──────. "Some Problems in Topology." In *Verhandlungen des Internationalen Mathematiker-Kongresses Zürich 1932.* 2 Vols. Zürich and Leipzig: Orell Füssli, 1932), 1: 249–257.

──────. "Topological Invariants of Knots and Links." *Transactions of the American Mathematical Society* 30 (1928), 275–306.

Alexandroff, Paul. "Untersuchungen über Gestalt und Lage abgeschlossener Mengen beliebiger Dimension." *Annals of Mathematics* 30 (1928), 101–187.

American Mathematical Society: List of Officers and Members, January 1916. New York: American Mathematical Society, 1916 (also *Bulletin of the American Mathematical Society* 22 (1916)).

American Mathematical Society: List of Officers and Members, 1923–1924. New York: American Mathematical Society, 1924 (also *Bulletin of the American Mathematical Society* 30 (1924)).

American Mathematical Society: List of Officers and Members, 1929–1930. New York: American Mathematical Society, 1930 (also *Bulletin of the American Mathematical Society* 36 (1930)).

Apushkinskaya, Darya E., Nazarov, Alexander I., and Sinkevich, Galina. "In Search of Shadows: The First Topological Conference, Moscow 1935." *The Mathematical Intelligencer* 41 (2019), 37–42.

Archibald, Raymond C. "R.G.D. Richardson 1878–1949." *Bulletin of the American Mathematical Society* 56 (1950), 256–265.

──────. "History of the American Mathematical Society 1888–1938." *Bulletin of the American Mathematical Society* 45 (1939), 31–46.

──────. *A Semicentennial History of the American Mathematical Society, 1888–1938.* New York: American Mathematical Society, 1938; Reprint Ed. New York: Arno Press, 1980.

Archibald, Thomas; Dumbaugh, Della; and Kent, Deborah. "A Mobilized Community: Mathematicians in the United States during World War I." In *The War of Guns and Mathematics:*

Mathematical Practices and Communities in France and Its Western Allies around World War I. Ed. David Aubin and Catherine Goldstein, pp. 229–271.

Artin, Emil. "Braids and Permutations, *Annals of Mathematics* 48 (1947), 643–649.

————. *Galois Theory.* Notre Dame Mathematical Lectures. Vol. 2. Notre Dame: Notre Dame University Press, 1942.

————. "Theorie der Zöpfe." *Abhandlungen aus dem Mathematischen Seminar der Universität Hamburg* 4 (1926), 47–72.

————. "Theory of Braids." *Annals of Mathematics* 48 (1947), 101–126.

Askey, Richard and Nevai, Paul. "Gábor Szegő: 1895–1985." *The Mathematical Intelligencer* 18 (3) (1996), 10–22.

Aspray, William. "The Emergence of Princeton as a World Center for Mathematical Research, 1896–1939." In *History and Philosophy of Modern Mathematics.* Ed. William Aspray and Philip Kitcher, pp. 346–366.

————. *John von Neumann and the Origins of Modern Computing.* Cambridge: The MIT Press, 1990.

————. "Oswald Veblen and the Origins of Mathematical Logic at Princeton." In *Perspectives on the History of Mathematical Logic.* Ed. Thomas Drucker, pp. 346–366.

Aspray, William and Kitcher, Philip, Ed. *History and Philosophy of Modern Mathematics.* Minneapolis: University of Minnesota Press, 1988.

Assmus, Alexi. "The Creation of Postdoctoral Fellowships and the Siting of American Scientific Research." *Minerva: A Review of Science, Learning, and Policy* 31 (1993), 151–183.

Atti del Congresso internazionale dei matematici. 6 Vols. Bologna: Nicola Zanichelli Editore, 1929.

Aubin, David. "George David Birkhoff, *Dynamical Systems* (1927)." In *Landmark Writings in Western Mathematics, 1640–1940.* Ed. Ivor Grattan-Guinness, pp. 871–881.

Aubin, David and Goldstein, Catherine, Ed. *The War of Guns and Mathematics: Mathematical Practices and Communities in France and Its Western Allies around World War I.* HMATH. Vol. 42. Providence: American Mathematical Society, 2014.

Aull, Charles E. and Lowen, Robert, Ed. *Handbook of the History of General Topology.* 3 Vols. Dordrecht: Kluwer Academic Publishers, 1997–2001.

Austin, Allan W. *Quaker Brotherhood: Interrracial Activism and the American Friends Service Committee, 1917–1950.* Urbana: University of Illinois Press, 2012.

Ayres, William L. "The Annual Meeting of the Society." *Bulletin of the American Mathematical Society* 47 (1941), 175–187.

————. "The September Meeting in Madison." *Bulletin of the American Mathematical Society* 45 (1939), 801–811.

————. "The Summer Meeting in Chicago." *Bulletin of the American Mathematical Society* 47 (1941), 832–845.

Baer, Reinhold."The Applicability of Lattice Theory to Group Theory." *Bulletin of the American Mathematical Society* 44 (1938), 817–820.

————. "The Significance of the System of Subgroups for the Structure of the Group." *American Journal of Mathematics* 61 (1939), 1–44.

Baez, Joan C. et al. "Irving Ezra Segal (1918–1998)." *Notices of the American Mathematical Society* 46 (1999), 659–668.

Banach, Stefan. *Théorie des opérations linéaires*. Warszawa: Z subwencji Funduszu kultury narodowej, 1932.

Banach, Stefan and Steinhaus, Hugo. "Sur le principe de la condensation de singularités." *Fundamenta mathematicae* 9 (1927), 50–61.

Barany, Michael J. "Distributions in Postwar Mathematics." Unpublished doctoral dissertation, Princeton University, 2016.

———. "The Myth and the Medal." *Notices of the American Mathematical Society* 62 (January 2015), 15–20.

———. "Remunerative Combinatorics: Mathematicians and Their Sponsors in the Midtwentieth Century." In *Mathematical Cultures: The LondonMeetings 2012–2014*, ed. Brendan Larmor, pp. 329–346.

———. "The World War II Origins of Mathematics Awareness." *Notices of the American Mathematical Society* 64 (2017), 363–367.

Barrows, Albert L. "General Organization and Activities." In *A History of the National Research Council, 1919–1933*. Washington, D.C.: National Research Council, 1933, pp. 7–11.

Bartlow, Thomas L. and Zitarelli, David E. "Who Was Miss Mullikin?" *American Mathematical Monthly* 116 (2009), 99–114.

Barton, Henry A. "American Institute of Physics: Report of Director for 1943." *Review of Scientific Instruments* 15 (1944), 130–133.

Bass, Hyman; Cartan, Henri; Freyd, Peter; Heller, Alex; and Mac Lane, Saunders. "Samuel Eilenberg (1913–1998)." *Notices of the American Mathematical Society* 45 (1998), 1344–1352.

Bass, Hyman and Lam, T. Y. "Irving Kaplansky (1917–2007)." *Notices of the American Mathematical Society* 54 (2007), 1477–1493.

Bateman, Harry. "The Control of Elastic Fluids." *Bulletin of the American Mathematical Society* 51 (1945), 601–646.

———. *Partial Differential Equations of Mathematical Physics*. Cambridge: Cambridge University Press and New York: The Macmillan Co., 1932.

Batterson, Steve. *American Mathematics 1890–1913: Catching Up to Europe*. Washington, D.C.: MAA Press, 2017.

———. *Pursuit of Genius: Flexner, Einstein, and the Early Faculty at the Institute for Advanced Study*. Wellesley: A K Peters, Ltd., 2006.

———. "The Vision, Insight, and Influence of Oswald Veblen, *Notices of the American Mathematical Society* 54 (2007), 606–618.

Beaulieu, Liliane. "A Parisian Café and Ten Proto-Bourbaki Meetings (1934–1935)." *The Mathematical Intelligencer* 15 (1993), 27–35.

Bell, Eric T. *Algebraic Arithmetic*. AMS Colloquium Publications. Vol. 7. New York: American Mathematical Society, 1927.

———. "Arithmetical Paraphrases," *Transactions of the American Mathematical Society* 22 (1921), 1–30, 198–219.

———. "Fifty Years of Algebra in America, 1888–1938." In *Semicentennial Addresses of the American Mathematical Society*. New York: American Mathematical Society, 1938, pp. 1–34.

———. *The Handmaiden of the Sciences*. Baltimore: Williams and Wilkins Company, 1937.

———. *Men of Mathematics*. New York: Simon and Schuster, 1937.

————. *The Queen of the Sciences*. Baltimore: Williams and Wilkins Company, 1931.

Bender, Harry. "Sylow Subgroups in the Group of Isomorphisms of Prime Power Abelian Groups." *American Journal of Mathematics* 45 (1923), 223–250.

Benkart, Georgia; Kaplansky, Irving: McCrimmon, Kevin; Saltman, David; and Seligman, George. "Nathan Jacobson (1910–1999)." *Notices of the American Mathematical Society* 47 (2000), 1061–1071.

Bergman, Stefan. *The Kernel Function and Conformal Mapping*. Mathematical Surveys. Vol. 5. New York: American Mathematical Society, 1950.

————. "Solutions of Linear Partial Differential Equations of the Fourth Order." *Duke Mathematical Journal* 11 (1944), 617–649.

Berkovitz, Leonard D. and Fleming, Wendell H. "Edward James McShane 1904–1989." *Biographical Memoirs*. Vol. 80. Washington, D.C.: National Academy of Sciences, 2001, pp. 227–239.

Berndt, Bruce. "Hans Rademacher (1892–1969)." *Acta Arithmetica* 61 (1992), 209–231.

Bernstein, Benjamin A. "Complete Sets of Representations of Two-Element Algebras." *Bulletin of the American Mathematical Society* 30 (1924), 24–30.

————. "The June Meeting in Berkeley." *Bulletin of the American Mathematical Society* 35 (1929), 593–606.

————. "Operations with Respect to Which the Elements of a Boolean Algebra Form a Group." *Transactions of the American Mathematical Society* 26 (1924), 171–175.

Bers, Lipman. "Editor's Introduction." In *Charles Loewner: Collected Papers*. Ed. Lipman Bers. Boston: Birkhäuser Boston, Inc., 1988, pp. vii–x.

————. "The Migration of European Mathematicians to America." In *A Century of Mathematics in America*. Ed. Peter L. Duren et al. 1: 231–243.

"Bibliotheca Mathematica." *Science* 68 (1928), 474–475.

Bing, R. H. "The Kline Sphere Characterization Problem." *Bulletin of the American Mathematical Society* 52 (1946), 644–653.

Birkhoff, Garrett. "Applications of Lattice Theory." *Proceedings of the Cambridge Philosophical Society* 30 (1934), 115–122.

————. *Lattice Theory*. AMS Colloquium Publications. Vol. 25. New York: American Mathematical Society, 1940.

————. "Lattices and Their Applications." *Bulletin of the American Mathematical Society* 44 (1938), 793–800.

————. "Mathematics at Harvard, 1836–1944." In *A Century of Mathematics in America*. Ed. Peter L. Duren et al. 2: 3–58.

————. "On the Combination of Subalgebras." *Proceedings of the Cambridge Philosophical Society* 29 (1933), 441–464.

Birkhoff, Garrett and Mac Lane, Saunders. *Survey of Modern Algebra*. New York: Macmillan, 1941.

————. "*A Survey of Modern Algebra*: The Fiftieth Anniversary of Its Publication," *The Mathematical Intelligencer* 14 (1992), 26–31.

Birkhoff, George D. *Aesthetic Measure*. Cambridge: Harvard University Press, 1933.

————. "Boundary Value and Expansion Problems of Ordinary Linear Differential Equations." *Transactions of the American Mathematical Society* 9 (1908), 373–395.

————. *Dynamical Systems*. AMS Colloquium Publications. Vol. 9. New York: American Mathematical Society, 1927.

————. "Dynamical Systems with Two Degrees of Freedom." *Transactions of the American Mathematical Society* 18 (1917), 199–300.

————. "Fifty Years of American Mathematics." In *Semicentennial Addresses of the American Mathematical Society*. New York: American Mathematical Society, 1938, pp. 270–315.

————. "General Theory of Linear Difference Equations." *Transactions of the American Mathematical Society* 12 (1911), 243–284.

————. "The Mathematical Work of Oliver Dimon Kellogg." *Bulletin of the American Mathematical Society* 39 (1933), 171–177.

————. "On the Asymptotic Character of the Solutions of Certain Linear Differential Equations Containing a Parameter." *Transactions of the American Mathematical Society* 9 (1908), 219–231.

————. "Proof of the Ergodic Theorem." *Proceedings of the National Academy of Sciences of the United States of America* 17 (1931), 656–660.

————. "Proof of Poincaré's Geometric Theorem." *Transactions of the American Mathematical Society* 14 (1913), 14–22.

————. "Quantum Mechanics and Asymptotic Series." *Bulletin of the American Mathematical Society* 39 (1933), 681–700.

————. "Quelques éléments mathématiques de l'art." In *Atti del Congresso internazionale dei matematici*. 6 Vols. Bologna: Nicola Zanichelli Editore, 1929, 1: 315–333.

————. "What Is the Ergodic Theorem." *AMM* 49 (1942), 222–226.

Birkhoff, George D.; Bliss, Gilbert A.; and Hedrick, Earle. "The Visiting Lectureship of the American Mathematical Society." *Bulletin of the American Mathematical Society* 34 (1928), 22.

Birkhoff, George D.; Bliss, Gilbert A.; and Veblen, Oswald. "The Colloquium Publications." *Bulletin of the American Mathematical Society* 32 (1926), 100.

Birkhoff, George D. and Hestenes, Magnus. "Generalized Minimax Principle in the Calculus of Variations." *Duke Mathematics Journal* 1 (1935), 413–432.

Birkhoff, George D. and Kellogg, Oliver D. "Invariant Points in Function Space." *Transactions of the American Mathematical Society* 23 (1922), 96–115.

Birkhoff, George D. and Koopman, Bernard. "Recent Contributions to the Ergodic Theory." *Proceedings of the National Academy of Sciences of the United States of America* 18 (1932), 279–282.

Birkhoff, George D. and Langer, Rudolph E. "The Boundary Problems and Developments Associated with a System of Linear Differential Equations of the First Order." *Proceedings of the American Academy of Arts and Sciences* 58 (1923), 49–128.

Blackwell, David. "Idempotent Markoff Chains." *Annals of Mathematics* 43 (1942), 560–567.

Bliss, Gilbert A. *Algebraic Functions*. AMS Colloquium Publications. Vol. 16. New York: American Mathematical Society, 1933.

————. "Algebraic Functions and Their Divisors." *Annals of Mathematics* 26 (1924), 95–124.

————. *The Calculus of Variations*. Carus Mathematical Monographs. No. 1. Chicago: The Open Court Publishing Company, 1925.

————. "The Evolution of Problems of the Calculus of Variations." *American Mathematical Monthly* 43 (1936), 598–609.

————. "A Letter from the President." *Bulletin of the American Mathematical Society* 28 (1922), 16.

————. *Mathematics for Exterior Ballistics.* New York: J. Wiley and Sons, Inc. and London: Chapman and Hall, Ltd., 1944.

————. "The Problem of Lagrange in the Calculus of Variations." *American Journal of Mathematics* 52 (1930), 674–744.

————. "Some Recent Developments in the Calculus of Variations." *Bulletin of the American Mathematical Society* 26 (1920), 343–361.

————. "The Transformation of Clebsch in the Calculus of Variations." *Proceedings of the International Mathematical Congress Held in Toronto, August 11–16, 1924.* Ed. John C. Fields. 2 Vols. Toronto: University of Toronto Press, 1928, 1: 589–603.

Bochner, Salomon. Review of "*Functional Analysis and Semi-groups.* By Einar Hille. American Mathematical Society Colloquium Publications, vol. 31. New York, American Mathematical Society, 1948, 12+528 pp. $7.50." *Bulletin of the American Mathematical Society* 55 (1949), 528–533.

————. "Vector Fields and Ricci Curvature." *Bulletin of the American Mathematical Society* 52 (1946), 776–797.

Bochner, Salomon and Martin, William Ted. *Several Complex Variables.* Princeton: Princeton University Press, 1948.

Bochner, Salomon and Phillips, Ralph. "Absolutely Convergent Fourier Expansions for Non-commutative Normed Rings." *Annals of Mathematics* 43 (1942), 409–418.

Bonner, Thomas. *Iconoclast: Abraham Flexner and a Life in Learning.* Baltimore: Johns Hopkins University Press, 2002.

Boone, William. "The Word Problem." *Proceedings of the National Academy of Sciences of the United States of America* 44 (1958), 1061–1065.

Borel, Armand. "The School of Mathematics at the Institute for Advanced Study." In *A Century of Mathematics in America.* Ed. Peter L. Duren et al., 3: 119–147.

Borel, Émile. "Aggregates of Zero Measure." *Rice Institute Pamphlets–Rice University Studies* 4 (1917), 1–21.

————. "Molecular Theories and Mathematics." *Rice Institute Pamphlets–Rice University Studies* 1 (1915), 163–193.

————. "Monogenic Uniform Non-analytic Functions." *Rice Institute Pamphlets–Rice University Studies* 4 (1917), 22–52.

Bott, Raoul. "Marston Morse and His Mathematical Works." *Bulletin of the American Mathematical Society* 3 (1980), 907–950.

Boyer, Carl B. "The First Twenty-five Years." In *The Mathematical Association of America: Its First Fifty Years.* Ed. Kenneth O. May, pp. 24–54.

Brahana, Henry R. "Certain Perfect Groups Generated by Two Operators of Orders Two and Three." *American Journal of Mathematics* 50 (1928), 345–356.

————. "George Abram Miller." *Biographical Memoirs.* Vol. 30. Washington, D.C.: National Academy of Sciences, 1957, pp. 257–312.

Brauer, Alfred. "Gedankrede auf Issai Schur." In Schur, Issai. *Gesammelte Abhandlungen*. Ed. Alfred Brauer and Hans Rohrbach. 3 Vols. Berlin: Springer-Verlag, 1973, 1: v-xiii.

_____. "On the Density of the Sum of Sets of Positive Integers II." *Annals of Mathematics* 42 (1941), 959–988.

_____. "Über die Dichte der Summe von Mengen positiver ganzer Zahlen I." *Annals of Mathematics* 39 (1938), 322–340.

Brauer, Richard. "Emil Artin." *Bulletin of the American Mathematical Society* 73 (1967), 27–43.

_____. "On Artin's *L*-Series with General Group Characters." *Annals of Mathematics* 48 (1947), 502–514.

_____. "On Modular and p-adic Representations of Algebras" (*Proceedings of the National Academy of Sciences of the United States of America* 25 (1939), 252–258.

Brauer, Richard; Hasse, Helmut; and Noether, Emmy. "Beweis eines Hauptsatzes in der Theorie der Algebren." *Journal für die reine und angewandte Mathematik* 167 (1931), 399–404.

Brauer, Richard and Nesbitt, Cecil. "On the Modular Representation of Groups of Finite Order I." *University of Toronto Studies* 4 (1937).

_____. "On the Regular Representations of Algebras." *Proceedings of the National Academy of Sciences of the United States of America* 23 (1937), 236–240.

Brelot, Marcel. "Norbert Wiener and Potential Theory." *Bulletin of the American Mathematical Society* 72 (1966), 39–41.

Brigaglia, Aldo and Ciliberto, Ciro. *Italian Algebraic Geometry between the Two World Wars*. Queen's Papers in Pure and Applied Mathematics. Vol. 100. Kingston: Queen's University Press, 1995.

Browder, Felix. "The Stone Age of Mathematics on the Midway." *The University of Chicago Magazine* 69 (Sept. 1976), 28–29; Reprinted in *A Century of Mathematics in America*. Ed. Peter L. Duren et al. 2: 191–193.

Brown, Ernest W. "Elements of the Theory of Resonances Illustrated by the Motion of a Pendulum." *Rice Institute Pamphlets–Rice University Studies* 19 (1932), 1–60.

_____. "The Relation of Mathematics to the Natural Sciences." *Bulletin of the American Mathematical Society* 23 (1917), 213–230.

Brown, J. Douglas. "The Princeton Bicentennial Conferences: A Retrospect." *The Journal of Higher Education* 19 (1948), 55–59.

Brown, Robert F. "Fixed Point Theory." In *History of Topology*. Ed. Ioan M. James, pp. 271–299.

Bruce, Philip A. *History of the University of Virginia: 1819-1919*. 5 Vols. New York: The Macmillian Company, 1920-1922.

Bruck, Richard H. "The Annual Meeting of the Society." *Bulletin of the American Mathematical Society* 52 (1946), 35–47.

_____. "The Summer Meeting in Madison." *Bulletin of the American Mathematical Society* 54 (1948), 1042–1086.

Bush, Vannevar. *Operational Circuit Analysis*. New York: Wiley, 1929.

_____. *Pieces of the Action*. New York: William Morrow and Company, Inc., 1970.

"Cabot Fellowship to Prof. Royce." *The Harvard Crimson* (15 April, 1911).

Cairns, William D. "The Mathematical Association of America." *American Mathematical Monthly* 23 (1916), 1–6.

_____. "The Sixteenth Annual Meeting of the Association." *American Mathematical Monthly* 39 (1932), 123–134.

Cameron, Robert H. "Some Introductory Exercises in the Manipulation of Fourier Transforms." *National Mathematics Magazine* 15 (1941), 331–356.

Carathéodory, Constantin. "On Dirichlet's Problem." *American Journal of Mathematics* 59 (1937), 709–731.

Cardozier, V. Ray. *Colleges and Universities in World War II*. Westport: Praeger, 1993.

Carmichael, Leonard. "The National Roster of Scientific and Specialized Personnel: A Progress Report." *Science* 93 (1941), 217–219.

Carmichael, Richard D. "Alfred T. Brauer: Teacher, Mathematician, and Developer of Libraries." *The Journal of the Elisha Mitchell Scientific Society* 103 (1986), 88–106.

Carmichael, Robert D. "On the Expansion of Certain Analytic Functions in Series." *Annals of Mathematics* 22 (1920), 29–34.

Cartan, Élie. *La géométrie des espaces de Riemann*. Paris: Gauthiers-Villars, 1925.

_____. "Sur les variétés à connexion affine et la théorie de la relativité généralisée." *Annales de l'École normale supérieure* 40 (1923), 325–412.

_____. "La théorie des groupes et les recherches récentes de géométrie différentielle." *Proceedings of the International Mathematical Congress Held in Toronto, August 11–16, 1924*. Ed. John Fields, 1: 85–94.

Case, Bettye Anne. *A Century of Mathematical Meetings*. Providence: American Mathematical Society, 1996.

Cattell, James McKeen, Ed. *American Men of Science*. 2d Ed. Lancaster: The Science Press, 1910.

Cattell, Jaques, Ed. *American Men of Science: A Biographical Dictionary*. 7th Ed. Lancaster: The Science Press, 1944.

_____. *American Men of Science: A Biographical Dictionary*. 9th Ed. Vol. 1 "Physical Sciences." Lancaster: The Science Press and New York: R. R. Bowker Company, 1955.

Čech, Eduard. "Multiplications on a Complex." *Annals of Mathematics* 37 (1936), 681–697.

Chern, Shiing-Shen. "On Integral Geometry in Klein Spaces." *Annals of Mathematics* 43 (1942), 178–189.

Chevalley, Claude. "La Théorie du corps de classes." *Annals of Mathematics* 41 (1940), 394–418.

Chittenden, Edward W. "Infinite Developments and the Composition Property $(K_{12}B_1)$ in General Analysis." *Rendiconti del Circolo matematico di Palermo* 39 (1915), 81–108.

Chung, Kai-Lai. "Probability and Doob." *American Mathematical Monthly* 105 (1998), 28–35.

Church, Alonzo. "Alternatives to Zermelo's Assumption." *Transactions of the American Mathematical Society* 29 (1927), 178–208.

_____. "A Bibliography of Symbolic Logic." *Journal of Symbolic Logic* 1 (1936), 121–123.

_____. *The Calculi of Lambda-Conversion*. Annals of Mathematics Studies. No. 6. Princeton: Princeton University Press and London: Humphrey Milford and Oxford University Press, 1941.

_____. "On Irredundant Sets of Postulates." *Transactions of the American Mathematical Society* 27 (1925), 318–328.

_____. "An Unsolvable Problem of Elementary Number Theory." *American Journal of Mathematics* 58 (1936), 345–363.

Churchill, Ruel V.; Reissner, Eric; Taub, Abraham H., Ed. *Elasticity.* Proceedings of Symposia in Applied Mathematics. Vol. 3. New York: American Mathematical Society, 1950.

Ciesielski, Krzysztof and Pogoda, Zdzislaw. "The Beginning of Polish Topology." *The Mathematical Intelligencer* 18 (1996), 32–39.

Claytor, William. "On Peanian Continua Not Imbeddable in a Spherical Surface." *Annals of Mathematics* 38 (1937), 631–646.

————. "Topological Immersion of Peanian Continua in a Spherical Surface." *Annals of Mathematics* 35 (1934), 809–835.

Clemens, Herbert. "Commentary on Algebraic Geometry." In *A Century of Mathematics in America.* Ed. Peter L. Duren et al. 2: 337–338.

Coble, Arthur. *Algebraic Geometry and Theta Functions.* AMS Colloquium Publications. Vol. 10. New York: American Mathematical Society, 1929.

Cohen, Leon W. "The Annual Meeting of the Society." *Bulletin of the American Mathematical Society* 58 (1952), 157–217.

Coifman Ronald R. and Strichartz, Robert S. "The School of Antoni Zygmund." In *A Century of Mathematics in America.* Ed. Peter L. Duren et al., 3: 343–368.

Cole, Frank N. "The April Meeting of the Society in New York." *Bulletin of the American Mathematical Society* 21 (1915), 481–493.

————. "The April Meeting of the American Mathematical Society in New York." *Bulletin of the American Mathematical Society* 26 (1920), 433–444.

————. "The October Meeting of the Society." *Bulletin of the American Mathematical Society* 20 (1914), 169–176.

————. "On a Certain Simple Group." In *Mathematical Papers Read at the International Mathematical Congress Held in Conjunction with the World's Columbian Exposition, Chicago 1893.* Ed. E. H. Moore et al., pp. 40–43.

————. "The Twelfth Summer Meeting of the American Mathematical Society," *Bulletin of the American Mathematical Society* 12 (1906), 53–63.

————. "The Twenty-sixth Annual Meeting of the American Mathematical Society." *Bulletin of the American Mathematical Society* 26 (1920), 241–259.

Committee on Publications. "Preface." *Semicentennial Addresses of the American Mathematical Society.* New York: American Mathematical Society, 1938, p. i.

————. *Semicentennial Addresses of the American Mathematical Society.* New York: American Mathematical Society, 1938.

Compton, Karl T. "Research Board for National Security." *Science* 101 (1945), 226–228.

Condon, E. C. "Duggan, Stephen Pierce." In *Biographical Dictionary of American Educators.* Ed. John Ohles, 1: 402–403.

Contributions to the Calculus of Variations 1930, 1931–1932, 1933–1937, 1938–1941. Chicago: University of Chicago Press, 1931, 1933, 1937, 1942; Reprint Ed. New York: Johnson Reprint Corp., 1965.

Conway, John B. "Paul Halmos and the Progress of Operator Theory." In *Paul Halmos: Celebrating 50 Years of Mathematics.* Ed. John Ewing and Fred Gehring, pp. 155–167.

Coolidge, Julian L. *A History of Geometrical Methods.* Oxford: University Press, 1940; Reprint Ed. New York: Dover Publications, Inc., 1963.

————. *An Introduction to Mathematical Probability.* Oxford: Clarendon Press, 1925.

_____. *A Treatise on Algebraic Plane Curves*. Oxford: Clarendon Press, 1931.

_____. "William Caspar Graustein–In Memoriam." *Bulletin of the American Mathematical Society* 47 (1941), 343–349.

Comptes rendus du Congrès international des mathématiciens Oslo 1936. 2 Vols. Oslo: A. W. Brøggers Boktrykkeri A/S, 1937.

Copeland, Arthur H. "Admissible Numbers in the Theory of Probability." *American Journal of Mathematics* 50 (1928), 535–552.

_____. "Consistency of the Conditions Determining Kollektivs." *Transactions of the American Mathematical Society* 42 (1937), 333–357.

_____. Review of "*Theory of Games and Economic Behavior*. By John von Neumann and Oskar Morgenstern. Princeton University Press, 1944. 18 + 625 pp. $10.00." *Bulletin of the American Mathematical Society* 51 (1945), 498–504.

Cordasco, Francesco. *Daniel Coit Gilman and the Protean Ph.D.: The Shaping of American Graduate Education*. Leiden: E. J. Brill, 1960.

Corry, Leo. *Modern Algebra and the Rise of Mathematical Structures*. Science Networks–Historical Studies. Vol. 17. Basel: Birkhäuser Verlag, 1996.

Courant, Richard, *Dirichlet's Principle, Conformal Mapping, and Minimal Surfaces*. New York: Interscience Publishers, 1950.

_____. "Plateau's Problem and Dirichlet's Principle." *Annals of Mathematics* 38 (1937), 679–724.

_____. "Über das Problem von Plateau." *Comptes rendus du Congrès international des mathématiciens Oslo 1936*. 2 Vols. Oslo: A. W. Brøggers Boktykkeri A/S, 1937, 2: 143.

Courant, Richard and Friedrichs, Kurt O. *Supersonic Flow and Shock Waves*. New York: Interscience Publishers, 1948.

Courant, Richard and Hilbert, David. *Methoden der Mathematischen Physik*. 2 Vols. Berlin: Julius Springer, 1924 and 1937.

"Cronica: Asamblea de la Unión Matemática Argentina." *Revista de la Unión Matématica Argentina* 9 (1943), 144.

Cunningham, Frank D. "Harry S. Truman and Universal Military Training, 1945." *The Historian* 46 (1984), 397–415.

Curtis, Charles W. *Pioneers of Representation Theory: Frobenius, Burnside, Schur, and Brauer*. HMATH. Vol. 15. Providence: American Mathematical Society and London: London Mathematical Society, 1999.

Curtiss, David R. *Analytic Functions of a Complex Variable*. Carus Mathematical Monographs. No. 2. Chicago: The Open Court Publishing Company, 1926.

Curtiss, John H. "A Federal Program in Applied Mathematics." *Science* 107 (1948), 257–262.

_____. "The Institute for Numerical Analysis of the National Bureau of Standards." *American Mathematical Monthly* 58 (1951), 372–379.

Dahan-Dalmedico, Amy. "L'essor des mathématiques appliquées aux États-Unis: L'impact de la Seconde Guerre Mondiale." *Revue d'histoire des mathématiques* 2 (1996), 149–213.

D'Ambrosio, Ubiratan; Dauben, Joseph W.; Parshall, Karen Hunger. "Mathematics Education in America in the Premodern Period." In *Handbook on the History of Mathematics Education*. Ed. Alexander Karp and Gert Schubring, pp. 175–199.

Dauben, Joseph W. *Georg Cantor: His Mathematics and Philosophy of the Infinite*. Cambridge: Harvard University Press, 1979.

Dawson, John W., Jr. "Max Dehn, Kurt Gödel, and the Trans-Siberian Escape Route." *Notices of the American Mathematical Society* 49 (2002), 1068–1075.

Decker, M. "Eberhard Hopf 04-17-1902 to 07-24-1983." *Jahresbericht der Deutschen Mathematiker-Vereinigung* 92 (1990), 47–57.

"Degrees Conferred June 12, 1900," *Johns Hopkins University Circulars* 19 (June 1900), 84–85.

Dehn, Max. "Mathematics, 200 B.C.-600 A.D." *American Mathematical Monthly* 51 (1944), 149–157.

————. "Über die Grundlagen der projektiven Geometrie und allgemeine Zahlsysteme." *Mathematische Annalen* 85 (1922), 184–194.

Denjoy, Arnaud. *L'énumération transfinie*. 4 Vols. Paris: Gauthier-Villars, 1946–1954.

Despeaux, Sloan Evans. "The Development of a Publication Community: Nineteenth-Century Mathematics in British Scientific Journals." Unpublished doctoral dissertation, University of Virginia, 2002.

————. "Fit to Print?: Referee Reports on Mathematics for the Nineteenth-century Journals of the Royal Society." *Notes and Records of the Royal Society* 65 (2011), 233–252.

Dewey, John. "The Supreme Intellectual Obligation." *Science* 49 (16 March, 1934), 240–243.

Dick, Auguste. *Emmy Noether 1882–1935*. Boston: Birkhäuser Verlag, 1981.

Dickson, Leonard. *Algebras and Their Arithmetics*. Chicago: University of Chicago Press, 1923; Reprint Ed. New York: Dover Publications, Inc., 1960.

————. *Algebren und ihre Zahlentheorie*. Zürich: Orell Füssli Verlag, 1927.

————. "Definitions of a Linear Associative Algebra by Independent Postulates." *Transactions of the American Mathematical Society* 4 (1903), 21–26.

————. "Further Development of the Theory of Arithmetics of Algebras." *Proceedings of the International Mathematical Congress Held in Toronto, August 11–16, 1924*. Ed. John Fields, 1: 173–184.

————. "Hans Frederik Blichfeldt, 1873–1945." *Bulletin of the American Mathematical Society* 53 (1947), 882–883.

————. *History of the Theory of Numbers*. 3 Vols. Washington, D.C.: Carnegie Institution of Washington, 1919, 1920, 1923; Reprint Ed. New York: Chelsea Publishing Company, 1992. "Recent

————. "Outline of the Theory to Date of the Arithmetics of Algebras." *Proceedings of the International Mathematical Congress Held in Toronto, August 11–16, 1924*. Ed. John Fields, 1: 95–102.

————. "Recent Progress on Waring's Theorem and Its Generalizations." *Bulletin of the American Mathematical Society* 39 (1933), 701–727.

————. "Some Relations Between the Theory of Numbers and Other Branches of Mathematics." In *Comptes rendus du Congrès international des Mathématiciens (Strasbourg, 22–30 septembre 1920)*. Ed. Henri Villat, pp. 41–56.

Dickson, Leonard E.; Mitchell, Howard H.; Vandiver, Harry S.; and Wahlin, Gustav E. *Algebraic Numbers*. Bulletin of the National Research Council. No. 28. Washington, D.C.: National Research Council of the National Academy of Sciences, 1923.

Dietz, David. "'Scientist in Exile' Now Aids U. S. Defense." *New York World Telegram*. 3 December, 1941.

Dieudonné, Jean. *A History of Algebraic and Differential Topology, 1900–1960*. Boston: Birkhäuser Boston, 1989.

———. *History of Algebraic Geometry*. Monterey: Wadsworth, Inc., 1985.

———. *History of Functional Analysis*. Amsterdam: North-Holland Publishing Company, 1981.

Dieudonné, Jean and Tits, Jacques. "Claude Chevalley (1909–1984)." *Bulletin of the American Mathematical Society* 17 (1987), 1–7.

Dinnerstein, Leonard. "Antisemitism in Crisis Times in the United States: The 1920s and 1930s." In *Anti-Semitism in Times of Crisis*. Ed. Sander L. Gilman and Steven T. Katz, pp. 212–226.

Donaldson, James A. and Fleming, Richard J. "Elbert F. Cox: An Early Pioneer." *American Mathematica Monthly* 107 (2000), 105–128.

Donoghue, Eileen F. "The Emergence of a Profession: Mathematics Education in the United States, 1890–1920." In *A History of School Mathematics*. Ed. George M. A. Stanic and Jeremy Kilpatrick, 1: 159–193.

Doob, Joseph. "The Brownian Movement and Stochastic Equations." *Annals of Mathematics* 43 (1942), 351–369.

———. "Commentary on Probability." In *A Century of Mathematics in America*. Ed. Peter L. Duren et al. 2: 353–354.

———. "Note on Probability." *Annals of Mathematics* 37 (1936), 363–367.

———. "Probability and Statistics." *Transactions of the American Mathematical Society* 36 (1934), 759–775.

———. "Probability as Measure." *Annals of Mathematical Statistics* 12 (1941), 206–214.

———. "Topics in the Theory of Markoff Chains." *Transactions of the American Mathematical Society* 52 (1942), 37–64.

———. "Wiener's Work in Probability Theory," *Bulletin of the American Mathematical Society* 72 (1966), 69–72.

Dorwart, Harold L. "Mathematics at Yale in the Nineteen Twenties." In *A Century of Mathematics in America*. Ed. Peter L. Duren et al., 2: 87–97.

Douglas, Jesse. "Edward Kasner, 1878–1955." *Biographical Memoirs*. Vol. 31. Washington, D.C.: National Academy of Sciences, 1958, pp. 179–209.

———. "The General Geometry of Paths." *Annals of Mathematics* 29 (1927–1928), 143–168.

———. "Green's Functions and the Problem of Plateau." *American Journal of Mathematics* 61 (1939), 545–589.

———. "The Most General Form of the Problem of Plateau." *American Journal of Mathematics* 61 (1939), 590–608.

———. "Solution of the Inverse Problem of the Calculus of Variations." *Proceedings of the National Academy of Sciences of the United States of America* 25 (1939), 631–637.

———. "Solution of the Problem of Plateau." *Transactions of the American Mathematical Society* 33 (1931), 263–321.

Dresden, Arnold. "The April Meeting of the Society." *Bulletin of the American Mathematical Society* 34 (1928), 419–432.

————. "The March Meeting in New York," *Bulletin of the American Mathematical Society* 35 (1929), 433–446.

————. "Mathematics in a Changing World." *The Scientific Monthly* 38 (1934), 568–570.

————. "The Migration of Mathematics." *American Mathematical Monthly* 49 (1942), 415–429.

————. "A Program for Mathematics." *American Mathematical Monthly* 42 (1935), 198–208.

————. "Report of the Committee on Aid to Devastated Libraries." *Bulletin of the American Mathematical Society* 55 (1949), 1–2.

————. Review of "*The Calculus of Variations in the Large*. By Marston Morse. Colloquium Publications of the American Mathematical Society, vol. 18. New York, 1934. ix+368 pp." *Bulletin of the American Mathematical Society* 42 (1936), 607–612.

————. Review of "*Contributions to the Calculus of Variations, 1933–1937*. (Theses submitted to the Department of Mathematics at the University of Chicago.) University of Chicago Press, 1937. 7+566 pp." *Bulletin of the American Mathematical Society* 44 (1938), 604–609.

————. Review of "*Fondamenti di Calcolo delle Variazioni*. By Leonida Tonelli. Bologna, Zanichelli. Vol. 1, 7+406 pp., 1921, 55 Lire; vol. 2, 8+660 pp., 1923, 80 Lire." *Bulletin of the American Mathematical Society* 32 (1926), 381–386.

————. Review of "*Semicentennial Addresses*. (American Mathematical Society Semicentennial Publications, vol. 2) New York, American Mathematical Society, 1938, 6 + 315 pp." *Bulletin of the American Mathematical Society* 45 (1939), 50–51.

————. "The Thirty-fourth Annual Meeting of the American Mathematical Society." *Bulletin of the American Mathematical Society* (1928), 129–154.

Drucker, Thomas, Ed. *Perspectives on the History of Mathematical Logic.* Boston: Birkhäuser, 1991.

Dryden, Hugh L., Murnaghan, Frances D., and Bateman, Harry. *Hydrodynamics.* Bulletin of the National Research Council. No. 84. Washington, D.C.: National Research Council of the National Academy of Sciences, 1932.

Duda, Roman. "*Fundamenta Mathematicae* and the Warsaw School of Mathematics," In *L'Europe mathématique/Mathematical Europe.* Ed. Catherine Goldstein, Jeremy Gray, and Jim Ritter, pp. 481–498.

Duggan, Stephen and Drury, Betty. *The Rescue of Science and Learning: The Story of the Emergency Committee in Aid of Displaced Foreign Scholars.* New York: The Macmillan Company, 1948.

Dumbaugh, Della and Schwermer, Joachim. "Creating a Life: Emil Artin in America." *Bulletin of the American Mathematical Society* 50 (2013), 321–330.

Duran, Samson. "Des géométries étatsuniennes à partir de l'étude de l'*American Mathematical Society*: 1888–1920." Unpublished doctoral dissertation, Université Paris-Sud (Orsay), 2019.

Durden, Robert F. *Bold Entrepreneur: A Life of James B. Duke.* Durham: Carolina Academic Press, 2003.

————. *The Launching of Duke University.* Durham: Duke University Press, 1993.

Duren, Peter L., et al., Ed. *A Century of Mathematics in America.* 3 Vols. Providence: American Mathematical Society, 1988–1989.

Du Val, Patrick. "Beniamino Segre." *Bulletin of the London Mathematical Society* 11 (1979), 215–235.

Eilenberg, Samuel and Mac Lane, Saunders. "Cohomology Theory in Abstract Groups I." *Annals of Mathematics* 48 (1947), 51–78.

———. "Cohomology Theory in Abstract Groups II: Group Extensions with a Non-Abelian Kernel." *Annals of Mathematics* 48 (1947), 326–341.

———. "General Theory of Natural Equivalences." *Transactions of the American Mathematical Society* 58 (1945), 231–294.

Eilenberg, Samuel and Steenrod, Norman. "Axiomatic Approach to Homology Theory." *Proceedings of the National Academy of Sciences of the United States of America* 31 (1945), 117–120.

———. *Foundations of Algebraic Topology* (Princeton: Princeton University Press, 1952.

Eisenhart, Luther P. *Continuous Groups of Transformations*. Princeton: Princeton University Press and Oxford: Oxford University Press, 1933.

———. "Infinitesimal Deformation of Surfaces." *American Journal of Mathematics* 24 (1902), 173–204.

———. *Non-Riemannian Geometry*. AMS Colloquium Publications. Vol 8. New York: American Mathematical Society, 1927.

———. "The Permanent Gravitational Field in the Einstein Theory." *Annals of Mathematics* 22 (1920), 86–94.

———. *Riemannian Geometry*. Princeton: Princeton University Press, 1926.

———. *Transformations of Surfaces*. Princeton: Princeton University Press, 1923.

Eisenhart, Luther P. and Knebelman, Morris. "Invariant Theory of Homogeneous Contact Transformations." *Annals of Mathematics* 37 (1936), 747–765.

Eisenhart, Luther P. and Veblen, Oswald. "The Riemannian Geometry and Its Generalizations." *Proceedings of the National Academy of Sciences of the United States of America* 8 (1922), 19–23.

Elliott, Clark A. "The Tercentenary of Harvard University in 1936: The Scientific Dimension." In *Commemorative Practices in Science: Historical Perspectives on the Politics of Collective Memory*. Ed. Pnina G. Abir-Am and Clark A. Elliott, *Osiris* 14 (1999), 153–175.

Enderton, Herbert. "Alonzo Church and the Reviews." *Bulletin of Symbolic Logic* 4 (1998), 172–180.

England, J. Merton. *A Patron of Pure Science: The National Science Foundation's Formative Years, 1945–57*. Washington, D.C.: National Science Foundation, 1982.

Epstein, Paul. "On the Evaluation of Certain Integrals in the Theory of Quanta." *Proceedings of the National Academy of Sciences of the United States of America* 12 (1926), 629–633.

Erdős, Paul. "On a New Method in Elementary Number Theory Which Leads to An Elementary Proof of the Prime Number Theorem." *Proceedings of the National Academy of Sciences of the United States of America* 35 (1949), 374–384.

Evans, Griffith C. "Dirichlet Problems." *Semicentennial Addresses of the American Mathematical Society*. New York: American Mathematical Society, 1938, pp. 185–226.

———. *Functionals and Their Applications: Selected Topics, Including Integral Equations*. AMS Colloquium Lectures. Vol. 5. New York: American Mathematical Society, 1918.

_____. *Mathematical Introduction to Economics*. New York: McGraw Hill Book Company, Inc., 1930.

_____. "Modern Methods of Analysis in Potential Theory." *Bulletin of the American Mathematical Society* 43 (1937), 481–502.

Ewing, John and Gehring, Fred. Ed. *Paul Halmos: Celebrating 50 Years of Mathematics*. New York: Springer-Verlag, 1991.

Feferman, Anita Burdman and Feferman, Solomon. *Alfred Tarski: Life and Logic*. Cambridge: Cambridge University Press, 2004.

Feffer, Loren Butler. "Oswald Veblen and the Capitalization of American Mathematics: Raising Money for Research 1923–1928." *Isis* 89 (1998), 474–497.

Feit, Walter. "Richard D. Brauer." *Bulletin of the American Mathematical Society*, n.s., 1 (1979), 1–20.

Fejér, Leopold. "On the Infinite Sequences Arising in the Theories of Harmonic Analysis, of Interpolation, and of Mechanical Quadratures." *Bulletin of the American Mathematical Society* 39 (1933), 521–534.

Feller, William. *An Introduction to Probability Theory and Its Applications*. New York: John Wiley, 1950.

Fenster, Della Dumbaugh. "American Initiatives toward Internationalization; The Case of Leonard Dickson." In *Mathematics Unbound: The Evolution of an International Mathematical Research Community, 1800–1945*. Ed. Karen Hunger Parshall and Adrian C. Rice, pp. 311–333.

_____. "Funds for Mathematics: Carnegie Institution of Washington Support for Mathematics from 1902 to 1921." *Historia Mathematica* 30 (2003), 195–216.

_____. "Leonard Dickson, History of the Theory of Numbers." In *Landmark Writings in Western Mathematics, 1640–1940*. Ed. Ivor Grattan-Guinness, pp. 833–843.

_____. "Leonard Eugene Dickson and His Work in the Arithmetics of Algebras." *Archive for History of Exact Sciences* 52 (1998), 119–159.

_____. "Role Modeling in Mathematics: The Case of Leonard Eugene Dickson (1874–1954)." *Historia Mathematica* 24 (1997), 7–24.

_____. "Why Dickson Left Quadratic Reciprocity Out of His History of the Theory of Numbers." *American Mathematical Monthly* 106 (1999), 618–627.

Fenster, Della Dumbaugh and Schwermer, Joachim. "A Delicate Collaboration: Adrian Albert and Helmut Hasse and the Principal Theorem in Division Algebras in the Early 1930s." *Archive for History of Exact Sciences* 59 (2005), 349–379.

Fermi, Laura. *Illustrious Immigrants: The Intellectual Migration from Europe 1930/41*. Chicago: University of Chicago Press, 1971.

Ficken, Frederick A. "*Reports of a Mathematical Colloquium*. Series 2, no. 1. Edited by Karl Menger. Notre Dame University Press, 1939, 64 pp." *Bulletin of the American Mathematical Society* 45 (1939), 813–814.

Fields, John C., Ed. *Proceedings of the International Mathematical Congress Held in Toronto, August 11–16, 1924*. 2 Vols. Toronto: University of Toronto Press, 1928.

Fiske, Thomas S. "The Buffalo Colloquium." *Bulletin of the American Mathematical Society* 3 (1896), 49–59.

Fitzpatrick, Ben. "Some Aspects of the Work and Influence of R. L. Moore." In *Handbook of the History of General Topology*. Ed. Charles E. Aull and Robert Lowen, 1: 41–61.

Fleming, Donald and Bailyn, Bernard, Ed. *The Intellectual Migration: Europe and America, 1930–1960. Perspectives in American History*. Vol. 2. Cambridge: Charles Warren Center for Studies in American History of Harvard University, 1968.

Fleming, John A, "The International Scientific Unions." *Proceedings of the American Philosophical Society* 91 (1947), 121–125.

Flexner, Abraham. *I Remember: The Autobiography of Abraham Flexner*. New York: Simon and Schuster, 1940.

Floyd, Edwin and Jones, F. Burton, "Gordon T. Whyburn (1904–1969)," *Bulletin of the American Mathematical Society* 77 (1971), 57–72.

Fokkink, Robbert. "A Forgotten Mathematician." *European Mathematical Society Newsletter* 52 (2004), 9–14.

Folkerts, Menso, et al., Ed. *Amphora: Festschrift für Hans Wussing zu seinem 65. Geburtstag*. Basel: Birkhäuser Verlag, 1992.

Fong, Paul and Wong, Warren. Ed. *Richard Brauer: Collected Papers*. 3 Vols. Cambridge: The MIT Press, 1980.

Ford, Walter B. "Earle Raymond Hedrick." *AMM* 50 (1943), 409–411.

Fort, Tomlinson. "The Annual Meeting in New Orleans." *Bulletin of the American Mathematical Society* 38 (1932), 145–154.

Fosdick, Raymond B. *The Story of the Rockefeller Foundation*. New York: Harper & Brothers, 1952.

Fréchet, Maurice. "La différentielle dans l'analyse générale." *Annales scientifiques de l'École Normale Supérieure*. 42 (1925), 293–323.

Frei, Günther and Stammbach, Urs. "Heinz Hopf." In *History of Topology*. Ed. Ioan M. James, pp. 991–1008.

Frewer, M. "Felix Bernstein." *Jahresbericht der Deutschen Mathematiker-Vereinigung* 83 (1981), 84–95.

Frink, Jr., Orrin. Review of *The Calculi of Lambda-Conversion*. By Alonzo Church. (Annals of Mathematics Studies, no. 6) Princeton, Princeton University Press; London, Humphrey Milford and Oxford University Press, 1941. 2+77 pp. $1.25. *Bulletin of the American Mathematical Society* 50 (1944), 169–172.

Fry, Thornton "Industrial Mathematics." *American Mathematical Monthly* 48 (June-July 1941), 1–38.

Gardner, David P. *The California Oath Controversy*. Berkeley: University of California Press, 1967.

Garrett, Alan W. and and Davis, Jr., O. L. "A Time of Uncertainty and Change: School Mathematics from World War II until the New Math." In *A History of School Mathematics*. Ed. George M. A. Stanic and Jeremy Kilpatrick, 1: 493–519.

Gehman, Harry M. "Moore on Point Sets." *Bulletin of the American Mathematical Society* 39 (1933), 479–483.

————. "The Thirtieth Summer Meeting of the Association." *American Mathematical Monthly* 55 (1948), 605–611.

_____. "The Thirty-Second Annual Meeting of the Association." *American Mathematical Monthly* 56 (1949), 200–207.

Geiger, Roger L. *To Advance Knowledge: The Growth of American Research Universities, 1900–1940*. New York: Oxford University Press, 1986.

Geiringer, Hilda. "On the Probability Theory of Linkage in Mendelian Heredity." *Annals of Mathematical Statistics* 15 (1944), 25–57.

_____. "On Some Mathematical Problems Arising in the Development of Mendelian Genetics." *Journal of the American Statistical Association* 44 (1949), 526–547.

Gemelli, Giuliana, Ed. *The "Unacceptables": American Foundations and Refugee Scholars between the Two World Wars and After*. Brussels: P.I.E.–Peter Lang, 2000.

Georgiadou, Maria. *Constantin Carathéodory: Mathematics and Politics in Turbulent Times*. Berlin: Springer-Verlag, 2004.

Gillispie, Charles C. Ed. *Dictionary of Scientific Biography*. 18 Vols. New York: Charles Scribner's Sons, 1970–1990.

Gilman, Sander L. and Katz, Steven T., Ed. *Anti-Semitism in Times of Crisis*. New York: New York University Press, 1991.

Gispert, Hélène. *La France mathématique: La Société mathématique de France (1872–1914)*. Paris: Société française d'histoire des sciences et des techniques & Société mathématique de France, 1991; Rev. Ed. *La France mathématique de la Troisième République avant la Grande Guerre*. Paris: Société mathématique de France, 2016.

Gleason, Andrew. "Groups without Small Subgroups." *Annals of Mathematics* 56 (1952), 193–212.

Glover, James W. "Statistical Teaching in American Colleges and Universities." *Journal of the American Statistical Association* 21 (1926), 419–446.

Gluchoff, Alan. "Pure Mathematics Applied in Early Twentieth-Century America: The Case of T. H. Gronwall, Consulting Mathematician." *Historia Mathematica* 32 (2005), 312–357.

Goldstein, Catherine; Gray, Jeremy; and Ritter, Jim, Ed. *L'Europe mathématique/Mathematical Europe*. Paris: Éditions de la Maison des sciences de l'homme, 1996.

Goldstine, Herman H. *The Computer from Pascal to von Neumann*. Princeton: Princeton University Press, 1972; Reprinted with a New Preface, 1993.

_____. "Remembrance of Things Past." In *A History of Scientific Computing*. Ed. Stephen Nash, pp. 5–16.

Goldstine, Herman H. and Goldstine, Adele. "The Electronic Numerical Integrator and Computer (ENIAC)." *Mathematical Tables and Other Aids to Computation* 2 (1946), 97–110.

Goldstine, Herman H. and von Neumann, John. "Numerical Inverting of Matrices of High Order." *Bulletin of the American Mathematical Society* 53 (1947), 1021–1099.

Golland, Louise and Sigmund, Karl. "Exact Thought in a Demented Time: Karl Menger and His Viennese Mathematical Colloquium." *The Mathematical Intelligencer* 22 (2000), 34–45.

Golomb, Solomon; Harris, Theodore; Seberry, Jennifer. "Albert Leon Whiteman (1915–1995)." *Notices of the American Mathematical Society* 44 (1997), 217–219.

Goodstein, Judith R. *Millikan's School: A History of The California Institute of Technology*. New York: W. W. Norton & Company, 1991.

————. "Olga Taussky-Todd." *Notices of the American Mathematical Society* 67 (2020), 345–353.

Goodstein, Judith R. and Babbitt, Donald. "E. T. Bell and Mathematics at Caltech between the Wars." *Notices of the American Mathematical Society* 60 (2013), 686–698.

Grattan-Guinness, Ivor. *Landmark Writings in Western Mathematics, 1640–1940*. Amsterdam: Elsevier Press, 2005.

————. *The Search for Mathematical Roots 1870–1940: Logics, Set Theories and the Foundations of Mathematics from Cantor through Russell to Gödel*. Princeton: Princeton University Press, 2000.

Graustein, William C. "Eisenhart's Transformation of Surfaces." *Bulletin of the American Mathematical Society* 30 (1924), 454–460.

————. "Invariant Methods in Classical Differential Geometry." *Bulletin of the American Mathematical Society* 36 (1930), 489–521.

————. "Méthodes invariantes dans la géométrie infinitésimale." *Mémoires de l'Académie Royale de Belgique (Classe des Sciences)* 11 (1929), 1–96.

————. "Schouten and Struik on Differential Geometry." *Bulletin of the American Mathematical Society* 45 (1939), 649–650.

Graves, Lawrence M. "The Existence of an Extremum in Problems of Mayer." *Transactions of the American Mathematical Society* 39 (1936), 456–471.

————. "On the Weierstrass Condition for the Problem of Bolza in the Calculus of Variations." *Annals of Mathematics* 33 (1932), 747–752.

Graves, Lawrence; Hille, Einar; Smith, Paul; and Zariski, Oscar, Ed. *Proceedings of the International Congress of Mathematicians: Cambridge, Massachusetts, U.S.A., August 30–September 6, 1950*. 2 Vols. Providence: American Mathematical Society, 1952.

Gray, Jeremy and Micallef, Mario. "About the Cover: The Work of Jesse Douglas on Minimal Surfaces." *Bulletin of the American Mathematical Society* 45 (2008), 293–302.

Gray, Jeremy J. and Parshall, Karen Hunger, Ed. *Episodes in the History of Modern Algebra (1800–1950)*. HMATH. Vol. 32. Providence: American Mathematical Society and London: London Mathematical Society, 2007.

Green, Judy and LaDuke, Jeanne. *Pioneering Women in American Mathematics: The Pre-1940 PhD's*. HMATH. Vol 34. Providence: American Mathematical Society and London: London Mathematical Society, 2009.

Greenberg, John L. and Goodstein, Judith R. "Theodore van Kármán and Applied Mathematics in America." *Science* 222 (1984), 1300–1304; Reprinted in *A Century of Mathematics in America*. Ed. Peter L. Duren et al., 2: 467–477.

Grier, David Alan. "Dr. Veblen Takes a Uniform: Mathematics in the First World War." *American Mathematical Monthly* 108 (2001), 922–931.

————. *When Computers Were Human*. Princeton: Princeton University Press, 2005.

Griffiths, Phillip; Spencer, Donald; and Whitehead, George. "Solomon Lefschetz: September 3, 1884–October 5, 1972." *Biographical Memoirs*. Vol. 61. Washington, D.C.: National Academy of Sciences, 1992, pp. 271–313.

Grinstein, Louise S. and Campbell, Paul J. Ed. *Women in Mathematics: A Biobibliographic Sourcebook*. Westport: Greenwood Press, 1982.

Gronwall, Thomas. "On the Zeros of the Function $\beta(z)$ Associated with the Gamma Function." *Transactions of the American Mathematical Society* 28 (1926), 391–399.

Gross, Benjamin and Tate, John. "Commentary on Algebra." In *A Century of Mathematics in America*. Ed. Peter L. Duren et al. 2: 335–336.

Guerraggio, Angelo and Nastasi, Pietro. *Italian Mathematics Between the Two World Wars*. Basel: Birkhäuser Verlag, 2005.

Guillaume, Marcel. "La Logique mathématique en sa jeunesse." In *Development of Mathematics 1900–1950*, Ed. Jean-Paul Pier, pp. 185–367.

Gunning, Robert C. "Bochner, Salomon." *Dictionary of Scientific Biography*. Ed. Charles C. Gillispie, 17: 88–90.

Hadamard, Jacques. "The Early Scientific Work of Henri Poincaré." *Rice Institute Pamphlet* 9 (3) (1922), 111–184.

_____. "The Later Scientific Work of Henri Poincaré." *Rice Institute Pamphlet* 20 (1) (1933), 1–86.

_____. *Lectures on Cauchy's Problem in Linear Partial Differential Equations*. New Haven: Yale University Press, 1922.

Haigh, Thomas; Priestley, Mark; Rope, Crispin. *ENIAC in Action: Making and Remaking the Modern Computer*. Cambridge: The MIT Press, 2016.

Hall, Marshall. "Projective Planes." *Transactions of the American Mathematical Society* 54 (1943), 229–277.

Halmos, Paul R. *Finite Dimensional Vector Spaces*. Annals of Mathematics Studies. No. 7. Princeton: Princeton University Press, 1942.

_____. "The Foundations of Probability," *American Mathematical Monthly* 51 (1944), 493–510.

_____. *I Want to Be a Mathematician: An Automathography*. New York: Springer-Verlag, 1985.

_____. *Introduction to Hilbert Space and the Theory of Spectral Multiplicity*. New York: Chelsea Publishing Company, 1951.

_____. "Invariants of Certain Stochastic Transformations: The Mathematical Theory of Gambling Systems," *Duke Mathematical Journal* 5 (1939), 461–478.

_____. *Measure Theory*. New York: Van Nostrand, 1950.

_____. "Normal Dilations and Extensions of Operators." *Summa Brasiliensis Mathematicae* 2 (1950), 125–134.

Hardy, G. H. "An Introduction to the Theory of Numbers." *Bulletin of the American Mathematical Society* 35 (1929), 778–818.

"Harold W. Stoke, College President." *New York Times* (7 April, 1982), A15.

Hart, William L. "On Education for the Service." *American Mathematical Monthly* 48 (1941), 353–362.

Hartman, Philip. "Aurel Wintner." *Journal of the London Mathematical Society* 37 (1962), 483–503.

"The Harvard Tercentenary Conference of Arts and Sciences." *Science* 83 (1936), 311–313.

Hasse, Helmut. "Theory of Cyclic Algebras over an Algebraic Number Field." *Transactions of the American Mathematical Society* 34 (1932), 171–214.

————. "Über p-adische Schiefkörper." *Mathematische Annalen* 104 (1931), 495–534.

Hawkins, Hugh. *Pioneer: A History of the Johns Hopkins University 1874–1889.* Ithaca: Cornell University Press, 1960.

Hawkins, Thomas W. *Lebesgue's Theory of Integration: Its Origin and Development.* Madison: University of Wisconsin Press, 1970; Reprint of 2d. Ed. New York: AMS Chelsea Books, 2001.

Hazlett, Olive. "The Arithmetic of a General Algebra." *Annals of Mathematics* 28 (1926), 92–102.

————. "Integers as Matrices." In *Atti del Congresso internazionale dei matematici, Bologna, 3–10 settembre 1928,* 2: 57–62.

————. "On the Arithmetic of a General Associative Algebra." *Proceedings of the International Mathematical Congress Held in Toronto, August 11–16, 1924.* Ed. John C. Fields, 1: 185–191.

————. "On the Classification and Invariantive Characterization of Nilpotent Algebras." *American Journal of Mathematics* 38 (1916), 109–110.

Hedlund, Gustav. "Dynamics of Geodesic Flows." *Bulletin of the American Mathematical Society* 45 (1939), 241–260.

Hedrick, Earle. "On the Derivatives of Non-analytic Functions." *Proceedings of the National Academy of Sciences of the United States of America* 14 (1928), 649–654.

————. "A Proposed Journal of Applied Mathematics." *Bulletin of the American Mathematical Society* 48 (1942), 791.

————. "A Tentative Platform for the Association." *American Mathematical Monthly* 23 (1916), 31–33.

Hedrick, Earle R. and Cairns, William D. "First Summer Meeting of the Association." *American Mathematical Monthly* 23 (1916), 273–288.

Heims, Steve J. *John von Neumann and Norbert Wiener: From Mathematics to the Technologies of Life and Death.* Cambridge: The MIT Press, 1980.

Henderson, Robert. "Life Insurance as a Social Science and as a Mathematical Problem." *Bulletin of the American Mathematical Society* 31 (1925), 227–252.

Herstein, Israel N. *Noncommutative Rings.* Carus Mathematical Monographs. No. 15. N.p.: Mathematical Association of America, 1968.

Hestenes, Magnus. "The Problem of Bolza in the Calculus of Variations." *Bulletin of the American Mathematical Society* 48 (1942), 57–75.

————. "Sufficient Conditions for the Problem of Bolza in the Calculus of Variations," *Transactions of the American Mathematical Society* (1934), 793–818.

Hestenes, Magnus and Todd, John. *NBS-INA—The Institute for Numerical Analysis—UCLA 1947–1954.* NIST Special Publication 730. Washington, D.C.: U. S. Government Printing Office, 1991.

Hilbert, David. "Mathematical Problems." Trans. Mary Winston Newson. *Bulletin of the American Mathematical Society* 8 (1902), 437–479.

————. "Die Theorie der algebraischen Zahlkörper," *Jahresbericht der Deutschen Mathematiker-Vereinigung* 4 (1894–95), 175–535.

————. "Über das Unendliche." *Mathematische Annalen* 95 (1926), 161–190.

Hildebrandt, Theophil H. "The Borel Theorem and Its Generalizations." *Bulletin of the American Mathematical Society* 32 (1926), 423–474.

_____. "On Uniform Limitedness of Sets of Functional Operations." *Bulletin of the American Mathematical Society* 29 (1923), 309–315.

_____. "The Second Madison Colloquium." *Bulletin of the American Mathematical Society* 33 (1927), 663–665.

Hille, Einar. "Algebraic Tendencies in Analysis." In *Proceedings of the International Congress of Mathematicians: Cambridge, Massachusetts, U.S.A., August 30–September 6, 1950*. Ed. Lawrence Graves, Einar Hille, Paul Smith, and Oscar Zariski, 2: 104–105.

_____. *Functional Analysis and Semi-groups*. AMS Colloquium Publications. Vol. 31. New York: American Mathematical Society, 1948.

Hille, Einar and Phillips, Ralph S. *Functional Analysis and Semi-groups*. AMS Colloquium Publications. Vol. 31. Rev. Ed. Providence: American Mathematical Society, 1957.

Hirsch, Morris W. "Edwin Henry Spanier (1921–1996)." *Notices of the American Mathematical Society* 45 (1998), 704–705.

A History of the National Research Council, 1919–1933. Washington, D.C.: National Research Council, 1933.

Hitchin, Nigel. "Shiing-Shen Chern, 26 October 1911–3 December 2004." *Biographical Memoirs of Fellows of the Royal Society* 60 (2014), 75–85.

Hochschild, Gerhard. Review of "*The Theory of Valuations*. By O.F.G. Schilling. Mathematical Surveys, no. 4 New York, American Mathematical Society, 1950. 8+253 pp. $6.00." *Bulletin of the American Mathematical Society* 57 (1951), 91–94.

Hollcroft, Temple R. "The Annual Meeting of the Society." *Bulletin of the American Mathematical Society* 45 (1939), 197–208.

_____. "The Annual Meeting of the Society," *Bulletin of the American Mathematical Society* 48 (1942), 183–198.

_____. "The Annual Meeting of the Society." *Bulletin of the American Mathematical Society* 56 (1950), 140–190.

_____. "The April Meeting in Charlottesville." *Bulletin of the American Mathematical Society* 44 (1938), 463–470.

_____. "The April Meeting in New York." *Bulletin of the American Mathematical Society* 52 (1946), 580–593.

_____. "The April Meeting in Philadelphia," *Bulletin of the American Mathematical Society* 55 (1949), 678–705.

_____. "The February Meeting in New York," *Bulletin of the American Mathematical Society* 47 (1941), 333–342.

_____. "The March Meeting in New York," *Bulletin of the American Mathematical Society* 43 (1937), 307–314.

_____. "The May Meeting in Washington," *Bulletin of the American Mathematical Society* 47 (1941), 523–531.

_____. "The Summer Meeting in Ithaca," *Bulletin of the American Mathematical Society* 52 (1946), 964–975.

_____. "The Summer Meeting in New Brunswick." *Bulletin of the American Mathematical Society* 49 (1943), 823–834.

_____. "The Summer Meeting in State College," *Bulletin of the American Mathematical Society* 43 (1937), 745–757.

Hollings, Christopher. "The Early Development of the Algebraic Theory of Semigroups." *Archive for History of Exact Sciences* 63 (2009), 497–536.

―――――. *Mathematics across the Iron Curtain: A History of the Algebraic Theory of Semigroups.* HMATH. Vol. 41. Providence: American Mathematical Society, 2014.

―――――. " 'Nobody Could Possibly Misunderstand What a Group Is': A Study in Early Twentieth-Century Group Axiomatics." *Archive for History of Exact Sciences* 71 (2017), 409–481.

―――――. "A Tale of Mathematical Myth-Making: E. T. Bell and the 'Arithmetization of Algebra.' " *BSHM Bulletin: Journal of the British Society for the History of Mathematics* 31 (2016), 69–80.

Hollings, Christopher and Siegmund-Schultze, Reinhard. *Meeting under the Integral Sign?: The Oslo Congress of Mathematicians on the Eve of the Second World War.* HMATH. Vol. 44. Providence: American Mathematical Society, 2020.

Hope, Arthur J. *Notre Dame: One Hundred Years.* Notre Dame: University of Notre Dame Press, 1943.

Hopf, Eberhard. "On the Time Average Theorem in Dynamics." *Proceedings of the National Academy of Sciences of the United States of America* 18 (1932), 93–100.

Hopf, Heinz. "A New Proof of the Lefschetz Theorem on Invariant Points." *Proceedings of the National Academy of Sciences of the United States of America* 14 (1928), 149–153.

―――――. "On Some Properties of One-valued Transformations of Manifolds." *Proceedings of the National Academy of Sciences of the United States of America* 14 (1928), 206–214.

―――――. "Über die Abbildungen der dreidimensionalen Sphäre auf der Kugelfläche." *Mathematische Annalen* 104 (1931), 637–665.

―――――. "Über die algebraische Anzahl von Fixpunkten." *Mathematische Zeitschrift* 29 (1929), 493–524.

Hotelling, Harold. "The Place of Statistics in the University." In *Proceedings of the Berkeley Symposium on Mathematical Statistics and Probability.* Ed. Jerzy Neyman, pp. 21–40.

―――――. Review of *Mathematical Introduction to Economics* by Griffith C. Evans. *Journal of Political Economy* 39 (1931), 107–109.

Hudson, Richard H. and Markham, Thomas L. "Alfred T. Brauer As a Mathematician and Teacher." *Linear Algebras and Its Applications* 59 (1984), 1–17.

Hunter, Patti W. "Connections, Context, Community: Abraham Wald and the Sequential Probability Ratio Test." *The Mathematical Intelligencer* 26 (2004), 25–33.

―――――. "Drawing the Boundaries: Mathematical Statistics in 20th-Century America." *Historia Mathematica* 23 (1996), 7–30.

―――――. "Foundations of Statistics in American Textbooks: Probability and Pedagogy in Historical Context." In *From Calculus to Computers: Using the Last 200 Years of Mathematics History in the Classroom.* Ed. Amy Shell-Gellasch and Richard Jardine, pp. 165–180.

―――――. "An Unofficial Community: American Mathematical Statisticians before 1935." *Annals of Science* 56 (1999), 47–68.

Huntington, Edward V. "The Pan-American Scientific Congress." *Bulletin of the American Mathematical Society* 31 (1925), 290.

―――――. "Sets of Completely Independent Postulates for Cyclic Order." *Proceedings of the National Academy of Sciences of the United States of America* 10 (2) (1924), 74–78.

Hyers, Donald. "Aristotle D. Michal 1899–1953." *Mathematics Magazine* 27 (1954), 237–244.

Ingraham, Mark. "The Summer Meeting in Chicago." *Bulletin of the American Mathematical Society* 39 (1933), 633–640.

—————. "The Trip of Associate Secretary M. H. Ingraham in the Interests of the Society." *Bulletin of the American Mathematical Society* 40 (1934), 641–643.

"In Memoriam Eberhard Hopf: 1902–1983." *Indiana University Mathematics Journal* 32 (1983), i-ii.

"In Memoriam: Yale Mathematician Shizuo Kakutani Known for His Work in Functional Analysis and Probability." *Yale Bulletin and Calendar* 33 (27 August, 2004).

Institute for Advanced Study Bulletin No. 3. Princeton: Princeton University Press, 1934.

Institute for Advanced Study Bulletin No. 4. Princeton: Princeton University Press, 1935.

Institute for Advanced Study Bulletin No. 9. Princeton: Princeton University Press, 1940.

"The International Congress of Mathematics: Committees of the Congress." In *Proceedings of the International Congress of Mathematicians: Cambridge, Massachusetts, U.S.A., August 30–September 6, 1950,* Ed. Lawrence Graves, Einar Hille, Paul Smith, and Oscar Zariski, 1: 1–4.

Jackson, Dunham. "Boundedness of Orthonormal Polynomials on Loci of the Second Degree." *Duke Mathematical Journal* 11 (1944), 351–365.

—————. *Fourier Series and Orthogonal Polynomials.* Carus Mathematical Monographs, No. 6. N.p: Mathematical Association of America, 1941.

—————. "The Harmonic Boundary Value Problem for an Ellipse or an Ellipsoid," *American Mathematical Monthly* 51 (1944), 555–563.

—————. *The Theory of Approximation.* AMS Colloquium Publications. Vol. 11. New York: American Mathematical Society, 1930.

Jackson, Kenneth, Ed. *The Encyclopedia of New York City.* New Haven: Yale University Press, 1995.

Jacobson, Nathan. "Abraham Adrian Albert 1905–1972." *Bulletin of the American Mathematical Society* 80 (1974), 1075–1100.

—————. "Isomorphisms of Jordan Rings." *American Journal of Mathematics* 70 (1948), 317–326.

—————. "Non-Commutative Polynomials and Cyclic Algebras." *Annals of Mathematics* 35 (1934), 197–208.

—————. "A Personal History and Commentary 1910–1943." In *Nathan Jacobson: Collected Mathematical Papers.* Ed. Nathan Jacobson, 1: 1–11.

—————. "A Personal History and Commentary, 1943–1946." In *Nathan Jacobson: Collected Mathematical Papers.* Ed. Nathan Jacobson, 1: 276–284.

—————. "A Personal History and Commentary 1947–1955." In *Nathan Jacobson: Collected Mathematical Papers.* Ed. Nathan Jacobson, 2: 1–16.

—————. "The Radical and Semi-simplicity for Arbitrary Rings." *American Journal of Mathematics* 67 (1945), 300–320.

—————. "Rational Methods in the Theory of Lie Algebras." *Annals of Mathematics* 36 (1935), 875–881.

—————. "Simple Lie Algebras of Type A." *Annals of Mathematics* 39 (1938), 181–189.

—————. "Simple Lie Algebras over a Field of Characteristic Zero." *Duke Mathematical Journal* 4 (1938), 534–551.

————. "Structure Theory for Algebraic Algebras of Bounded Degree." *Annals of Mathematics* 46 (1945), 695–707.

————. *Structure Theory of Jordan Algebras*. Lecture Notes in Mathematics. University of Arkansas, 1981.

————. "Structure Theory of Simple Rings Without Finiteness Assumptions." *Transactions of the American Mathematical Society* 57 (1945), 228–245.

————. *The Theory of Rings*. Mathematical Surveys. No. 2. New York: American Mathematical Society, 1943.

James, Ioan M. "Combinatorial Topology Versus Point-set Topology." In *Handbook of the History of General Topology*. Ed. Charles E. Aull and Robert Lowen, 3: 809–834.

James, Ioan M., Ed. *History of Topology*. Amsterdam: Elsevier Science B.V., 1999.

Janet, Maurice. "Les systèmes d'équations aux dérivées partielles." *Journal de mathématiques pures et appliquées*. 8th Ser. 3 (1920), 65–151.

Jewett, Frank B. "The Mobilization of Science in National Defense." *Science* 95 (1942), 235–241.

Jones, Bessie Z. "To the Rescue of the Learned: The Asylum Fellowship Plan at Harvard, 1938–1940." *Harvard Library Bulletin* 32 (1984), 205–238.

Jones, F. Burton. "The Beginning of Topology in the United States and the Moore School." In *Handbook of the History of General Topology*. Ed. Charles E. Aull and Robert Lowen, 1: 97–103.

Jordan, Pascual; von Neumann, John; and Wigner, Eugene. "On an Algebraic Generalization of the Quantum Mechanical Formalism." *Annals of Mathematics* 35 (1934), 29–64.

Judt, Tony. *Postwar: A History of Europe Since 1945*. London: Penguin Books, 2005.

Kac, Mark. "Random Walks and the Theory of Brownian Motion." *American Mathematical Monthly* 54 (1947), 369–391.

————. Review of "*Finite Dimensional Vector Spaces*. By Paul R. Halmos. (Annals of Mathematics Studies, no. 7.) Princeton University Press, 1942. 5 + 196 pp." *Bulletin of the American Mathematical Society* 49 (1943), 349–350.

————. Review of "*Proceedings of the Berkeley Symposium on Mathematical Statistics and Probability*. Ed. by J. Neyman. Berkeley: University of California Press, 194[9]). [8+501] pp. $7.50." *BAMS* 56 (1950), 267–268

Kahane, Jean-Pierre. "Norbert Wiener et l'analyse de Fourier." *Bulletin of the American Mathematical Society* 72 (1966), 42–47.

Kakutani, Shizuo. "Ergodic Theory." In *Proceedings of the International Congress of Mathematicians: Cambridge, Massachusetts, U.S.A., August 30–September 6, 1950*. Ed. Lawrence Graves, Einar Hille, Paul Smith, and Oscar Zariski, 2: 128–142.

Kamminga, Harmke and Somsen, Geert, Ed. *Pursuing the Unity of Science: Ideology and Scientific Practice from the Great War to the Cold War*. Abingdon and New York: Routledge/Taylor and Francis Group, 2016.

Kaplansky, Irving. "Abraham Adrian Albert: November 9, 1905–June 6, 1972." *Biographical Memoirs*. Vol. 51. Washington, D.C.: National Academy of Sciences, 1980, pp. 3–22.

————. "Maximal Fields with Valuations." *Duke Mathematical Journal* 9 (1942), 303–321.

————. "Rings with a Polynomial Identity." *Bulletin of the American Mathematical Society* 54 (1948), 575–580.

Karabel, Jerome. *The Chosen: The Hidden History of Admission and Exclusion at Harvard, Yale, and Princeton*. New York: Houghton Mifflin and Company, 2005.

Karp, Alexander and Schubring, Gert. Ed. *Handbook on the History of Mathematics Education*. New York: Springer Verlag, 2014.

Kasner, Edward. "The Impossibility of Einstein Fields Immersed in Flat Space of Five Dimensions." *American Journal of Mathematics* 43 (1921), 126–129.

Kass, Seymour. "Karl Menger." *Notices of the American Mathematical Society* 43 (1996), 558–561.

Katz, Victor and Parshall, Karen Hunger. *Taming the Unknown: A History of Algebra from Antiquity to the Early Twentieth Century*. Princeton: Princeton University Press, 2014.

Keller, Morton and Phyllis. *Making Harvard Modern: The Rise of America's University*. New York: Oxford University Press, 2001.

Kellogg, Oliver D. *Foundations of Potential Theory*. New York: Frederick Ungar Publishing Company, 1929.

————. "Recent Progress on the Dirichlet Problem." *Bulletin of the American Mathematical Society* 32 (1926), 601–625.

Kempner, Aubrey. "Polynomials and Their Residue Systems," *Transactions of the American Mathematical Society* 22 (1921), 240–266 and 267–288.

Kendall, David C.; Bartlett, Maurice S.; and Page, Thornton L. "Jerzy Neyman, 16 April, 1894–5 August, 1981." *Biographical Memoirs*. Vol. 28. Washington, D.C.: National Academy of Sciences, 1982, pp. 379–412.

Kendig, Keith. "Hassler Whitney: 1907–1989." *Celebratio Mathematica* (2013), 1–19 at https://celebratio.org/Whitney_H/245/.

————. *Never a Dull Moment: Hassler Whitney, Mathematics Pioneer*. AMS/ MAA Spectrum. Vol. 93. Providence: American Mathematical Society, 2018.

Kennedy, Stephen et al., Ed. *A Century of Advancing Mathematics*. Washington, D.C.: Mathematical Association of America, 2015.

Kenschaft, Patricia. "Charlotte Angas Scott (1858–1931)." In *Women in Mathematics: A Biobibliographic Sourcebook*. Ed. Louise S. Grinstein and Paul J. Campbell, pp. 193–203.

Kevles, Daniel J. *The Physicists: The History of a Scientific Community in Modern America*. Cambridge: Harvard University Press, 1987.

————. "Scientists, the Military, and the Control of Postwar Defense Research: The Case of the Research Board for National Security, 1944–46." *Technology and Culture* 16 (1975), 20–47.

Kinderlehrer, David. "Hans Lewy: A Brief Biographical Sketch." In *Hans Lewy Selecta*. Ed. David Kinderlehrer, 2: xv-xx.

Kinderlehrer, David, Ed. *Hans Lewy Selecta*. 2 Vols. Boston: Birkhäuser Verlag, 2002.

Kirby, Linda. "David H. Blackwell: Biography." *Celebratio Mathematica* (2011) at https://celebratio.org/Blackwell_DH/article/27/.

Klein, Felix. *The Evanston Colloquium Lectures on Mathematics*. New York: Macmillan & Co., 1894.

————. *Vorlesungen über das Ikosaeder und die Auflösung der Gleichungen vom fünften Grade*. Leipzig: B. G. Teubner Verlag, 1884.

Kliebard, Herbert and Franklin, Barry. "The Ascendance of Practical and Vocational Mathematics, 1893–1945: Academic Mathematics Under Siege." In *A History of School Mathematics*. Ed. George Stanic and Jeremy Kilpatrick, 1: 399–440.

Kline, John R. "The Annual Meeting of the Society." *Bulletin of the American Mathematical Society* 54 (1948), 260–297.

————. "The April Meeting in New York." *Bulletin of the American Mathematical Society* (1936), 449–455.

————. "The April Meeting of the Society," *Bulletin of the American Mathematical Society* 54 (1948), 622–647.

————. "The February Meeting in New York." *Bulletin of the American Mathematical Society* 39 (1933), 313–315.

————. "The February Meeting in New York," *Bulletin of the American Mathematical Society* 54 (1948), 461–479.

————. "Rehabilitation of Graduate Work." *American Mathematical Monthly* 53 (1946), 121–131.

————. "Report of the Meeting of the Council of the Society in New York City on February 24, 1945." *Bulletin of the American Mathematical Society* 51 (1945), 351–355.

————. "Secretary's Report: The International Congress of Mathematicians." In *Proceedings of the International Congress of Mathematicians: Cambridge, Massachusetts, U.S.A., August 30–September 6, 1950*. Ed. Lawrence Graves, Einar Hille, Paul Smith, and Oscar Zariski, 1: 121–145.

————. "Separation Theorems and Their Relation to Recent Developments in Analysis Situs." *Bulletin of the American Mathematical Society* 34 (1928), 155–192.

————. "A Theorem Concerning Connected Point Sets." *Fundamenta Mathematicae* 3 (1922), 238–239.

Knaster, Bronisław and Kuratowski, Casimir. "A Connected and Connected im kleinen Point Set Which Contains No Perfect Subset." *Bulletin of the American Mathematical Society* 33 (1927), 106–109.

————. "Sur les ensembles connexes." *Fundamenta Mathematicae* 2 (1921), 206–255.

Knoche, Michael. "Scientific Journals under National Socialism." *Libraries & Culture* 26 (1991), 415–426.

Koetsier, Teun and van Mill, Jan. " 'By their fruits ye shall know them': Some Remarks on the Introduction of General Topology and Other Areas of Mathematics." In *History of Topology*. Ed. Ioan M. James, pp. 199–239.

Kohler, Robert. *Partners in Science: Foundations and Natural Scientists 1900–1945*. Chicago: University of Chicago Press, 1991.

Kolmogoroff, Andrei. *Grundbegriffe der Wahrscheinlichkeitsrechnung*. Ergebnisse der Mathematik und ihrer Grenzgebiete. Vol. 2. No. 3. Berlin: Julius Springer, 1933.

————. "Homologiering des Komplexes und des lokal-bikompakten Raumes." *Matematicheskiĭ Sbornik* 43 (5) (1936), 701–706.

————. "Über die Dualität im Aufbau der kombinatorischen Topologie." *Matematicheskiĭ Sbornik* 43 (1) (1936), 97–102.

Koopman, Bernard O. "Hamiltonian Systems and Transformations in Hilbert Space." *Proceedings of the National Academy of Sciences of the United States of America* 17 (1931), 315–318.

Koopman, Bernard O. and von Neumann, John. "Dynamical Systems with Continuous Spectra." *Proceedings of the National Academy of Sciences of the United States of America* 18 (1932), 255–263.

Kreyszig, Erwin. "Remarks on the Mathematical Work of Tibor Radó." In *The Problem of Plateau: A Tribute to Jesse Douglas & Tibor Radó.* Ed. Themistocles M. Rassias, pp. 13–32.

Krige, John. *American Hegemony and the Postwar Reconstruction of Science in Europe.* Cambridge: The MIT Press, 2006.

Krohn, Claus-Dieter. *Intellectuals in Exile: Refugee Scholars and the New School for Social Research.* Trans. Rita and Robert Kimber. Amherst: University of Massachusetts Press, 1993.

Kuratowski, Casimir. "Quelques applications d'éléments cycliques de M. Whyburn." *Fundamenta Mathematicae* 14 (1929), 138–144.

———. "Sur la méthode d'inversion." *Fundamenta Mathematicae* 4 (1923), 151–163.

Kuratowski, Casimir and Whyburn, Gordon T. "Sur les éléments cycliques et leurs applications." *Fundamenta Mathematicae* 16 (1930), 305–331.

Langer, Rudolph E. "Developments Associated with a Boundary Problem not Linear in the Parameter." *Transactions of the American Mathematical Society* 25 (1923), 155–172.

Lankford, John. *American Astronomy: Community, Careers, and Power, 1859–1940.* Chicago: University of Chicago Press, 1997.

Larmor, Brendan, Ed. *Mathematical Cultures: The London Meetings 2012–2014.* Cham: Springer Basel, 2016.

Laurence, William L. "New 10-Billion Volt Atom Smasher Is Planned by California Scientists." *New York Times* (25 September, 1946), 46.

Lax, Peter. "The Flowering of Applied Mathematics in America." In *A Century of Mathematics in America.* Ed. Peter L. Duren et al. 2: 455–466.

———. "The Old Days." In *A Century of Mathematical Meetings.* Ed. Bettye Anne Case. Providence: American Mathematical Society, 1996, pp. 281–283.

Lefschetz, Solomon. *Algebraic Topology.* AMS Colloquium Publications. Vol. 27. New York: American Mathematical Society, 1942.

———. *L'Analysis situs et la géométrie algébrique.* Paris: Gauthier-Villars et Cie, 1924.

———. "Intersections and Transformations of Complexes and Manifolds." *Transactions of the American Mathematical Society* 28 (1926), 1–49.

———. "On Certain Numerical Invariants of Algebraic Varieties with Application to Abelian Varieties." *Transactions of the American Mathematical Society* 22 (1921), 327–406.

———. Review of Oswald Veblen's *Analysis Situs. Bulletin des sciences mathématiques.* Ser. 2. 46 (1922), 421–424.

———. "The Role of Algebra in Topology." *Bulletin of the American Mathematical Society* 43 (1937), 345–359.

———. *Topology.* AMS Colloquium Publications. Vol. 12. New York: American Mathematical Society, 1930.

———. "Witold Hurewicz, in Memoriam." *Bulletin of the American Mathematical Society* 63 (1957), 77–82.

———. "Zariski on Algebraic Surfaces." *Bulletin of the American Mathematical Society* 42 (1936), 13–14.

Lefschetz, Solomon and Whitehead, J. Henry C. "On Analytical Complexes." *Transactions of the American Mathematical Society* 35 (1933), 510–517.

Lehto, Olli. *Mathematics without Borders: A History of the International Mathematical Union.* New York: Springer-Verlag, 1998.

Lesley, Frank D. "Biography of S. E. Warschawski." *Complex Variables, Theory and Application* 5 (1986), 95–109.

Levi-Civita, Tullio. "Some Mathematical Aspects of the New Mechanics." *Bulletin of the American Mathematical Society* 39 (1933), 535–563.

Levinson, Norman. *Gap and Density Theorems.* AMS Colloquium Publications. Vol. 26. New York: American Mathematical Society, 1940.

Lewis, Albert C. "The Beginnings of the R. L. Moore School of Topology." *Historia Mathematica* 31 (2004), 279–295.

––––––. "The Building of the University of Texas Mathematics Faculty, 1883–1938." In *A Century of Mathematics in America.* Ed. Peter L. Duren et al. 3: 205–239.

Lewy, Hans. *Aspects of the Calculus of Variations.* Berkeley: University of California Press, 1939.

Lin, Chia-Chiao. Review of "*Supersonic Flow and Shock Waves.* By R. Courant and K. O. Friedrichs. New York, Interscience, 1948. 16+464 pp. $7.00." *Bulletin of the American Mathematical Society* 57 (1951), 85–87.

Linehan, Paul H. "Review of *The Queen of the Sciences* by E. T. Bell." *American Mathematical Monthly* 39 (1932), 296–297.

"List of Members." In *Proceedings of the International Congress of Mathematicians: Cambridge, Massachusetts, U.S.A., August 30–September 6, 1950.* Ed. Lawrence Graves, Einar Hille, Paul Smith, and Oscar Zariski, 1: 36–87.

"List of Officers and Members of the Association for Symbolic Logic." *Journal of Symbolic Logic* 1 (1936), 106–109.

Lorch, Edgar R. "Joseph Fels Ritt." *Bulletin of the American Mathematical Society* 57 (1951), 307–318.

––––––. "Mathematics at Columbia during Adolescence." In *A Century of Mathematics in America.* Ed. Peter L. Duren et al., 3: 149–161.

Lorenat, Jemma. "'Actual Accomplishments in This World': The Other Students of Charlotte Angas Scott." *The Mathematical Intelligencer* 42 (2020), 56–65.

Lurie, Edward. *Louis Agassiz: A Life in Science.* Chicago: University of Chicago Press, 1960.

Macaulay, Francis. "Dr. Charlotte Angas Scott." *Journal of the London Mathematical Society* 7 (1932), 230–240.

MacDuffee, Cyrus C. "The Coordinating Committee." *American Mathematical Monthly* 52 (1945), 294.

––––––. "An Introduction to the Theory of Ideals in Linear Associative Algebras." *Transactions of the American Mathematical Society* 31 (1929), 71–90.

––––––. *The Theory of Matrices.* Ergebnisse der Mathematik und ihrer Grenzgebiete. Vol. 2. No. 5. Berlin: Julius Springer, 1933.

Mackey, George W. "Marshall Harvey Stone 1903–1989." *Notices of the American Mathematical Society* 36 (3) (1989), 221–223.

Mac Lane, Saunders. "The Applied Mathematics Group at Columbia in World War II." In *A Century of Mathematics in America.* Ed. Peter L. Duren et al. 3: 495–515.

————. "Cohomology Theory in Abstract Groups III: Operator Homomorphisms of Kernels." *Annals of Mathematics* 50 (1949), 736–761.

————. "Mathematics at the University of Chicago: A Brief History." In *A Century of Mathematics in America*. Ed. Peter L. Duren et al. 2: 127–154.

————. "Oswald Veblen: 1880–1960," *Biographical Memoirs*. Washington, D.C.: National Academy of Sciences, 1964, pp. 324–371.

————. *Saunders Mac Lane: A Mathematical Autobiography*. Wellesley: A K Peters, 2005.

————. "Some Recent Advances in Algebra." *American Mathematical Monthly* 46 (1939), 3–19.

————. "Topology and Logic at Princeton." In *A Century of Mathematics in America*. Ed. Peter L. Duren et al. 2: 217–221.

MacNeille, Holbrook M. "The Application of Lattice Theory to Integration." *Bulletin of the American Mathematical Society* 44 (1938), 825–827.

————. "Extension of a Distributive Lattice to a Boolean Ring." *Bulletin of the American Mathematical Society* 45 (1939), 452–455.

Macrae, Norman. *John von Neumann*. New York: Pantheon Books, 1992.

Maddox, Robert F. "The Politics of World War II Science: Senator Harley M. Kilgore and the Legislative Origins of the National Science Foundation." *West Virginia History* 41 (1979), 20–39.

Manheim, Jerome H. *The Genesis of Point Set Topology*. London: The Macmillan Company, 1964.

Mann, Henry B. "A Proof of the Fundamental Theorem on the Density of Sums of Sets of Positive Integers." *Annals of Mathematics* 43 (1942), 523–527.

Manning, William. "On the Primitive Groups of Class $3p$." *Transactions of the American Mathematical Society* 6 (1905), 42–47.

————. "The Primitive Groups of Class $2p$ Which Contain a Substitution of Order p and Degree $2p$. *Transactions of the American Mathematical Society* 4 (1903), 351–357.

————. "The Primitive Groups of Class 14." *American Journal of Mathematics* 51 (1929), 619–652.

Marden, Morris. *Geometry of Polynomials*. Mathematical Surveys. Vol. 3. New York: American Mathematical Society, 1949.

Markov, Andrei. "The Insolubility of the Problem of Homeomorphy" (in Russian). *Proceedings of the USSR Academy of Sciences* 121 (1958), 218–220.

Martinelli, Enzo. "Beniamino Segre: His Life, His Work." *Rendiconti dell'Accademia nationale delle scienze detta dei XL* 4 (1979–1980), 1–12.

Masani, Pesi. "Wiener's Contributions to Generalized Harmonic Analysis, Prediction Theory and Filter Theory." *Bulletin of the American Mathematical Society* 72 (1966), 73–125.

Massey, William S. "A History of Cohomology Theory." In *History of Topology*. Ed. Ioan M. James, pp. 579–603.

"Mathematics and the National Science Foundation." *Bulletin of the American Mathematical Society* 50 (1950), 285–287.

Mattingly, Paul H. "New York University." In *The Encyclopedia of New York City*. Ed. Kenneth T. Jackson.

Mattuck, Arthur. "Arthur Byron Coble." *Bulletin of the American Mathematical Society* 76 (1970), 693–699.

May, Kenneth O., Ed. *The Mathematical Association of America: Its First Fifty Years.* Washington, D.C.: The Mathematical Association of America, 1972.

Maz'ya, Vladimir and Shaposhmikova, Tatyana. *Jacques Hadamard: A Universal Mathematician.* HMATH. Vol. 14. Providence: American Mathematical Society and London: London Mathematical Society, 1998.

McArthur, Charles W. *Operations Analysis in the U. S. Army Eighth Air Force in World War II.* HMATH. Vol. 4. Providence: American Mathematical Society and London: London Mathematical Society, 1990.

McClintock, Emory. "The Past and Future of the Society." *Bulletin of the American Mathematical Society* 1 (1895), 85–94.

McKinsey, J.C.C. "Chen." "The Decision Problem for Some Classes of Sentences Without Quantifiers." *Journal of Symbolic Logic* 8 (1943), 61–76.

————. "A Solution of the Decision Problem for the Lewis systems S2 and S4, with an Application to Topology." *Journal of Symbolic Logic* 6 (1941), 117–134.

McNeill, William H. *Hutchins' University: A Memoir of the University of Chicago, 1929–1950.* Chicago: University of Chicago Press, 1991.

McShane, Edward J. *Integration.* Princeton Mathematical Series. Vol. 7. Princeton: Princeton University Press, 1944.

————. "Recent Developments in the Calculus of Variations." *Semicentennial Addresses of the American Mathematical Society.* New York: American Mathematical Society, 1938, pp. 69–97.

————. Review of *"Length and Area.* By Tibor Radó. American Mathematical Society Colloquium Publications, vol. 30. New York: American Mathematical Society, 1948. 6+572 pp. $6.75." *Bulletin of the American Mathematical Society* 54 (1948), 861–863.

————. "Semi-continuity of Integrals in the Calculus of Variations." *Duke Mathematical Journal* 2 (1936), 597–616.

————. "Some Existence Theorems in the Calculus of Variations. I. The Dresden Corner Condition." *TAMS* 44 (1938), 429–438.

————. "Some Existence Theorems in the Calculus of Variations. II. Existence Theorems for Isoperimetric Problems in the Plane." *TAMS* 44 (1938), 439–453.

————. "Some Existence Theorems in the Calculus of Variations. III. Existence Theorems for Nonregular Problems." *TAMS* 45 (1939), 151–171.

————. "Some Existence Theorems in the Calculus of Variations. IV. Isoperimetric Problems in Non-parametric Form." *TAMS* 45 (1939), 173–196.

————. "Some Existence Theorems in the Calculus of Variations. V. The Isoperimetric Problem in Parametric Form." *TAMS* 45 (1939), 197–216.

McShane, Edward J.; Kelley, John L; and Reno, Franklin V. *Exterior Ballistics.* Denver: University of Denver Press, 1953.

Mehrtens, Herbert. *Die Entstehung der Verbandstheorie.* Hildesheim: Gerstenberg Verlag, 1979.

"Members of the Society." *Bulletin of the American Mathematical Society* 7 (1901), 9–33.

"A Memorial to a Scholar-Teacher." *Princeton Alumni Weekly* 32 (30 October, 1931), pp. 111–113.

Menger, Karl. "New Foundations of Projective and Affine Geometry." *Annals of Mathematics* 37 (1936), 456–482.

——. "Non-Euclidean Geometry of Joining and Intersecting." *Bulletin of the American Mathematical Society* 44 (1938), 821–824.

——. *Reminiscences of the Vienna Circle and the Mathematical Colloquium.* Ed. Louise Golland, Brian McGuinness, and Abe Sklar. Dordrecht: Kluwer Academic Publishers, 1994.

Meyerhoff, Howard. "The National Science Foundation: S. 1850, Final Senate Bill." *Science* 103 (1946), 270–273.

Michal, Aristotle. "The April Meeting in Berkeley." *Bulletin of the American Mathematical Society* 50 (1944), 476–477.

——. "General Differential Geometries and Related Topics." *Bulletin of the American Mathematical Society* 45 (1939), 529–563.

Miller, George A. *The Collected Works of George Abram Miller.* 5 Vols. Urbana: University of Illinois, 1935–1959.

——. "Determination of All the Abstract Groups of Order 72," *American Journal of Mathematics* 51 (1929), 491–494.

——. "Groups Generated by Two Operators of Order Three Whose Product Is of Order Six," *Proceedings of the National Academy of Sciences of the United States of America* 13 (1927), 170–174.

——. "Groups Generated by Two Operators of Order Three Whose Product Is of Order Three." *Proceedings of the National Academy of Sciences of the United States of America* 13 (1927), 24–26.

——. "The Simple Group of Order 2520." *Bulletin of the American Mathematical Society* 28 (1922), 98–102.

——. "Subgroups of Index p^2 Contained in a Group of Order p^m." *American Journal of Mathematics* 48 (1926), 253–256.

Miller, George A., Blichfeldt, Hans F., and Dickson, Leonard E. *Theory and Applications of Finite Groups.* New York: John Wiley & Sons, 1916.

Miller, George A. and Ling, George H. "Proof That There Is No Simple Group Whose Order Lies Between 1092 and 2001." *American Journal of Mathematics* 22 (1900), 13–26.

Minorsky, Nicolas. Review of "*Nonlinear Vibrations in Mechanical and Electrical Systems.* By J. J. Stoker. New York, Interscience, 1950. 20+273 pp. $5.00." *Bulletin of the American Mathematical Society* 56 (1950), 519–521.

Mitchell, Janet A., Ed. *A Community of Scholars: The Institute for Advanced Study: Faculty and Members 1930–1980.* Princeton: The Institute for Advanced Study, 1980.

Montgomery, Dean and Zippin, Leo. "Small Groups of Finite-Dimensional Groups." *Annals of Mathematics* (1952), 213–241.

Moore, Charles N. "On the Application of Borel's Method to the Summation of Fourier's Series." *Proceedings of the National Academy of Sciences of the United States of America* 11 (1925), 284–287.

——. *Summable Series and Convergence Factors.* AMS Colloquium Publications. Vol. 22. New York: American Mathematical Society, 1938.

Moore, Eliakim H. "Introduction to a Form of General Analysis." In *The New Haven Mathematical Colloquium.* New Haven, CT: Yale University Press, 1910, pp. 1–150.

Moore, Eliakim H. et al., Ed. *Mathematical Papers Read at the International Mathematical Congress Held in Conjunction with the World's Columbian Exposition, Chicago 1893*. New York: Macmillan & Co., 1896.

Moore, Eliakim H. and Smith, Herman L. "A General Theory of Convergence." *American Journal of Mathematics* 44 (1922), 102–121.

Moore, Gregory. *Zermelo's Axiom of Choice: Its Origins, Development, and Influence*. New York: Springer-Verlag, 1982.

Moore, Robert L. "Concerning Connectedness im kleinen and a Related Property." *Fundamenta Mathematicae* 3 (1922), 232–237.

————. "A Connected and Regular Point Set Which Contains No Arc." *Bulletin of the American Mathematical Society* 32 (1926), 331–332.

————. *Foundations of Point Set Theory*. AMS Colloquium Publications. Vol. 13. New York: American Mathematical Society, 1932.

————. "On the Foundations of Plane Analysis Situs," *Transactions of the American Mathematical Society* 17 (1916), 131–164.

Moore, Robert L. and Kline, John R. "On the Most General Closed Point-set Through Which It Is Possible To Pass a Simple Continuous Arc" *Annals of Mathematics* 20 (1919), 218–223.

Morawetz, Cathleen Synge. "Kurt Otto Friedrichs 1901–1983." *Biographical Memoirs*. Vol. 40. Washington, D.C.: National Academy of Sciences, 1969, pp. 69–90.

Morrey, Charles B. "Griffith Conrad Evans." *Biographical Memoirs*. Vol. 54. Washington, D.C.: National Academy of Sciences, 1983, pp. 126–155.

————. "Multiple Integral Problems in the Calculus of Variations and Related Topics." *University of California Publications in Mathematics*. New Ser. 1 (1943), 1–130.

Morse, Marston. "The Calculus of Variations in the Large." *Verhandlungen der Internationalen Mathematiker-Kongresses Zürich 1932*. Ed. Walter Saxer, 1: 173–188.

————. *The Calculus of Variations in the Large*. AMS Colloquium Publications. Vol. 18. New York: American Mathematical Society, 1934.

————. "The Foundations of a Theory of the Calculus of Variations in the Large." *Transactions of the American Mathematical Society* 30 (1928), 213–274.

————. "The Foundations of a Theory the Calculus of Variations in the Large in m-Space (First Paper)." *Transactions of the American Mathematical Society* 31 (1929), 379–404.

————. "The Foundations of a Theory of the Calculus of Variations in the Large in m-Space (Second Paper)." *Transactions of the American Mathematical Society* 32 (1930), 599–631.

————. "Functional Topology and Abstract Variational Theory." *Annals of Mathematics* 38 (1937), 386–44.

————. "George David Birkhoff and His Mathematical Work." *Bulletin of the American Mathematical Society* 52 (1946), 357–391.

————. "The International Congress in Oslo." *Bulletin of the American Mathematical Society* 42 (1936), 777–781.

————. "A One-to-one Representation of Geodesics on a Surface of Negative Curvature." *American Journal of Mathematics* 43 (1921), 33–51.

————. "Relations Between the Critical Points of a Real Function of n Independent Variables." *Transactions of the American Mathematical Society* 27 (1925), 345–396.

————. "Report of the War Preparedness Committee of the American Mathematical Society and Mathematical Association of America at the Chicago Meeting." *Bulletin of the American Mathematical Society* 47 (1941), 829–831.

————. "Report of the War Preparedness Committee of the American Mathematical Society and Mathematical Association of America at the Hanover Meeting." *Bulletin of the American Mathematical Society* 46 (1940), 711–714.

————. "Sufficient Conditions in the Problem of Lagrange without Assumptions of Normalcy." *Transactions of the American Mathematical Society* 37 (1935), 147–160.

————. "What Is Analysis in the Large?" *American Mathematical Monthly* 49 (1942), 358–364.

Morse, Marston and Hart, William L. "Mathematics in the Defense Program." *American Mathematical Monthly* 48 (1941), 293–302.

Morse, Marston and Hedlund, Gustav. "Unending Chess, Symbolic Dynamics and a Problem in Semigroups." *Duke Mathematical Journal* 11 (1944), 1–7.

Morse, Marston and Myers, Sumner. "The Problems of Lagrange and Mayer with Variable End Points." *Proceedings of the American Academy of Arts and Sciences* 66 (1931), 235–253.

Morse, Philip M. "Mathematical Problems in Operations Research." *Bulletin of the American Mathematical Society* 54 (1948), 602–620.

Morse, Philip M. and Kimball, George E. *Methods of Operations Research*. Cambridge: Technology Press of the Massachusetts Institute of Technology and New York: John Wiley & Sons, Inc., 1945. Rev. First Ed., 1951.

Moschovakis, Yiannis N. "Commentary on Mathematical Logic." In *A Century of Mathematics in America*. Ed. Peter L. Duren et al. 2: 343–346.

Moser, Jürgen. "Fritz John, 1910–1994." *Notices of the American Mathematical Society* 42 (1995), 256–257.

Moulton, Elton J. "Mathematical Reviews." *American Mathematical Monthly* 46 (1939), 523.

————. "Report on the Training of Teachers of Mathematics." *American Mathematical Monthly* 42 (1935), 263–277.

————. "The Unemployment for Ph.D.'s in Mathematics." *American Mathematical Monthly* 42 (1935), 143–144.

Moulton, Forest R. *Differential Equations*. New York: The Macmillan Company, 1930.

Moyer, Albert E. *Joseph Henry: The Rise of an American Scientist*. Washington, D.C.: Smithsonian Institution Press, 1997.

Mundy, Liza. *Code Girls: The Untold Story of the American Women Code Breakers of World War II*. New York: Hachette Books, 2017.

Murray, Francis J. and von Neumann, John. "On Rings of Operators." *Annals of Mathematics* 37 (1936), 116–229.

————. "On Rings of Operators II." *Transactions of the American Mathematical Society* 41 (1937), 208–248.

————. "On Rings of Operators IV." *Annals of Mathematics* 44 (1943), 716–808.

Myers, Sumner. "*Variationsrechnung im Grossen (Theorie von Marston Morse)*. By H. Seifert and W. Threlfall. Leipzig and Berlin, Teubner, 1938. 115 pp." *Bulletin of the American Mathematical Society* 46 (1940), 390.

Nash, Stephen, Ed. *A History of Scientific Computing*. Reading: Addison-Wesley Publishing Co., 1990.

"The National Research Council." *Science* 49 (1919), 458–462.

National Research Fellowships 1919–1938. Washington, D.C.: National Research Council, 1938.

Navy College Training Program: V-12: Curricula, Schedules, Course Descriptions. N.p.: Training Division, Bureau of Naval Personnel, U.S. Navy, 1943.

Neumark, Fritz. *Zuflucht am Bosporus: Deutsche Gelehrte, Politiker und Künstler in der Emigration, 1933–1953*. Frankfurt am Main: Verlag Josef Knecht, 1980.

New Haven Mathematical Colloquium. New Haven: Yale University Press, 1910.

Newman, Maxwell H. A. "John Henry Constantine Whitehead." *Biographical Memoirs of Fellows of the Royal Society* 7 (1961), 349–363.

Newsom, Carroll V., Ed. "General Information." *American Mathematical Monthly* 52 (1945), 409–414.

————. "War Information." *American Mathematical Monthly* 50 (1943), 464–470.

Neyman, Jerzy. "Contribution to the Theory of Sampling Human Populations." *Journal of the American Statistical Association* 33 (1938), 101–116.

————. *First Course in Probability and Statistics*. New York: Henry Holt and Co., 1950.

————. "Foreword." In *Proceedings of the Berkeley Symposium on Mathematical Statistics and Probability*. Ed. Jerzy Neyman, p. vi.

————. "On a Statistical Problem Arising in Routine Analyses and in Sampling Inspections of Mass Production." *Annals of Mathematical Statistics* 12 (1941), 46–76.

Neyman, Jerzy, Ed. *Proceedings of the Berkeley Symposium on Mathematical Statistics and Probability*. Berkeley: University of California Press, 1949.

Nirenberg, Louis. "Partial Differential Equations in the First Half of the Century." In *Development of Mathematics 1900–1950*. Ed. Jean-Paul Pier, pp. 479–515.

Niven, Ivan. "The Threadbare Thirties." In *A Century of Mathematics in America*. Ed. Peter L. Duren *et al.*, 1: 209–229.

Noether, Max. "Über einen Satz aus der Theorie der algebraischen Funktionen." *Mathematische Annalen* 6 (1873), 351–359.

Norwood, Stephen H. *The Third Reich in the Ivory Tower: Complicity and Conflict on American Campuses*. New York: Cambridge University Press, 2009.

"Notes." *Bulletin of the American Mathematical Society* 26 (1920), 184–188.

"Notes." *Bulletin of the American Mathematical Society* 26 (1920), 425–430.

"Notes." *Bulletin of the American Mathematical Society* 26 (1920), 464–469.

"Notes." *Bulletin of the American Mathematical Society* 27 (1920), 39–46.

"Notes." *Bulletin of the American Mathematical Society* 28 (1922), 72–82.

"Notes." *Bulletin of the American Mathematical Society* 30 (1924), 376–378.

"Notes." *Bulletin of the American Mathematical Society* 32 (1926), 176–186.

"Notes." *Bulletin of the American Mathematical Society* 32 (1926), 565–570.

"Notes." *Bulletin of the American Mathematical Society* 33 (1927), 499–508.

"Notes." *Bulletin of the American Mathematical Society* 34 (1928), 121–124.

"Notes." *Bulletin of the American Mathematical Society* 34 (1928), 385–394.

"Notes." *Bulletin of the American Mathematical Society* 34 (1928), 533–538.

"Notes." *Bulletin of the American Mathematical Society* 35 (1929), 279–283.

"Notes." *Bulletin of the American Mathematical Society* 35 (1929), 740–746.

"Notes." *Bulletin of the American Mathematical Society* 47 (1941), 548–552.

"Notes." *Bulletin of the American Mathematical Society* 48 (1942), 206–209.

"Notes." *Bulletin of the American Mathematical Society* 50 (1944), 173–178.

"Notes." *Bulletin of the American Mathematical Society* 51 (1945), 868–873.

"Notes." *Bulletin of the American Mathematical Society* 52 (1946), 979–1002.

"Notes." *Bulletin of the American Mathematical Society* 53 (1947), 268–271.

"Notes." *Bulletin of the American Mathematical Society* 55 (1949), 321–324.

Novikov, Pyotr. "On the Algorithmic Unsolvability of the Word Problem in Group Theory" (in Russian). *Proceedings of the Steklov Institute of Mathematics* 44 (1955), 1–143.

Official Guide Book of the Fair 1933: With 1934 Supplement. Chicago: The Cuneo Press, Inc., 1934.

Ohles, John, Ed. *Biographical Dictionary of American Educators.* 3 Vols. Westport: Greenwood Publishing Group, 1978.

"One Hundred Percent Membership." *American Mathematical Monthly* (1930), 563.

Ore, Oystein. "Abriß einer arithmetischen Theorie der Galoisschen Körper." *Mathematische Annalen* 100 (1928), 650–673.

————. *Les corps algébriques et la théorie des idéaux.* Paris: Gauthier-Villars, 1934.

————. "On the Decomposition Theorems of Algebra." *Comptes rendus du Congrès international des mathématiciens Oslo 1936.* 2 Vols. Oslo: A. W. Brøggers Boktrykkeri A/S, 1937, 1: 297–307.

————. "On the Foundation of Abstract Algebra." *Annals of Mathematics* 36 (1935), 406–437.

————. "Theory of Non-Commutative Polynomials." *Verhandlungen der Internationalen Mathematiker-Kongresses Zürich 1932.* Ed. Walter Saxer, 2: 19–20.

Ortiz, Eduardo. "La politica interamericana de Roosevelt: George D. Birkhoff y la inclusión de América Latina en las redes matemáticas internacionales: Primera Parte." *Saber y tiempo: Revista de historia de la ciencia* 15 (2003), 53–111.

————. "La politica interamericana de Roosevelt: George D. Birkhoff y la inclusión de América Latina en las redes matemáticas internacionales: Segunda Parte," *Saber y tiempo: Revista de historia de la ciencia* 16 (2003), 21–70.

Osserman, Robert. "Commentary on Differential Geometry." In *A Century of Mathematics in America.* Ed. Peter L. Duren et al. 2: 339–341.

Owens, Larry. "Mathematicians and the War: Warren Weaver and the Applied Mathematics Panel, 1942–1945." In *The History of Modern Mathematics.* Ed. David Rowe and John McCleary, 2: 287–305.

"Oystein Ore 1899–1968." *Journal of Combinatorial Theory* 8 (1970), i–iii.

Page, James. *Ordinary Differential Equations: An Elementary Text-Book with an Introduction to Lie's Theory of the Group of One Parameter.* London and New York: Macmillan & Co., 1897.

Paley, Raymond E. A. and Wiener, Norbert. *Fourier Transforms in the Complex Domain.* AMS Colloquium Publications. Vol. 19. New York: American Mathematical Society, 1934.

Parker, John. *R. L. Moore: Mathematician and Teacher.* Washington, D.C.: Mathematical Association of America, 2005.

Parikh, Carol. *The Unreal Life of Oscar Zariski.* Boston: Academic Press, Inc., 1991.

Parshall, Karen Hunger. "*The American Mathematical Monthly* (1894–1919): A New Journal in the Service of Mathematics and Its Educators." In *Research in History and Philosophy of Mathematics*. Ed. Maria Zach and Elaine Landry, pp. 193–204.

──────. "America's First School of Mathematical Research: James Joseph Sylvester at the Johns Hopkins University 1876–1883." *Archive for History of Exact Sciences* 38 (1988), 153–196.

──────. "Defining a Mathematical Research School: The Case of Algebra at the University of Chicago, 1892–1945." *Historia Mathematica* 31 (2004), 263–278.

──────. "E. H. Moore and the Founding of a Mathematical Community in America: 1892–1902." *Annals of Science* 41 (1984), 313–333; Reprinted in *A Century of Mathematics in America*. Ed. Peter L. Duren et al. 2: 155–175.

──────. *James Joseph Sylvester: Jewish Mathematician in a Victorian World*. Baltimore: Johns Hopkins University Press, 2006.

──────. "Joseph H. M. Wedderburn and the Structure Theory of Algebras." *Archive for History of Exact Sciences* 32 (1985), 223–349.

──────. "In Pursuit of the Finite Division Algebra Theorem and Beyond: Joseph H. M. Wedderburn, Leonard E. Dickson, and Oswald Veblen." *Archives internationales d'Histoires des Sciences* 35 (1983), 274–299.

──────. " 'Increasing the Utility of the Society': The Colloquium Lectures of the American Mathematical Society." *Philosophia Scientaie* 19 (2015), 153–169.

──────. "Journals in the Evolution of a National Research Community: The Case of Mathematics in the United States (1776–1940)." To appear.

──────. "Marshall Stone and the Internationalization of the American Mathematical Research Community." *Bulletin of the American Mathematical Society* 46 (2009), 459–482.

──────. "A Mathematical 'Good Neighbor': Marshall Stone in Latin America (1943)." *Revista brasileira de história da matemática: Especial No 1—Festschrift Ubiratan D'Ambrosio* (Dec. 2007), 19–31.

──────. "Mathematics and the Politics of Race: The Case of William Claytor (Ph.D., University of Pennsylvania, 1933)." *American Mathematical Monthly* 123 (2016), 214–240.

──────. "Mathematics in National Contexts (1875–1900): An International Overview." In *Proceedings of the International Congress of Mathematicians: Zürich*. 2 Vols. Basel: Birkhäuser Verlag, 1995, 2: 1581–1591.

──────. " 'A New Era in the Development of Our Science': The American Mathematical Research Community, 1920–1950." In *A Delicate Balance: Global Perspectives on Innovation and Tradition in the History of Mathematics: A Festschrift in Honor of Joseph W. Dauben*. Ed. David E. Rowe and Wann-Sheng Horng, pp. 275–308.

──────. "New Light on the Life and Work of Joseph Henry Maclagan Wedderburn (1882–1948)." In *Amphora: Festschrift für Hans Wussing zu seinem 65. Geburtstag*. Ed. Menso Folkerts et al., pp. 523–537.

──────. "Perspectives on American Mathematics." *Bulletin of the American Mathematical Society* 37 (2000), 381–405.

──────. "The Stratification of the American Mathematical Community: The Mathematical Association of America and the American Mathematical Society, 1915–1925." In *A Century of Advancing Mathematics*. Ed. Stephen Kennedy, et al., pp. 159–175.

————. "Training Research Mathematicians circa 1900: The Cases of the United States, Germany, France, and Great Britain," in *A Global History of Research Education: Disciplines, Institutions and Nations*, ed. Ku-ming "Kevin" Chang and Alan Rocke. Vol. 34(1). Oxford: Oxford University Press, 2021, pp. 65–83.

————. "Training Women in Mathematical Research: The First Fifty Years of Bryn Mawr College (1885–1935)," *The Mathematical Intelligencer* 37 (2) (2015), 71–83.

Parshall, Karen Hunger and Rice, Adrian C., Ed. *Mathematics Unbound: The Evolution of an International Mathematical Research Community, 1800–1945*. HMATH. Vol. 23. Providence: American Mathematical Society and London: London Mathematical Society, 2002.

Parshall, Karen Hunger and Rowe, David E. "Embedded in the Culture: Mathematics at the World's Columbian Exposition." *The Mathematical Intelligencer* 15 (2) (1993), 40–45.

————. *The Emergence of the American Mathematical Research Community, 1876–1900: J. J. Sylvester, Felix Klein, and E. H. Moore*. HMATH 8. Providence: American Mathematical Society and London: London Mathematical Society, 1994.

Phillips, Ralph. "Reminiscences about the 1930s." *The Mathematical Intelligencer* 16 (3) (1994), 6–8.

Picard, Émile. "Allocution" (Séance de Cloture du Congrès). In *Comptes rendus du Congrès international des mathématiciens (Strasbourg, 22–30 septembre 1920)*. Ed. Henri Villat, pp. xxxi-xxxiii.

————. "Allocution" (Séance d'Ouverture du Congrès). In *Comptes rendus du Congrès international des mathématiciens (Strasbourg, 22–30 septembre 1920)*. Ed. Henri Villat, pp. xxvi-xxix.

Pier, Jean-Paul, Ed. *Development of Mathematics 1900–1950*. Basel: Birkhäuser Verlag, 1994.

Pitcher, Everett. *A History of the Second Fifty Years: American Mathematical Society 1939–1988*. Providence: American Mathematical Society, 1988.

————. "Marston Morse, March 24, 1890–June 22, 1977." *Biographical Memoirs*. Vol. 65. Washington, D.C.: National Academy of Sciences, 1994, pp. 222–240.

Post, Emil. "Recursively Enumerable Sets of Positive Integers and Their Decision Problem." *Bulletin of the American Mathematical Society* 50 (1944), 284–316.

————. "A Variant of a Recursively Unsolvable Problem." *Bulletin of the American Mathematical Society* 52 (1946), 264–269.

Prager, William and Hollcroft, Temple R. "First Symposium in Applied Mathematics." *Bulletin of the American Mathematical Society* 53 (1947), 878–881.

"Pre-Training of Aviation Cadets." *American Mathematical Monthly* 49 (1942), 274–276.

Price, G. Baley. "Adjustments in Mathematics to the Impact of War." *American Mathematical Monthly* 50 (1943), 31–34.

————. "The Mathematical Scene, 1940–1965." In *A Century of Mathematics in America*. Ed. Peter L. Duren et al. 2: 379–404.

"The Princeton University Bicentennial Conference on the Problems of Mathematics." In *A Century of Mathematics in America*. Ed. Peter Duren et al. 2: 309–333.

Problems for the Numerical Analysis of the Future. Applied Mathematics Series. Vol. 15. Washington, D.C.: Department of Commerce's National Bureau of Standards, 1951.

Problems of Mathematics. Princeton: Princeton University Press, 1946.

Pupin, Michael. "From Chaos to Cosmos." *Scribner's Magazine* 76 (1924), 3–10.

Radó, Tibor. *Length and Area*. AMS Colloquium Publications. Vol. 30. New York: American Mathematical Society, 1948.

_____. *On the Problem of Plateau*. Ergebnisse der Mathematik und ihrer Grenzgebiete. Vol. 2. No. 2. Berlin: Springer-Verlag, 1933.

_____. "The Problem of Least Area and the Problem of Plateau." *Mathematische Zeitschrift* 32 (1930), 763–796.

Rainich, George. "Electrodynamics in General Relativity." *Transactions of the American Mathematical Society* 17 (1925), 106–136.

Rassias, Themistocles M., Ed. *The Problem of Plateau: A Tribute to Jesse Douglas & Tibor Radó*. Singapore: World Scientific, 1992.

Raynoud, Michel. "André Weil and the Foundations of Algebraic Geometry." *Notices of the American Mathematical Society* 46 (1999), 864–867.

Rédei, Miklós, Ed. *John von Neumann: Selected Letters*. Vol. 27. Providence: American Mathematical Society and London: London Mathematical Society, 2005.

Rees, Mina S. "The Mathematical Sciences and World War II." *American Mathematical Monthly* 87 (1980), 607–621. Reprinted in *A Century of Mathematics in America*. Ed. Peter L. Duren et al. 1: 275–289.

_____. "Mathematics and the Government: The Post-War Years As Augury of the Future." In *The Bicentennial Tribute to American Mathematics, 1776–1976*. Ed. Dalton Tarwater, pp. 101–116.

_____. "The Mathematics Program of the Office of Naval Research." *Bulletin of the American Mathematical Society* 54 (1948), 1–5.

Register of Officers and Members for the Academic Year 1922–1923. Lancaster and Providence: Mathematical Association of America, 1922 (also *American Mathematical Monthly* 29 (1922)).

Reid, Constance. *Courant in Göttingen and New York: The Story of an Improbable Mathematician*. New York: Springer-Verlag, 1976.

_____. *Neyman–From Life*. New York: Springer-Verlag, 1982.

_____. *The Search for E. T. Bell Also Known as John Taine*. Washington, D.C.: The Mathematical Association of America, 1993.

Reid, William T. "*Aspects of the Calculus of Variations*. Notes by J. W. Green after lectures by Hans Lewy. Berkeley, University of California Press, 1939. 6+96 pp." *Bulletin of the American Mathematical Society* 46 (1940), 595–596.

_____. "A Direct Sufficiency Proof for the Problem of Bolza in the Calculus of Variations. " *Transactions of the American Mathematical Society* 42 (1937), 141–154.

Reingold, Nathan. "Refugee Mathematicians in the United States of America, 1933–1941: Reception and Reaction." *Annals of Science* 38 (1981), 313–338; Reprinted in *A Century of Mathematics in America*. Ed. Peter L. Duren et al. 1: 175–200.

Reingold, Nathan, Ed. *The Sciences in the American Context: New Perspectives*. Washington, D.C.: Smithsonian Institution, 1979.

Reingold, Nathan and Reingold, Ida H., Ed. *Science in America: A Documentary History 1900–1939*. Chicago: University of Chicago Press, 1981.

Reissner, Eric; Prager, William; and Stoker, James J., Ed. *Nonlinear Problems in Mechanics of Continua*. Proceedings of Symposia in Applied Mathematics. Vol. 1. New York: American Mathematical Society, 1949.

Remmert, Volkert. "Mathematical Publishing in the Third Reich: Springer-Verlag and the Deutsche Mathematiker-Vereinigung." *The Mathematical Intelligencer* 22 (2000), 22–30.

Report of the Delegates of the United States of America to the Third Pan-American Scientific Congress. Washington: Government Printing Office, 1925.

"Research Fellowships in Mathematics." *American Mathematical Monthly* 31 (1924), 168–169.

Rhodes, Richard. *The Making of the Atomic Bomb.* New York: Simon and Schuster, 1986.

Rice, Adrian C. and Wilson, Robin J. "From National to International Society: The London Mathematical Society 1867–1900." *Historia Mathematica* 25 (1998), 185–217.

Richards, Joan L. "Hilda Geiringer von Mises (1893–1973)." In *Women of Mathematics: A Biobibliographical Sourcebook.* Ed. Louise S. Grinstein and Paul J. Campbell, pp. 41–46.

Richardson, Roland G. D. "Applied Mathematics and the Present Crisis." *American Mathematical Monthly* 50 (1943), 415–423.

————. "The February Meeting of the American Mathematical Society." *Bulletin of the American Mathematical Society* 28 (1922), 233–244.

————. "The Frank Nelson Cole Prize in Algebra." *Bulletin of the American Mathematical Society* 29 (1923), 14.

————. "Incorporation of the American Mathematical Society." *Bulletin of the American Mathematical Society* 30 (1924), 1–3.

————. "International Congress of Mathematicians, Zurich, 1932." *Bulletin of the American Mathematical Society* 38 (1932), 769–774.

————. "The Josiah Willard Gibbs Lectureship." *Bulletin of the American Mathematical Society* 29 (1923), 385.

————. "The October Meeting of the Society." *Bulletin of the American Mathematical Society* 30 (1924), 4–11.

————. "The Ph.D. Degree and Mathematical Research." *American Mathematical Monthly* 43 (1936), 199–215; Reprinted in *A Century of Mathematics in America.* Ed. Peter L. Duren et al. 2: 361–378.

————. "A Problem in the Calculus of Variations with an Infinite Number of Auxiliary Conditions." *Transactions of the American Mathematical Society* 30 (1928), 155–189.

————. "Report of the Secretary to the Council for the Years 1921–1925." *Bulletin of the American Mathematical Society* 32 (1926), 203–211.

————. "The Society Establishes a New Periodical." *Bulletin of the American Mathematical Society* 45 (1939), 641–643.

————. "The Society's Prizes." *Bulletin of the American Mathematical Society* 36 (1930), 3–4.

————. "The Summer Meeting in Cambridge." *Bulletin of the American Mathematical Society* 42 (1936), 761–776.

————. "The Thirtieth Annual Meeting of the Society." *Bulletin of the American Mathematical Society* 30 (1924), 199–216.

————. "The Thirty-third Annual Meeting of the Society." *Bulletin of the American Mathematical Society* 33 (1927), 129–152.

————. "The Twenty-ninth Annual Meeting of the Society." *Bulletin of the American Mathematical Society* 29 (1923), 97–116.

————. "The Twenty-seventh Annual Meeting of the American Mathematical Society." *Bulletin of the American Mathematical Society* 27 (1921), 245–265.

Richardson, Roland G. D. et al. "The Semicentennial Celebration: September 6–9, 1938." *Bulletin of the American Mathematical Society* 45 (1939), 1–30.

Richtmyer, Floyd K. "Division of Physical Sciences." In *A History of the National Research Council, 1919–1933*. Washington, D.C.: National Research Council, 1933, pp. 12–16.

Rider, Robin E. "An Opportune Time: Griffith C. Evans and Mathematics at Berkeley." In *A Century of Mathematics in America*. Ed. Peter L. Duren et al. 2: 283–302.

Riehm, Elaine McKinnon and Hoffman, Frances. *Turbulent Times in Mathematics: The Life of J. C. Fields and the History of the Fields Medal*. Providence: American Mathematical Society, 2011.

Rietz, Henry L. "Coolidge on Probability." *Bulletin of the American Mathematical Society* 32 (1926), 83–85.

_____. *Mathematical Statistics*. Carus Mathematical Monographs. No. 3. Chicago: The Open Court Publishing Company, 1927.

Riquier, Charles. *Les systèmes d'équations aux dérivées partielles*. Paris, Gauthier-Villars, 1910.

Ritt, Joseph F. "Algebraic Aspects of the Theory of Differential Equations," *Semicentennial Addresses of the American Mathematical Society*. New York: American Mathematical Society, 1938, pp. 35–55.

_____. "*Algebraic Functions*. By Gilbert Ames Bliss. American Mathematical Society Colloquium Publications, Volume 16, New York, 1933. vi+218 pp." *Bulletin of the American Mathematical Society* 41 (1935), 9–10.

_____. *Differential Algebra*. AMS Colloquium Publications. Vol. 33. New York: American Mathematical Society, 1950.

_____. *Differential Equations from the Algebraic Standpoint*. AMS Colloquium Publications. Vol. 14. New York: American Mathematical Society, 1932.

_____. "Elementary Functions and Their Inverses." *Transactions of the American Mathematical Society* 27 (1925), 68–90.

Ritter, Jim. "Geometry as Physics: Oswald Veblen and the Princeton School." In *Mathematics Meets Physics: A Contribution to Their Interaction in the 19th and the First Half of the 20th Century*. Ed. Karl-Heinz Schlote and Martina Schneider, pp. 146–179.

Rhodes, Richard. *The Making of the Atomic Bomb*. New York: Simon and Schuster, 1986.

Roberts, David L. "Albert Harry Wheeler (1873–1950): A Case Study in the Stratification of American Mathematical Activity." *Historia Mathematica* 23 (1996), 269–287.

Robertson, Howard P. "Dynamical Space-Times Which Contain a Conformal Euclidean-Space." *Transactions of the American Mathematical Society* 29 (1927), 481–496.

_____. "Note on the Preceding Paper: The Two Body Problem in General Relativity." *Annals of Mathematics* 39 (1938), 101–104.

Robertson, Howard P. and Weyl, Hermann. "On a Problem in the Theory of Groups Arising in the Foundations of Infinitesimal Geometry." *BAMS* 35 (1929), 686–690.

Robinson, Julia. "Definability and Decision Problems in Arithmetic." *Journal of Symbolic Logic* 14 (1949), 98–114.

Roosevelt, Franklin. *Roosevelt's Foreign Policy 1933–1941: Franklin D. Roosevelt's Unedited Speeches and Messages*. New York: Wilfred Funk, Inc., 1942.

Rovnyak, James. "Ernst David Hellinger 1883–1950: Göttingen, Frankfurt Idyll, and the New World." In *Topics in Operator Theory: Ernst D. Hellinger Memorial Volume*. Operator Theory, Advances and Applications. Vol. 48. Basel: Birkhäuser Verlag, 1990, pp. 1–44.

Rowe, David E. "Mathematics in Wartime: Private Reflections of Clifford Truesdell." *The Mathematical Intelligencer* 34 (2012), 29–39.

————. "On Projecting the Future and Assessing the Past: The 1946 Princeton Bicentennial Conference." *The Mathematical Intelligencer* 25 (2003), 8–15.

Rowe, David E. and Horng, Wann-Sheng, Ed. *A Delicate Balance: Global Perspectives on Innovation and Tradition in the History of Mathematics: A Festschrift in Honor of Joseph W. Dauben.* Basel: Birkhäuser Verlag, 2015.

Rowe, David E. and McCleary, John, Ed. *The History of Modern Mathematics.* 2 Vols. Boston: Academic Press, Inc., 1989.

Royden, Halsey. "A History of Mathematics at Stanford." In *A Century of Mathematics in America.* Ed. Peter L. Duren et al. 2: 237–277.

Rydell, Robert W. *Worlds of Fairs: The Century-of-Progress Expositions.* Chicago: University of Chicago Press, 1993.

Sapolsky, Harvey M. *Science and the Navy: The History of the Office of Naval Research.* Princeton: Princeton University Press, 1990.

Sarkowski, Heinz. *Springer-Verlag: History of a Scientific Publishing House.* Part 1. Trans. Gerald Graham. Berlin: Springer-Verlag, 1996.

Sarnak, Peter. "Ralph Phillips (1913–1998)." *Notices of the American Mathematical Society* 47 (2000), 561–563.

Saxer, Walter, Ed. *Verhandlungen des Internationalen Mathematiker-Kongresses Zürich 1932.* 2 Vols. Zürich and Leipzig: Orell Füssli Verlag, n.d.

Saxon, Wolfgang. "James Stoker Jr., 87, Ex-Director of N.Y.U. Mathematical Institute." *New York Times.* 22 October, 1992, B12.

Scanlon, Michael. "Who Were the American Postulate Theorists?" *Journal of Symbolic Logic* 56 (1991), 981–1002.

Schaeffer, Albert C. and Spencer, Donald C. *Coefficient Regions for Schlicht Functions.* AMS Colloquium Publications. Vol. 35. New York: American Mathematical Society, 1950.

————. "Coefficients of Schlicht Functions, I, II, III, IV." *Duke Mathematical Journal* 10 (1943), 611–635; *Duke Mathematical Journal* 12 (1945), 107–125; *Proceedings of the National Academy of Sciences of the United States of America* 23 (1946), 111–116; and *Proceedings of the National Academy of Sciences of the United States of America* 35 (1949), 143–150.

Schafer, Alice. "Two Singularities of Space Curves." *Duke Mathematical Journal* 11 (1944), 655–670.

Schappacher, Norbert. "A Historical Sketch of B. L. van der Waerden's Work in Algebraic Geometry: 1926–1946." In *Episodes in the History of Modern Algebra (1800–1950).* Ed. Jeremy J. Gray and Karen Hunger Parshall, pp. 245–283.

Scherk, Peter. "Two Estimates Connected with the (α, β) Hypothesis." *Annals of Mathematics* 42 (1941), 538–546.

————. "Über reelle geschlossene Raumkurven vierter Ordnung." *Mathematische Annalen* 112 (1936), 743–766.

Schilling, Otto F. G. *The Theory of Valuations.* Mathematical Surveys. Vol. 4. New York: American Mathematical Society, 1950.

Schlote, Karl-Heinz and Schneider, Martina, Ed. *Mathematics Meets Physics: A Contribution to Their Interaction in the 19th and the First Half of the 20th Century.* Frankfurt: Verlag Harri Deutsch, 2011.

Schorling, Raleigh. "A Program for Improving the Teaching of Science and Mathematics." *American Mathematical Monthly* 55 (1948), 221–237.

Schouten, Jan and Struik, Dirk. *Einführung in die neuren Methoden der Differentialgeometrie.* 2 Vols. Groningen: Noordhoff, 1935 and 1938.

Schwartz, Laurent. *A Mathematician Grappling with His Century.* Trans. Leila Schneps. Basel: Birkhäuser Verlag, 2001.

Scott, Charlotte Angas. "A Proof of Noether's Fundamental Theorem." *Mathematische Annalen* 52 (1899), 593–597.

Seifert, Herbert and Threlfall, Wilhelm. *Variationsrechnung im Grossen (Theorie von Marston Morse).* Leipzig and Berlin: B. G. Teubner Verlag, 1938.

Selberg, Atle. "An Elementary Proof of Dirichlet's Theorem About Primes in Arithmetic Progression." *Annals of Mathematics* 50 (1948), 297–304.

———. "An Elementary Proof of the Prime-Number Theorem." *Annals of Mathematics* 50 (1948), 305–313.

Shell-Gellasch, Amy. *In Service to Mathematics: The Life and Work of Mina Rees.* Boston: Docent Press, 2000.

———. "Mina Rees and the Funding of the Mathematical Sciences." *American Mathematical Monthly* 109 (2002), 873–889.

Shell-Gellasch, Amy and Jardine, Richard, Ed. *From Calculus to Computers: Using the Last 200 Years of Mathematics History in the Classroom.* Washington, D.C.: Mathematical Association of America, 2005.

Shields, Brittany. "A Mathematical Life: Richard Courant, New York University, and Scientific Diplomacy in Twentieth-Century America." Unpublished doctoral dissertation, University of Pennsylvania, 2015.

Shohat, James A. and Tamarkin, Jacob D. *The Problem of Moments.* Mathematical Surveys. No. 1. New York: American Mathematical Society, 1943.

Siegel, Carl Ludwig. "Algebraic Integers Whose Conjugates Lie in the Unit Circle." *Duke Mathematical Journal* 11 (1944), 597–602.

———. "On the History of the Frankfurt Mathematics Seminar." *The Mathematical Intelligencer* 1 (1979), 223–230.

Siegmund-Schultze, Reinhard. "Eliakim Hastings Moore's 'General Analysis.'" *Archive for History of Exact Sciences* 52 (1998), 51–89.

———. "The Emancipation of Mathematical Research Publishing in the United States from German Dominance (1878–1945)." *Historia Mathematica* 24 (1997), 135–166.

———. "The Ideology of Applied Mathematics within Mathematics in Germany and the U.S. until the End of World War II." *LLULL* 27 (2004), 791–811.

———. "The Late Arrival of Academic Applied Mathematics in the United States: A Paradox, Theses, and Literature." *NTM International Journal of the History and Ethics of Natural Sciences, Technology, and Medicine* 11 (2003), 116–127.

———. *Mathematicians Fleeing from Nazi Germany: Individual Fates and Global Impact.* Princeton: Princeton University Press, 2009.

———. *Mathematische Berichterstattung in Hitlerdeutschland: Der Niedergang des 'Jahrbuchs über die Fortschritte der Mathematik'.* Göttingen: Vandenhoeck & Ruprecht, 1993.

————. "Probability in 1919/20: The von Mises-Polya-Controversy." *Archive for History of Exact Sciences* 60 (2006), 431–515.

————. *Rockefeller and the Internationalization of Mathematics between the Two World Wars.* Basel: Birkhäuser Verlag, 2001.

————. "Rockefeller Support for Mathematicians Fleeing from the Nazi Purge." In *The "Unacceptables": American Foundations and Refugee Scholars between the Two World Wars and After.* Ed. Giuliana Gemelli, pp. 83–106.

————. " 'Scientific Control' in Mathematical Reviewing and German-U.S.-American Relations between the Two World Wars." *Historia Mathematica* 21 (1994), 306–329.

Sierpiński, Wacław. "Sur une condition pour qu'un continu soit une courbe jordanienne." *Fundamenta Mathematicae* 1 (1920), 44–60.

Sigmund, Karl. *Exact Thinking in Demented Times: The Vienna Circle and the Epic Quest for the Foundations of Science.* New York: Basic Books, 2017.

Sisam, Charles H. "White on Cubic Curves." *Bulletin of the American Mathematical Society*, 32 (1926) 555–556.

Slater, John C. "Physics and the Wave Equation." *Bulletin of the American Mathematical Society* 52 (1946), 392–400.

Slaught, Herbert E. "The Carus Mathematical Monographs." *American Mathematical Monthly* 30 (1923), 151–155.

Slembek, Silke. "On the Arithmetization of Algebraic Geometry." In *Episodes in the History of Modern Algebra (1800–1950).* Ed. Jeremy J. Gray and Karen Hunger Parshall, pp. 285–300.

Smith, David Eugene. "Mary Hegeler Carus." *American Mathematical Monthly* 44 (1937), 280–283.

Smith, David Eugene and Ginsburg, Jekuthiel. *A History of Mathematics in America before 1900.* No. 5. Chicago: Open Court Publishing Company, 1934.

Smith, Paul A. "Lefschetz on Topology." *Bulletin of the American Mathematical Society* 37 (1931), 645–648.

Snyder, Virgil. "Coolidge on Algebraic Curves." *Bulletin of the American Mathematical Society* 38 (1932), 163–165.

————. "Some Recent Contributions to Algebraic Geometry." *Bulletin of the American Mathematical Society* 40 (1934), 673–687.

Snyder, Virgil; Black, Amos H.; and Dye, Leaman A. *Selected Topics in Algebraic Geometry–II.* Bulletin of the National Research Council. No. 96. Washington, D.C.: National Research Council of the National Academy of Sciences, 1934.

Snyder, Virgil; Coble, Arthur; Emch, Arnold; Lefschetz, Solomon; Sharpe, Francis R.; and Sisam, Charles H. *Selected Topics in Algebraic Geometry.* Bulletin of the National Research Council. No. 63. Washington, D.C.: National Research Council of the National Academy of Sciences, 1928.

Soderberg, C. Richard. "Stephen P. Timoshenko 1878–1972." *Biographical Memoirs.* Vol. 53. Washington, D.C.: National Academy of Sciences, 1982, pp. 323–350.

Sokolnikoff, Ivan S. *Mathematical Theory of Elasticity.* New York: McGraw-Hill, 1946.

Sokolnikoff, Ivan S. and Sokolnikoff, Elizabeth Stafford. *Higher Mathematics for Engineers and Physicists.* 2d. ed. New York and London: McGraw-Hill, 1941.

Sørensen, Henrik. "Confluences of Agendas: Emigrant Mathematicians in Transit in Denmark, 1933–1945." *Historia Mathematica* 41 (2016), 157–187.

Spector, Ronald H. *Listening to the Enemy: Key Documents on the Role of Communications Intelligence in the War with Japan*. Wilmington: Scholarly Resources Inc., 1988.

Stadman, Verne A. *The University of California, 1868–1968*. New York: McGraw Hill Book Company, 1970.

Stanford, Edna Cleo, "The History of the Department of Mathematics at the University of Illinois." University of Illinois: Master's Thesis in Mathematics, 1940.

Stanic, George M. A. and Kilpatrick, Jeremy, Ed. *A History of School Mathematics*. 2 Vols. Reston: National Council of Teachers of Mathematics, 2003.

"A Statement of Policy." *Journal of Symbolic Logic* 1 (1936), 1.

Statistical Research Group. *Sequential Analysis of Statistical Data: Applications*. New York: Columbia University Press, 1945.

Steelman, John R. *Manpower for Research: Science and Public Policy: A Report to the President*. 5 Vols. Washington, D.C.: Government Printing Office, 1947.

Steen, Lynn A. "Conjectures and Counterexamples in Metrization Theory," *American Mathematical Monthly* 79 (1972), 113–132.

Stewart, George R. *The Year of the Oath: The Fight for Academic Freedom at the University of California*. Garden City: Doubleday, 1950.

Stewart, Irvin. *Organizing Scientific Research for War: The Administrative History of the Office of Scientific Research and Development*. Boston: Little, Brown and Company, 1948.

Stillwell, John. "Max Dehn." In *History of Topology*. Ed. Ioan M. James. Amsterdam: Elsevier Science B. V., 1999, pp. 965–978.

Stoker, James J. *Nonlinear Vibrations in Mechanical and Electrical Systems*. New York: Interscience Publishers, 1950.

Stone, Marshall. "American Mathematics in the Present War." *Science* 100 (1944), 529–535.

——— . "A Comparison of the Series of Fourier and Birkhoff." *Transactions of the American Mathematical Society* 28 (1926), 695–761.

——— . "Linear Transformations in Hilbert Space. I. Geometrical Aspects." *Proceedings of the National Academy of Sciences of the United States of America* 15 (1929), 198–200.

——— . "Linear Transformations in Hilbert Space. II. Analytical Aspects." *Proceedings of the National Academy of Sciences of the United States of America* 15 (1929), 423–425.

——— . "Linear Transformations in Hilbert Space. III. Operational Methods and Group Theory." *Proceedings of the National Academy of Sciences of the United States of America* 16 (1930), 172–175.

——— . *Linear Transformations in Hilbert Space and Their Applications to Analysis*. AMS Colloquium Publications. Vol. 15. New York: American Mathematical Society, 1932.

——— . "Reminiscences of Mathematics at Chicago." *The University of Chicago Magazine* 69 (Sept. 1976), 27–31; Reprinted in *A Century of Mathematics in America*, ed. Peter Duren et al. 2: 183–190.

——— . "The Representation of Boolean Algebras." *Bulletin of the American Mathematical Society* 44 (1938), 807–816.

——— . "The Theory of Representations for Boolean Algebras." *Transactions of the American Mathematical Society* 40 (1936), 37–111.

Stouffer, Ellis. "Singular Ruled Surfaces in Space of Five Dimensions." *Transactions of the American Mathematical Society* 29 (1927), 80–95.

Struik, Dirk J. "Julian Lowell Coolidge." *American Mathematical Monthly* 62, (1955), 669–682.

————. "The MIT Department of Mathematics During Its First Seventy-Five Years: Some Recollections." In *A Century of Mathematics in America*. Ed. Peter L. Duren et al. 3: 163–177.

————. "On Sets of Principle Directions in a Riemannian Manifold of Four Dimensions." *Journal of Mathematics and Physics* 7 (1927–1928), 193–197.

————. "Review of *The Differential Invariants of Generalized Spaces*. By T. Y. Thomas. Cambridge, University Press, 1934." *Bulletin of the American Mathematical Society* 41 (1935), 477–478.

Struik, Dirk and Wiener, Norbert. "A Relativistic Theory of Quanta." *Journal of Mathematics and Physics* 7 (1927–1928), 1–23.

Swerdlow, Noel M. "Otto E. Neugebauer (26 May 1899–19 February, 1990)." *Proceedings of the American Philosophical Society* 137 (1993), 138–165.

Synge, John. "Focal Properties of Optical and Electromagnetic Systems." *American Mathematical Monthly* 51 (1944), 185–200.

Szegő, Gábor. *Orthogonal Polynomials*. AMS Colloquium Publications. Vol. 23. New York: American Mathematical Society, 1939.

————. "Otto Szász." *Bulletin of the American Mathematical Society* 60 (1954), 261–63.

Tarwater, Dalton, Ed. *The Bicentennial Tribute to American Mathematics, 1776–1976*. N.p.: Mathematical Association of America, 1977.

Taub, Abraham H.; Reissner, Eric; and Churchill, Ruel V., Ed. *Electromagnetic Theory*. Proceedings of Symposia in Applied Mathematics. Vol. 2. New York: American Mathematical Society, 1950.

Taub, Abraham; Veblen, Oswald; and von Neumann, John. "The Dirac Equation in Projective Relativity." *Proceedings of the National Academy of Sciences of the United States of America* 20 (1934), 383–388.

Taylor, Angus E. "Differentiation of Fourier Series and Integrals." *American Mathematical Monthly* 51 (1944), 19–25.

————. "A Life in Mathematics Remembered." *American Mathematical Monthly* 91 (1984), 605–618.

"Text of the New Kilgore-Magnuson Bill." *Science* 103 (1946), 225–230.

Thomas, Joseph M. "Conformal Invariants." *Proceedings of the National Academy of Sciences of the United States of America* 12 (1926), 389–393.

————. *Differential Systems*. AMS Colloquium Publications. Vol. 21. New York: American Mathematical Society, 1937.

————. "Riquier's Existence Theorems." *Annals of Mathematics* 30 (1928–1929), 285–310.

Thomas, Joseph M. and Veblen, Oswald. "Projective Invariants of Affine Geometry of Paths." *Annals of Mathematics* 27 (1926), 279–296.

Thomas, Tracy Y. *The Differential Invariants of Generalized Space*. Cambridge: Cambridge University Press, 1934.

————. "On Conformal Geometry," *Proceedings of the National Academy of Sciences of the United States of America* 12 (1926), 352–359.

————. "On the Projective and Equi-projective Geometries of Paths." *Proceedings of the National Academy of Sciences of the United States of America* 11 (1925), 199–203.

————. "Recent Trends in Geometry." In *Semicentennial Addresses of the American Mathematical Society.* New York: American Mathematical Society, 1938, pp. 98–135.

Thrall, Robert M. Review of "*A Survey of Modern Algebra.* By Garrett Birkhoff and Saunders Mac Lane. New York, Macmillan 1941. 11+450 pp. $3.75." *Bulletin of the American Mathematical Society* 48 (1942), 342–345.

Tonelli, Leonida. "The Calculus of Variations." *Bulletin of the American Mathematical Society* 31 (1925), 163–172.

"Training in Meteorology." *American Mathematical Monthly* 49 (1942), 697–698.

Tucker, Albert. "An Abstract Approach to Manifolds." *Annals of Mathematics* 34 (1933), 191–234.

————. "The Topological Congress in Moscow." *Bulletin of the American Mathematical Society* 41 (1935), 764.

Turing, Alan. "On Computable Numbers, with an Application to the Entscheidungsproblem." *Proceedings of the London Mathematical Society* 42 (1937), 230–265.

Uhlenbeck, Karen. "Commentary on 'Analysis in the Large.'" In *A Century of Mathematics in America.* Ed. Peter L. Duren et al. 2: 357–359.

Ulam, Stanislaw. *Adventures of a Mathematician.* New York: Charles Scribner's Sons, 1976.

————. "John von Neumann 1903–1957." *Bulletin of the American Mathematical Society* 64 (1958), 1–49.

Van der Waerden, Bartel. *Moderne Algebra.* 2 Vols. Berlin: Julius Springer, 1930–1931.

————. Review of "*Differential Algebra.* By Joseph Fels Ritt. American Mathematical Society Colloquium Publications, vol. 33. New York, American Mathematical Society, 1950. 8+181 pp. $4.40." *Bulletin of the American Mathematical Society* 56 (1950), 521–523.

Vandiver, Harry S. "On Fermat's Last Theorem," *Transactions of the American Mathematical Society* 31 (1929), 613–642.

Vandiver, Harry S. and Wahlin, Gustav E. *Algebraic Numbers II.* Bulletin of the National Research Council. No. 62. Washington, D.C.: National Research Council of the National Academy of Sciences, 1928.

Van Est, Willem T. "Hans Freudenthal, 17 September 1905 – 13 October 1990." In *History of Topology.* Ed. Ioan M. James. Amsterdam: Elsevier Science B.V., 1999, pp. 1009–1019.

Van Vleck, Edward; White, Henry; and Woods, Frederick. *The Boston Colloquium: Lectures on Mathematics.* Vol. 1. London: Macmillan & Co., 1905.

Veblen, Oswald. *The Cambridge Colloquium: Part II: Analysis Situs.* AMS Colloquium Publications. Vol 5. New York: American Mathematical Society, 1922. 2d Ed., 1931.

————. "Differential Invariants in Geometry." In *Atti del Congresso internazionale dei matematici,* 1: 181–189.

————. "Generalized Projective Geometry." *Journal of the London Mathematical Society* 4 (1929), 140–160.

————. "George David Birkhoff (1884–1944)." *Yearbook of the American Philosophical Society* (1946), 279–285.

————. *Invariants of Quadratic Differential Forms.* Cambridge Tracts in Mathematics and Mathematical Physics. No. 24. Cambridge: University Press, 1927.

————. "Opening Address." In *Proceedings of the International Congress of Mathematicians: Cambridge, Massachusetts, U.S.A., August 30–September 6, 1950*. Ed. Lawrence Graves, Einar Hille, Paul Smith, and Oscar Zariski, 1: 124–125.

————. *Projective Relativitätstheorie*. Ergebnisse der Mathematik und ihrer Grenzgebiete. Vol. 2. No. 1. Berlin: Verlag von Julius Springer, 1933.

————. "Remarks on the Foundations of Geometry." *Bulletin of the American Mathematical Society* 31 (1925), 121–141.

Veblen, Oswald and Alexander, James. "Manifolds of N Dimensions." *Annals of Mathematics* 14 (1912–1913), 163–178.

Veblen, Oswald and Thomas, Joseph. "Projective Normal Coordinates for the Geometry of Paths." *Proceedings of the National Academy of Sciences of the United States of America* 11 (1925), 204–207.

Veblen, Oswald and Whitehead, J. Henry C. *The Foundations of Differential Geometry*. Cambridge Tracts in Mathematics and Mathematical Physics. No. 29. Cambridge: University Press, 1932.

Veblen, Thorstein. *The Higher Learning in America: A Memorandum on the Conduct of Universities by Business Men*. New York: B. W. Huebsch, 1918; Reprint Ed. New York: The Viking Press, 1935.

Veysey, Lawrence. *The Emergence of the American University*. Chicago: University of Chicago Press, 1965.

"Vie de la Société" (Supplément spécial). *Bulletin de la Société mathématique de France* 52 (1924), 1–67.

Villat, Henri, Ed. *Comptes rendus du Congrès international des mathématiciens (Strasbourg, 22–30 septembre 1920)*. Toulouse: Imprimerie et Librairie Éduoard Privat, 1921.

Visher, Stephen. *Scientists Starred, 1903–1943, in American Men of Science: A Study of Collegiate and Doctoral Training, Birthplace, Distribution, Background, and Developmental Influences*. Baltimore: Johns Hopkins University Press, 1947.

Von Kármán, Theodore and Biot, Maurice A. *Mathematical Methods in Engineering: An Introduction to the Mathematical Treatment of Engineering Problems*. New York: McGraw-Hill, 1940.

Von Mises, Richard. *Fluglehre: Vorträge über Theorie und Berechnung der Flugzeuge in elementarer Darstellung*. Berlin: Verlag von Julius Springer, 1918.

————. "Fundamentalsätze der Wahrscheinlichkeitsrechnung." *Mathematische Zeitschrift* 4 (1919), 1–97.

————. "Grundlagen der Wahrscheinlichkeitsrechnung." *Mathematische Zeitschrift* 5 (1919), 52–99.

————. "On the Foundations of Probability and Statistics." *Annals of Mathematical Statistics* 12 (1941), 191–205.

Von Mises, Richard and Doob, Joseph. "Discussion of Papers on Probability Theory." *Annals of Mathematical Statistics* 12 (1941), 215–217.

Von Mises, Richard with Prager, William and Kuerti, Gustav. *Theory of Flight*. New York: McGraw-Hill, 1945; Reprint Ed. New York: Dover Publications, 1959.

Von Neumann, John. *Collected Works*. Ed. Abraham H. Taub. 6 Vols. New York: Pergamon Press, 1961.

————. "Mathematische Begründung der Quantenmechanik." *Nachrichten von der Gesellschaft der Wissenschaften zu Göttingen, Mathematisch-Physikalische Klasse* (1927), 1–57.

————. "On Regular Rings." *Proceedings of the National Academy of Sciences of the United States of America* 22 (1936), 707–713.

————. "On Rings of Operators III." *Annals of Mathematics* 41 (1940), 94–161.

————. "Proof of the Quasi-Ergodic Hypothesis." *Proceedings of the National Academy of Sciences of the United States of America* 18 (1932), 70–82.

Von Neumann, John and Morgenstern, Oskar. *Theory of Games and Economic Behavior*. Princeton: Princeton University Press, 1944.

Von Neumann, John and Stone, Marshall H. "The Determination of Representative Elements in the Residual Classes of a Boolean Algebra." *Fundamenta Mathematicae* 25 (1935), 353–378.

Von Plato, Jan. *Creating Modern Probability: Its Mathematics, Physics and Philosophy in Historical Perspective*. Cambridge: Cambridge University Press, 1994.

Wade, Thomas L. and Bruck, Richard H. "Types of Symmetries." *American Mathematical Monthly* 51 (1944), 123–129.

Wald, Abraham. "Contributions to the Theory of Statistical Estimation and Testing Hypotheses." *Annals of Mathematical Statistics* 10 (1939), 299–326.

————. "Die Widerspruchtsfreiheit des Kollectivbegriffes in der Wahrscheinlichkeitsrechnung." *Ergebnisse eines mathematischen Kolloquiums* 8 (1937), 38–72.

————. *On the Principles of Statistical Inference*. Notre Dame Mathematical Lectures. Vol. 1. Notre Dame: University of Notre Dame Press, 1942.

————. "Sequential Method of Sampling for Deciding between Two Courses of Action." *Journal of the American Statistical Association* 40 (1945), 277–306.

————. "Sequential Tests of Statistical Hypotheses." *Annals of Mathematical Statistics* 16 (1945), 117–186.

————. "Statistical Decision Functions Which Minimize the Maximum Risk." *Annals of Mathematics* 46 (1945), 265–280.

Walker, Robert J. "Reduction of the Singularities of an Algebraic Surface." *Annals of Mathematics* 36 (1935), 336–365.

Wallis, W. Allen. "The Statistical Research Group, 1942–1945." *Journal of the American Statistical Association* 75 (1980), 320–330.

Walsh, Joseph L. "The Approximation of Harmonic Functions by Harmonic Polynomials and by Harmonic Rational Functions." *Bulletin of the American Mathematical Society* 35 (1929), 499–544.

————. *Interpolation and Approximation by Rational Functions in the Complex Domain*. AMS Colloquium Publications. Vol. 20. New York: American Mathematical Society, 1935.

————. *The Location of Critical Points of Analytic and Harmonic Functions*. AMS Colloquium Publications. Vol. 34. New York: American Mathematical Society, 1950.

————. "On Approximation by Rational Functions to an Arbitrary Function of a Complex Variable." *Transactions of the American Mathematical Society* 31 (1929), 477–502.

Weart, Spencer R. "The Physics Business in America, 1919–1940: A Statistical Reconnaissance." In *The Sciences in the American Context: New Perspectives*. Ed. Nathan Reingold, pp. 295–358.

Wedderburn, Joseph H. M. "Algebras Which Do Not Possess a Finite Basis." *Transactions of the American Mathematical Society* 26 (1924), 395–426.

_____. *Lectures on Matrices*. AMS Colloquium Publications. Vol. 17. New York: American Mathematical Society, 1934.

_____. "On Division Algebras." *Transactions of the American Mathematical Society* 22 (1921), 129–135.

_____. "On Hypercomplex Number Systems." *Proceedings of the London Mathematical Society*. 2d Ser., 6 (November 1907), 77–118.

_____. "A Type of Primitive Algebra." *Transactions of the American Mathematical Society* 15 (1914), 162–166.

Weil, André. *The Apprenticeship of a Mathematician*. Boston: Birkhäuser Verlag, 1992.

_____. *Foundations of Algebraic Geometry*. AMS Colloquium Publications. Vol. 29. New York: American Mathematical Society, 1946.

_____. "Numbers and Solutions of Equations in Finite Fields." *Bulletin of the American Mathematical Society* 55 (1949), 497–508.

Weiner, Charles. "A New Site for the Seminar: The Refugees and American Physics in the Thirties." in *The Intellectual Migration: Europe and America, 1930–1960*. Ed. Donald Fleming and Bernard Bailyn, pp. 190–234.

Weiss, Marie J. "Primitive Groups Which Contain Substitutions of Prime Order p and of Degree $6p$ or $7p$." *Transactions of the American Mathematical Society* 30 (1928), 333–359.

Weyl, F. Joachim. "Mina Rees, President-Elect 1970." *Science* 167 (1970), 1149–1151.

Weyl, Hermann. *Classical Groups: Their Invariants and Representations*. Princeton: Princeton University Press, 1939.

_____. "Generalized Riemann Matrices and Factor Sets." *Annals of Mathematics* 37 (1936), 709–745.

_____. "Note on Matric Algebras." *Annals of Mathematics* 38 (1937), 477–483.

_____. "On the Foundations of General Infinitesimal Geometry." *Bulletin of the American Mathematical Society* 35 (1929), 716–725.

_____. "On Generalized Riemann Matrices." *Annals of Mathematics* 35 (1934), 714–729.

_____. "Riemannsche Matrizen und Faktorsysteme." *Comptes rendus du Congrès international des mathématiciens Oslo 1936*. 2 Vols. Oslo: A. W. Brøggers Boktykkeri A/S, 1937, 2: 3.

Wheeler, Anna Pell. "Linear Ordinary Self-adjoint Differential Equations of the Second Order." *American Journal of Mathematics* 49 (1927), 309–320.

Whidden, Georgia. "Anna Stafford Henriques." *Attributions: Newsletter from the Development Office of the Institute for Advanced Study* 1 (2001).

White, Henry S. "The Frank Nelson Cole Prize in Algebra." *Bulletin of the American Mathematical Society* 21 (1925), 289.

_____. *Plane Curves of the Third Order*. Cambridge: Harvard University Press, 1925.

White, Lyman Cromwell. *300,000 New American: The Epic of a Modern Immigrant-Aid Service*. New York: Harper & Brothers, 1957.

Whitehead, J. Henry C. "The Representation of Projective Spaces." *Annals of Mathematics* 32 (1931), 327–360.

Whitney, Hassler. "Analytic Extensions of Differentiable Functions Defined in Closed Sets." *Transactions of the American Mathematical Society* 36 (1934), 63–89.

_____. "The Coloring of Graphs." *Proceedings of the National Academy of Sciences of the United States of America* 17 (1931), 122–125.

_____. "Differentiable Manifolds." *Annals of Mathematics* 37 (1936), 645–680.

_____. "Moscow 1935: Topology Moving toward America." In *A Century of Mathematics in America.* Ed. Peter L. Duren et al., 1: 97–117.

_____. "On Products in a Complex," *Annals of Mathematics* 39 (1938), 397–432.

_____. "The Self-Intersections of a Smooth *n*-Manifold in 2*n*-Space," *Annals of Mathematics* 45 (1944), 220–246.

_____. "The Singularities of a Smooth *n*-Manifold in (2*n* − 1)-Space." *Annals of Mathematics* 45 (1944), 247–293.

Whyburn, Gordon T. *Analytic Topology.* AMS Colloquium Publications. Vol. 28. New York: American Mathematical Society, 1942.

_____. "Concerning Accessibility in the Plane and Regular Accessibility in *n* Dimensions." *Bulletin of the American Mathematical Society* 34 (1928), 504–510.

_____. "Concerning the Complementary Domains of Continua." *Annals of Mathematics* 29 (1927–1928), 399–411.

_____. "Cyclically Connected Continuous Curves." *Proceedings of the National Academy of Sciences of the United States of America* 13 (1927), 31–38.

_____. "Non-alternating Interior Retracting Transformations." *Annals of Mathematics* 40 (1939), 914–921.

_____. Review of Beatrice Aitchison, "Concerning Regular Accessibility," *Fundamenta Mathematicae* 20 (1933), 117–125. In *Zentralblatt für Mathematik und ihrer Grenzgebiete* 7 (1933), 82–83.

Whyburn, William M. "The Annual Meeting of the Society." *Bulletin of the American Mathematical Society* 57 (1951), 109–152.

Wiener, Norbert. "Differential Space." *Journal of Mathematics and Physics* 2 (1923), 131–174.

_____. "The Dirichlet Problem." *Journal of Mathematics and Physics* 3 (1924), 127–146.

_____. *Extrapolation, Interpolation and Smoothing of Stationary Time Series with Engineering Applications.* Cambridge: The MIT Press; New York: Wiley; and London: Chapman and Hall, 1949.

_____. "Fourier Analysis and Asymptotic Series." In Vannevar Bush. *Operational Circuit Analysis.* New York: Wiley, 1929, pp. 366–379.

_____. *The Fourier Integral and Certain of Its Applications.* Cambridge: Cambridge University Press, 1933.

_____. "Generalized Harmonic Analysis." *Acta Mathematica* 55 (1930), 117–258.

_____. "The Historical Background of Harmonic Analysis." *Semicentennial Addresses of the American Mathematical Society.* New York: American Mathematical Society, 1938, pp. 56–68.

_____. *I Am a Mathematician: An Autobiography.* Garden City: Doubleday & Co., 1956.

_____. "R. E. C. Paley—In Memoriam Jan. 7, 1907-April 7, 1933." *Bulletin of the American Mathematical Society* 39 (1933), 476.

_____. "Tauberian Theorems." *Annals of Mathematics* 33 (1932), 1–100.

Wilcox, Lee R. "*Lattice Theory.* By Garrett Birkhoff. (American Mathematical Society Colloquium Publications, vol. 25) New York, American Mathematical Society, 1940. 6 + 155 pp. $2.50," *Bulletin of the American Mathematical Society* 47 (1941), 194–196.

Wilczynski, Ernest. *Projective Differential Geometry of Curves and Ruled Surfaces*. Leipzig: B. G. Teubner Verlag, 1906.

Wilder, Raymond L. "The Mathematical Work of R. L. Moore." *Archive for History of Exact Sciences* 26 (1982) 73–97. Reprinted in *A Century of Mathematics in America*. Ed. Peter L. Duren et al. 3: 265–291.

————. "The Nature of Mathematical Proof." *American Mathematical Monthly* 51 (1944), 309–323.

————. "On the Linking of Jordan Continua in E_n by $(n-2)$-Cycles." *Annals of Mathematics* 34 (1933), 441–449.

————. "Point Sets in Three and Higher Dimensions and Their Investigation by Means of a Unified Analysis Situs." *Bulletin of the American Mathematical Society* 38 (1932), 649–692.

————. "Reminiscences of Mathematics at Michigan." In *A Century of Mathematics in America*. Ed. Peter L. Duren et al. 3: 191–204.

————. "Some Unsolved Problems of Topology." *American Mathematical Monthly* 44 (1937), 61–70.

————. *Topology of Manifolds*. AMS Colloquium Publications. Vol. 32. New York: American Mathematical Society, 1949.

Wilder, Raymond L. and Ayres, William L. Ed. *Lectures on Topology*. Ann Arbor: University of Michigan Press, 1941.

Wilkins, J. Ernest. "Definitely Self-Conjugate Adjoint Integral Equations." *Duke Mathematical Journal* 11 (1944), 155–166.

Wilks, Samuel. *Mathematical Statistics*. Princeton: Princeton University Press, 1943.

Williams, Kathleen Broome. *Improbable Warriors: Women Scientists and the U. S. Navy in World War II*. Annapolis: Naval Institute Press, 2001.

Wilson, Woodrow. "The Preceptorial System at Princeton." *Educational Review* 39 (1910), 385–390.

Wolfowitz, Jacob. Review of *"An Introduction to Probability Theory and Its Applications*. Vol. I. By William Feller. New York, Wiley, 1950. 12+419 pp. $6.00," *Bulletin of the American Mathematical Society* 57 (1951), 156–159.

Woodward, Robert S. "The Century's Progress in Applied Mathematics." *Bulletin of the American Mathematical Society* 6 (1900), 133–163.

Xu, Yibao. "Chinese–U. S. Mathematical Relations, 1859–1949." In *Mathematics Unbound: The Evolution of an International Mathematical Research Community, 1800–1945*. Ed. Karen Hunger Parshall and Adrian Rice, pp. 287–309.

Young, John W. "Functions of the Mathematical Association of America." *American Mathematical Monthly* 39 (1932), 6–15.

————. *Projective Geometry*. Carus Mathematical Monographs. No. 4. Chicago: The Open Court Publishing Company, 1930.

Youngs, John W. T. "The Annual Meeting of the Society." *Bulletin of the American Mathematical Society* 55 (1949), 261–311.

Zach, Maria and Landry, Elaine, Ed. *Research in History and Philosophy of Mathematics*. Basel: Birkhäuser Verlag, 2016.

Zariski, Oscar. *Algebraic Surfaces*. Ergebnisse der Mathematik und ihrer Grenzgebiete. Vol. 3. No. 5. Berlin: Springer-Verlag, 1935.

———. "Algebraic Varieties over Ground Fields of Characteristic Zero." *American Journal of Mathematics* 62 (1940), 187–221.

———. "Local Uniformization on Algebraic Varieties." *Annals of Mathematics* 41 (1940), 852–896.

———. "The Reduction of the Singularities of an Algebraic Surface." *Annals of Mathematics* 40 (1939), 639–689.

———. Response in "1981 Steele Prizes." *Notices of the American Mathematical Society* 28 (1981), 504–507.

———. "Some Results in the Arithmetic Theory of Algebraic Varieties." *American Journal of Mathematics* 61 (1939), 249–294.

———. *Theory and Applications of Holomorphic Functions on Algebraic Varieties over Arbitrary Ground Fields.* Memoirs of the American Mathematical Society. No. 5. New York: American Mathematical Society, 1951.

Zassenhaus, Hans. "Emil Artin, His Life and His Work." *Notre Dame Journal of Formal Logic* 5 (1964), 1–9.

Zelinsky, Daniel. "A. A. Albert." *American Mathematical Monthly* 80 (1973), 661–665.

Zippin, Leo. "Independent Arcs of a Continuous Curve." *Annals of Mathematics* 34 (1933), 95–113.

Zitarelli, David E. "The Mathematical Association of America: Its First 100 Years." In *A Century of Advancing Mathematics.* Ed. Steve Kennedy, et al., pp. 135–157.

———. "The Origin and Early Impact of the Moore Method." *American Mathematical Monthly* 111 (2004), 465–486.

———. "Who Was Miss Mullikin?" *American Mathematical Monthly* 116 (2009), 99–114.

Zund, Joseph D. "George David Birkhoff and John von Neumann: A Question of Priority and the Ergodic Theorems, 1931–1932." *Historia Mathematica* 29 (2002), 138–156.

Zunz, Olivier. *Philanthropy in America: A History.* Princeton: Princeton University Press, 2012.

Zygmund, Antoni. *Trigonometrical Series.* Warszawa: Z subwencji Funduszu kultury narodowej, 1935.

INDEX

Abel, Niels Henrik, 275

Aberdeen Proving Ground, 365, 396; Ballistics Research Laboratory at (*see* Ballistics Research Laboratory (Aberdeen)); mathematicians at the, 228, 370; von Neumann as consultant to, 369; Ordnance Training School of the, 365; Veblen's association with, 365

Academic Assistance Council (later Society for the Protection of Science and Learning), 201, 207, 227, 298, 300

Académie des Sciences (Paris), prize competitions of, 9

Acoustical Society of America, 406

Acta Arithmetica, 221n96

actuarial science, 72, 73; associated with the ICMs, 109, 292, 486

Adams, C. Raymond, 329; as chair of AMS committee to consider creation of a new abstracting journal, 327, 331–32

Adelbert College. *See* Case Western Reserve University

aeronautics, 110, 160, 343, 351, 355, 468, 472

Agassiz, Louis, 63

Ahlfors, Lars, 441, 493, 501; Fields Medal (1936), 198

airborne fire control, work of the Applied Mathematics Group-Columbia on, 375

Aitchison, Beatrice, 242

Aitken, Alexander, 295

Albanese, Giacomo, 255

Albert, A. Adrian, 45–46, 272, 317, 327, 400, 414, 425, 436, 443; AMS Colloquium volume, 267; competition with Hasse, 262–64; as editor of *Transactions of the American Mathematical Society*; and the ICM (Cambridge, MA) (1950), 500, 501; at the Institute for Advanced Study, 177, 266–67; photo, 263; and the planned ICM (Cambridge, MA) (1940), 292, 294, 295; as a test case for the notion of paid leaves of absence, 175–76; war work of, 383; as winner of the Cole Prize in Algebra (1939), 266; work on division algebras, 45–46, 262–64; work on Riemann matrices, 264–67, 272; work on the structure theory of non-associative algebras, 443–44

Alderman, Edwin, 164

Alexander, James, 14, 76n41, 97, 124, 136, 137, 177, 179, 248, 294, 295, 377n107, 401; independent discovery of cohomology, 249; plenary lecture at the ICM (Zürich) (1932), 193, 194–95; work in algebraic topology, 15–18, 246

Alexander duality, 19

Alexander horned sphere, 19n37

Alexandroff, Paul, 294, 295, 487; IEB fellowship, 124n64, 125–27, 253; work in algebraic topology, 125, 246, 449

algebra, xvii, 34–46, 261–72, 285, 286, 292, 335, 443–46; Artin's work in, 446; Baer's work in, 271; Chicago as American

A NOTE ON THE TYPE

This book has been composed in Arno, an Old-style serif typeface in the
classic Venetian tradition, designed by Robert Slimbach at Adobe.

CPSIA information can be obtained
at www.ICGtesting.com
Printed in the USA
LVHW080952290122
709394LV00008B/2